MASTER PARTS LIST

COVERING UNITS

BUILT FOR

UNITED STATES ARMY

BY

THE AUTOCAR COMPANY

ARDMORE, PA., U. S. A.

FROM

JAN. 1, 1940 to AUG. 1, 1942

WILDSIDE PRESS

TM—10—1373
Released August 1, 1942

Engine	01
Clutch	02
Fuel System	03
Exhaust System	04
Cooling System	05
Electrical System	06
Transmission	07
Transfer Case	08
Prop. Shafts / Univ. Joints	09
Front Axle	10
Rear Axle	11
Brakes	12
Wheels, Hubs and Drums	13
Steering	14
Frame	15
Springs / Shock Absorbers	16
Body	18
Winch	19
Bumpers Guard	21
Access. Group	22
Standard Parts	23
INDEX	

FOREWORD

In the preparation of this Parts List, the grouping used is that specified by the Quartermaster General's Office, as listed on the following page. Group numbers are composed of four figures, the first two of which index the main group, while the two following figures indicate those sub-groups of parts which comprise the main groups. For example, the Engine Group is **01**; the Electrical Group **06**; the Front Axle Group **10**, etc. Subdivisions such as Connecting Rods and Bearings are 0104; Light Switches and Cables, 0606; Steering Knuckle, Flange and Arm, 1006, etc.

Part numbers are separately listed in numerical order and indexed to the groups in which they may be found, while an alphabetical index to the parts keying them to their group is also provided at the end of this book.

(Prepared in accordance with ES-737)

THE AUTOCAR COMPANY
ARDMORE, PENNSYLVANIA
U. S. A.

Standard Set-up for Parts Books and Setting Up of Parts Depots

01 Engine Group
		PAGE
0100	Engine Assembly	01-1
0101	Cylinder Block and Head	01-1
0102	Crankshaft, Bearings and Caps	01-3
0103	Pistons, Rings and Piston Pins	01-5
0104	Connecting Rods and Bearings	01-6
0105	Valves, Springs, Guides, Tappets	01-7
0106	Camshaft, Timing, Bearings, Etc.	01-7
0107	Oil Pump, Oil Pan, Gauge	01-8
0108	Manifold	01-12
0109	Flywheel, Ring Gear, Etc.	01-13
0110	Mountings	01-14

02 Clutch Group
0200	Clutch Assembly	02-1
0201	Clutch Disc	02-1
0202	Cover, Pressure Plate and Springs	02-1
0203	Release Lever, Bearings, Forks, Etc.	02-1
0204	Pedal	02-2
0205	Pilot Bearing	02-3

03 Fuel System Group
0300	Gasoline Tank	03-0
0301	Carburetor, Air Cleaner	03-0
0302	Fuel Pump, Fuel Pump Tube, Hose, Etc.	03-2
0303	Accelerator, Throttle and Choke	03-3
0304	Fuel Tank Lines to Gauge	03-4
0305	Governor	03-4

04 Exhaust Group
0401	Muffler	04-1
0402	Pipe and Tail Pipe	04-1

05 Cooling Group
0501	Radiator Shell, Core and Mounting	05-1
0502	Thermostat	05-1
0503	Water Pump, Fan, Belt	05-1
0505	Engine Water Fittings and Hose	05-4

06 Electrical Group
0601	Generator	06-1
0602	Starting Motor	06-2
0603	Distributor	06-6
0604	Ignition Coil and Wiring (Spark Plugs, Ignition Lock)	06-8
0605	Instruments and Carrier	06-9
0606	Light Switches and Cables	06-9
0607	Headlamps, Sealed Beams, Bulbs	06-9
0608	Tail and Auxiliary Lamp	06-10
0609	Horn	06-10
0610	Battery, Starting Cables and Connections	06-10
0611	Radio Suppression	06-11

07 Transmission Group
0700	Transmission	07-0
0701	Case	07-0
0702	Gears	07-1
0703	Main Drive Pinion and Bearings	07-2
0704	Main Shaft, Countershaft, and Reverse Idler Countershaft	07-4
0705	Speedometer Drive Gears	07-5
0706	Shift Forks, Levers	07-5

08 Transfer Case Group
0800	Transmission Transfer Assembly	08-1
0801	Case	08-1
0802	Drive Gear, Shaft, Bearings	08-1
0803	Driven Gear, Shaft, Bearings	08-2
0804	Idler Gear, Shaft, Bearings and Caps	08-3
0805	Shifter Shafts, Yokes, and Shift Levers	08-4

09 Propeller Shaft and Universal Joint Group
0901	Propeller Shaft Assembly	09-1
0902	Universal Joints	09-1

10 Front Axle Group
		PAGE
1000	Front Axle Assembly	10-1
1001	Housing	10-1
1002	Differential and Carrier Assembly	10-1
1003	Drive Gear, Pinion and Bearings	10-2
1006	Steering Knuckle, Flange and Arm	10-4
1007	Axle Shaft and Universal Joint	10-6

11 Rear Axle Group
1100	Rear Axle Assembly	11-0
1101	Housing Assembly	11-0
1102	Axle Drive Shafts	11-0
1103	Differential and Carrier Assembly	11-1
1104	Differential Side Gears and Pinions	11-3
1105	Drive Gear and Bearings	11-3

12 Brake Group
1201	Hand Brake	12-1
1202	Shoes and Facing	12-4
1203	Brake Shoe Support, Guide, Springs, Adjusting Pins, and Anchor Plate	12-5
1204	Pedal	12-7
1205	Air Compressor	12-7
1206	Tubes and Clips, Brackets, Springs	12-9
1207	Brake Chamber	12-9
1208	Brake Dust Shield	12-10
1209	Brake Lines, Pipes, Hoses	12-10

13 Wheel, Hub and Drum Group
1301	Wheel Assembly, Bearings, Retainers, Etc.	13-1
1302	Hubs and Drums	13-2

14 Steering Group
1401	Steering Connecting Rod (Drag Link)	14-0
1402	Tie Rod	14-0
1403	Gear Assembly	14-1
1404	Wheel Assembly	14-3
1405	Brackets	14-3

15 Frame and Bracket Group
1500	Frame and Brackets	15-1
1501	Towing Attachment	15-1
1502	Pintle Hooks	15-1

16 Spring and Shock Absorber Group
1601	Front and Rear Springs	16-0
1602	Shackles and Spring Attaching Parts	16-3
1603	Shock Absorbers and Mountings	16-5

18 Body Group
1800	Cab Assembly and Parts	18-0
	Body Parts List	18-0
1801	Windshield Wiper and Parts	18-5

19 Winch Assy.—Winch and Control Levers
1900	Winch and Winch Drive Shafts	19-0
1910	Control Levers and Rods	19-1
1911	P.T.O. Assembly for Winch	19-2

21 Bumper and Guard Group
2101	Bumpers	21-1
2103	Radiator Guard	21-1

22 Miscellaneous Body, Chassis and Accessory Group
2200	Identification and Caution Plates	22-1
2201	Rear View Mirrors	22-1
2202	Tarpaulins, Bows, and End Curtains	22-1
2203	Speedometer and Parts	22-1

23 General Use, Standardized Parts Group
2300	Miscellaneous Tools, Tire Chains, Extinguishers	23-1
	Summary of Ball and Roller Bearings	23-3
2304	Miscellaneous Nuts, Bolts, Screws, and Washers	23-6

THE AUTOCAR COMPANY
ARDMORE, PA.

Master Parts List

00-1
GENERAL

00

AUTOCAR MODEL U-2044
2½ Ton 4 x 4 Oil Servicing Truck

GENERAL DATA

Wheelbase	128"
Back of Cab to C/L of Rear Axle	87"
Back of Cab to End of Frame	119"
ENGINE	Hercules
Model	JXD
Cylinders	6
Bore	4"
Stroke	4¼"
Displacement—Cu. In.	320
A.M.A. Horsepower	38.4

WEIGHT	FRONT	REAR	TOTAL
Chassis and Cab	4750	3910	8660

TIRES—9.00/20—10 Ply

Contract	No. Units	Date	Autocar Serial No.	U. S. Registration No.	Parts List
W-398-QM-7902	50	5/27/40	1883 2250 to 2298 incl.	W-80238 to W-80287	TM-10-1392
W-398-QM-8534	183	11/12/40	3449 to 3628 incl. 3801 to 3803 incl.	W-80568 to W-80750	TM-10-1394

AUTOCAR MODEL U-4044
2½ TON 4 x 4 Tractor Truck

GENERAL DATA

Wheelbase	131"
Back of Cab to C/L of Rear Axle	90"
Back of Cab to End of Frame	122"
ENGINE	Autocar
Model	358
Cylinders	6
Bore	4"
Stroke	4¾"
Displacement—Cu. In.	358
A.M.A. Horsepower	38.4

WEIGHT	FRONT	REAR	TOTAL
Chassis and Cab	5730	4310	10040

TIRES—9.00/20—10 Ply

Contract	No. Units	Date	Autocar Serial No.	U. S. Registration No.	Parts List
W-398-QM-7902	99	5/27/40	1842 2750 2100 to 2170 incl. 2172 to 2183 incl. 2184 to 2197 incl.	W-413527 W-413599 W-413528 to W-413598 W-413716 to W-413727 W-413600 to W-413613	TM-10-1390
W-398-QM-8551	1	10/18/40	2171	W-417599	
W-398-QM-8534	448	11/12/40	3001 to 3448 incl.	W-428106 to W-428338 W-428374 to W-428588	TM-10-1396

THE AUTOCAR COMPANY
ARDMORE, PA.

Master Parts List

00-3
GENERAL

AUTOCAR MODEL U-4144
2½ Ton 4 x 4 Oil Servicing Truck

GENERAL DATA

Wheelbase	128″
Back of Cab to C/L of Rear Axle	87″
Back of Cab to End of Frame	119″
ENGINE	Autocar
Model	358
Cylinders	6
Bore	4″
Stroke	4¾″
Displacement—Cu. In.	358
A.M.A. Horsepower	38.4

WEIGHT	FRONT	REAR	TOTAL
Chassis and Cab	5730	4310	10040

TIRES—9.00/20

Contract	No. Units	Date	Autocar Serial No.	U. S. Registration No.	Parts List
W-398-QM-9433	138	2/6/41	6001 to 6138 incl.	W-80770 to W-80907	

00-4 GENERAL

Master Parts List

THE AUTOCAR COMPANY ARDMORE, PA.

AUTOCAR MODEL U-4144-T
2½ Ton 4 x 4 Tractor Truck

GENERAL DATA

Wheelbase	131″
Back of Cab to C/L of Rear Axle	90″
Back of Cab to End of Frame	122″
ENGINE	Autocar
Model	358
Cylinders	6
Bore	4″
Stroke	4¾″
Displacement—Cu. In.	358
A.M.A. Horsepower	38.4

WEIGHT	FRONT	REAR	TOTAL
Chassis and Cab	5730	4310	10040

TIRES—9.00/20—10 Ply

Contract	No. Units	Date	Autocar Serial No.	U. S. Registration No.	Parts List
W-398-QM-9433	274	2/6/41	6139 to 6412 incl.	W-460205 to W-460478	

THE AUTOCAR COMPANY
ARDMORE, PA.

Master Parts List

00-5
GENERAL

AUTOCAR MODEL C-50
5 Ton 4 x 2 Dump Truck

GENERAL DATA

Wheelbase	175"
Back of Cab to C/L of Rear Axle	95"
Back of Cab to End of Frame	129½"
ENGINE	Autocar
Model	358
Cylinders	6
Bore	4"
Stroke	4¾"
Displacement—Cu. In.	358
A.M.A. Horsepower	38.4

WEIGHT	FRONT	REAR	TOTAL
Chassis and Cab	4820	4370	9190

TIRES—9.75/20

Contract	No. Units	Date	Autocar Serial No.	U. S. Registration No.	Parts List
W-398-QM-8290	7	10/4/40	2491-2493-2494-2496 2498-2500-2501	W-51015 to W-51021	

AUTOCAR MODEL U-5044
5 Ton 4 x 4 Tractor Truck

GENERAL DATA

Wheelbase	131"
Back of Cab to C/L of Rear Axle	90"
Back of Cab to End of Frame	122"
ENGINE	Autocar
Model	377
Cylinders	6
Bore	4"
Stroke	5"
Displacement—Cu. In.	377
A.M.A. Horsepower	38.4

WEIGHT	FRONT	REAR	TOTAL
Chassis and Cab	5670	4170	9840

TIRES—9.00/20—10 Ply

Contract	No. Units	Date	Autocar Serial No.	U. S. Registration No.	Parts List
W-398-QM-8687	97	9/24/40	3629 to 3725 incl.	W-51313 to W-51409	TM-10-1160

THE AUTOCAR COMPANY
ARDMORE, PA.

Master Parts List

00-7
GENERAL

AUTOCAR MODEL U-7144-T
4-5 Ton 4 x 4 Tractor Truck

GENERAL DATA

Wheelbase	134½"
Back of Cab to C/L of Rear Axle	90"
Back of Cab to End of Frame	121"
ENGINE	Hercules
Model	RXC
Cylinders	6
Bore	4⅝"
Stroke	5¼"
Displacement—Cu. In.	529
A.M.A. Horsepower	51.3

WEIGHT	FRONT	REAR	TOTAL
Chassis and Cab, Complete as Illustrated	6900	5300	12200

TIRES—9.00/20—10 Ply, Bus-Balloon, Mud and Snow Type Tread.

Contract	No. Units	Date	Autocar Serial No.	U. S. Registration No.	Parts List
W-398-QM-10141	1182	6/13/41	A-1 to A-1169 incl. A-1552 to A-1564 incl.	W-461794 to W-462750 W-482421 to W-482632 W-482662 to W-482674	TM-10-1116
W-398-QM-10802	100	10/16/41	A-1565 to A-1664 incl.	W-487221 to W-487320	
W-398-QM-11410	1790	11/27/41	A-1710 to A-3459 incl. A-4419 to A-4458 incl.	490699 to 492448 4108532 to 4108571	
W-398-QM-12332	357	3/6/42	A-4459 to A-4815 incl.	4110585 to 4110735 4113260 to 4113465	
W-398-QM-12974	8346				

AUTOCAR MODEL U-8144
5-6 Ton 4 x 4 Van Body Truck

GENERAL DATA

Wheelbase	163½"
Back of Cab to C/L of Rear Axle	119"
Back of Cab to End of Frame	209 5/16"
ENGINE	Hercules
Model	RXC
Cylinders	6
Bore	4⅝"
Stroke	5¼"
Displacement—Cu. In.	529
A.M.A. Horsepower	51.3

WEIGHT	FRONT	REAR	TOTAL
Chassis and Cab	7060	6550	13610
Chassis, Cab, Body	7320	11960	19280

TIRES—12.00/20—14 Ply

Contract	No. Units	Date	Autocar Serial No.	U. S. Registration No.	Parts List
W-398-QM-11624	302		A-3461 to A-3762 incl.	55953 to 56254	TM-10-1498

THE AUTOCAR COMPANY
ARDMORE, PA.

Master Parts List

00—9
GENERAL

AUTOCAR MODEL U-8144-T
5-6 Ton 4 x 4 Ponton Tractor Truck

GENERAL DATA

Wheelbase	163½"
Back of Cab to C/L of Rear Axle	119"
Back of Cab to End of Frame	150"
ENGINE	Hercules
Model	RXC
Cylinders	6
Bore	4⅝"
Stroke	5¼"
Displacement—Cu. In.	529
A.M.A. Horsepower	51.3

WEIGHT	FRONT	REAR	TOTAL
Chassis and Cab	8280	5935	14215
Chassis, Cab, Body, 5th Wheel, Complete	8860	6865	15725

TIRES—12.00/20—14 Ply

Contract	No. Units	Date	Autocar Serial No.	U. S. Registration No.	Parts List
W-398-QM-10325	240	6/11/41	A-1170 to A-1409 incl.	W-52930 to W-53089 W-53117 to W-53196	TM-10-1118
DA-W-398-QM-7	142	6/11/41	A-1410 to A-1551 incl.		
W-398-QM-10803	45	9/ 6/41	A-1665 to A-1709 incl.	W-53526 to W-53570	TM-10-1118
W-398-QM-11562	1		A-3460	55274	
W-398-QM-11624			A-3763 to A-4326	56255 to 56818	TM-10-1496
DA-W-398-QM-526	72				

THE AUTOCAR COMPANY,
ARDMORE, PA.

Master Parts List

GROUP 01

ILLUSTRATIONS

GROUP 01

ENGINE

GROUP 01

Master Parts List

THE AUTOCAR COMPANY, ARDMORE, PA.

As supplied on Autocar Models
U-7144-T U-8144 U-8144-T

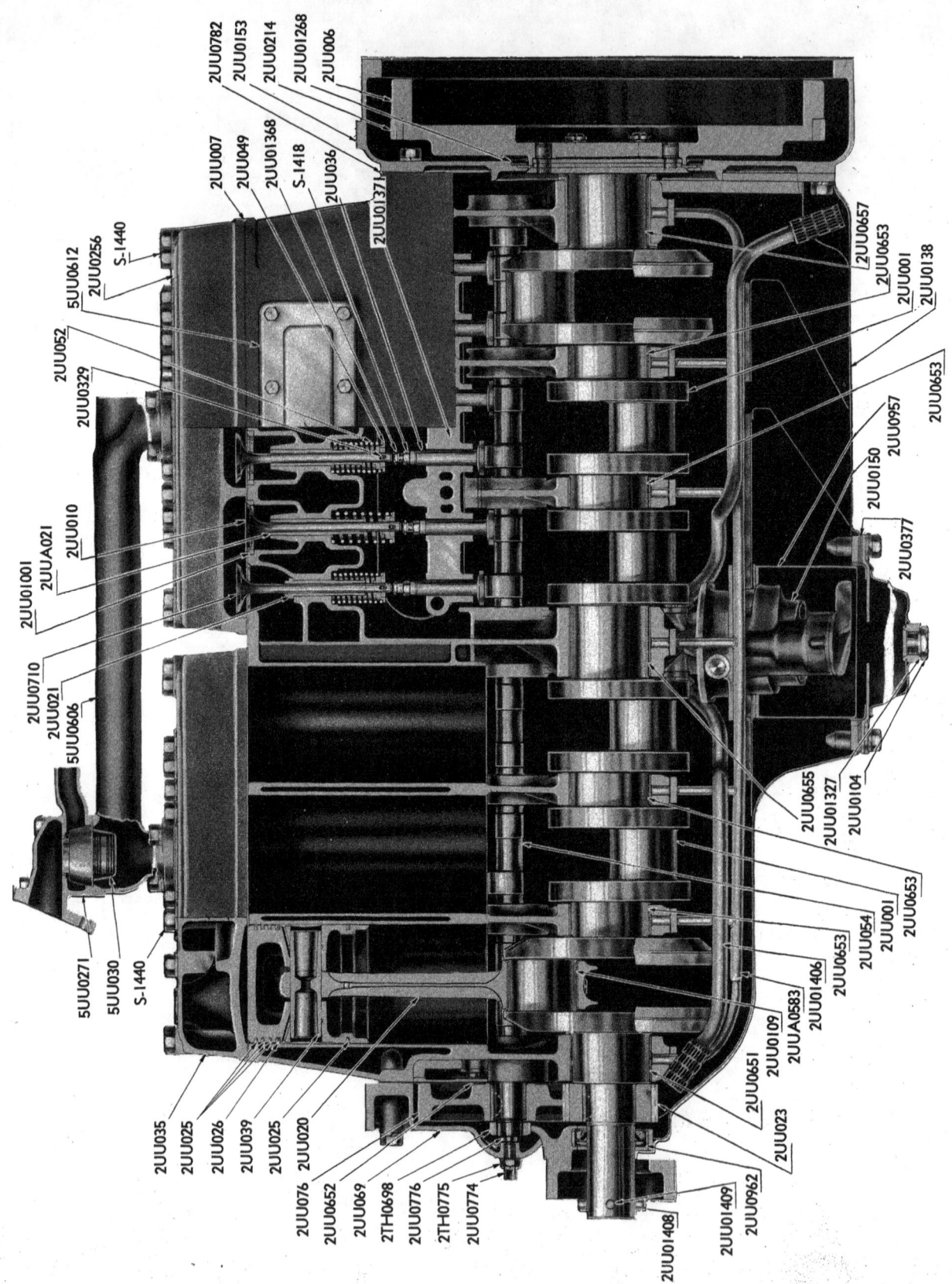

0100—Hercules Model RXC Engine—Autocar Assembly No. 2UU080

THE AUTOCAR COMPANY, ARDMORE, PA.

Master Parts List

GROUP 01

As supplied on Autocar Models
U-7144-T   U-8144   U-8144-T

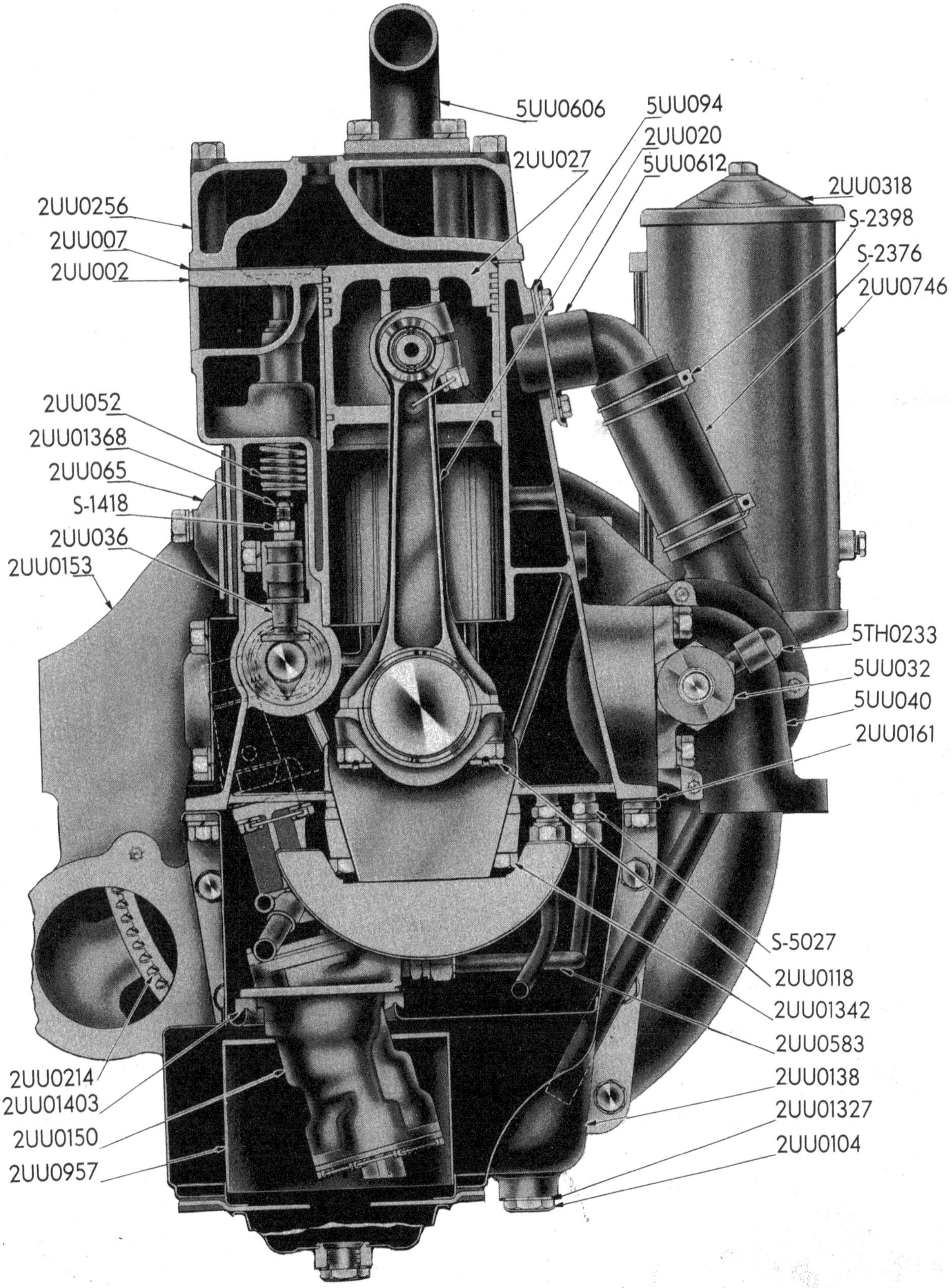

0100—Hercules Model RXC Engine—Autocar Assembly No. 2UU080

GROUP 01

Master Parts List

THE AUTOCAR COMPANY, ARDMORE, PA.

As supplied on Autocar Models
U-8144 U-8144-T

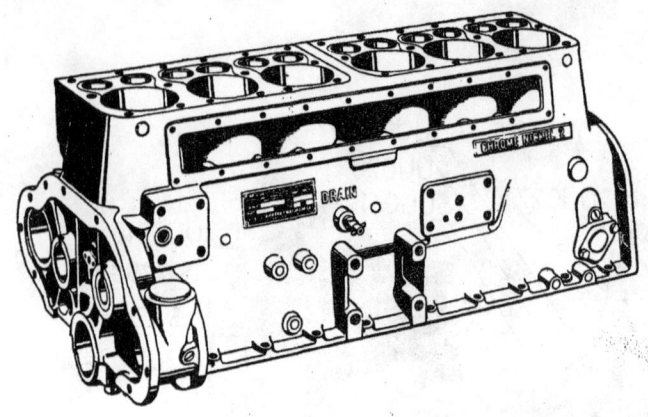
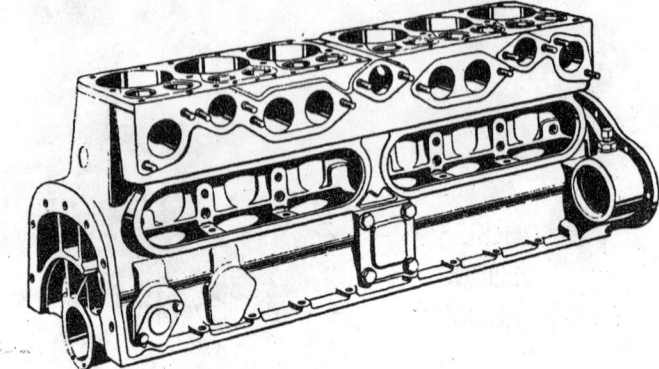
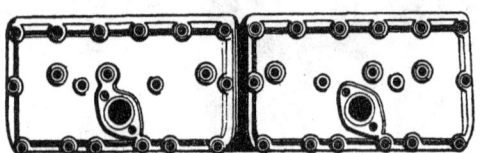
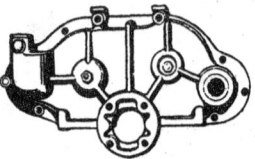

0101—Cylinder Block and Head

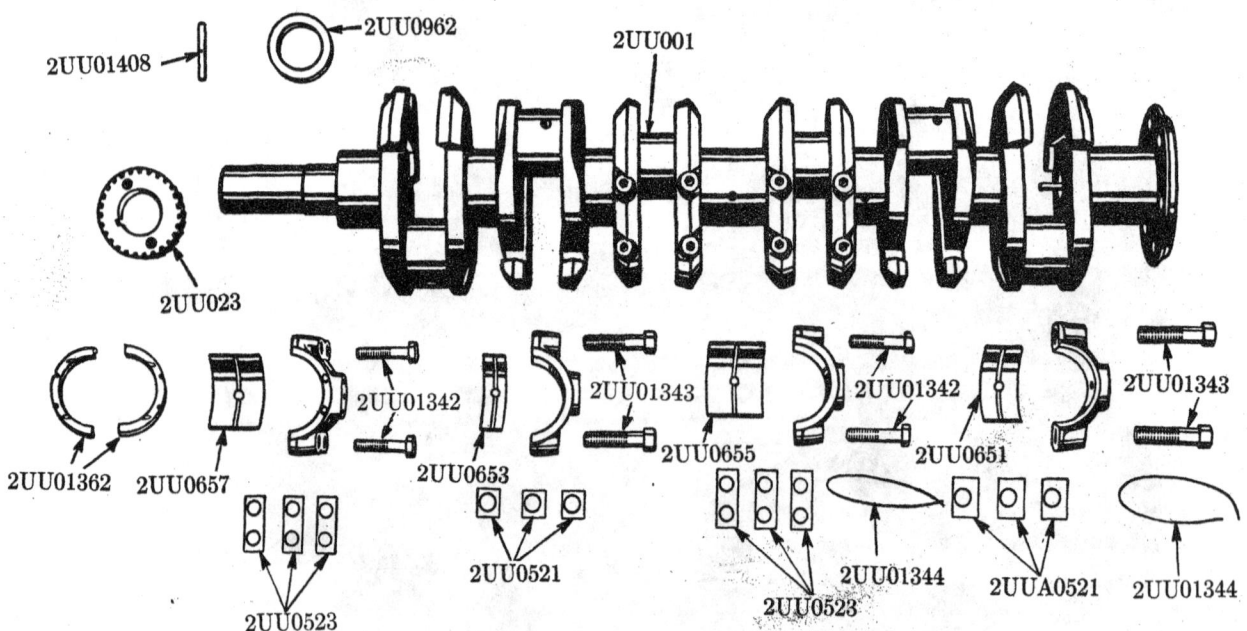

0102—Crankshaft, Bearings and Caps

0103—Pistons, Rings, and Piston Pins

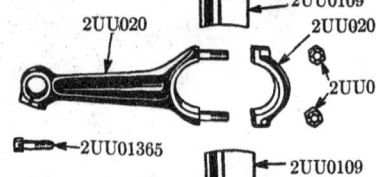
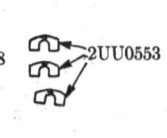

0104—Connecting Rods and Bearings

THE AUTOCAR COMPANY, ARDMORE, PA.

Master Parts List

GROUP 01

As supplied on Autocar Models
U-8144   U-8144-T

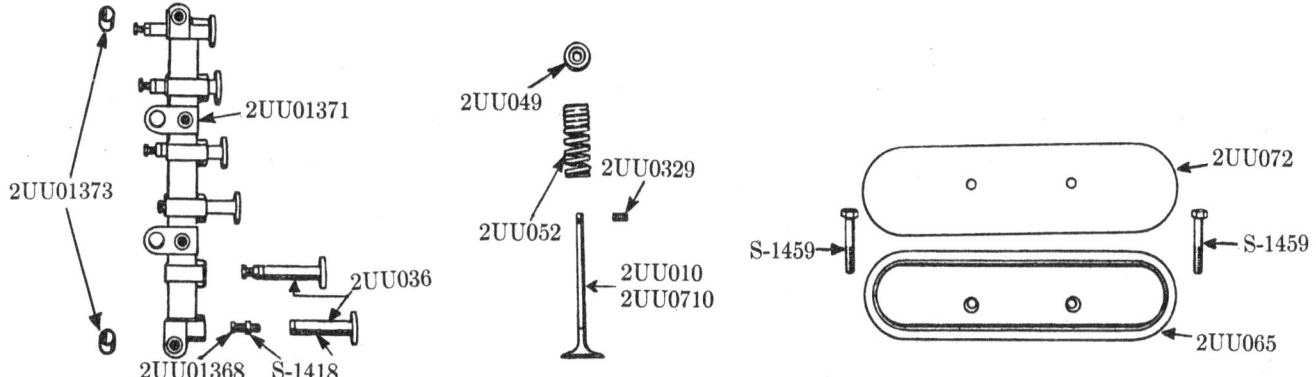

0105—Valves, Springs, Guides and Tappets

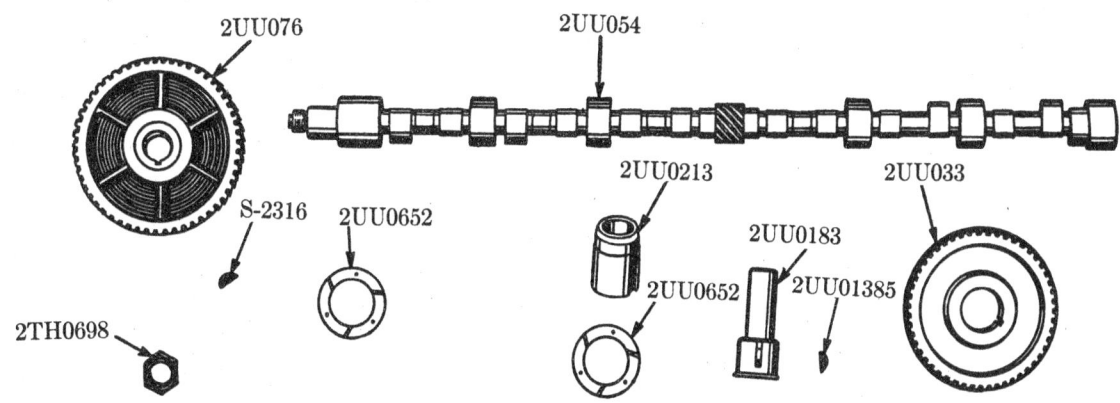

0106—Camshaft and Bearings

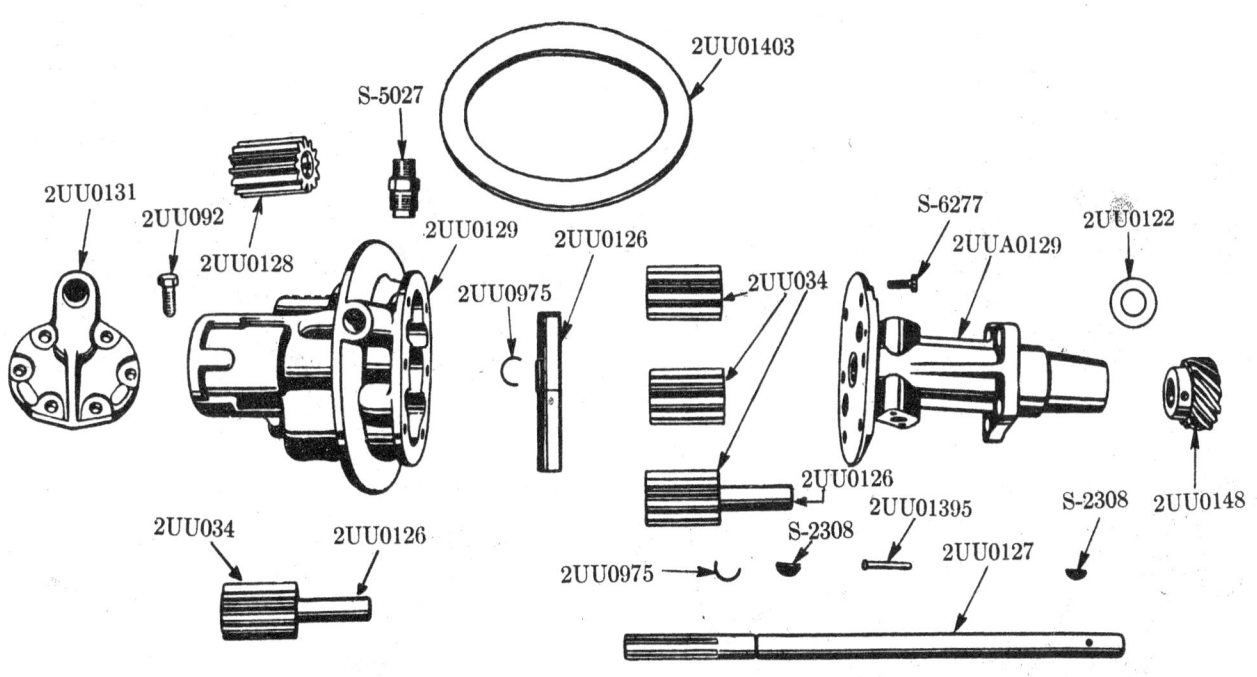

0107—Oil Pump

GROUP 01

Master Parts List

THE AUTOCAR COMPANY, ARDMORE, PA.

As supplied on Autocar Models
U-8144   U-8144-T

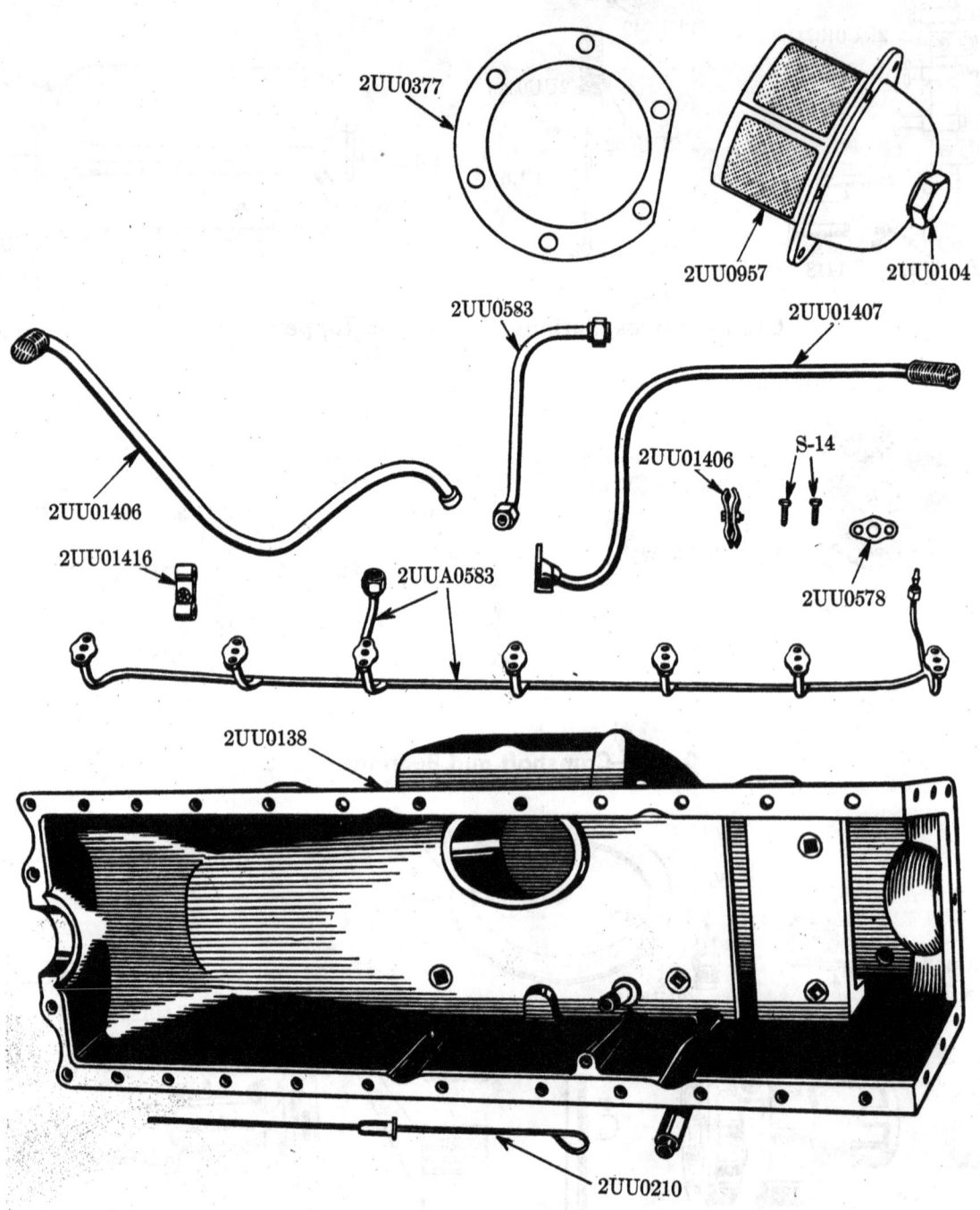

0107—Oil Pan and Filter

THE AUTOCAR COMPANY, ARDMORE, PA.

Master Parts List

GROUP 01

As supplied on Autocar Models
U-8144   U-8144-T

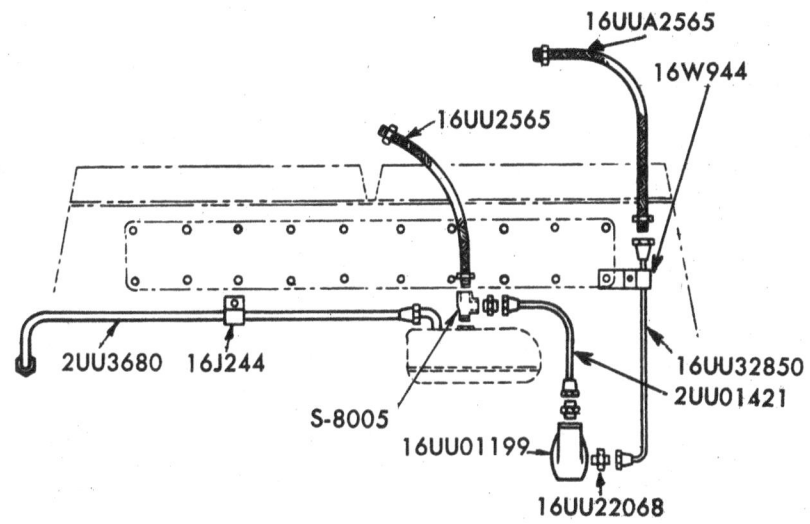

0107—Oil Line

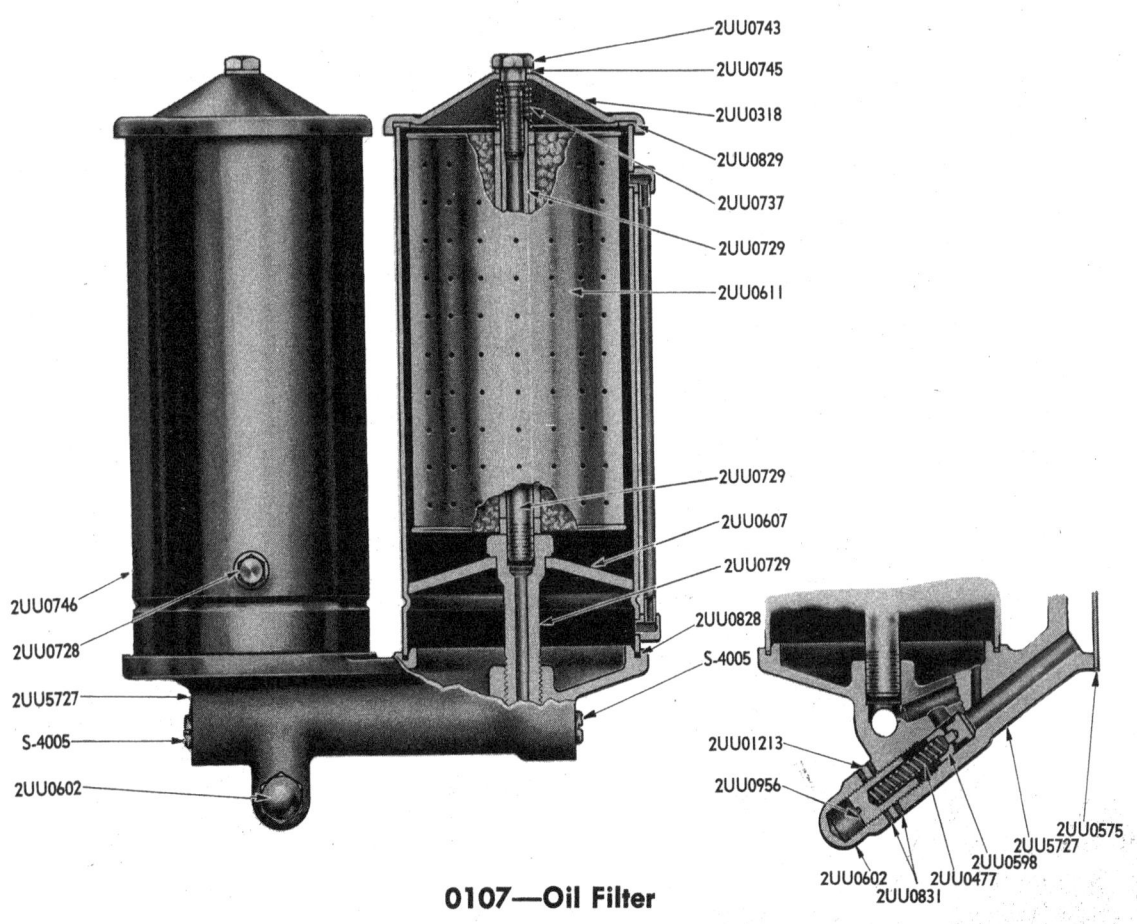

0107—Oil Filter

GROUP 01

Master Parts List

THE AUTOCAR COMPANY, ARDMORE, PA.

As supplied on Autocar Models
U-8144 U-8144-T

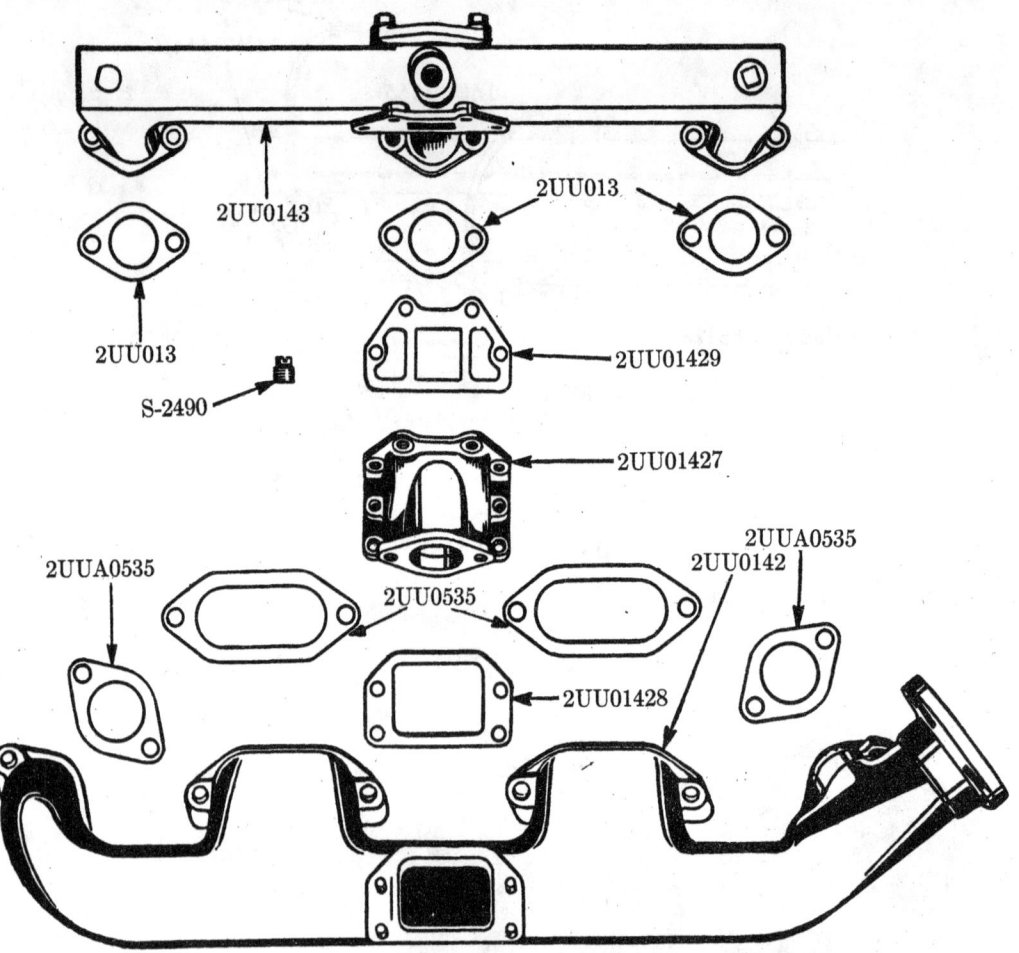

0108—Manifold

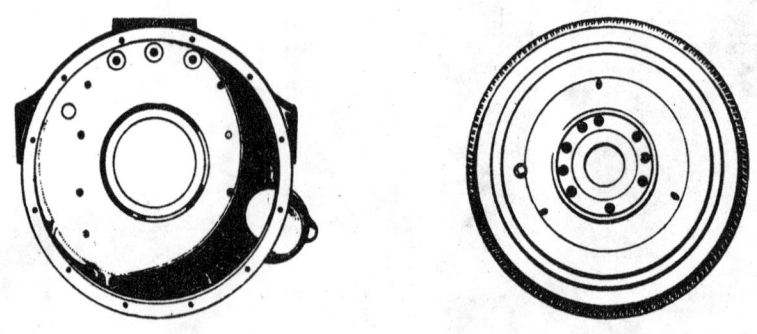

0109—Flywheel, Ring Gear

THE AUTOCAR COMPANY
ARDMORE, PA.

Master Parts List

01-1 ENGINE

CODE	MFR. PART No.	LIST PRICE	NAME OF PART	AUTOCAR PART No.	U-2044	U-4044	U-4144	U-4144-T	C-50	U-5044	U-7144-T	U-8144	U-8144-T	No. REQ.
0100	**Engine Assembly**													
	JXD (320)	....	Engine Assembly (Hercules)...................	JXD (320)	✓									1
	358	....	Engine Assembly (Autocar)....................	358		✓	✓	✓	✓					1
	377	....	Engine Assembly (Autocar)....................	377						✓				1
	RXC	$900.00	Engine Assembly (Hercules)...................	2UU0-80							✓	✓	✓	1
0101	**Cylinder Block and Head**													
	JXD (320)	161.64	Cylinder and Crankcase Assembly (Including Valve Seats, Main Bearings & Valve Guides).....................	2UGA0-809	✓									1
	2D0130	141.60	Cylinder Block Assembly (Including Valve Seats & all Studs).	2D0-130		✓	✓	✓	✓					1
	2J0130	141.60	Cylinder Block Assembly (Including Valve Seats & all Studs).	2J0-130						✓				1
	16356E	350.50	Cylinder Block (Includes Bearing Caps, Valve Guides, Camshaft Bearings & Idler Bearing)..................	2UU0-02							✓	✓	✓	1
	......	230.40	Crankcase Assembly............................	2UG0-320		✓	✓	✓						1
	......	230.40	Crankcase Assembly............................	2SA0-320						✓				1
	......	258.00	Crankcase Assembly............................	2B0-320						✓				1
	......	.12	Gasket, Cylinder Block.........................	2SA4-181B		✓	✓	✓	✓					1
	......	.12	Gasket, Cylinder Block.........................	2SA4-181A						✓				1
	......	.20	Stud, Cylinder Block..........................	2T2-48		✓	✓	✓	✓	✓				6
	......	.20	Stud, Cylinder Block..........................	2SA2-48		✓	✓	✓	✓	✓				12
	......	.20	Screw, Generator Lock.........................	2T2-93		✓	✓	✓	✓					1
	......	.05	Nut, Generator Lock Screw (⅝"—11)...........	S-2004		✓	✓	✓	✓					1
	......	.01	Washer, ⅝" Spring............................	S-1351		✓	✓	✓	✓					1
	......	.56	Washer, Generator—Felt......................	16A2-2535		✓	✓	✓	✓					1
	......	.56	Ring, Generator Oil Seal (Duprene).............	16A2-2535						✓				1
	......	.15	Shield, Generator Oil (Steel Washer)............	16A2-2534						✓				1
	......	.75	Plate, Engine Serial Number....................	2RE2-571B		✓	✓	✓	✓					1
	......	.35	Plate, Engine Serial Number....................	2B2-571						✓				1
	......	.01	Screw, Engine Serial Number Plate Drive (#8 x ⅜")....	S-3263		✓	✓	✓	✓	✓				4
	23123-A	.28	Inserts, Valve Seat............................	2BL0-1001	✓									6
	......	2.00	Inserts, Valve Seat............................	2D2-1001		✓	✓	✓	✓					6
	......	1.70	Inserts, Valve Seat............................	2B2-1001						✓				6
	16511-A	.70	Inserts, Exhaust Valve Seat....................	2UU0-1001							✓	✓	✓	6
	8-A	.45	Cock, Cylinder Drain (⅛").....................	S-2500	✓									1
	......	.25	Screw, Cylinder Water Jacket Pet Cock Cap....	2SA2-96A		✓	✓	✓	✓	✓				1
	8082-A	.38	Cock, Cylinder Block Drain (⅜") Pipe Thread...	2UU0-1339							✓	✓	✓	1
	......	.10	Washer, Cylinder Water Jacket Pet Cock Cap Screw......	2SA2-164		✓	✓	✓	✓	✓				1
	......	.20	Washer, Cylinder Water Jacket Pet Cock Cap Screw—Lead..	2SA2-169		✓	✓	✓	✓	✓				1
	1609-A	.04	Plug, Cylinder Welch (1¼")...................	S-4212	✓									9
	2534A	.03	Plug, Cylinder Block Expansion (1¼")...........	2UU0-1338							✓	✓	✓	2
	665A	.02	Plug, Cylinder Block Expansion (⅝")...........	2UUA0-1338							✓	✓	✓	5
	565A	.04	Plug, Cylinder Block Pipe (¾").................	2UU0-1341							✓	✓	✓	4
	7165A	.06	Plug, Cylinder Block Pipe (1").................	2UUA0-1341							✓	✓	✓	1
	15739A	.70	Orifice, Cylinder Block Oil.....................	2UU0-1345							✓	✓	✓	1
	4251A	.02	Plug, Cylinder Block Oil Orifice................	2UU0-1346							✓	✓	✓	1
	15133A	.02	Gasket, Cylinder Block Oil Orifice..............	2UU0-578							✓	✓	✓	1
	4243A	.02	Screw, Cylinder Block Oil Orifice (5/16"—18 x ¾")....	2UU0-1348							✓	✓	✓	2
	4242A	.02	Screw, Cylinder Block Oil Orifice L.W. (5/16")....	2UU0-1349							✓	✓	✓	2
	2129A	.04	Dowel, Cylinder Block Pressure Regulator.......	2UU0-1356							✓	✓	✓	1
	4790A	.04	Oil Seal, Cylinder Block Generator.............	2UU0-1359							✓	✓	✓	1
	17105A	.50	Pin, Cylinder Block Fuel Pump Drive...........	2UU0-1361							✓	✓	✓	1
	60A	.04	Plug, Cylinder Block Fuel Pump Drive Pin Hole Pipe.....	S-4748							✓	✓	✓	2
	......	.30	Cover, Fuel Pump Opening.....................	2T2-74		✓	✓	✓	✓	✓				1
	......	.12	Gasket, Fuel Pump Opening Cover..............	2T2-67		✓	✓	✓	✓	✓				1
	......	.05	Screw, Fuel Pump Opening Cover (⅜"—16 x ⅞").	S-50		✓	✓	✓	✓	✓				4
	......	.01	Lockwasher, ⅜" Spring........................	S-1347		✓	✓	✓	✓	✓				4
	4879A	.04	Cover, Cylinder Block Fuel Pump Opening......	2UU0-74							✓	✓	✓	1
	22564A	.03	Gasket, Cylinder Block Fuel Pump Opening Cover	2BL0-67							✓	✓	✓	1
	1048A	.03	Screw, Cylinder Block Fuel Pump Opening Cover (5/16"—18 x ½")	2UU0-1357							✓	✓	✓	2
	615A	.02	Screw, Cylinder Block Fuel Pump Opening Cover L.W. (5/16")..	2UU0-1358							✓	✓	✓	2
	45921D	21.50	Cylinder Head................................	2UG0-35	✓									1
	4118A	.09	Screw, Cylinder Head (½"—13 x 3⅛")...........	S-3106	✓									26

Master Parts List

CODE	MFR. PART No.	LIST PRICE	NAME OF PART	AUTOCAR PART No.	MODEL									No. REQ.
					U-2044	U-4044	U-4144	U-4144-T	C-50	U-5044	U-7144-T	U-8144	U-8144-T	
0101	1075A	$0.04	**Plug,** Cylinder Head Pipe (½″)	S-4172	✓									1
	......	31.00	**Cylinder** Head Assembly	2D0-912		✓			✓					1
	......	31.75	**Cylinder** Head Assembly	2URM0-912			✓	✓						1
	......	30.15	**Cylinder** Head Assembly	2J0-912						✓				..
	......	29.50	**Cylinder** Head	2URM6-35		✓	✓	✓						1
	......	29.50	**Cylinder** Head	2D6-35					✓					1
	......	29.50	**Cylinder** Head	2B6-35						✓				1
	17214D	18.12	**Cylinder** Head—Front	2UU0-35							✓	✓	✓	1
	17215D	18.12	**Cylinder** Head—Rear	2UU0-256							✓	✓	✓	1
	......	2.50	**Gasket,** Cylinder Head	2RM5-07		✓	✓	✓		✓				1
	......	2.25	**Gasket,** Cylinder Head	2SA5-07					✓					1
	16155C	1.20	**Gasket,** Cylinder Head	2UU0-07							✓	✓	✓	2
	......	.35	**Stud,** Cylinder Head	2T2-151		✓	✓	✓	✓	✓				23
	......	.30	**Stud,** Cylinder Head	2T2-149		✓	✓	✓						1
	......	.12	**Nut,** Cylinder Head Acorn (½″—20 Nickel Plated)	S-3742		✓	✓	✓	✓	✓				23
	......	1.50	**Elbow,** Cylinder Head Water Outlet	5URM3-62		✓	✓	✓						1
	......	.12	**Gasket,** Cylinder Head Water Outlet Elbow	S-4083		✓	✓	✓		✓				1
	......	1.20	**Elbow,** Cylinder Head Water Outlet	5RH3-62					✓					1
	......	3.50	**Elbow,** Cylinder Head Water Outlet	5UP3-62						✓				1
	1075A	.04	**Plug,** Cylinder Head Pipe	2UUA0-1376							✓	✓	✓	2
	1608A	.08	**Screw,** Cylinder Head Cap	2UU0-1377							✓	✓	✓	32
	17030D	18.76	**Cover,** Cylinder Block Gear Case	2UU0-69							✓	✓	✓	1
	15024C	.12	**Gasket,** Cylinder Block Gear Case Cover	2UU0-178							✓	✓	✓	1
	352A	.04	**Screw,** Cylinder Block Gear Case Cover Attach. (½″—13 x 1¼″)	S-25							✓	✓	✓	5
	8552A	.06	**Screw,** Cylinder Block Gear Case Cover Attach. (½″—13 x 2¾″)	S-4077							✓	✓	✓	2
	4038A	.06	**Screw,** Cylinder Block Gear Case Cover Attach. (½″—13 x 3½″)	S-4063							✓	✓	✓	1
	312A	.02	**Lockwasher,** Cylinder Block Gear Case Cover	2UU0-1355							✓	✓	✓	8
	14236AS	1.10	**Oil Seal,** Cylinder Block Gear Case Cover (Accessory Shaft)	2UUA0-962							✓	✓	✓	1
	......	8.09	**Set,** Engine Gasket	2UU0-1200							✓	✓	✓	1
	40006A	.24	**Cover,** Cylinder Valve	2BL0-65	✓									2
	4068A	.11	**Screw,** Cylinder Valve Cover (½″—13 x 2½″)	S-1871	✓									4
	4643A	.03	**Plug,** Cylinder Oil Tube	S-4208	✓									1
	60A	.02	**Plug,** Cylinder Oil Tube Pipe (¼″)	S-629	✓									1
	59A	.02	**Plug,** Cylinder Oil Tube Pipe (⅛″)	S-619	✓									7
	4312A	.06	**Plug,** Cylinder Pipe (½″)	S-2490	✓									4
	22158A	.04	**Plug,** Cylinder Oil Header Pipe	22158A	✓									1
	......	2.00	**Plate,** Cylinder Water Jacket (Front)	2SA3-61		✓	✓	✓	✓	✓				1
	......	3.00	**Plate,** Cylinder Water Jacket (Rear)	2URL4-61A		✓	✓	✓		✓				1
	......	.05	**Screw,** Cylinder Water Jacket Plate (⅜″—16 x ¾″)	S-12		✓	✓	✓		✓				27
	......	.01	**Washer,** ⅜″ Spring	S-1347		✓	✓	✓		✓				27
	......	.20	**Stud**	2CF2-159A		✓	✓	✓						1
	......	.15	**Gasket,** Cylinder Water Jacket Plate	2URL3-81		✓	✓	✓		✓				2
	......	1.20	**Plate,** Valve Cover	2SA3-65C		✓	✓	✓	✓	✓				2
	......	.24	**Gasket,** Valve Cover Plate	2SA3-72A		✓	✓	✓	✓	✓				2
	......	.60	**Bar,** Valve Cover Plate Clamp	2D3-1131		✓	✓	✓	✓	✓				2
	......	.06	**Screw,** Valve Cover Plate Clamp Bar (⅜″—16 x 1¼″)	S-64		✓	✓	✓		✓				4
	......	.06	**Washer,** Valve Cover Plate Clamp ($\frac{13}{32}$″ I.D.—¾″ O.D. x $\frac{1}{16}$″)	S-735		✓	✓	✓		✓				4
	......	.20	**Stud,** Cylinder	2T2-48		✓	✓	✓	✓	✓				6
	......	.20	**Stud,** Cylinder	2SA2-48		✓	✓	✓	✓	✓				12
	......	.35	**Stud,** Cylinder Head	2T2-151		✓	✓	✓	✓	✓				23
	......	.20	**Stud,** Hand Throttle	2TE2-1008A		✓	✓	✓	✓					1
	......	.20	**Stud,** Clutch Oiler & Coil	2TE2-1008B		✓	✓	✓	✓					2
	......	.20	**Stud,** Clutch Oil Line Bracket	2TE2-1008B		✓	✓	✓	✓	✓				2
	......	.20	**Stud,** Clutch Oil Line Bracket	2TE2-1008B			✓	✓						2
	......	.01	**Nut,** Clutch Oil Line Bracket Stud (⅜″—16 x ¼″)	S-829		✓	✓	✓		✓				2
	......	.01	**Washer,** ⅜″ Spring	S-1347		✓	✓	✓		✓				4
	......	1.20	**Plate,** Cylinder Water Jacket (Pressed Steel)	2SA3-61B					✓					2
	......	.12	**Gasket,** Cylinder Water Jacket Plate	2SA3-81					✓					2
	......	.30	**Stud,** Cylinder Head	2T2-149						✓				1
	......	.06	**Nut,** Cylinder Head Stud (½″—20)	S-3723						✓				1

THE AUTOCAR COMPANY
ARDMORE, PA.

Master Parts List

01-3
ENGINE

CODE	MFR. PART No.	LIST PRICE	NAME OF PART	AUTOCAR PART No.	U-2044	U-4044	U-4144	U-4144-T	C-50	U-5044	U-7144-T	U-8144	U-8144-T	No. REQ.
0102	Crankshaft, Bearings and Caps													
	45853E	$51.50	Crankshaft	2BL0-01	✓									1
	......	....	*Crankshaft	2SA7-01C		✓	✓	✓						1
	......	....	*Crankshaft	2SA7-01B					✓					1
	......	....	*Crankshaft	2B7-01						✓				1
	17527E	220.00	Crankshaft	2UU0-01							✓	✓	✓	1
	22104A	.14	Bolt, Flywheel	2BL0-397	✓									4
	......	.15	Bolt, Flywheel	2T2-397A		✓	✓	✓	✓	✓				6
	7104A	.40	Bolt, Crankshaft Flywheel	2UU0-397							✓	✓	✓	4
	1656A	.05	Nut, Flywheel Bolt (½"—20)	S-811	✓	✓	✓	✓	✓	✓				6
	1382A	.06	Nut, Crankshaft Flywheel Bolt	2UU0-1411							✓	✓	✓	4
	......	.01	Pin, Flywheel Bolt Nut Cotter (3/32" x 1¼")	S-2200		✓	✓	✓	✓	✓				6
	1710A	.02	Cotter, Crankshaft Flywheel Bolt	2UU0-1364							✓	✓	✓	4
	1707A	.12	Dowel, Flywheel	2TH0-837	✓									2
	1707A	.12	Dowel, Crankshaft Flywheel	2TH0-837							✓	✓	✓	2
	665A	.03	Plug, Flywheel Welch (⅝")	S-4207	✓									2
	665A	.02	Plug, Crankshaft Flywheel Expansion	S-4207							✓	✓	✓	2
	22039B	2.60	Gear, Crankshaft	2BL0-23	✓									1
	......	2.70	Gear, Crankshaft	2SA2-23B		✓	✓	✓	✓	✓				1
	15039B	3.76	Gear, Crankshaft	2UU0-23							✓	✓	✓	1
	4265A	.06	Key, Crankshaft Gear (#91 Woodruff)	S-2319	✓									1
	1247A	.02	Key, Crankshaft Gear	2UU0-1387							✓	✓	✓	1
	40806A	.62	Ratchet, Crankshaft Starting	1BL0-37	✓									1
	......	3.30	Ratchet, Crankshaft Starting	2TE3-1068						✓				1
	......	3.50	Jaw, Crankshaft Starting Crank	1UU2-37							✓	✓	✓	1
	4642A	.04	Dowel, Crankshaft	2BL0-1221	✓									1
	1675A	.14	Pin, Crankshaft Starting Crank	2UU0-1408							✓	✓	✓	1
	1739A	.02	Screw, Crankshaft Starting Crank Pin Set	2UU0-1409							✓	✓	✓	1
	2129A	.04	Pin, Crankshaft Knurled	2UU0-1356							✓	✓	✓	1
	40076A	.10	Thrower, Crankshaft Oil	2BL0-1225	✓									1
	14626AS	.38	Wick, Crankshaft Oil, Assembly	2UU0-1225							✓	✓	✓	1
	......	6.00	Seal, Crankshaft Rear Bearing & Gasket Assembly	2T0-410		✓	✓	✓	✓					1
	......	.12	Gasket, Crankshaft Rear Bearing Seal	2T3-642		✓	✓	✓	✓					1
	......	.08	Gasket, Crankshaft Rear Bearing Seal—Center	2T2-988		✓	✓	✓	✓					2
	11137AS	1.60	Oil Seal, Crankshaft	2UU0-962							✓	✓	✓	1
	19053A	.36	Oil Thrower, Crankshaft	2UU0-1448							✓	✓	✓	1
	45693B	.60	Bearing, Crankshaft Front Main (Standard)	2UG0-651	✓									2
	......	4.00	Bearing, Crankshaft Front Main (Standard)	2D2-651		✓	✓	✓						2
	......	5.00	Bearing, Crankshaft Front Main (.010 U/S)	2D2-402		✓	✓	✓						2
	......	3.55	Bearing, Crankshaft Front Main (.020 U/S)	2D2-512		✓	✓	✓						2
	......	2.71	Bearing, Crankshaft Front Main (.030 U/S)	2D2-513		✓	✓	✓						2
	......	5.00	Bearing, Crankshaft Front Main (.040 U/S)	2D2-371		✓	✓	✓						2
	......	2.35	Bearing, Crankshaft Front Main (Standard)	2T2-651					✓					2
	......	3.55	Bearing, Crankshaft Front Main (.010 U/S)	2T2-402					✓					2
	......	3.55	Bearing, Crankshaft Front Main (.020 U/S)	2T2-512					✓					2
	......	3.55	Bearing, Crankshaft Front Main (.030 U/S)	2T2-513					✓					2
	......	4.55	Bearing, Crankshaft Front Main (.040 U/S)	2T2-371					✓					2
	......	2.00	Bearing, Crankshaft Front Main (Standard)	2B3-651						✓				2
	......	3.55	Bearing, Crankshaft Front Main (.010 U/S)	2B2-402						✓				2
	......	3.25	Bearing, Crankshaft Front Main (.020 U/S)	2B2-512						✓				2
	......	3.25	Bearing, Crankshaft Front Main (.030 U/S)	2B2-513						✓				2
	......	2.00	Bearing, Crankshaft Front Main (.040 U/S)	2B2-371						✓				2
	17071B—Std.	1.20	Bearing, Crankshaft Front Main (Standard)	2UU0-651							✓	✓	✓	2
	17071B—.010 U/S	1.68	Bearing, Crankshaft Front Main (.010 U/S)	2UU0-402							✓	✓	✓	2
	17071B—.020 U/S	1.68	Bearing, Crankshaft Front Main (.020 U/S)	2UU0-512							✓	✓	✓	2
	17071B—.030 U/S	1.68	Bearing, Crankshaft Front Main (.030 U/S)	2UU0-513							✓	✓	✓	2
	45694B	.82	Bearing, Crankshaft Center Main (Standard)	2UG0-655	✓									2
	......	4.00	Bearing, Crankshaft Center Main (Standard)	2D2-655		✓	✓	✓						2
	......	5.00	Bearing, Crankshaft Center Main (.010 U/S)	2D2-489		✓	✓	✓						2
	......	4.81	Bearing, Crankshaft Center Main (.020 U/S)	2D2-516		✓	✓	✓						2
	......	5.81	Bearing, Crankshaft Center Main (.030 U/S)	2D2-517		✓	✓	✓						2
	......	6.75	Bearing, Crankshaft Center Main (.040 U/S)	2D2-372		✓	✓	✓						2

* Not Sold Separately

01-4 ENGINE — Master Parts List

THE AUTOCAR COMPANY, ARDMORE, PA.

CODE	MFR. PART No.	LIST PRICE	NAME OF PART	AUTOCAR PART No.	U-2044	U-4044	U-4144	U-4144-T	C-50	U-5044	U-7144-T	U-8144	U-8144-T	No. REQ.
0102	......	$3.53	Bearing, Crankshaft Center Main (Standard)	2SA2-655						✓				2
	......	4.81	Bearing, Crankshaft Center Main (.010 U/S)	2SA2-489						✓				2
	......	5.81	Bearing, Crankshaft Center Main (.020 U/S)	2SA2-516						✓				2
	......	4.81	Bearing, Crankshaft Center Main (.030 U/S)	2SA2-517						✓				2
	......	5.81	Bearing, Crankshaft Center Main (.040 U/S)	2SA2-372						✓				2
	......	3.08	Bearing, Crankshaft Center Main (Standard)	2B3-655						✓				2
	......	3.55	Bearing, Crankshaft Center Main (.010 U/S)	2B2-489						✓				2
	......	3.55	Bearing, Crankshaft Center Main (.020 U/S)	2B2-516						✓				2
	......	3.60	Bearing, Crankshaft Center Main (.030 U/S)	2B2-517						✓				2
	......	3.08	Bearing, Crankshaft Center Main (.040 U/S)	2B2-372						✓				2
	17072B—Std.	1.60	Bearing, Crankshaft Center Main (Standard)	2UU0-655							✓	✓	✓	2
	17072B—.010 U/S	2.24	Bearing, Crankshaft Center Main (.010 U/S)	2UU0-489							✓	✓	✓	2
	17072B—.020 U/S	2.24	Bearing, Crankshaft Center Main (.020 U/S)	2UU0-516							✓	✓	✓	2
	17072B—.030 U/S	2.24	Bearing, Crankshaft Center Main (.030 U/S)	2UU0-517							✓	✓	✓	2
	45696B	.58	Bearing, Crankshaft Intermediate Main (Standard)	2UG0-653	✓									8
	......	3.00	Bearing, Crankshaft Intermediate Main (Standard)	2D2-653		✓	✓	✓						8
	......	4.00	Bearing, Crankshaft Intermediate Main (.010 U/S)	2D2-401		✓	✓	✓						8
	......	2.71	Bearing, Crankshaft Intermediate Main (.020 U/S)	2D2-514		✓	✓	✓						8
	......	2.71	Bearing, Crankshaft Intermediate Main (.030 U/S)	2D2-515		✓	✓	✓						8
	......	3.00	Bearing, Crankshaft Intermediate Main (.040 U/S)	2D2-374		✓	✓	✓						8
	......	1.98	Bearing, Crankshaft Intermediate Main (Standard)	2SA2-653						✓				8
	......	3.71	Bearing, Crankshaft Intermediate Main (.010 U/S)	2SA2-401						✓				8
	......	2.71	Bearing, Crankshaft Intermediate Main (.020 U/S)	2SA2-514						✓				8
	......	2.71	Bearing, Crankshaft Intermediate Main (.030 U/S)	2SA2-515						✓				8
	......	3.71	Bearing, Crankshaft Intermediate Main (.040 U/S)	2SA2-374						✓				8
	......	1.98	Bearing, Crankshaft Intermediate Main (Standard)	2B3-653						✓				8
	......	1.90	Bearing, Crankshaft Intermediate Main (.010 U/S)	2B2-401						✓				8
	......	1.90	Bearing, Crankshaft Intermediate Main (.020 U/S)	2B2-514						✓				8
	......	1.90	Bearing, Crankshaft Intermediate Main (.030 U/S)	2B2-515						✓				8
	......	1.98	Bearing, Crankshaft Intermediate Main (.040 U/S)	2B2-374						✓				8
	17074B—Std.	1.00	Bearing, Crankshaft Intermediate Main (Standard)	2UU0-653							✓	✓	✓	8
	17074B—.010 U/S	1.40	Bearing, Crankshaft Intermediate Main (.010 U/S)	2UU0-401							✓	✓	✓	8
	17074B—.020 U/S	1.40	Bearing, Crankshaft Intermediate Main (.020 U/S)	2UU0-514							✓	✓	✓	8
	17074B—.030 U/S	1.40	Bearing, Crankshaft Intermediate Main (.030 U/S)	2UU0-515							✓	✓	✓	8
	45695B	1.28	Bearing, Crankshaft Rear Main (Standard)	2UG0-657	✓									2
	......	5.00	Bearing, Crankshaft Rear Main (Standard)	2D2-657		✓	✓	✓						2
	......	6.00	Bearing, Crankshaft Rear Main (.010 U/S)	2D2-499		✓	✓	✓						2
	......	4.35	Bearing, Crankshaft Rear Main (.020 U/S)	2D2-509		✓	✓	✓						2
	......	3.41	Bearing, Crankshaft Rear Main (.030 U/S)	2D2-511		✓	✓	✓						2
	......	7.15	Bearing, Crankshaft Rear Main (.040 U/S)	2D2-373		✓	✓	✓						2
	......	3.53	Bearing, Crankshaft Rear Main (Standard)	2SA2-657						✓				2
	......	4.35	Bearing, Crankshaft Rear Main (.010 U/S)	2SA2-499						✓				2
	......	5.35	Bearing, Crankshaft Rear Main (.020 U/S)	2SA2-509						✓				2
	......	4.35	Bearing, Crankshaft Rear Main (.030 U/S)	2SA2-511						✓				2
	......	5.35	Bearing, Crankshaft Rear Main (.040 U/S)	2SA2-373						✓				2
	......	3.23	Bearing, Crankshaft Rear Main (Standard)	2B3-657						✓				2
	......	3.55	Bearing, Crankshaft Rear Main (.010 U/S)	2B2-499						✓				2
	......	3.55	Bearing, Crankshaft Rear Main (.020 U/S)	2B2-509						✓				2
	......	3.60	Bearing, Crankshaft Rear Main (.030 U/S)	2B2-511						✓				2
	......	3.23	Bearing, Crankshaft Rear Main (.040 U/S)	2B2-373						✓				2
	17073B—Std.	1.40	Bearing, Crankshaft Rear Main (Standard)	2UU0-657							✓	✓	✓	6
	17073B—.010 U/S	1.96	Bearing, Crankshaft Rear Main (.010 U/S)	17073B							✓	✓	✓	..
	17073B—.020 U/S	1.96	Bearing, Crankshaft Rear Main (.020 U/S)	2UU0-509							✓	✓	✓	2
	17073B—.030 U/S	1.96	Bearing, Crankshaft Rear Main (.030 U/S)	2UU0-511							✓	✓	✓	2
	......	.95	Washer, Main Bearing Thrust—Center	2D3-382		✓	✓	✓						4
	......	1.15	Washer, Main Bearing Center Thrust	2B3-382						✓				4
	......	.30	Stud, Main Bearing	2SA2-644		✓	✓	✓	✓	✓				14
	......	.30	Stud, Main Bearing—Center	2SA2-687		✓	✓	✓	✓	✓				4
	......	.35	Nut, Main Bearing Stud	4H1-227		✓	✓	✓	✓	✓				18
	......	.12	Pin, Main Bearing Dowel	2T2-643		✓	✓	✓						7
	......	.12	Pin, Main Bearing Dowel	2CB2-643						✓				7
	80172A	1.50	Flange, Crankshaft Rear Main Bearing	2UU0-1362							✓	✓	✓	4

THE AUTOCAR COMPANY
ARDMORE, PA.

Master Parts List

01-5
ENGINE

CODE	MFR. PART No.	LIST PRICE	NAME OF PART	AUTOCAR PART No.	U-2044	U-4044	U-4144	U-4144-T	C-50	U-5044	U-7144-T	U-8144	U-8144-T	No. REQ.
0102	4765A	$0.10	Dowel, Crankshaft Rear Main Bearing Flange	2UU0-1363							✓	✓	✓	4
	40550A	.02	Shim, Crankshaft Main Bearing—Front & Intermediate (.002)	2BL0-521	✓									As
	40552A	.02	Shim, Crankshaft Main Bearing—Front & Intermediate (.003)	2BL0-522	✓									30
	40549A	.02	Shim, Crankshaft Main Bearing—Center & Rear (.002)	2BL0-523	✓									As
	40551A	.02	Shim, Crankshaft Main Bearing—Center & Rear (.003)	2BL0-552	✓									12
	......	.06	Shim, Crankshaft Bearing—Front (.010)	2T2-521		✓	✓	✓	✓	✓				2
	......	.06	Shim, Crankshaft Bearing—Intermediate (.010)	2T2-522		✓	✓	✓	✓	✓				8
	......	.06	Shim, Crankshaft Bearing—Center (.010)	2SA2-523		✓	✓	✓	✓	✓				4
	80164A	.02	Shim, Crankshaft Front Main Bearing (.002)	2UU0-521							✓	✓	✓	As
	80165A	.02	Shim, Crankshaft Front Main Bearing (.003)	2UU0-522							✓	✓	✓	6
	80168A	.02	Shim, Crankshaft Center & Rear Main Bearing (.002)	2UU0-523							✓	✓	✓	As
	80169A	.02	Shim, Crankshaft Center & Rear Main Bearing (.003)	2UUA0-523							✓	✓	✓	12
	80166A	.02	Shim, Crankshaft Intermediate Main Bearing (.002)	2UUA0-521							✓	✓	✓	As
	80167A	.02	Shim, Crankshaft Intermediate Main Bearing (.003)	2UUA0-522							✓	✓	✓	24
	15103A	.22	Screw, Crankshaft Front & Intermediate Main Bearing Cap	2UU0-1343							✓	✓	✓	10
	1703A	.20	Screw, Crankshaft Center & Rear Main Bearing Cap	2UU0-1342							✓	✓	✓	8
	267A	.02	Lockwire, Crankshaft Main Bearing Screw	2UU0-1344							✓	✓	✓	16
0103	**Pistons, Rings, and Piston Pins**													
	40120-C	7.75	Piston (Standard) (4″ Dia. Aluminum)	2BN0-27	✓									6
	40120-C	7.75	Piston (.010 O/S) (4″ Dia. Aluminum)	2BN0-891	✓									6
	40120-C	7.75	Piston (.020 O/S) (4″ Dia. Aluminum)	2BN0-873	✓									6
	40120-C	7.75	Piston (.030 O/S) (4″ Dia. Aluminum)	2BN0-864	✓									6
	......	8.00	Piston (Standard)	2D4-27A		✓	✓	✓		✓				6
	......	8.00	Piston (.020 O/S)	2D4-873A		✓	✓	✓		✓				6
	......	8.00	Piston (.030 O/S)	2D4-864A		✓	✓	✓		✓				6
	......	8.00	Piston (.040 O/S)	2D4-862A		✓	✓	✓		✓				6
	......	8.00	Piston (Standard)	2D4-27					✓					6
	......	8.00	Piston (.020 O/S)	2D4-873					✓					6
	......	8.00	Piston (.030 O/S)	2D4-864					✓					6
	......	8.00	Piston (.040 O/S)	2D4-862					✓					6
	16519-C—Std.	10.50	Piston (Standard)	2UU0-27							✓	✓	✓	6
	16519-C—.010 O/S	10.50	Piston (.010 O/S)	2UU0-891							✓	✓	✓	6
	16519-C—.020 O/S	10.50	Piston (.020 O/S)	2UU0-873							✓	✓	✓	6
	16519-C—S.F.	8.24	Piston (Semi-Finished)	2UU0-1393							✓	✓	✓	6
	22238-B	.56	Pin, Piston	2BN0-39	✓									6
	......	.60	Pin, Piston	2H2-39		✓	✓	✓		✓				6
	......	.50	Pin, Piston	2B2-39					✓					6
	15929-B—Std.	1.36	Pin, Piston (Standard)	2UU0-39							✓	✓	✓	6
	15929-B—.003 O/S	1.36	Pin, Piston (.003 O/S)	2UU0-447							✓	✓	✓	6
	15929-B—.005 O/S	1.36	Pin, Piston (.005 O/S)	2UUA0-447							✓	✓	✓	6
	15929-B—.010 O/S	1.36	Pin, Piston (.010 O/S)	2UUB0-447							✓	✓	✓	6
	42297-A	.28	Ring, Piston Compression (Standard) (4″) Top Groove	42297-A	✓									6
	42297-A	.28	Ring, Piston Compression (.010 O/S) (4″) Top Groove	42297-A	✓									6
	42297-A	.28	Ring, Piston Compression (.020 O/S) (4″) Top Groove	42297-A	✓									6
	42297-A	.28	Ring, Piston Compression (.030 O/S) (4″) Top Groove	42297-A	✓									6
	3813-A	.26	Ring, Piston Compression (Standard) (4″) 2nd & 3rd Groove	2BN0-25	✓									12
	3813-A	.26	Ring, Piston Compression (.010 O/S) (4″) 2nd & 3rd Groove	2BN0-888	✓									12
	3813-A	.26	Ring, Piston Compression (.020 O/S) (4″) 2nd & 3rd Groove	2BN0-875	✓									12
	3813-A	.26	Ring, Piston Compression (.030 O/S) (4″) 2nd & 3rd Groove	2BN0-353	✓									12
	......	.25	Ring, Piston Compression (Standard)	2D2-25		✓	✓	✓	✓	✓				18
	......	.25	Ring, Piston Compression (.020 O/S)	2D2-875		✓	✓	✓	✓	✓				18
	......	.25	Ring, Piston Compression (.030 O/S)	2D2-353		✓	✓	✓	✓	✓				18
	......	.25	Ring, Piston Compression (.040 O/S)	2D2-863		✓	✓	✓	✓	✓				18
	3824-A—Std.	.32	Ring, Piston Compression (Standard)	2UU0-25							✓	✓	✓	24
	......	7.68	Ring, Piston Compression (Standard) Complete Set	2UU0-25S							✓	✓	✓	1
	3824-A—.010 O/S	.32	Ring, Piston Compression (.010 O/S)	2UU0-888							✓	✓	✓	24
	......	7.68	Ring, Piston Compression (.010 O/S) Complete Set	2UU0-888S							✓	✓	✓	1
	3824-A—.020 O/S	.32	Ring, Piston Compression (.020 O/S)	2UU0-875							✓	✓	✓	24
	......	7.68	Ring, Piston Compression (.020 O/S) Complete Set	2UU0-875S							✓	✓	✓	1
	3824-A—.030 O/S	.32	Ring, Piston Compression (.030 O/S)	2UU0-353							✓	✓	✓	24
	......	7.68	Ring, Piston Compression (.030 O/S) Complete Set	2UU0-353S							✓	✓	✓	1

Master Parts List

THE AUTOCAR COMPANY, ARDMORE, PA.

CODE	MFR. PART No.	LIST PRICE	NAME OF PART	AUTOCAR PART No.	U-2044	U-4044	U-4144	U-4144-T	C-50	U-5044	U-7144-T	U-8144	U-8144-T	No. REQ.
0103	3824-A—.040 O/S	$0.32	Ring, Piston Compression (.040 O/S)....................	2UU0-863							✓	✓	✓	24
	......	7.68	Ring, Piston Compression (.040 O/S) Complete Set.........	2UU0-863S							✓	✓	✓	1
	3824-A—.050 O/S	.32	Ring, Piston Compression (.050 O/S)....................	2UU0-554							✓	✓	✓	24
	......	7.68	Ring, Piston Compression (.050 O/S) Complete Set.........	2UU0-554S							✓	✓	✓	1
	3824-A—.060 O/S	.32	Ring, Piston Compression (.060 O/S)....................	2UU0-869							✓	✓	✓	24
	......	7.68	Ring, Piston Compression (.060 O/S) Complete Set.........	2UU0-869S							✓	✓	✓	1
	3913-A	.48	Ring, Piston Oil Regulator (Standard) (4″)...............	2BN0-26	✓									6
	3913-A	.48	Ring, Piston Oil Regulator (.010 O/S) (4″)...............	2BN0-889	✓									6
	3913-A	.48	Ring, Piston Oil Regulator (.020 O/S) (4″)...............	2BN0-877	✓									6
	3913-A	.48	Ring, Piston Oil Regulator (.030 O/S) (4″)...............	2BN0-878	✓									6
	......	.59	Ring, Piston Oil Regulator (Standard)...................	2D2-26A		✓	✓	✓		✓				6
	......	.48	Ring, Piston Oil Regulator (.020 O/S)...................	2D2-877		✓	✓	✓		✓				6
	......	.50	Ring, Piston Oil Regulator (.030 O/S)...................	2D2-878		✓	✓	✓		✓				6
	......	.50	Ring, Piston Oil Regulator (.040 O/S)...................	2D2-879		✓	✓	✓		✓				6
	......	.50	Ring, Piston Oil Regulator (Standard)...................	2D2-26					✓					6
	......	.48	Ring, Piston Oil Regulator (.020 O/S)...................	2D2-877					✓					6
	......	.50	Ring, Piston Oil Regulator (.030 O/S)...................	2D2-878					✓					6
	......	.50	Ring, Piston Oil Regulator (.040 O/S)...................	2D2-879					✓					6
	3923-A—Std.	.54	Ring, Piston Oil Regulator (Standard)...................	2UU0-26							✓	✓	✓	6
	......	3.24	Ring, Piston Oil Regulator (Standard) Complete Set........	2UU0-26S							✓	✓	✓	6
	15946-A—.010 O/S	.54	Ring, Piston Oil Regulator (.010 O/S)...................	2UU0-889							✓	✓	✓	6
	......	3.24	Ring, Piston Oil Regulator (.010 O/S) Complete Set........	2UU0-889S							✓	✓	✓	1
	15946-A—.020 O/S	.54	Ring, Piston Oil Regulator (.020 O/S)...................	2UU0-877							✓	✓	✓	6
	......	3.24	Ring, Piston Oil Regulator (.020 O/S) Complete Set........	2UU0-877S							✓	✓	✓	1
	15946-A—.030 O/S	.54	Ring, Piston Oil Regulator (.030 O/S)...................	2UU0-878							✓	✓	✓	6
	......	3.24	Ring, Piston Oil Regulator (.030 O/S) Complete Set........	2UU0-878S							✓	✓	✓	1
	15946-A—.040 O/S	.54	Ring, Piston Oil Regulator (.040 O/S)...................	2UU0-879							✓	✓	✓	6
	......	3.24	Ring, Piston Oil Regulator (.040 O/S) Complete Set........	2UU0-879S							✓	✓	✓	1
	15946-A—.050 O/S	.54	Ring, Piston Oil Regulator (.050 O/S)...................	2UU0-1188							✓	✓	✓	6
	......	3.24	Ring, Piston Oil Regulator (.050 O/S) Complete Set........	2UU0-1188S							✓	✓	✓	1
	15946-A—.060 O/S	.54	Ring, Piston Oil Regulator (.060 O/S)...................	2UU0-881							✓	✓	✓	6
	......	3.24	Ring, Piston Oil Regulator (.060 O/S) Complete Set........	2UU0-881S							✓	✓	✓	1
0104	**Connecting Rods and Bearings**													
	40390-CS	9.70	Rod, Connecting—Assembly............................	2BL0-20	✓									6
	45707B or 45692B	.48	Shell, Connecting Rod Bearing..........................	2BL0-109	✓									12
	22059-A	.28	Bolt, Connecting Rod Bearing...........................	2BL0-116	✓									12
	21056-A	.04	Nut, Connecting Rod Bearing Bolt.......................	2BL0-118	✓									12
	40556-A	.02	Shim, Connecting Rod Bearing (.003)....................	2BL0-1189					✓					18
	40553-A	.02	Shim, Connecting Rod Bearing (.002)....................	2BL0-553	✓									As
	22111-A	.10	Screw, Piston Pin Clamp................................	2BL0-162	✓									6
	14761-A	.01	Washer, 3/8″ Lock.....................................	S-3185	✓									6
	......	16.20	Rod, Connecting—Assembly (Standard)..................	2D4-20		✓	✓	✓	✓					6
	......	14.40	Rod, Connecting—Assembly (Standard)..................	2B4-20						✓				6
	......	16.20	Rod, Connecting—Assembly (.010 U/S Bearings).........	2D4-510		✓	✓	✓	✓					6
	......	14.40	Rod, Connecting—Assembly (.010 U/S Bearings).........	2B4-510						✓				6
	......	16.20	Rod, Connecting—Assembly (.020 U/S Bearings).........	2D4-520		✓	✓	✓	✓					6
	......	16.80	Rod, Connecting—Assembly (.020 U/S Bearings).........	2B4-520						✓				6
	......	16.20	Rod, Connecting—Assembly (.030 U/S Bearings).........	2D4-530		✓	✓	✓	✓					6
	......	14.40	Rod, Connecting—Assembly (.030 U/S Bearings).........	2B4-530						✓				6
	......	16.20	Rod, Connecting—Assembly (.040 U/S Bearings).........	2D0-540		✓	✓	✓	✓					6
	......	14.40	Rod, Connecting—Assembly (.040 U/S Bearings).........	2B4-540						✓				6
	......	1.50	Bearing, Connecting Rod (Standard)....................	2D2-248		✓	✓	✓	✓					12
	......	1.50	Bearing, Connecting Rod (Standard)....................	2B2-248						✓				12
	......	2.21	Bearing, Connecting Rod (.010 U/S)....................	2D2-231		✓	✓	✓	✓					12
	......	1.50	Bearing, Connecting Rod (.010 U/S)....................	2B2-231						✓				12
	......	2.21	Bearing, Connecting Rod (.020 U/S)....................	2DE2-231		✓	✓	✓	✓					12
	......	1.50	Bearing, Connecting Rod (.020 U/S)....................	2BE2-231						✓				12
	......	2.21	Bearing, Connecting Rod (.030 U/S)....................	2DF2-231		✓	✓	✓	✓					12
	......	1.50	Bearing, Connecting Rod (.030 U/S)....................	2BF2-231						✓				12
	......	2.21	Bearing, Connecting Rod (.040 U/S)....................	2DG2-231		✓	✓	✓	✓					12
	......	1.50	Bearing, Connecting Rod (.040 U/S)....................	2BG2-231						✓				12

THE AUTOCAR COMPANY
ARDMORE, PA.

Master Parts List

01-7
ENGINE

CODE	MFR. PART No.	LIST PRICE	NAME OF PART	AUTOCAR PART No.	MODEL U-2044	U-4044	U-4144	U-4144-T	C-50	U-5044	U-7144-T	U-8144	U-8144-T	No. REQ.
0104	......	$0.35	Bolt, Connecting Rod Bearing....................	2SA2-116A		✓	✓	✓	✓					12
	......	.35	Bolt, Connecting Rod Bearing....................	2B4-116						✓				12
	......	.20	Nut, Connecting Rod Bearing Bolt...............	2T2-118		✓	✓	✓	✓					12
	......	.20	Nut, Connecting Rod Bearing Bolt...............	2B2-118						✓				12
	......	.06	Shim, Connecting Rod Bearing...................	2D2-942		✓	✓	✓						6
	......	.06	Shim, Connecting Rod Bearing (.006)............	2CB2-942						✓				12
	......	.06	Shim, Connecting Rod Bearing (.006)............	2B2-942						✓				6
	......	.40	Bushing, Piston Pin............................	2T2-15		✓	✓	✓						6
	......	.40	Bushing, Piston Pin............................	2T2-15A						✓				6
	......	.20	Bushing, Piston Pin............................	2B2-15						✓				6
	......	.03	Dowel, Connecting Rod..........................	2T2-559A						✓				12
	17076AS	14.00	Rod, Connecting, with Bearings..................	2UU0-20							✓	✓	✓	6
	17075B—Std.	.80	Bearing, Connecting Rod (Standard)..............	2UU0-1109							✓	✓	✓	12
	17075B—.010 U/S	1.12	Bearing, Connecting Rod (.010 U/S)..............	2UU0-231A							✓	✓	✓	12
	17075B—.020 U/S	1.12	Bearing, Connecting Rod (.020 U/S)..............	2UUA0-231							✓	✓	✓	12
	17075B—.030 U/S	1.12	Bearing, Connecting Rod (.030 U/S)..............	2UUB0-231							✓	✓	✓	12
	16692A—.002	.02	Shim, Connecting Rod (.002)....................	2UU0-553							✓	✓	✓	As
	16693A—.003	.02	Shim, Connecting Rod (.003)....................	2UUA0-553							✓	✓	✓	18
	7056A	.12	Nut, Connecting Rod Cap.......................	2UU0-118							✓	✓	✓	12
	1710A	.01	Cotter, Connecting Rod Cap Nut (1/8″ x 1″).....	S-2204							✓	✓	✓	12
	1711A	.08	Screw, Connecting Rod Piston Pin Clamp.........	2UU0-1365							✓	✓	✓	6
	267A	.02	Lockwire, Connecting Rod Piston Pin Clamp Screw..	2UU0-1344							✓	✓	✓	6
0105	**Valves, Springs, Guides and Tappets**													
	45814-A	.90	Valve, Exhaust................................	2BL0-10	✓									6
	......	2.40	Valve, Exhaust................................	2B2-10		✓	✓	✓	✓					6
	......	2.40	Valve, Exhaust (Round Groove Type Latest Design)...	2D2-10A						✓				6
	......	2.40	Valve, Exhaust (Tapered Groove Type Original Design)......	2SA2-10C						✓				6
	15510-A	1.40	Valve, Exhaust................................	2UU0-10							✓	✓	✓	6
	40107-A	.60	Valve, Intake.................................	2BL0-710	✓									6
	......	2.00	Valve, Intake.................................	2D2-710		✓	✓	✓	✓					6
	......	2.00	Valve, Intake (Round Groove Type Latest Design)....	2D2-710						✓				6
	......	2.00	Valve, Intake (Tapered Groove Type Original Design).....	2SA2-710A						✓				6
	15520-A	1.10	Valve, Intake.................................	2UU0-710							✓	✓	✓	6
	40008-A	.10	Spring, Valve.................................	2BL0-52	✓									12
	......	.40	Spring, Valve.................................	2T2-52		✓	✓	✓	✓					12
	19772-A	.24	Spring, Valve.................................	2UU0-52							✓	✓	✓	12
	1613-A	.12	Seat, Valve Spring............................	2UU0-49							✓	✓	✓	12
	21011-A	.08	Pin, Valve Spring Seat.........................	2UU0-329							✓	✓	✓	12
	22011-A	.18	Guide, Intake Valve...........................	2BL0-21	✓									6
	......	.35	Guide, Intake Valve...........................	2M2-21		✓	✓	✓	✓					6
	1611-A	.30	Guide, Intake Valve...........................	2UU0-21							✓	✓	✓	6
	22011-A	.18	Guide, Exhaust Valve..........................	2BL0-21	✓									6
	......	.35	Guide, Exhaust Valve..........................	2M2-21		✓	✓	✓	✓					6
	15009-A	.20	Guide, Exhaust Valve..........................	2UUA0-21							✓	✓	✓	6
	15088-B	2.96	Cluster, Valve Tappet.........................	2UU0-1371							✓	✓	✓	2
	1701-A	.06	Screw, Valve Tappet Cluster...................	2UU0-1372							✓	✓	✓	8
	2210-A	.06	Dowel, Valve Tappet Cluster...................	2UU0-1373							✓	✓	✓	8
	312-A	.02	Lockwasher, Valve Tappet Cluster (1/2″)........	S-1349							✓	✓	✓	8
	16506-A	1.10	Tappet, Valve................................	2UU0-36							✓	✓	✓	12
	4682-A	.10	Screw, Valve Tappet Adjusting.................	2UU0-1368							✓	✓	✓	12
	2186-A	.03	Nut, Valve Tappet Adjusting Screw.............	2UU0-1369							✓	✓	✓	12
	15006-B	.42	Cover, Valve.................................	2UU0-65							✓	✓	✓	2
	15005-B	.14	Gasket, Valve Cover..........................	2UU0-72							✓	✓	✓	2
	4038-A	.06	Screw, Valve Cover...........................	2UU0-1367							✓	✓	✓	4
	794-A	.02	Washer, Valve Cover Screw....................	2UU0-1375							✓	✓	✓	4
	795-A	.10	Leadwasher, Valve Cover Screw................	2UUA0-1375							✓	✓	✓	4
0106	**Camshaft and Bearings**													
	45686-D	15.20	Camshaft.....................................	2UG0-54	✓									1
	......	45.00	Camshaft.....................................	2SA6-54B		✓	✓	✓	✓	✓				1
	16567-D	26.50	Camshaft.....................................	2UU0-54							✓	✓	✓	1

01-8 ENGINE

Master Parts List — The Autocar Company, Ardmore, PA.

CODE	MFR. PART No.	LIST PRICE	NAME OF PART	AUTOCAR PART No.	U-2044	U-4044	U-4144	U-4144-T	C-50	U-5044	U-7144-T	U-8144	U-8144-T	No. REQ.
0106	40063-A	$0.40	Bearing, Camshaft Front & Rear	2BL0-44	✓									2
	15043-B	1.52	Bearing, Camshaft Front	2UU0-44							✓	✓	✓	1
	40065-B	.32	Bearing, Camshaft Intermediate	2BL0-43	✓									1
	15044-B	1.12	Bearing, Camshaft Center	2UU0-43							✓	✓	✓	4
	16543-A	1.30	Bearing, Camshaft Rear	2UUA0-44							✓	✓	✓	1
	22049-B	2.84	Gear, Camshaft	2BL0-76	✓									1
	......	7.20	Gear, Camshaft	2SA3-76		✓	✓	✓	✓	✓				1
	15049-B	4.10	Gear, Camshaft	2UU0-76							✓	✓	✓	1
	4265-A	.06	Key, Camshaft Gear (#91 Woodruff)	S-2319	✓									1
	1247-A	.02	Key, Camshaft Gear	2UU0-1387							✓	✓	✓	1
	11023-A	.12	Nut, Camshaft Gear	2UG0-698	✓									1
	......	.30	Nut, Camshaft Gear Clamp	2T2-616		✓	✓	✓	✓	✓				1
	1698-A	.20	Nut, Camshaft Gear	2TH0-698							✓	✓	✓	1
	2045-A	.55	Washer, Camshaft Gear Thrust	2TH0-699	✓									1
	......	1.50	Washer, Camshaft Gear Thrust	2SA2-492		✓	✓	✓	✓	✓				1
	4342-A	.34	Washer, Camshaft Gear	2UU0-652							✓	✓	✓	1
	40068-A	.20	Plunger, Camshaft	2BL0-776	✓									1
	14209-A	.30	Plunger, Camshaft Thrust	2UU0-776							✓	✓	✓	1
	2048-A	.05	Nut, Camshaft Thrust Plunger Adjusting	2TH0-775							✓	✓	✓	1
	14596-A	.32	Screw, Camshaft Thrust Plunger Adjusting	2UU0-774							✓	✓	✓	1
	40136-A	1.82	Gear, Camshaft Idler Shaft, Plunger and Plug Assembly	2BL0-183	✓									1
	......	14.55	Gear, Camshaft Idler Assembly with Shaft	2SA3-50		✓	✓	✓	✓	✓				1
	17878-B	3.50	Shaft, Idler	2UU0-183							✓	✓	✓	1
	40135-B	3.16	Gear, Camshaft Idler	2BL0-33	✓									1
	......	5.40	Gear, Camshaft Idler	2SA3-33		✓	✓	✓	✓	✓				1
	15105-B	3.90	Gear, Idler Shaft	2UU0-33							✓	✓	✓	1
	920-A	.02	Key, Idler Shaft Gear	2UU0-1385							✓	✓	✓	1
	40137-A	1.10	Bearing, Camshaft Idler Gear Shaft	2BL0-213	✓									1
	18098-A	2.15	Bearing, Idler Shaft	2UU0-213							✓	✓	✓	1
	22107-A	.08	Washer, Camshaft Idler Gear Thrust	2BL0-652	✓									1
	4342-A	0.34	Washer, Idler Shaft Gear Thrust	2UU0-652							✓	✓	✓	1
	40068-A	.20	Plunger, Camshaft Idler Gear	2BL0-776	✓									1
	14209-A	.30	Plunger, Camshaft Idler Gear Thrust	2UU0-776							✓	✓	✓	1
	14594-AS	.20	Screw, Idler Shaft Gear Thrust Plunger Adjusting, Assy.	2BL0-774							✓	✓	✓	1
	2048-A	.05	Nut, Idler Shaft Gear Thrust Plunger Adjusting	2TH0-775							✓	✓	✓	1
	59-A	.02	Plug, Camshaft Idler Gear Shaft Pipe (⅛")	S-619	✓									1
	......	.12	Screw, Camshaft Idler Gear Cap	2CB2-92		✓	✓	✓	✓					4
	......	.12	Screw, Camshaft Idler Gear Cap	2T2-92A						✓				4
	......	8.40	Shaft, Camshaft Idler Gear	2T3-183		✓	✓	✓	✓	✓				1
	......	4.00	Bushing, Camshaft Idler Gear Shaft	2SA2-213		✓	✓	✓	✓					1
	......	4.00	Bushing, Camshaft Idler Gear Shaft	2SA2-213A						✓				1
	......	54.65	Camshaft Assembly with Gear	2SA0-310A		✓	✓	✓	✓	✓				1
	......	.06	Washer, Camshaft Gear Nut Lock	2T2-618		✓	✓	✓	✓	✓				1
	......	4.80	Bushing, Camshaft, #1	2T2-44B		✓	✓	✓	✓	✓				1
	......	4.80	Bushing, Camshaft, #5	2T2-45A		✓	✓	✓	✓	✓				1
	......	3.60	Bushing, Camshaft, #2	2T2-648A		✓	✓	✓	✓	✓				1
	......	3.60	Bushing, Camshaft, #4	2T2-649A		✓	✓	✓	✓	✓				1
	......	.12	Screw, Camshaft Bushing	2SA2-62		✓	✓	✓	✓	✓				2
	......	.12	Screw, Camshaft Bushing	2SCH2-62		✓	✓	✓	✓	✓				1
0107	**Oil Pump, Oil Pan, Oil Gauge, Oil Filter**													
	45290-CS	8.50	Pump, Oil, Assembly	2UG0-150	✓									1
	......	30.00	Pump, Oil, Assembly	2N0-150		✓	✓	✓	✓	✓				1
	17523-ES	33.00	Pump, Oil, Assembly	2UU0-150							✓	✓	✓	1
	41124-C	3.00	Body, Oil Pump	2UG0-129	✓									1
	......	7.20	Body, Oil Pump	2N4-129		✓	✓	✓	✓	✓				1
	17524-D	8.60	Body, Oil Pump (Lower)	2UU0-129							✓	✓	✓	1
	16538-D	8.10	Body, Oil Pump (Upper)	2UUA0-129							✓	✓	✓	1
	19017-A	1.36	Bushing, Oil Pump Body (Lower)	2UU0-464							✓	✓	✓	1
	16539-A	1.22	Bushing, Oil Pump Body (Upper)	2UUA0-464							✓	✓	✓	2
	4605-A	.24	Screw, Oil Pump Body Attaching	2UU0-1391							✓	✓	✓	6
	267-A	.02	Lockwire, Oil Pump Body Attaching Screw	2UU0-1344							✓	✓	✓	1

THE AUTOCAR COMPANY — ARDMORE, PA.

Master Parts List

01-9 ENGINE

CODE	MFR. PART No.	LIST PRICE	NAME OF PART	AUTOCAR PART No.	MODEL U-2044	U-4044	U-4144	U-4144-T	C-50	U-5044	U-7144-T	U-8144	U-8144-T	No. REQ.
0107	40153-A	$1.60	Cover, Oil Pump	2BL0-131	✓									1
	......	5.70	Cover, Oil Pump	2N4-131		✓	✓	✓	✓	✓				1
	19028-B	.96	Cover, Oil Pump	2UU0-131							✓	✓	✓	1
	14069-A	.02	Screw, Oil Pump Cover (¼"—20 x ¾")	S-30	✓									6
	4604-A	.16	Screw, Oil Pump Cover	2UU0-92							✓	✓	✓	6
	267-A	.02	Lockwire, Oil Pump Cover Screw	2UU0-1344							✓	✓	✓	1
	22124-A	1.28	Shaft, Oil Pump	2BL0-127	✓									1
	16452-B	4.90	Shaft, Oil Pump	2UU0-127							✓	✓	✓	1
	45266-A	1.22	Gear, Oil Pump Drive	2BL0-148	✓									1
	15131-A	2.60	Gear, Oil Pump Drive	2UU0-148							✓	✓	✓	1
	2165-A	.02	Key, Oil Pump Drive Gear	2UU0-1394							✓	✓	✓	1
	1157-A	.06	Pin, Oil Pump Drive Gear	2UU0-1395							✓	✓	✓	1
	4809-A	.01	Pin, Oil Pump Drive Gear ($\frac{5}{32}$")	S-3192	✓									1
	1157-A	.06	Pin, Oil Pump Drive Gear	2UU0-1395							✓	✓	✓	1
	2047-A	.04	Washer, Oil Pump Drive Gear	2BL0-122	✓									1
	15126-A	.12	Washer, Oil Pump Drive Gear	2UU0-122							✓	✓	✓	1
	22155-A	.64	Shaft, Oil Pump Idler	2BL0-126	✓									1
	15125-A	1.30	Shaft, Oil Pump Idler	2UU0-126							✓	✓	✓	3
	15122-A	2.80	Gear, Oil Pump Idler Shaft	2UU0-34							✓	✓	✓	4
	2165-A	.02	Key, Oil Pump Idler Shaft Gear	2UU0-1394							✓	✓	✓	1
	4362-A	.04	Ring, Oil Pump Idler Shaft Snap	2UU0-975							✓	✓	✓	1
	......	2.40	Gear, Oil Pump	2N2-128		✓	✓	✓	✓	✓				1
	......	2.40	Gear, Oil Pump	2N2-29					✓					1
	16453-A	4.80	Gear, Oil Pump	2UU0-128							✓	✓	✓	1
	1864-A	.01	Screw, Oil Pump Attaching (⅜"—16 x 1")	S-23	✓									3
	4629-A	.04	Screw, Oil Pump Attaching	2UUA0-92							✓	✓	✓	4
	342-A	.02	Lockwasher, Oil Pump Attaching Screw	2UU0-1399							✓	✓	✓	4
	267-A	.02	Lockwire, Oil Pump Attaching Screw	2UU0-1344							✓	✓	✓	1
	19019-A	.50	Seal, Oil Pump to Oil Pan Felt	2UU0-1403							✓	✓	✓	1
	2396-A	.12	Union, Oil Pump Pipe	2UU0-1404							✓	✓	✓	3
	4387-A	.08	Ring, Oil Pump Snap	2BL0-975	✓									2
	22154-A	1.20	Gears, Oil Pump Shaft	2BL0-195	✓									2
	......	2.40	Gear, Oil Pump	2N2-29		✓	✓	✓	✓	✓				1
	1179-A	.01	Key, Oil Pump Shaft Gear (#2 Whitney)	S-2325	✓									2
	......	.30	Shaft, Oil Pump Pinion	2N2-126		✓	✓	✓	✓	✓				1
	......	6.60	Cover, Oil Pump, and Bushing Assembly	2N0-845		✓	✓	✓	✓	✓				1
	......	.35	Bushing, Oil Pump Cover	2N2-976		✓	✓	✓	✓	✓				2
	......	2.90	Gear, Oil Pump Idler, and Bushing Assembly	2N0-922		✓	✓	✓	✓	✓				1
	......	2.40	Gear, Oil Pump Idler	2N2-34		✓	✓	✓	✓	✓				1
	......	.60	Bushing, Oil Pump Idler Gear	2N2-19		✓	✓	✓	✓	✓				1
	......	1.20	Shaft, Oil Pump Gear	2N2-127		✓	✓	✓	✓	✓				1
	......	.30	Collar, Oil Pump	2N2-975		✓	✓	✓	✓	✓				1
	22129-A	.10	Valve, Oil Pressure Regulator	2BL0-598	✓									1
	1347-A	.04	Spring, Oil Pressure Regulator	2BL0-477	✓									1
	2058-A	.06	Spring, Oil Pressure Regulator	2BL0-956	✓									1
	28-A	.01	Nut, Oil Pressure Regulator Screw ($\frac{5}{16}$")	S-3715	✓									1
	1660-A	.10	Nut, Oil Pressure Regulator	2BL0-1213	✓									1
	1385-A	.06	Button, Oil Pressure Regulator Spring	2BL0-1212	✓									1
	342-A	.01	Washer, Spring (⅜")	S-1347	✓									3
	2268-A	1.50	Wrench, Oil Pressure Crow Foot	2BL0-1215	✓									1
	X-5800	2.14	Wrench, Oil Pressure "T"	2BL0-1214	✓									1
	......	.30	Spring, Oil Relief Valve	2N2-477		✓	✓	✓	✓	✓				1
	......	.30	Guide, Oil Relief Valve Spring	2N2-295		✓	✓	✓	✓	✓				1
	......	.20	Ball, Oil Relief Valve	S-1678		✓	✓	✓	✓	✓				1
	......	.40	Screw, Oil Relief Valve	2N2-956C		✓	✓	✓	✓	✓				1
	......	.30	Washer, Oil Pump Body Felt (2$\frac{13}{16}$" I.D.—3⅝" O.D. x ⅜")	S-2544		✓	✓	✓	✓	✓				1
	45180-BS	.60	Gauge, Oil Level, Assembly	2UG0-210	✓									1
	......	.95	Gauge, Oil Level, Assembly	2UP3-210		✓			✓					1
	......	.95	Gauge, Oil Level, Assembly	2UP3-210			✓	✓						1
	......	.90	Gauge, Oil Level, Assembly	2URL3-210A						✓				1
	......	1.50	Gauge, Crankcase Oil Level, Unit	2UD3-1150		✓	✓	✓	✓					1
	......	.90	Gauge, Oil Level, Assembly	2URL3-210A						✓				1

01-10 ENGINE — Master Parts List

THE AUTOCAR COMPANY, ARDMORE, PA.

CODE	MFR. PART No.	LIST PRICE	NAME OF PART	AUTOCAR PART No.	U-2044	U-4044	U-4144	U-4144-T	C-50	U-5044	U-7144-T	U-8144	U-8144-T	No. REQ.
0107	17147-AS	$0.96	Gauge, Oil Pan Bayonet, Assembly	2UU0-210							✓	✓	✓	1
	......	2.15	Cap, Oil Filler, Assembly	2T2-946		✓	✓	✓	✓	✓				1
	......	1.80	Body, Oil Filler	2URL3-105B		✓	✓	✓		✓				1
	......	1.80	Body, Oil Filler	2SA2-105					✓					1
	......	2.15	Cap, Oil Filler Breather, Assembly	2T2-946A							✓	✓	✓	1
	......	.15	Brace, Oil Filler Breather Pipe	2UU2-1272							✓	✓	✓	1
	......	3.90	Pipe, Oil Filler Breather, Assembly	2UU3-1274							✓	✓	✓	1
	42176-D	24.36	Pan, Oil	2UG0-138	✓									1
	......	7.20	Pan, Oil (Small)	2UG4-592		✓	✓	✓		✓				1
	......	7.20	Pan, Oil (Small)	2T4-592					✓					1
	17183-D	39.30	Pan, Oil	2UU0-138							✓	✓	✓	1
	42183-B	.10	Gasket, Oil Pan (Half)	2UG0-161	✓									1
	......	.15	Gasket, Oil Pan (Small)	2SK3-634		✓	✓	✓		✓				1
	......	.39	Gasket, Oil Pan (Small)	2T2-634					✓					1
	17185-C	.08	Gasket, Oil Pan	2UU0-161							✓	✓	✓	2
	315-A	.05	Screw, Oil Pan (3/8″—16 x 7/8″)	S-50	✓									16
	14770-A	.02	Screw, Oil Pan (Front End)	14770-A	✓									4
	......	.05	Screw, Oil Pan Assembly Cap (3/8″—16 x 1″)	S-23		✓	✓	✓		✓				33
	2100-A	.04	Screw, Oil Pan Attaching (1/2″—13 x 1″)	2UU0-1388							✓	✓	✓	26
	342-A	.01	Washer, 3/8″ Spring	S-1347		✓	✓	✓		✓				33
	312-A	.02	Screw, Oil Pan Attaching, L.W.	2UU0-1355							✓	✓	✓	26
	17885-CS	4.70	Line, Oil Pump Front Suction, Assembly	2UU0-1406							✓	✓	✓	1
	8588-A	.04	Clamp, Oil Pump Suction Line Flange	2UU0-1416							✓	✓	✓	4
	4607-A	.03	Screw, Oil Pump Suction Line Flange Clamp	2UU0-1417							✓	✓	✓	2
	1-A	.04	Nut, Oil Pump Suction Line Flange Clamp Screw	2UUA0-1213							✓	✓	✓	2
	10-A	.02	Cotter, Oil Pump Suction Line Flange Clamp Screw	2UU0-1418							✓	✓	✓	2
	14742-AS	.36	Pipe, Oil Pump, Union Assembly	2UU0-1424							✓	✓	✓	1
	16425-CS	1.50	Pipe, Oil Pump Outlet, Assembly	2UU0-583							✓	✓	✓	1
	16454-CS	5.20	Line, Oil Pump Rear Suction	2UU0-1407							✓	✓	✓	1
	19011-A	.02	Gasket, Oil Pump Suction Line Flange	2UUA0-578							✓	✓	✓	2
	4606-A	.04	Screw, Oil Pump Suction Line Flange	2UU0-956							✓	✓	✓	2
	4772-A	.04	Screw, Oil Pump Suction Line Flange	2UUA0-956							✓	✓	✓	2
	17182-CS	7.00	Pipe, Oil Pump Main Discharge, Assembly	2UUA0-583							✓	✓	✓	1
	15133-A	.02	Gasket, Oil Pump Main Discharge Pipe	2UU0-578							✓	✓	✓	7
	4039-A	.04	Screw, Oil Pump Main Discharge Pipe	2UU0-1405							✓	✓	✓	14
	17899-BS	6.00	Strainer, Oil Pan Oil, and Water Trap Assembly	2UU0-957							✓	✓	✓	1
	8146-A	.12	Gasket, Oil Pan Oil Strainer Cap	2UU0-377							✓	✓	✓	1
	2100-A	.04	Screw, Oil Pan Oil Strainer Cap	2UU0-1388							✓	✓	✓	26
	312-A	.02	Screw, Oil Pan Oil Strainer Cap, L.W.	2UU0-1355							✓	✓	✓	26
	......	.09	Plug, Oil Pan Pipe (3/4″)	S-617	✓									2
	......	.12	Plug, Oil Pan Drain (3/8″)	2T2-104		✓								1
	......	.10	Plug, Oil Pan Drain (7/8″)	2A2-104			✓	✓						1
	14867-A	.66	Plug, Oil Pan Drain	2UU0-104							✓	✓	✓	2
	296-A	.02	Gasket, Oil Pan Drain Plug	2UU0-1327							✓	✓	✓	2
	16185-A	.10	Pipe, Oil Pan Sump Vent	2UU0-1402							✓	✓	✓	1
	16423-AS	1.36	Pipe, Oil Overflow, Assembly	5UU0-594							✓	✓	✓	1
	1536-A	.02	Screw, Oil Overflow Pipe Attaching	5UU0-595							✓	✓	✓	1
	628-A	.02	Screw, Oil Overflow Pipe Attaching, L.W.	5UU0-596							✓	✓	✓	1
	......	11.20	Viscometer (Oil Pan Unit)	16UU0-1199							✓	✓	✓	1
	......	.25	Adapter, Viscometer	16UU0-2068							✓	✓	✓	1
	......	.45	Tube, Viscometer Gauge, Assembly	16UU3-2850A							✓	✓	✓	1
	17884-A	1.00	Line, Oil, Assembly—Visco. Unit to Oil Pressure Line	2UU0-1421							✓	✓	✓	1
	......	.85	Line, Oil, Assembly—Crankcase to Governor	2UU3-680A							✓	✓	✓	1
	14406-A	.30	Elbow, Oil Line (1/4″)	2UU0-463							✓	✓	✓	1
	14407-A	.06	Nut, Oil Line (1/4″)	2UU0-1213							✓	✓	✓	1
	14408-A	.06	Ferrule, Oil Line (1/4″)	2UU0-1425							✓	✓	✓	1
	......	12.80	Filter, Oil, Assy.—RPM Type with 4″ O.D. at Base of Filter Cyl.	2UBB0-808	✓									1
	C-108	.80	Cartridge, Oil Filter	2A0-611	✓	✓								1
	C-107	.12	Gasket, Top	2A0-829	✓									1
	C-103	3.00	Cover, Top	2A0-318	✓									1
	C-102	.90	Bridge, Top Cover	2A0-608	✓									1
	C-101	.45	Screw, Bridge	2A0-319	✓									1

THE AUTOCAR COMPANY
ARDMORE, PA.

Master Parts List

01-11
ENGINE

CODE	MFR. PART No.	LIST PRICE	NAME OF PART	AUTOCAR PART No.	U-2044	U-4044	U-4144	U-4144-T	C-50	U-5044	U-7144-T	U-8144	U-8144-T	No. REQ.
0107	DM-104	$6.30	Cylinder, Filter	2E0-746	✓									1
	DM-110	.50	Plate, Convex Bottom	2A0-607	✓									1
	DM-130	.12	Gasket, Bottom Cylinder	2E0-828	✓									1
	DM-115	1.00	Screw, Bottom Cylinder	2A0-606	✓									1
	EM-116A	4.20	Base, Oil Filter	2URL5-727	✓									1
	EM-131	.10	Gasket, Oil Filter Base	2A0-575	✓									1
	......	.20	Valve, Oil Relief	2T2-598C	✓									1
	......	.12	Spring, Oil Relief Valve	2T2-477B	✓									1
	......	.25	Screw, Valve Adjusting	2T2-599A	✓									1
	......	.05	Nut, Valve Adjusting Screw Lock	S-2004	✓									1
	......	.20	Nut, Valve Adjusting Screw Cap	2T2-602	✓									1
	......	.03	Washer, Oil Relief Valve	S-1911	✓									2
	......	.15	Screw, Oil Filter Base	S-3108	✓									4
	......	.01	Washer, Spring	S-1349	✓									4
	DM-106	.02	Plug, Oil Filter Base	S-619	✓									1
	......	15.00	Filter, Oil, Assy.—TRM Type with 4″ O.D. at Base of Filter Cyl.	2J0-807		✓			✓					1
	T-108	1.15	Cartridge, Oil Filter	2J0-611		✓			✓					1
	T-107	.15	Gasket, Top	2J0-829		✓			✓					1
	T-103	3.00	Cover, Top	2J0-318		✓			✓					1
	T-102	.80	Bridge, Top Cover	2J0-608		✓			✓					1
	T-101	.40	Screw, Bridge	2J0-319		✓			✓					1
	TRM-104	....	Cylinder, Filter (4″ O.D. at Base)	2J0-746		✓			✓					1
	TRM-110	.50	Plate, Convex Bottom	2A0-607		✓			✓					1
	TRM-130	.12	Gasket, Bottom Cylinder (4″ O.D.)	2E0-828		✓			✓					1
	TRM-115	1.00	Screw, Bottom Cylinder	2A0-606		✓			✓					1
	......	4.20	Base, Oil Filter	2URL5-727		✓			✓					1
	......	.06	Gasket, Oil Filter Base	2URL2-575		✓			✓					1
	......	.20	Valve, Oil Relief	2T2-598C		✓			✓					1
	......	.12	Spring, Oil Relief Valve	2T2-477B		✓			✓					1
	......	.25	Screw, Valve Adjusting	2T2-599A		✓			✓					1
	......	.05	Nut, Valve Adjusting Screw Lock (5/8″—11)	S-2004		✓			✓					1
	......	.20	Nut, Valve Adjusting Screw Cap	2T2-602		✓			✓					1
	......	.03	Washer, Oil Relief Valve (5/8″ I.D.—7/8″ O.D. x 3/32″)	S-1911		✓			✓					2
	......	.15	Screw, Oil Filter Base (1/2″—13 x 3″)	S-3108		✓			✓					4
	......	.01	Washer, 1/2″ Spring	S-1349		✓			✓					4
	PP-2	.02	Plug, Oil Filter Base (1/8″)	S-619		✓			✓					1
	19000	16.56	Filter, Oil, Assembly—Complete	2UG0-1335			✓	✓						1
	SA-12800	2.00	Cartridge, Oil Filter	2UG0-611			✓	✓						2
	24	.35	Plug, Oil Filter Drain (1/4″)	2UG0-728			✓	✓						2
	560-S	.12	Plug, Oil Filter Base Pipe (1/4″)	S-4748			✓	✓						1
	31	.38	Rod, Oil Filter Center	2UG0-729			✓	✓						2
	37	.30	Valve, Oil Pressure Relief	2UG0-598			✓	✓						1
	38	.15	Spring, Oil Pressure Relief Valve	2UG0-477			✓	✓						1
	39	.45	Screw, Oil Pressure Relief Valve Adjusting	2UG0-599			✓	✓						1
	40	.06	Gasket, Oil Pressure Relief Valve	2UG0-831			✓	✓						2
	41	.35	Nut, Oil Pressure Relief Valve Lock	2UG0-736			✓	✓						1
	42	.75	Nut, Oil Pressure Relief Valve Cap	2UG0-602			✓	✓						1
	SA-33	13.64	Base, Oil Filter, and Shell Assembly	2UG0-746			✓	✓						1
	30	.23	Cover, Oil Filter Shell	2UG0-318			✓	✓						2
	17	.15	Gasket, Oil Filter Shell Cover	2UG0-829			✓	✓						2
	35	.05	Gasket, Oil Filter Shell Cover Fitting	2UG0-745			✓	✓						2
	32	.25	Fitting, Oil Filter Shell Cover	2UG0-743			✓	✓						2
	36	.15	Spring, Oil Filter	2UG0-737			✓	✓						2
	......	.05	Gasket, Oil Filter Mounting	2H2-575			✓	✓						1
	......	.15	Bracket, Oil Filter Steady	2A2-546			✓	✓						1
	......	.05	Screw, Oil Filter Steady Bracket (3/8″—16 x 7/8″)	S-50			✓	✓						1
	......	.03	Nut, Oil Filter Steady Bracket Screw (3/8″—16)	S-75			✓	✓						1
	......	.03	Washer, Oil Filter Steady Br'c't Screw (1 1/2″ I.D.—3/4″ O.D. x 1/16″)	S-387			✓	✓						1
	......	.01	Washer, 3/8″ Spring	S-1347			✓	✓						1
	......	12.00	Filter, Oil, Assy.—RPM Type with 4″ O.D. at Base of Filter Cyl.	2E0-807						✓				1
	C-108	.80	Cartridge, Oil Filter	2A0-611						✓				1
	C-107	.12	Gasket, Top	2A0-829						✓				1

01-12 ENGINE

Master Parts List

THE AUTOCAR COMPANY, ARDMORE, PA.

CODE	MFR. PART No.	LIST PRICE	NAME OF PART	AUTOCAR PART No.	U-2044	U-4044	U-4144	U-4144-T	C-50	U-5044	U-7144-T	U-8144	U-8144-T	NO. REQ.
0107	C-103	$3.00	Cover, Top	2A0-318					✓					1
	C-102	.90	Bridge, Top Cover	2A0-608					✓					1
	C-101	.45	Screw, Bridge	2A0-319					✓					1
	DM-104	6.30	Cylinder, Filter—4″ O.D. at Base (Part of 2E0807)	2E0-746					✓					1
	DM-104	6.30	Cylinder, Filter—4⅛″ O.D. at Base (Part of 2A0807)	2A0-746					✓					1
	DM-110	.50	Plate, Convex Bottom	2A0-607					✓					1
	DM-130	.12	Gasket, Bottom Cylinder (4″ O.D.)	2E0-828					✓					1
	DA-130	.06	Gasket, Bottom Cylinder (4⅛″ O.D.)	2A0-828					✓					1
	DM-115	1.00	Screw, Bottom Cylinder	2A0-606					✓					1
	......	4.20	Base, Oil Filter	2URL5-727					✓					1
	......	.06	Gasket, Oil Filter Base	2URL2-575					✓					1
	......	.20	Valve, Oil Relief	2T2-598C					✓					1
	......	.12	Spring, Oil Relief Valve	2T2-477B					✓					1
	......	.25	Screw, Valve Adjusting	2T2-599A					✓					1
	......	.05	Nut, Valve Adjusting Screw Lock	S-2004					✓					1
	......	.20	Nut, Valve Adjusting Screw Cap	2T2-602					✓					1
	......	.03	Washer, Oil Relief Valve	S-1911					✓					2
	......	.04	Screw, Oil Filter Base	S-3108					✓					4
	......	.01	Washer, Spring	S-1349					✓					4
	......	.02	Plug, Oil Filter Base	S-619					✓					1
	17246-A	.06	Gasket, Oil Filter Attaching	2UU0-575							✓	✓	✓	1
	17269-D	15.00	Bracket, Oil Filter	2UU5-727							✓	✓	✓	1
	4625-A	.04	Screw, Oil Filter Attaching (½″ x 1¼″)	2UU0-319							✓	✓	✓	3
	8552-A	.06	Screw, Oil Filter Attaching (½″ x 2¾″)	2UUA0-319							✓	✓	✓	1
	312-A	.02	Screw, Oil Filter Attaching—L.W. (17/32″)	S-1349							✓	✓	✓	4
	42843-CS	15.00	Filter, Oil—Unit Assembly (Fram Model F-36)	2UU0-1335							✓	✓	✓	2
	15243-A	.10	Valve, Oil Filter Pressure Relief	2UU0-598							✓	✓	✓	1
	8571-A	.06	Spring, Oil Filter Pressure Relief Valve	2UU0-477							✓	✓	✓	1
	8572-A	.30	Screw, Oil Filter Pressure Relief Valve Adjusting	2UU0-956							✓	✓	✓	1
	8573-A	.06	Nut, Oil Filter Pressure Relief Valve Adjusting Screw Lock	2UU0-1213							✓	✓	✓	1
	8574-A	.18	Cap, Oil Filter Pressure Relief Valve Adjusting Screw	2UU0-602							✓	✓	✓	1
	8575-A	.04	Gasket, Oil Filter Pressure Relief Valve Adjusting Screw	2UU0-831							✓	✓	✓	2
	61-A	.04	Plug, Oil Filter Bracket Pipe (⅜″)	S-4005							✓	✓	✓	2
	59-A	.02	Plug, Oil Filter Bracket Pipe (⅛″)	2UUB0-1447							✓	✓	✓	2
	60-A	.04	Plug, Oil Filter Bracket Pipe (¼″)	S-4748							✓	✓	✓	1
	17256-CS	2.80	Filter, Oil—Element Assembly (Fram Model C-31)	2UU0-611							✓	✓	✓	2
	42844-CS	7.00	Case, Oil Filter—Assembly (Fram #5267)	2UU0-746							✓	✓	✓	2
	17265-A	.10	Gasket, Oil Filter Case (Fram #11639)	2UU0-828							✓	✓	✓	2
	42845-B	2.00	Tube, Oil Filter Center—Assembly (Fram #11733)	2UU0-729							✓	✓	✓	2
	42847-A	1.00	Plate, Oil Filter Case Retaining (Fram #11734)	2UU0-607							✓	✓	✓	2
	11374-A	.10	Plug, Oil Filter Case Drain (Fram #11584)	2UU0-728							✓	✓	✓	2
	17260-A	.22	Gasket, Oil Filter Case Cover (Fram #11582)	2UU0-829							✓	✓	✓	2
	17261-A	1.10	Cover, Oil Filter Case (Fram #11559)	2UU0-318							✓	✓	✓	2
	11383-A	.05	Gasket, Oil Filter Case Cover Screw (Fram #11581)	2UU0-745							✓	✓	✓	2
	17262-A	.30	Screw, Oil Filter Case Cover (Fram #11580)	2UU0-743							✓	✓	✓	2
	17263-A	.05	Spring, Oil Filter Case Cover (Fram #11583)	2UU0-737							✓	✓	✓	2
	42846-A	1.25	Bolt, Oil Filter Case Clamp (Fram #11735)	2UU0-729							✓	✓	✓	2
0108	**Manifold**													
	45806-E	20.96	Manifold, Intake & Exhaust	2UG0-860	✓									1
	14895-A	.10	Stud, Intake & Exhaust Manifold	2UG0-159	✓									10
	848-A	.05	Nut, Intake & Exhaust Manifold Stud (7/16″—20)	S-1419	✓									10
	1388-A	.01	Washer, Intake & Ex. Man. Stud (15/32″ I.D.—⅞″ O.D. x 1/16″)	S-3326	✓									10
	4640-A	.03	Screw, Intake & Exhaust Manifold Set	2BL0-1216	✓									1
	305-A	.02	Plug, Intake & Exhaust Manifold Pipe	S-4720	✓									1
	40314-A	2.82	Flange, Intake & Exhaust Manifold	7BL0-04	✓									1
	......	62.10	Manifold, Exhaust & Intake Assembly	2URM0-860		✓	✓	✓	✓	✓				1
	......	62.10	Manifold, Exhaust & Intake—Assembly	2URM0-860		✓	✓	✓	✓	✓				1
	......	30.00	Manifold, Exhaust	2SA7-142B		✓	✓	✓	✓					1
	......	30.00	Manifold, Exhaust	2SA7-142C						✓				1
	......	.25	Gasket, Exhaust Manifold—Small	2SA2-535B		✓	✓	✓		✓				2
	......	.25	Gasket, Exhaust Manifold—End	2SA2-535A						✓				2

THE AUTOCAR COMPANY
ARDMORE, PA.

Master Parts List

01-13 ENGINE

CODE	MFR. PART No.	LIST PRICE	NAME OF PART	AUTOCAR PART No.	U-2044	U-4044	U-4144	U-4144-T	C-50	U-5044	U-7144-T	U-8144	U-8144-T	No. REQ.
0108	......	$0.40	Gasket, Exhaust Manifold—Large	2SA2-633A	✓	✓	✓	✓	✓					2
	......	18.00	Manifold, Intake	2SA5-141B	✓	✓	✓	✓	✓					1
	......	.15	Gasket, Intake Manifold	2SA2-461	✓	✓	✓	✓	✓					3
	......	3.25	Stove, Intake Manifold	2URM4-1161	✓	✓	✓	✓		✓				1
	......	4.60	Stove, Intake Manifold	2SA4-1161					✓					1
	......	.10	Screw, Intake Manifold Stove (7/16"—14 x 2¾")	S-1901	✓	✓	✓	✓		✓				2
	......	.05	Screw, Intake Manifold Stove (7/16"—20 x 1¾")	S-4042A	✓	✓	✓	✓		✓				2
	......	.04	Screw, Intake Manifold Stove Nut (7/16"—14)	S-850	✓	✓	✓	✓		✓				2
	......	.01	Washer, 7/16" Spring	S-1348	✓	✓	✓	✓		✓				4
	......	.35	Gasket, Intake Manifold Stove—Bottom	2D2-1162	✓	✓	✓	✓	✓					1
	......	.40	Gasket, Manifold Hot Spot	2C2-13	✓	✓	✓	✓	✓					2
	......	.20	Gasket, Intake Pipe	2SA2-1002	✓	✓	✓	✓	✓					1
	......	1.35	Pipe, Intake	2RM3-1003A	✓	✓	✓							1
	......	1.75	Pipe, Intake	2C3-1003				✓	✓	✓				1
	......	.20	Stud, Manifold	2T2-221	✓	✓	✓	✓		✓				18
	......	.20	Stud, Intake & Exhaust Manifold—Short	2URM2-159	✓	✓	✓	✓		✓				6
	......	.20	Stud, Intake & Exhaust Manifold	2SA2-159A					✓					4
	......	.20	Stud, Intake & Exhaust Manifold—Long	2C2-334	✓	✓	✓	✓	✓					2
	......	.20	Stud, Exhaust Manifold Front Flange	2SA2-963A	✓	✓	✓	✓	✓					1
	......	.20	Stud, Exhaust Manifold Rear Flange	2SA2-964	✓	✓	✓	✓	✓					1
	......	.15	Nut, Exhaust Manifold Stud	2T2-145	✓	✓	✓	✓	✓	✓				10
	......	.01	Washer, Ex. Man. Stud Nut (17/32" I.D.—7/8" O.D. x 3/32")	S-704	✓	✓	✓	✓	✓	✓				10
	......	.30	Nut, Intake & Exhaust Manifold Stud	2C2-158A	✓	✓	✓	✓	✓					8
	......	.40	Plate, Intake & Exhaust Filler	2C2-12A	✓	✓	✓	✓	✓					1
	......	.40	Shield, Intake Manifold	2T2-98	✓	✓	✓	✓						1
	......	.40	Shield, Intake Manifold	2T3-98						✓				1
	......	.05	Screw, Intake Manifold Shield (¼"—20 x 1¼")	S-67	✓	✓	✓	✓		✓				2
	......	.01	Nut, Intake Manifold Shield Screw (¼"—20)	S-76	✓	✓	✓	✓		✓				2
	......	.01	Washer, ¼" Spring	S-1345	✓	✓	✓	✓		✓				4
	......	1.10	Valve, Manifold Hot Spot	2URM3-342	✓	✓	✓	✓						1
	......	1.20	Valve, Manifold Hot Spot	2SA2-342					✓					1
	......	.60	Spindle, Manifold Hot Spot Valve	2SA2-08	✓	✓	✓	✓	✓					1
	......	.35	Joint, Ball	1OUF1-07	✓	✓	✓	✓						1
	......	.40	Clevis, Control	1OY1-136	✓	✓	✓	✓						1
	......	.30	Lever, Heat Control	2SD2-1011B	✓	✓	✓	✓		✓				1
	......	.55	Lever, Heat Control	2SD2-1011					✓					1
	......	.01	Pin, Heat Control Lever (1/8" Dia. x ¾")	S-1499	✓	✓	✓	✓		✓				1
	......	.06	Screw, Heat Control Lever Set (¼"—20 x 1¼")	S-425	✓	✓	✓	✓		✓				1
	......	.60	Lever, Choke Control	1ORL3-181					✓					1
	17048-D	18.80	Manifold, Exhaust	2UU0-142							✓	✓	✓	1
	15023-A	.32	Gasket, Exhaust Manifold Attaching—Inner	2UU0-535							✓	✓	✓	2
	15026-A	.20	Gasket, Exhaust Manifold Attaching—Outer	2UUA0-535							✓	✓	✓	2
	17047-D	12.80	Manifold, Intake	2UU0-143							✓	✓	✓	1
	15031-A	.18	Gasket, Intake Manifold Attaching	2UU0-13							✓	✓	✓	3
	11484-A	.16	Stud, Intake and Exhaust Manifold Attaching	2UU0-159							✓	✓	✓	14
	11485-A	.04	Washer, Intake and Exhaust Manifold Attaching Stud	2UU0-1426							✓	✓	✓	14
	4079-A	.04	Nut, Intake and Exhaust Manifold Attaching Stud	2UU0-1432							✓	✓	✓	14
	20230-C	3.00	Box, Intake Manifold Heat	2UU0-1427							✓	✓	✓	1
	19764-A	.84	Gasket, Intake Manifold Heat Box (To Exhaust Manifold)	2UU0-1428							✓	✓	✓	1
	19763-A	.84	Gasket, Intake Mnaifold Heat Box (To Intake Manifold)	2UU0-1429							✓	✓	✓	1
	14759-A	.12	Stud, Intake Manifold Heat Box Attaching (To Ex. Manifold)	2UU0-1431							✓	✓	✓	4
	4079-A	.04	Nut, Intake Manifold Heat Box Attaching Stud	2UU0-1432							✓	✓	✓	4
	315-A	.03	Screw, Intake Manifold Heat Box Attaching (To Intake Man.)	2UU0-1436							✓	✓	✓	4
	7219-A	.02	Gasket, Intake Manifold to Governor	2UU0-1433							✓	✓	✓	1
	305-A	.02	Plug, Intake Manifold Pipe (1/8")	2UU0-1434							✓	✓	✓	1
	4312-A	.04	Plug, Intake Manifold Pipe (½")	2UU0-1435							✓	✓	✓	1
0109	**Flywheel, Ring Gear, Etc.**													
	23705-C	25.10	Flywheel	2BN0-06	✓									1
	17486-C	34.30	Flywheel	2UU0-06							✓	✓	✓	1
	15471-C	3.80	Gear, Flywheel (Teeth) 126	2BL0-214	✓									1
	15480-C	3.80	Gear, Flywheel Ring	2UU0-214							✓	✓	✓	1

01-14 ENGINE

Master Parts List — The Autocar Company, Ardmore, PA.

CODE	MFR. PART NO.	LIST PRICE	NAME OF PART	AUTOCAR PART NO.	U-2044	U-4044	U-4144	U-4144-T	C-50	U-5044	U-7144-T	U-8144	U-8144-T	NO. REQ.
0109	16900-D	$32.00	Housing, Flywheel Bell	2UU0-153							✓	✓	✓	1
	352-A	.04	Screw, Flywheel Bell Housing to Crankcase (½" x 1¼")	2UU0-1378							✓	✓	✓	3
	2100-A	.04	Screw, Flywheel Bell Housing to Crankcase (½" x 1")	2UU0-1388							✓	✓	✓	11
	312-A	.02	Screw, Flywheel Bell Housing to Crankcase—L.W.	2UU0-1355							✓	✓	✓	14
	15065-B	.25	Gasket, Flywheel Bell Housing	2UU0-782							✓	✓	✓	1
	80634-A	5.80	Seal, Flywheel Bell Housing Oil	2UU0-1268							✓	✓	✓	1
	14221-A	.02	Screw, Flywheel Bell Housing Oil Seal	2UU0-1442							✓	✓	✓	3
	7064-B	3.20	Adapter, Flywheel Bell Housing Starter	2TH0-783							✓	✓	✓	1
	18421-A	.06	Plate, Flywheel Bell Housing Timing Hole Cover	2UU0-544							✓	✓	✓	1
	1048-A	.03	Screw, Flywheel Bell Housing Timing Hole Cover Plate	2UU0-1357							✓	✓	✓	3
	615-A	.02	Plate, Flywheel Bell Housing Timing Hole Cover—L.W.	2UU0-1358							✓	✓	✓	3
0110	**Mountings**													
	24476-D	26.00	Support, Rear Engine (Bellhousing #3 S.A.E.)	2UG0-153	✓									1
	3543-B	1.90	Seal, Rear Engine Support Oil	2BL0-1268	✓									1
	4123-A	.05	Screw, Rear Engine Support and Crankcase (½" x 13 x ⅞")	S-2086	✓									8
	2100-A	.08	Screw, Rear Engine Support (½"—13 x 1")	S-20	✓									3
	312-A	.01	Washer, ½" Spring	S-1349	✓									10
	......	.15	Plate, Rear Engine Support Cover—Front	2BL2-542	✓									2
	......	.04	Screw, Rear Engine Support Cover Plate (⅜"—16 x ½")	S-1461	✓									6
	......	.01	Washer, ⅜" Spring	S-1347	✓									6
	......	.20	Plate, Rear Engine Support Cover—Rear—R. H.	2BL3-543	✓									1
	......	.20	Plate, Rear Engine Support Cover—Rear—L. H.	2BL2-544	✓									1
	40700-B	2.90	Bracket, Front Engine Support	2BL3-132	✓									1
	1701-A	.10	Bolt, Front Engine Support Bracket (½"—13 x 2")	S-922	✓									1
	3293-A	.04	Nut, Front Engine Support Bracket Bolt (½"—13)	S-77	✓									1
	312-A	.01	Washer, ½" Spring	S-1349	✓									1
	......	8.60	Cross Member, Front Engine Support	2BL4-152A	✓									1
	......	.95	Rubber, Front Engine Support	2BL2-277	✓									1
	......	.15	Channel, Front Engine Support	2BL2-278	✓									1
	......	.15	Bolt, Front Engine Support Channel (½"—13 x 2½")	S-1877	✓									2
	......	.05	Nut, Front Engine Support Channel Bolt (½"—13)	S-1310	✓									2
	......	.50	Biscuit, Rear Engine Support Rubber	12NJ2-402	✓									2
	......	.25	Bushing, Rear Engine Support Rubber	12NJ2-403	✓									2
	......	.20	Bolt, Rear Engine Support	2BL2-239A	✓									2
	......	.12	Nut, Rear Engine Support Bolt (⅝"—11)	S-808	✓									2
	......	.04	Washer, Rear Engine Support Bolt (⅝" I.D.—1⅝" O.D. x ⅛")	S-927	✓									2
	......	44.65	Support, Rear Engine	2DK6-153		✓	✓	✓		✓				1
	......	38.40	Support, Rear Engine	2B6-153					✓					1
	......	.09	Stud, Rear Engine Support	2T2-239		✓	✓	✓		✓				10
	......	.04	Nut, Rear Engine Support (½"—13)	S-77		✓	✓	✓		✓				10
	......	.01	Washer, ½" Spring	S-1349		✓	✓	✓		✓				10
	......	4.80	Bracket, Front Engine Support	2NJ4-132		✓	✓	✓	✓	✓				1
	......	5.40	Bracket, Rear Engine Support	12C4-86		✓	✓	✓	✓	✓				1
	......	1.00	Bushing, Front Engine Support Bracket Rubber	12NJ3-401		✓	✓	✓	✓	✓				1
	......	.50	Biscuit, Rear Engine Support Bracket Rubber	12NJ2-402		✓	✓	✓	✓	✓				4
	......	.25	Bushing, Rear Engine Support Bracket Rubber	12NJ2-403		✓	✓	✓	✓	✓				4
	......	13.60	Support, Front Engine	2UU4-152							✓	✓	✓	1
	......	.09	Screw, Hex. Cap (½"—13 x 1⅜")	S-1428							✓	✓	✓	3
	......	.10	Screw, Hex. Cap (½"—13 x 1⅝")	S-1839							✓	✓	✓	3
	......	.04	Nut, Hex. (½"—13)	S-77							✓	✓	✓	6
	......	.01	Lockwasher (½")	S-1349							✓	✓	✓	12
	......	4.80	Bracket, Front Engine Support	2NJ4-132							✓	✓	✓	1
	......	1.00	Bushing, Front Engine Support Bracket Rubber	12NJ3-401							✓	✓	✓	1
	......	.15	Screw, Hex. Cap (½"—13 x 3¼")	S-3378							✓	✓	✓	1
	......	.04	Nut, Hex. (½"—13)	S-77							✓	✓	✓	1
	......	.01	Lockwasher (½")	S-1349							✓	✓	✓	2
	......	1.40	Spacer, Front Engine Support Bracket	2UU2-1185B							✓	✓	✓	1
	......	.15	Screw, Hex. Cap (½"—13 x 3⅛")	S-6280							✓	✓	✓	2
	......	.06	Nut, Hex.—Slotted (½"—13)	S-4199							✓	✓	✓	2
	......	.01	Pin, Cotter (⅛" x 1¼")	S-2205							✓	✓	✓	2
	......	1.75	Support Extension, Rear Engine	12UU3-399							✓	✓	✓	2

THE AUTOCAR COMPANY, ARDMORE, PA.

Master Parts List

01-15 ENGINE

CODE	MFR. PART No.	LIST PRICE	NAME OF PART	AUTOCAR PART No.	U-2044	U-4044	U-4144	U-4144-T	C-50	U-5044	U-7144-T	U-8144	U-8144-T	NO. REQ.
0110	......	$0.12	Screw, Hex. Cap (½"—13 x 1¾")....................	S-3101							✓	✓	✓	8
	......	.01	Lockwasher (½")..................................	S-1349							✓	✓	✓	8
	......	7.00	Support, Rear Engine—R. H........................	12UU4-86							✓	✓	✓	1
	......	.15	Screw, Hex. Cap (⅝"—11 x 2¾")...................	S-1874							✓	✓	✓	2
	......	.12	Screw, Hex. Cap (⅝"—11 x 3½")...................	S-4336							✓	✓	✓	2
	......	.05	Nut, Hex. (⅝"—11)................................	S-93							✓	✓	✓	4
	......	.01	Lockwasher (⅝")...................................	S-1351							✓	✓	✓	8
	......	7.15	Support, Rear Engine—L. H........................	12UU5-88							✓	✓	✓	1
	......	.15	Screw, Hex. Cap (⅝"—11 x 2¾")...................	S-1874							✓	✓	✓	2
	......	.12	Screw, Hex. Cap (⅝"—11 x 3½")...................	S-4336							✓	✓	✓	2
	......	.05	Nut, Hex. (⅝"—11)................................	S-93							✓	✓	✓	4
	......	.01	Lockwasher (⅝")...................................	S-1351							✓	✓	✓	8
	......	.50	Insulator, Rubber—Top............................	12NJ2-402							✓	✓	✓	4
	......	.25	Insulator, Rubber—Bottom.........................	12NJ2-403							✓	✓	✓	4
	......	.15	Screw, Hex. Cap (⅝"—11 x 3½")...................	S-4346							✓	✓	✓	4
	......	.05	Nut, Hex.—Slotted (⅝"—11).......................	S-4201							✓	✓	✓	4
	......	.04	Washer, Flat Steel (²¹⁄₃₂" x 1⅝" x ⅛")............	S-927							✓	✓	✓	4
	......	.01	Pin, Cotter (³⁄₃₂" x 1¼")...........................	S-2221							✓	✓	✓	4
	......	.04	Screw, Oil Jet Set.................................	2SA2-165A	✓	✓	✓	✓		✓				2
	......	.01	Washer, ³⁄₁₆" Spring...............................	S-1344	✓	✓	✓			✓				2
	......	.12	Spring, Oil Relief Valve............................	2T2-477B	✓					✓				1
	......	.15	Spring, Oil Relief Valve............................	2UG0-477		✓	✓							1
	......	.25	Screw, Oil Relief Valve.............................	2T2-599A	✓				✓	✓				1
	......	.45	Screw, Oil Relief Valve.............................	2UG0-599		✓	✓							1
	......	.20	Valve, Oil Relief...................................	2T2-598C	✓					✓				1
	......	.30	Valve, Oil Relief...................................	2UG0-598		✓	✓							1
	......	.05	Nut, Oil Relief Valve Adjusting Screw (⅝"—11)...	S-2004	✓					✓				1
	......	.35	Nut, Oil Relief Valve Lock..........................	2UG0-736		✓	✓							1
	......	.20	Nut, Oil Relief Valve Cap...........................	2T2-602	✓				✓	✓				1
	......	.75	Nut, Oil Relief Valve Cap...........................	2UG0-602		✓	✓							1
	......	.03	Gasket, Oil Relief Valve Cap........................	S-1911	✓					✓				2
	......	.06	Gasket, Oil Relief Valve Cap........................	2UG0-831		✓	✓							2
	......	.25	Tube, Oil Drain....................................	2D3-244	✓	✓	✓	✓		✓				1
	......	.40	Clip, Drain Tube—Front...........................	2G2-245	✓	✓	✓	✓		✓				1
	......	.55	Clip, Drain Tube—Rear............................	2G2-246	✓	✓	✓	✓		✓				1
	......	.08	Chain, Brass—4½" Long...........................	2T0-986	✓	✓	✓	✓		✓				1
	......	.04	Pin, Crankshaft Rear Bearing Dowel................	2T2-951						✓				2
	......	.06	Clip, Cable..	2SA2-937						✓	✓			1
	......	.04	Seal, Crankshaft Rear Bearing......................	2T2-951							✓			2
	......	.03	Screw, Cable Clip (¼"—20 x ½")...................	S-796							✓			1
	......	.01	Washer, ¼" Spring................................	S-1345							✓			1
	......	.09	Stud, Rear Motor Support..........................	2T2-239							✓			10
	22013-A	.10	Cup, Valve Spring..................................	2BL0-49	✓									12
	......	.12	Cup, Valve Spring..................................	2NB2-49A		✓	✓	✓		✓				12
	......	.12	Washer, Valve Spring..............................	2TE2-508		✓	✓	✓						12
	......	....	Washer, Valve Spring..............................	2T2-508						✓				12
	......	.12	Cup, Valve Spring—Latest Design (Use with 2D210A—2D2710)	2NB2-49A						✓				12
	......	.20	Cup, Valve Spring—Orig. Design (Use with 2SA210C—2SA2710A)	2M1-49						✓				12
	40011-A	.06	Pin, Valve Stem....................................	2BL0-329	✓									12
	......	.03	Key, Valve Stem...................................	2NB2-329A		✓	✓	✓		✓				24
	......	.03	Key, Valve Stem (Use with 2D210A—2D2710)......	2NB2-329A						✓				24
	......	.05	Key, Valve Stem (Use with 2SA210C—2SA2710A)...	2M1-329						✓				24
	45812-A	.48	Plunger, Valve.....................................	2UG0-36	✓									12
	......	1.75	Plunger, Valve, Assembly...........................	2C2-40		✓	✓	✓	✓	✓				12
	22089-A	.20	Guide, Valve Plunger...............................	2BL0-24	✓									12
	......	.90	Guide, Valve Plunger...............................	2T2-24A		✓	✓	✓	✓	✓				12
	......	.40	Yoke, Valve Plunger................................	2T2-28		✓	✓	✓	✓	✓				6
	2185-A	.10	Screw, Valve Plunger Adjusting.....................	2TH0-168	✓									12
	......	.20	Screw, Valve Plunger Adjusting.....................	2T2-168A		✓	✓	✓	✓					12
	......	.10	Screw, Valve Plunger Adjusting.....................	2CB2-168B						✓				12
	2186-A	.03	Nut, Valve Plunger Adjusting Screw Lock............	2TH0-748	✓									12
	......	.06	Nut, Valve Plunger Adjusting Screw Lock............	2T2-38		✓	✓	✓	✓	✓				12

01-16 ENGINE — Master Parts List
THE AUTOCAR COMPANY, ARDMORE, PA.

CODE	MFR. PART No.	LIST PRICE	NAME OF PART	AUTOCAR PART No.	U-2044	U-4044	U-4144	U-4144-T	C-50	U-5044	U-7144-T	U-8144	U-8144-T	No. REQ.
0110	......	$0.20	Stud, Valve Plunger Guide Yoke................	2T2-31		✓	✓	✓	✓	✓				6
	......	.03	Nut, Valve Plunger Yoke (3/8"—24).............	S-1418		✓	✓	✓	✓	✓				6
	......	.01	Washer, 3/8" Spring...........................	S-1347		✓	✓	✓	✓	✓				6
	......	.12	Ring, Piston Pin Lock.........................	2Y1-368		✓	✓	✓	✓	✓				12
	......	.40	Bushing, Piston Pin...........................	2T2-15A					✓					6
	......	51.62	Crankshaft and Gear Assembly..................	2BL0-911	✓									1
	......	130.00	Crankshaft and Gear Assembly..................	2SA0-911B		✓	✓	✓	✓					1
	......	130.00	Crankshaft and Gear Assembly..................	2J0-911						✓				1
	......	.12	Screw, Air Compressor Coupling Cap............	2G2-99		✓	✓	✓	✓	✓				1
	......	30.25	Flywheel and Ring Gear Assembly...............	2G0-910		✓	✓	✓						1
	......	30.00	Flywheel and Ring Gear Assembly...............	2RL0-910						✓				1
	......	30.00	Flywheel and Ring Gear Assembly...............	2ZB0-910							✓			1
	......	5.70	Gear, Flywheel................................	2T3-214		✓	✓	✓	✓	✓				1
	......	6.00	Seal, Crankshaft Rear Bearing and Gasket Assembly........	2T0-410		✓	✓	✓	✓	✓				1
	......	.06	Screw, Crankshaft Rear Bearing Seal (3/8"—16 x 1¼")....	S-64		✓	✓	✓	✓	✓				8
	......	.01	Washer, 3/8" Spring...........................	S-1347		✓	✓	✓	✓	✓				8
	......	.12	Gasket, Crankshaft Rear Bearing Seal...........	2T3-642		✓	✓	✓	✓	✓				1
	......	.06	Gasket, Crankshaft Rear Bearing Seal—Center....	2T2-988		✓	✓	✓	✓	✓				2
	......	44.60	Pan, Oil—Large and Small Assembly.............	2UG0-160		✓	✓	✓						1
	......	39.00	Pan, Oil—Large and Small Assembly (Conventional Models)..	SD0-160						✓				1
	......	24.75	Pan, Oil—Large Assembly (Conventional Models)..........	2SA0-1160						✓				1
	......	31.00	Pan, Oil—Large Assembly......................	2HD0-1160A		✓	✓	✓	✓					1
	......	.50	Gasket, Oil Pan—Large.........................	2SA4-161A		✓	✓	✓	✓					1
	......	.10	Stud, Oil Pan.................................	2SK2-998		✓	✓	✓						20
	......	.12	Stud, Oil Pan.................................	2T2-998					✓	✓				18
	......	3.00	Strainer, Oil Pump, Assembly..................	2UG0-957		✓	✓	✓	✓					1
	......	1.40	Pipe, Oil Pump Intake.........................	2DK2-1051		✓	✓	✓	✓					1
	......	.12	Plug, Oil Pump Drain (3/8")....................	2T2-104		✓			✓	✓				1
	......	.07	Gasket, Oil Pan Drain Plug.....................	S-1905			✓	✓						1
Oil Pan Group														
	......	.20	Valve, Oil Relief..............................	2T2-598C						✓				1
	......	.40	Screw, Oil Relief Valve........................	2N2-956C						✓				1
	......	.12	Spring, Oil Relief Valve.......................	2T2-477B						✓				1
	......	2.70	Strainer, Oil Pump, Assembly..................	2SD0-957						✓				1
	......	....	Pipe, Oil Pump Intake.........................	2N2-1051						✓				1
Cylinder and Crankcase Group														
40070-A		.10	Screw, Main Bearing (½"—13 x 2")...............	S-922	✓									10
	......	.01	Washer, ½" Spring............................	S-1349	✓									10
	......	.09	Screw, Main Bearing..........................	S-3373	✓									8
	......	.01	Washer, 7/16" Spring..........................	S-1348	✓									8
	......	.30	Cap, Oil Filler, Assembly......................	2BL0-946	✓									1
	......	2.40	Body, Oil Filler...............................	2BLW0-105	✓									1

ILLUSTRATIONS

GROUP 02

CLUTCH

GROUP 02

Master Parts List

THE AUTOCAR COMPANY, ARDMORE, PA.

As supplied on Autocar Models
U-7144-T U-8144 U-8144-T

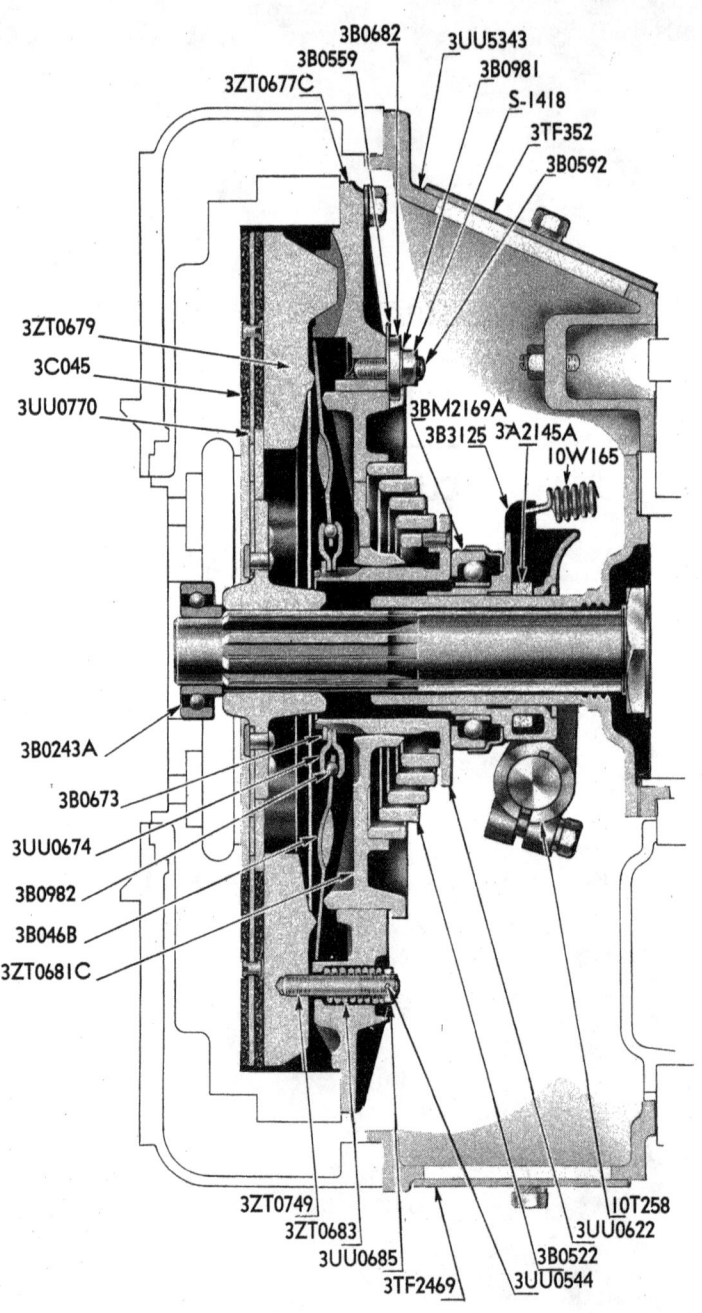

0200—Clutch Assembly

THE AUTOCAR COMPANY, ARDMORE, PA.

Master Parts List

GROUP 02

As supplied on Autocar Models
U-8144 U-8144-T

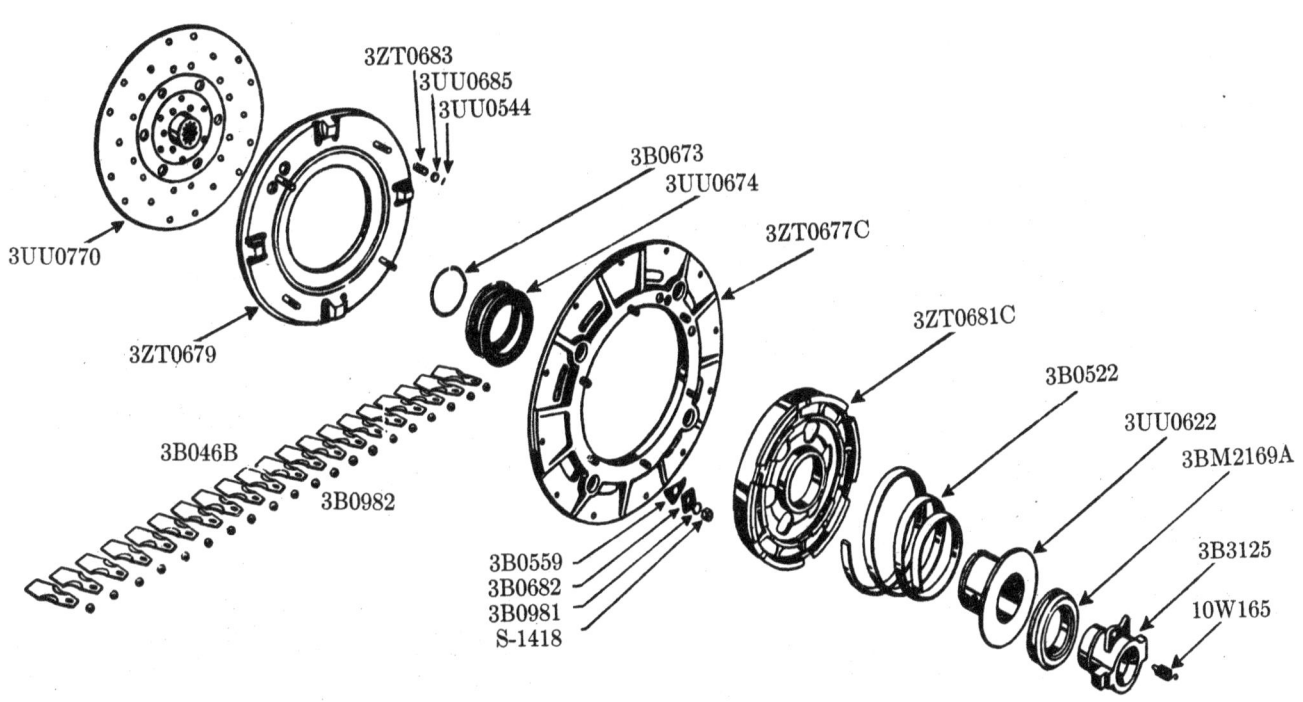

0200—Clutch Assembly

GROUP 02

Master Parts List

THE AUTOCAR COMPANY, ARDMORE, PA.

As supplied on Autocar Models
U-8144 U-8144-T

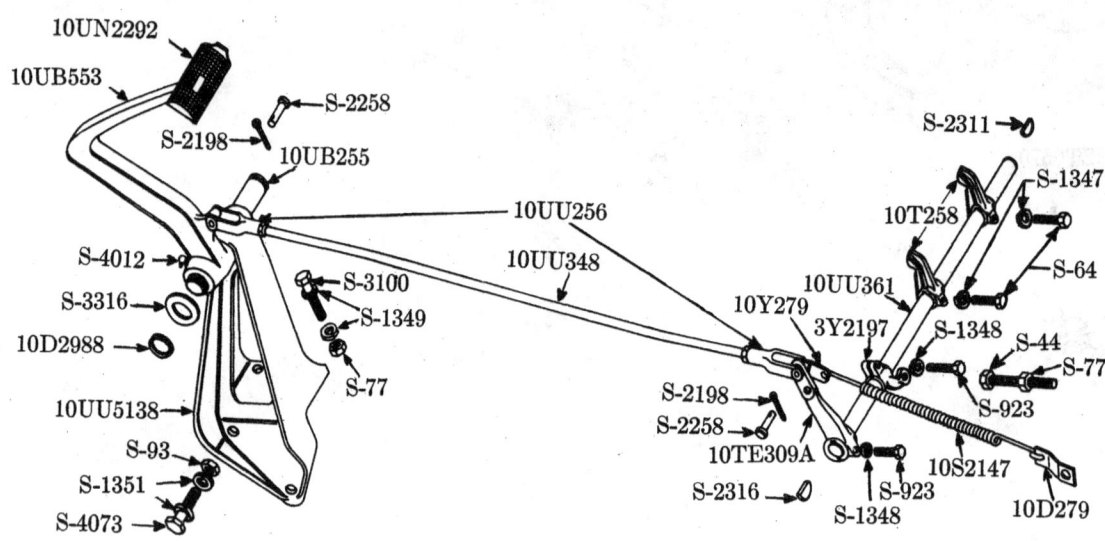

0200—Clutch Assembly

THE AUTOCAR COMPANY
ARDMORE, PA.

Master Parts List

02-1 CLUTCH

CODE	MFR. PART No.	LIST PRICE	NAME OF PART	AUTOCAR PART No.	U-2044	U-4044	U-4144	U-4144-T	C-50	U-5044	U-7144-T	U-8144	U-8144-T	NO. REQ.
0200	5854	....	**Clutch** Assembly (Long)(Does not include Housing & Control)	3RL4-30A	✓	✓	✓	✓	✓					1
	Z48-S	$55.00	**Clutch** Assy. (W. C. Lipe) (Does not include Housing & Control)	3DA5-30						✓				1
	Z42-S-15″	100.00	**Clutch** Assembly (W. C. Lipe)	3UU5-30							✓	✓	✓	1
	......	.08	**Screw**, Hex. cap (⅜″ x 1⅛″) NS	S-6257							✓	✓	✓	12
	......	.01	**Lockwasher** (⅜″)	S-1347							✓	✓	✓	12
0201	**Clutch Disc**													
	CM-4232	13.00	**Driven Member** Assembly (Including Hub and Plate Assy.)	3RL0-770	✓	✓	✓	✓	✓					1
	Z10-1	13.25	**Disc** Assembly	3DA0-770						✓				1
	Z15-2	25.00	**Disc**, Clutch with Facing	3UU0-770							✓	✓	✓	1
	CS-3599	3.40	**Plates**, Cushion and Facing Unit Assembly	3RL0-1470	✓	✓	✓	✓	✓					1
	C-3006	.01	**Rivet**, Cushion Plates and Facing Unit	3RL0-785	✓	✓	✓	✓	✓					12
	C-3593	1.80	**Facing**, Friction	3RL0-45	✓	✓	✓	✓	✓					2
	C-18-92	2.52	**Facing**, Disc	3DA0-45						✓				2
	C18-86	4.46	**Facing**, Clutch Disc	3C0-45							✓	✓	✓	2
	C-3429	.01	**Rivet**, Friction Facing	3RL0-114	✓	✓	✓	✓	✓					48
	X14-10	.01	**Rivets**, Disc Facing (⅛″ Dia. x 5/16″)	S-1018						✓				18
	X14-16	.01	**Rivet**, Clutch Disc Facing	3ZT0-114							✓	✓	✓	30
0202	**Cover, Pressure Plate and Springs**													
	CM-4348	30.00	**Cover** & Press. Plate Assy. (Inc. Cov. Plate, Spgs. & Press. Plate)	3RL0-800A	✓	✓	✓	✓	✓					1
	C1-16	10.00	**Plate**, Pressure	3DA0-679						✓				1
	C1-9	25.00	**Plate**, Clutch Pressure	3ZT0-679							✓	✓	✓	1
	X21-11	.25	**Stud**, Clutch Pressure Plate	3ZT0-749							✓	✓	✓	4
	C6-2	.08	**Spring**, Pull Back	3B0-683						✓				4
	C6-4	.10	**Spring**, Clutch Pressure Plate Retracting	3ZT0-683							✓	✓	✓	4
	C66-1	.10	**Washer**, Clutch Pressure Plate Retracting Spring—Ret.	3UU0-685							✓	✓	✓	4
	C7-3	.02	**Pin**, Clutch Pressure Plate Retracting Spring—Ret.	3UU0-544							✓	✓	✓	4
	C7-2	.02	**Pin**, Pull Back Spring	3B0-544						✓				4
	C2-32	13.50	**Ring**, Flywheel	3DA0-677						✓				1
	C2-27	20.95	**Ring**, Clutch Flywheel	3ZT0-677C							✓	✓	✓	1
	X21-12	.16	**Stud**, Clutch Flywheel Ring	3B0-592							✓	✓	✓	6
	C20-1	.02	**Shim**, Clutch Flywheel Ring Adjusting (.015)	3B0-559							✓	✓	✓	42
	C10-2	.08	**Strap**, Clutch Flywheel Ring Adjusting	3B0-682							✓	✓	✓	6
	X4-20	.03	**Nut**, Clutch Flywheel Ring Stud	S-1418							✓	✓	✓	6
	X10-23	.01	**Washer**, Clutch Flywheel Ring Stud Nut Lock	3B0-981							✓	✓	✓	6
	C3-22	5.50	**Plate**, Adjuster	3DA0-681						✓				1
	C3-36	10.00	**Plate**, Clutch Adjusting	3ZT0-681C							✓	✓	✓	1
	C5-10	4.50	**Spring**, Clutch—Inner	3NY0-522						✓				1
	C5-4	4.50	**Spring**, Clutch Pressure	3B0-522							✓	✓	✓	1
0203	**Release Lever, Bearings, Forks, Etc.**													
	C-1888	3.00	**Sleeve**, Release—Bare	3A0-125	✓				✓	✓				1
	......	3.60	**Sleeve**, Release—Bare	3RM0-125		✓	✓	✓						1
	C4-35	8.00	**Sleeve**, Clutch Release	3UU0-622							✓	✓	✓	1
	......	3.60	**Sleeve**, Clutch Release	3B3-125						✓				1
	C40-1	.02	**Pin**, Clutch Release Sleeve Spring Stop	3UU0-1004							✓	✓	✓	1
	C13-8	.60	**Ring**, Fulcrum	3NY0-674						✓				2
	C13-11	.60	**Ring**, Clutch Release Sleeve Fulcrum	3UU0-674							✓	✓	✓	2
	X17-3	.02	**Ball**, Pressure Lever Locking	3NY0-982						✓				20
	X17-7	.02	**Ball**, Clutch Release Sleeve Fulcrum Ring	3B0-982							✓	✓	✓	20
	C8-20	.20	**Lever**, Pressure	3NY0-46						✓				20
	C8-21	.24	**Lever**, Clutch Pressure	3B0-46B							✓	✓	✓	20
	C11-3	.15	**Ring**, Snap	3NY0-673						✓				1
	C11-4	.15	**Ring**, Clutch Release Sleeve Snap	3B0-673							✓	✓	✓	1
	A959-1	2.05	**Bearing**, Clutch Release (Aetna)	3BM2-169A	✓	✓	✓	✓	✓	✓				1
	......	3.60	**Block**, Clutch Release Trunnion	3B3-125										1
	......	.20	**Spring**, Clutch Release Trunnion Return	10W1-65	✓	✓	✓	✓	✓					1
	......	.12	**Link**, Clutch Release Trunnion Return Spring	10R1-148	✓	✓	✓	✓	✓					1
	......	.10	**Wick**, Clutch Release Trunnion Oil	3A2-145A	✓	✓	✓	✓	✓					1
	......	1.20	**Shaft**, Clutch Control Cross	10DF3-61A	✓	✓	✓	✓	✓					1
	......	1.15	**Shaft**, Clutch Throwout	10UU3-61							✓	✓	✓	1

02-2 CLUTCH — Master Parts List — THE AUTOCAR COMPANY, ARDMORE, PA.

CODE	MFR. PART No.	LIST PRICE	NAME OF PART	AUTOCAR PART No.	U-2044	U-4044	U-4144	U-4144-T	C-50	U-5044	U-7144-T	U-8144	U-8144-T	No. REQ.
0203	.....	$0.25	Bushing, Clutch Throwout Shaft	3UN2-119B							✓	✓	✓	..
	.....	.90	Lever, Clutch Trunnion	10T2-58	✓	✓	✓	✓	✓		✓	✓	✓	2
	.....	.06	Key, Woodruff (#11)	S-2311	✓	✓	✓	✓	✓		✓	✓	✓	3
	.....	.06	Screw, Hex. Cap (3/8"—16 x 1¼")	S-64	✓	✓	✓	✓	✓		✓	✓	✓	2
	.....	.01	Washer, 3/8" Spring Lock	S-1347	✓	✓	✓	✓	✓		✓	✓	✓	2
	.....	3.00	Lever, Clutch Control Shaft and Control, Assembly	10L0-1330	✓	✓			✓					1
	.....	1.50	Lever, Clutch Throwout Shaft Stop	3Y2-197							✓	✓	✓	1
	.....	.08	Screw, Hex. Cap (7/16"—14 x 1½")	S-923							✓	✓	✓	1
	.....	.01	Washer, 1+/32" Spring Lock	S-1348							✓	✓	✓	1
	.....	.10	Screw, Hex. Cap (½"—13 x 2¼")	S-44							✓	✓	✓	1
	.....	.04	Nut, Hex. (½"—13)	S-77							✓	✓	✓	1
	.....	1.60	Lever, Clutch Control	10NL3-09	✓	✓			✓					1
	.....	1.80	Lever, Clutch Control	10UGA3-09			✓	✓						1
	.....	1.90	Lever, Clutch Throwout Shaft	10TE3-09A							✓	✓	✓	1
	.....	.06	Key, Woodruff (#15)	S-2316	✓	✓	✓	✓	✓		✓	✓	✓	1
	.....	.08	Screw, Hex. Cap (7/16"—14 x 1½")	S-923	✓	✓	✓	✓	✓		✓	✓	✓	1
	.....	.01	Washer, 1+/32" Spring Lock	S-1348	✓	✓	✓	✓	✓		✓	✓	✓	1
	.....	.01	Washer, 3/8" Spring	S-1347	✓	✓	✓	✓	✓					2
	.....	.40	Spring, Clutch Pedal Return	10S2-147	✓	✓	✓	✓	✓					1
	.....	.40	Spring, Clutch Throwout Shaft Lever	10S2-147							✓	✓	✓	1
	.....	.12	Link, Clutch Pedal Return Spring—Upper	10T1-79A	✓	✓	✓	✓	✓					1
	.....	.12	Spring, Clutch Throwout Shaft Lever, Link	10D2-79							✓	✓	✓	1
	.....	.12	Link, Clutch Pedal Return Spring—Lower	10CH2-79A	✓	✓	✓	✓	✓					1
	.....	.12	Link, Clutch Throwout Shaft Lever Spring	10Y2-79							✓	✓	✓	1
	.....	1.80	Rod, Clutch Control, Assembly	10DFL0-470	✓	✓			✓					1
	.....	1.80	Rod, Clutch Control, Assembly	10UB0-470			✓	✓						1
	.....	.25	Rod, Clutch Control (6½" Long)	10DFL2-48	✓	✓			✓					1
	.....	.40	Rod, Clutch Control	10UB3-48			✓	✓						1
	.....	.45	Rod, Clutch Control	10UU3-48							✓	✓	✓	1
	.....	.03	Nut, Clutch Control Rod (3/8"—24)	S-1418	✓	✓	✓	✓	✓					1
	.....	.30	Clevis, Clutch Control Rod	10A2-56	✓	✓			✓					1
	.....	.50	Clevis, Clutch Control Rod	10F2-56			✓	✓						1
	.....	.85	Clevis, Clutch Control Rod	10UU2-56							✓	✓	✓	1
	.....	.06	Pin, Clutch Control Rod Clevis (3/8" x 1 7/32")	S-2259	✓	✓	✓	✓	✓					2
	.....	.06	Pin, Clutch Control Rod Clevis	S-2258							✓	✓	✓	2
	.....	.01	Pin, Cotter (3/32" x ¾")	S-2198	✓	✓	✓	✓	✓		✓	✓	✓	2
0204	.....	1.20	Clevis, Clutch Control Rod Adjusting	10A2-383	✓	✓			✓					1
	.....	1.20	Clevis, Clutch Control Rod Adjusting	10UE2-04			✓	✓						1
	.....	.06	Pin, Clutch Control Rod Adjusting Clevis (3/8" x 1 1/16")	S-2257	✓	✓	✓	✓	✓					1
	.....	.01	Pin, Cotter (3/32" x ¾")	S-2198	✓	✓	✓	✓	✓					1
	.....	4.30	Bracket, Pedal, and Bushing Assembly	10NL0-828	✓	✓			✓					1
	.....	3.60	Bracket, Pedal	10NL4-138	✓	✓			✓					1
	.....	6.00	Bracket, Pedal	10UG5-138			✓	✓						1
	.....	10.00	Bracket, Clutch and Brake Pedal	10UU5-138							✓	✓	✓	2
	.....	.15	Screw, Hex. Cap (5/8"—11 x 1¾")	S-4073							✓	✓	✓	2
	.....	.05	Nut, Hex. (5/8"—11)	S-93							✓	✓	✓	2
	.....	.01	Washer, 21/32" Spring Lock	S-1351							✓	✓	✓	4
	.....	.05	Screw, Hex. Cap (½"—13 x 1¾")	S-3100							✓	✓	✓	2
	.....	.04	Nut, Hex. (½"—13)	S-77							✓	✓	✓	2
	.....	.01	Washer, 1+/32" Spring Lock	S-1349							✓	✓	✓	4
	.....	.35	Bushing, Pedal Bracket	10NL2-15	✓	✓			✓					2
	.....	.25	Bushing, Pedal Bracket	10S2-15			✓	✓						2
	.....	.09	Bolt, Pedal Bracket (½"—13 x 1½")	S-1872	✓	✓	✓	✓	✓					4
	.....	.04	Nut, Pedal Bracket Bolt (½"—13)	S-77	✓	✓	✓	✓	✓					4
	.....	.01	Washer, ½" Spring	S-1349	✓	✓	✓	✓	✓					4
	.....	.70	Shaft, Pedal Lever	10NL2-55A	✓	✓			✓					1
	.....	.35	Shaft, Pedal Lever	10UB2-55			✓	✓						1
	.....	.25	Collar, Pedal Lever Shaft	10Y1-135	✓	✓	✓	✓	✓					1
	.....	.01	Pin, Cotter (1/8" x 2")	S-2208	✓	✓	✓	✓	✓					1
	.....	4.80	Pedal, Clutch	10NL4-53A	✓	✓			✓					1
	.....	7.55	Pedal, Clutch	10UB5-53			✓	✓						1

CODE	MFR. PART No.	LIST PRICE	NAME OF PART	AUTOCAR PART No.	U-2044	U-4044	U-4144	U-4144-T	C-50	U-5044	U-7144-T	U-8144	U-8144-T	No. REQ.
					MODEL									
0204	......	$0.06	**Key,** Clutch Pedal (#15 Whitney)	S-2316	✓	✓	✓	✓	✓					1
	......	.08	**Bolt,** Clutch Pedal (⅜″—16 x 2¼″)	S-40	✓	✓	✓	✓	✓					1
	......	.03	**Nut,** Clutch Pedal Bolt (⅜″—16)	S-75	✓	✓	✓	✓	✓					1
	......	.01	**Washer,** ⅜″ Spring	S-1347	✓	✓	✓	✓	✓					1
	......	.35	**Shaft,** Clutch and Brake Pedal	10UB2-55							✓	✓	✓	1
	......	.01	**Washer,** Flat (1 1/32″ x 1¾″ x 3/32″)	S-3316							✓	✓	✓	1
	......	.25	**Ring,** Lock	10D2-988							✓	✓	✓	1
	......	1.80	**Bushing,** Clutch Pedal Lever, and Assembly	10NL0-833	✓	✓	✓	✓	✓					1
	......	1.60	**Lever,** Clutch Pedal	10NL2-08	✓	✓	✓	✓	✓					1
	......	.12	**Bushing,** Clutch Pedal Lever	10UA1-15	✓	✓	✓	✓	✓					1
	......	.06	**Key,** Clutch Pedal Lever (#15 Whitney)	S-2316	✓	✓	✓	✓	✓					1
	......	.08	**Screw,** Clutch Pedal Lever (7/16″—14 x 1½″)	S-923	✓	✓	✓	✓	✓					1
	......	.01	**Washer,** 7/16″ Spring	S-1348	✓	✓	✓	✓	✓					1
	......	.11	**Screw,** Clutch Pedal Lever Stop (½″—13 x 2¼″)	S-1880	✓	✓	✓	✓	✓					1
	......	.04	**Nut,** Clutch Pedal Lever Stop Screw (½″—13)	S-77	✓	✓	✓	✓	✓					1
	......	7.55	**Lever,** Clutch Pedal	10UB5-53							✓	✓	✓	1
	......	.20	**Fitting,** 67½° Alemite	S-4012							✓	✓	✓	1
	......	.55	**Pad,** Clutch Pedal	10UN2-292	✓	✓	✓	✓	✓					1
	......	.03	**Bolt,** Clutch Pedal Pad (¼″ x 1¼″)	S-3086	✓	✓	✓	✓	✓					2
	......	.01	**Washer,** ¼″ Spring	S-1345	✓	✓	✓	✓	✓					2
	......	.55	**Pad,** Clutch Pedal	10UN2-292							✓	✓	✓	2
	......	.90	**Line,** Clutch Trunnion Oil, Assembly (1¼″ I.D.)	3CA9-1020	✓	✓	✓	✓	✓					1
	......	.30	**Nut,** Clutch Trunnion Oil Line Clamp	3UN2-625	✓	✓	✓	✓	✓					1
	......	.17	**Ell,** Clutch Trunnion Oil Line	S-4501	✓	✓	✓	✓	✓					1
	......	.12	**Union,** Clutch Trunnion Oil Line Single (⅛″ Pipe)	S-5023	✓	✓	✓	✓	✓					1
	......	.12	**Nut,** Clutch Trunnion Oil Line Union	S-4302	✓	✓	✓	✓	✓					1
	......	1.20	**Oiler,** Magazine	3B3-298	✓	✓	✓	✓	✓					1
	......	.25	**Wick,** Magazine Oiler	3M2-301A	✓	✓	✓	✓	✓					1
	......	.20	**Stud,** Magazine Oiler	2TE2-1008C	✓	✓			✓					2
	......	.20	**Stud,** Magazine Oiler	2TE2-1008B			✓	✓						2
	......	.01	**Nut,** Magazine Oiler Stud (⅜″—16)	S-829	✓	✓	✓	✓	✓					2
	......	.01	**Washer,** ⅜″ Spring	S-1347	✓	✓	✓	✓	✓					4
	......	.03	**Plug,** Magazine Oiler Filler (¾″)	S-4131	✓	✓	✓	✓	✓					1
	......	.02	**Plug,** Magazine Oiler (⅛″)	S-619	✓	✓	✓	✓	✓					3
	......	.50	**Oil Line,** Clutch, Assembly	3UU4-1020A							✓	✓	✓	1
0205	FLB30-C-003	4.60	**Bearing,** Clutch Pilot (S.K.F.)	3B0-243A						✓	✓	✓	✓	1
	......	.10	**Wick,** Clutch Release Sleeve Oil	3A2-145						✓				1
	CS-4349	7.50	**Plate,** Cover, Unit (Including Riveted Spring Cups)	3RL0-889A	✓	✓	✓	✓	✓					1
	C-4343	6.00	**Plate,** Cover, Bare	3RL0-48A	✓	✓	✓	✓	✓					1
	C-3437	.25	**Cup,** Spring	3RL0-642	✓	✓	✓	✓	✓					12
	C-3436	.01	**Rivet,** Spring Cup	3RL0-644	✓	✓	✓	✓	✓					12
	CS-4347	18.50	**Lever,** Pressure Plate and, Unit	3RL0-790A	✓	✓	✓	✓	✓					1
	C-3594	7.00	**Plate,** Pressure, Bare	3RL0-43	✓	✓	✓	✓	✓					1
	C-3141	.12	**Pin,** Release Lever	3RL0-782	✓	✓	✓	✓	✓					6
	C-3142	.10	**Pin,** Release Lever Adjusting Yoke	3RL0-783	✓	✓	✓	✓	✓					6
	C-3077	.01	**Roller,** Needle	3RL0-781	✓	✓	✓	✓	✓					156
	C-3390	.60	**Lever,** Release	3RL0-46	✓	✓	✓	✓	✓					6
	C-4344	.50	**Yoke,** Release Lever Adjusting	3RL0-784A	✓	✓	✓	✓	✓					6
	C-4345	.06	**Nut,** Release Lever Adjusting	3RL0-105A	✓	✓	✓	✓	✓					6
	C-4346	.10	**Spring,** Release Lever Tension	3RL0-132A	✓	✓	✓	✓	✓					6
	C-3431	.20	**Spring,** Pressure	3RL0-128	✓	✓	✓	✓	✓					12
	C-2085	.02	**Button,** Pressure Spring Insulator	3T0-124B	✓	✓	✓	✓	✓					12
	CM-1012	3.20	**Sleeve,** Release, Assembly (Including Wick & Spring Post)	3A0-780	✓	✓	✓	✓	✓					1
	C-959	.20	**Post,** Return Spring	3A0-136	✓	✓	✓	✓	✓					1
	6205-Z	3.60	**Bearing,** Clutch Pilot (S.K.F.)	3A0-243	✓	✓	✓	✓	✓					1
	......	.05	**Screw,** Cover Plate (⅜″—16 x ⅞″)	S-50	✓	✓	✓	✓	✓					12
	......	.01	**Washer,** ⅜″ Spring	S-1347	✓	✓	✓	✓	✓					12

THE AUTOCAR COMPANY, ARDMORE, PA.

Master Parts List

GROUP 03

ILLUSTRATIONS

GROUP 03

FUEL SYSTEM

GROUP 03

Master Parts List

THE AUTOCAR COMPANY, ARDMORE, PA.

As supplied on Autocar Models
U-8144-T

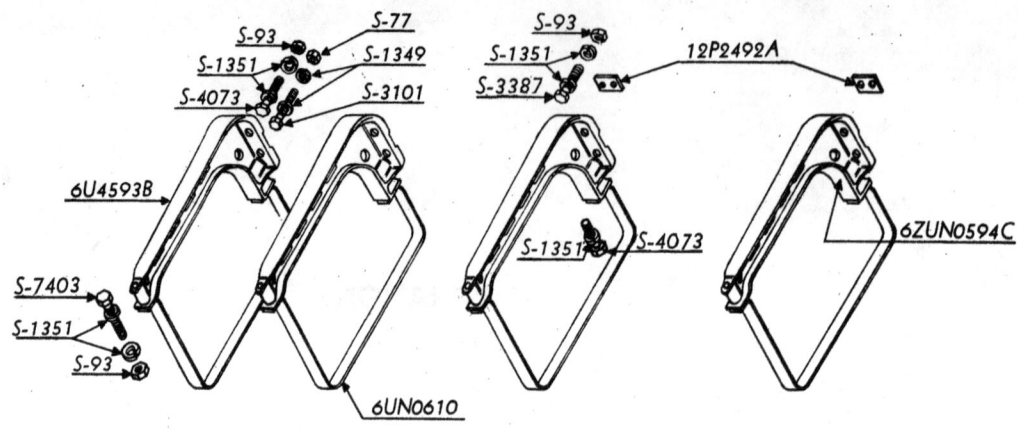

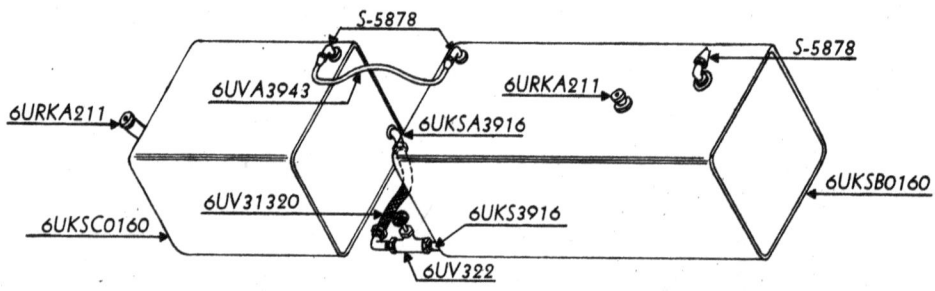

0300—Gasoline Tank

THE AUTOCAR COMPANY, ARDMORE, PA.

Master Parts List

GROUP 03

As supplied on Autocar Models U-8144 U-8144-T

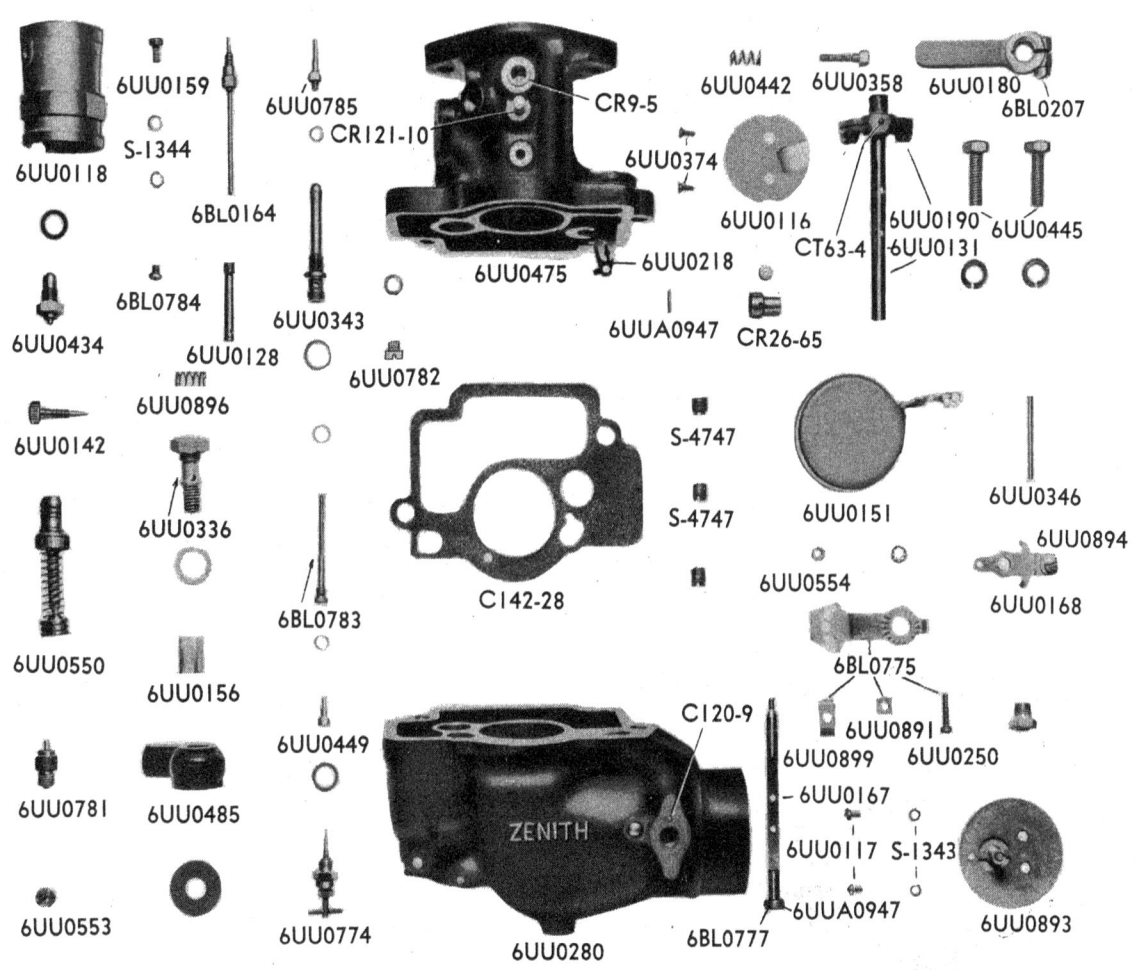

0301—Carburetor, Air Cleaner

GROUP 03

Master Parts List

THE AUTOCAR COMPANY, ARDMORE, PA.

As supplied on Autocar Models
U-8144   U-8144-T

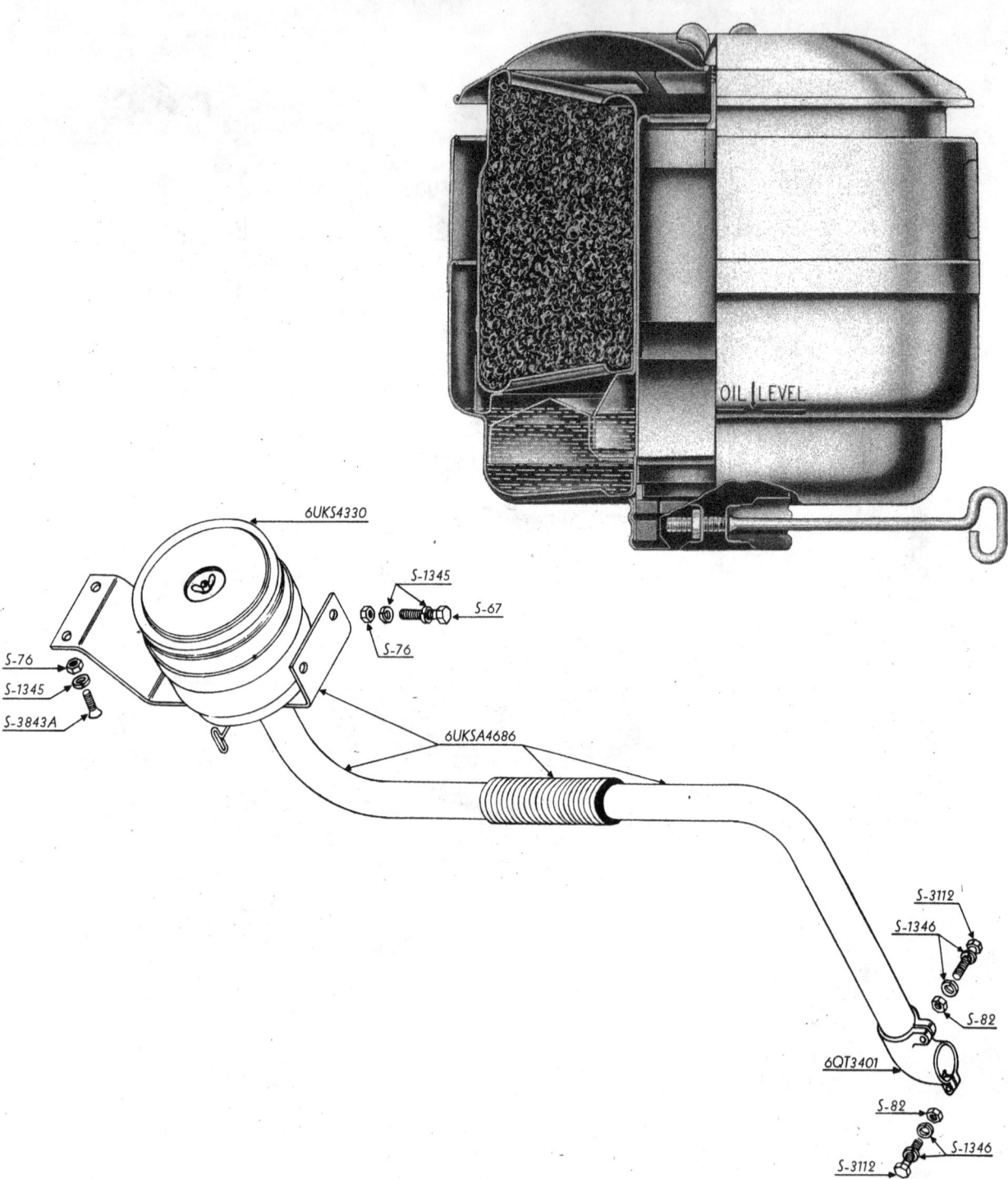

0301—Carburetor, Air Cleaner

THE AUTOCAR COMPANY, ARDMORE, PA.
Master Parts List
GROUP 03

As supplied on Autocar Models
U-7144-T   U-8144   U-8144-T

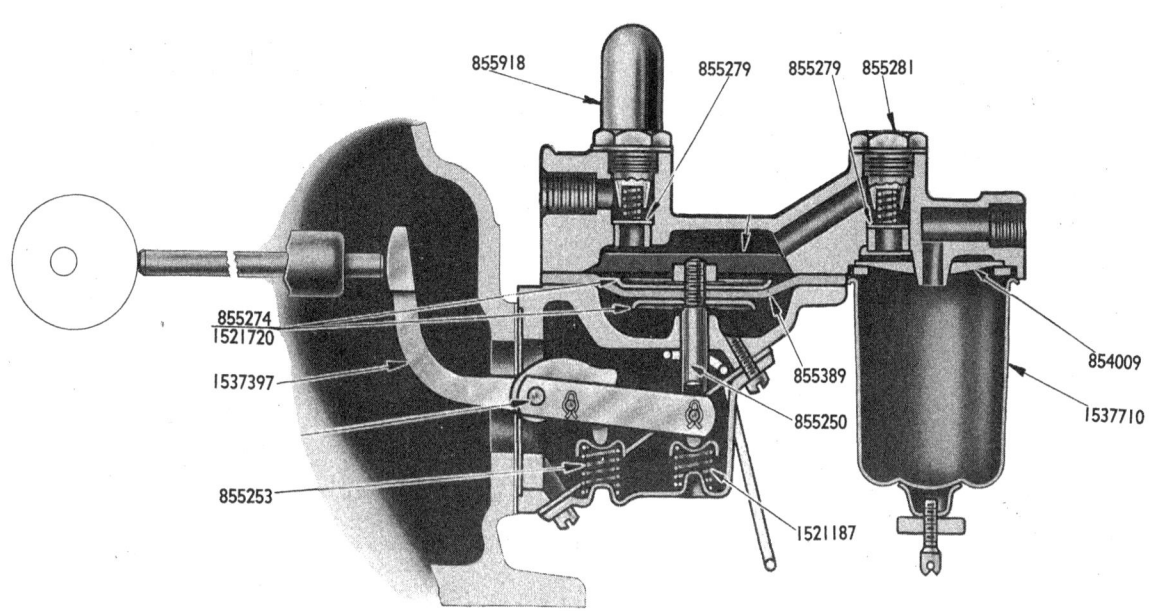

0302—Fuel Pump and Filter

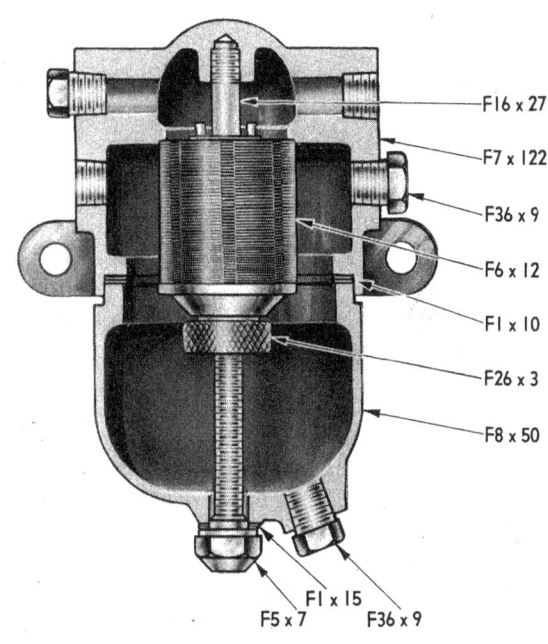

0302—Fuel Filter

GROUP 03

Master Parts List

As supplied on Autocar Models
U-8144-T

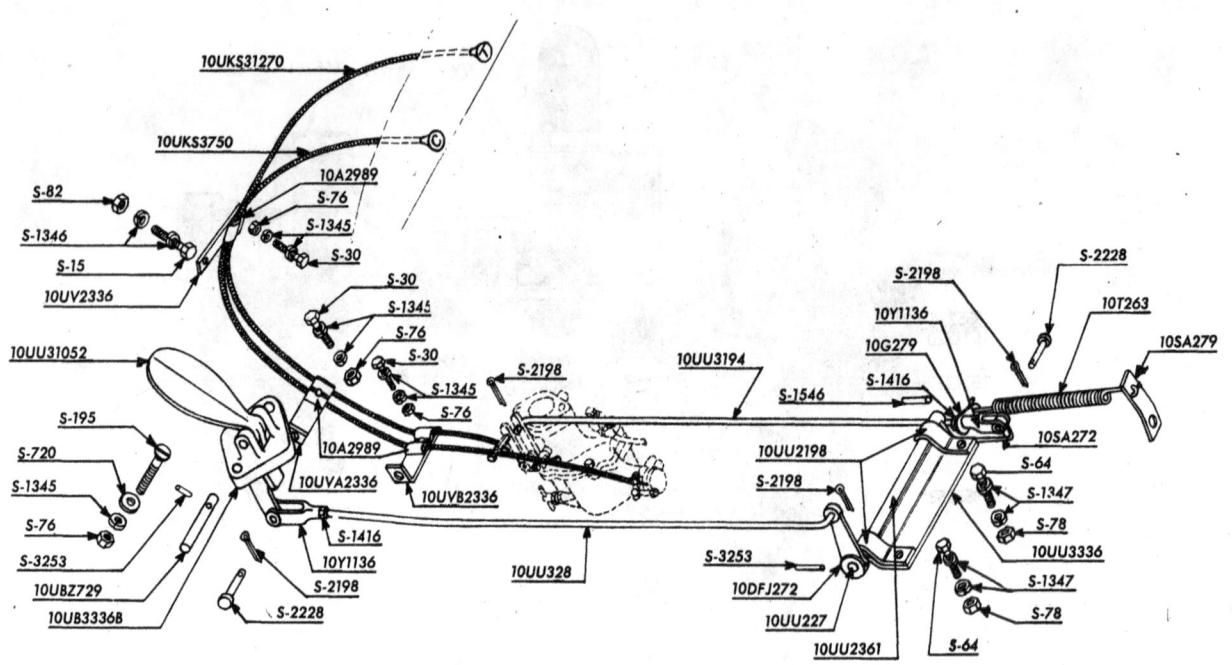

0303—Accelerator, Throttle and Choke

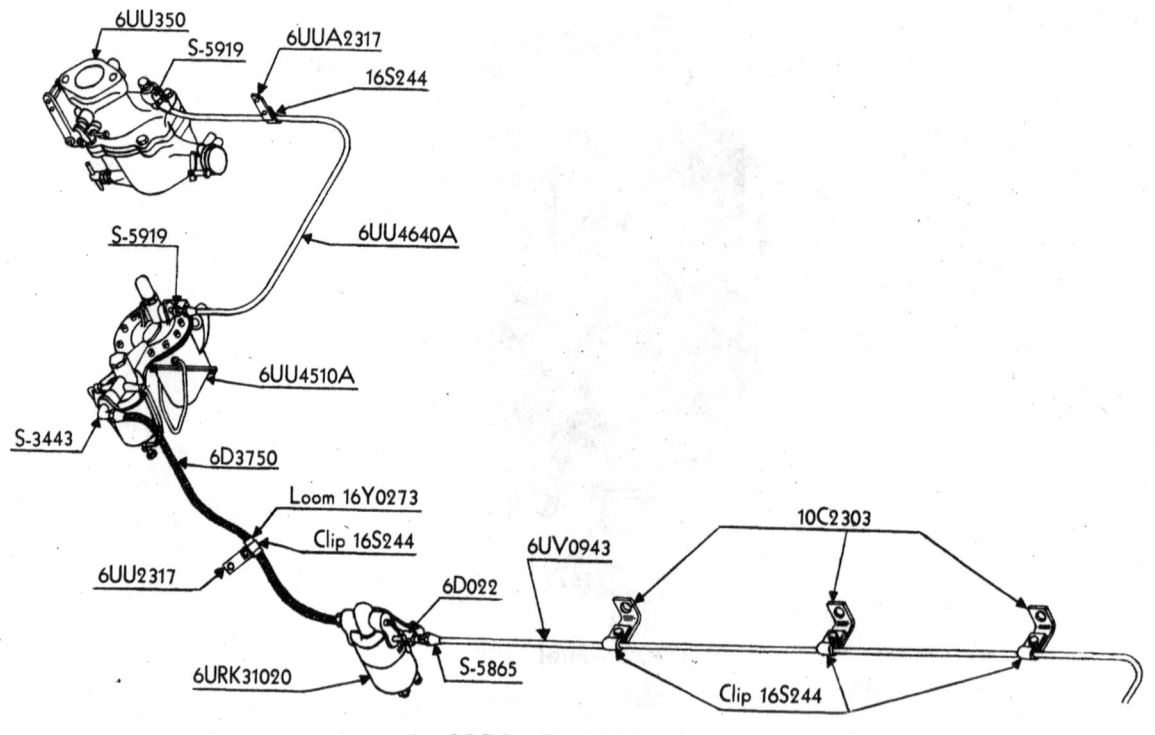

0304—Fuel Lines

Master Parts List

GROUP 03

THE AUTOCAR COMPANY, ARDMORE, PA.

As supplied on Autocar Models
U-7144-T   U-8144   U-8144-T

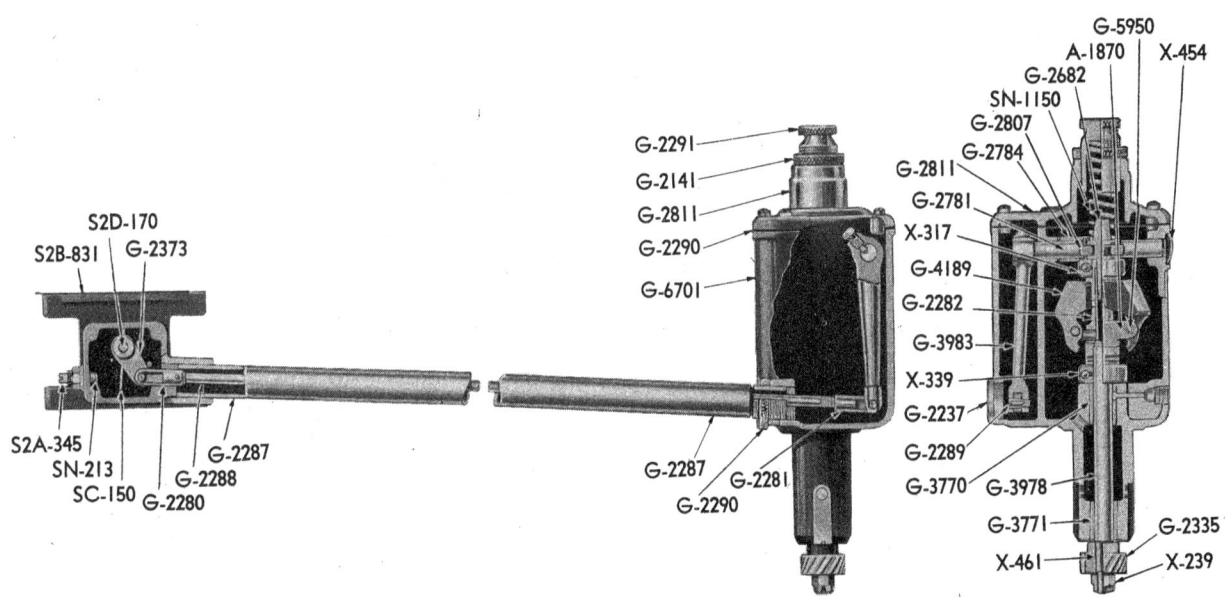

0305—Governor

03-0 FUEL SYSTEM

Master Parts List

THE AUTOCAR COMPANY, ARDMORE, PA.

CODE	MFR. PART No.	LIST PRICE	NAME OF PART	AUTOCAR PART No.	U-2044	U-4044	U-4144	U-4144-T	C-50	U-5044	U-7144-T	U-8144	U-8144-T	NO. REQ.
0300			**Gasoline Tank**											
		$40.00	Tank, Gas, Assembly	6URK4-720	✓	✓	✓	✓		✓				1
		42.00	Tank, Gas, 60 Gal.	6UKSB0-160							✓	✓	✓	1
		31.75	Tank, Gas, 30 Gal.	6UKSC0-160							✓	✓	✓	1
		.45	Elbow, Long Nut Compression (⅜"—¼")	S-5878							✓	✓	✓	3
		2.00	Cap, Gas Tank, and Chain	6URK0-730	✓	✓	✓	✓		✓				1
		1.50	Cap, Gas Tank, Assembly	6URKA2-11							✓	✓	✓	2
		3.00	Bracket, Gas Tank—Front	6UBK3-610	✓									1
		3.75	Bracket, Gas Tank—Front	6URK3-610		✓	✓	✓		✓				1
		3.00	Bracket, Gas Tank—Rear	6UBKA3-610	✓									1
		3.75	Bracket, Gas Tank—Rear	6URKA3-610		✓	✓	✓		✓				1
		5.25	Bracket, Gas Tank	6U4-593B							✓	✓	✓	4
		.12	Screw, Hex. Cap (½"—13 x 1¾")	S-3101							✓	✓	✓	8
		.15	Screw, Hex. Cap (⅝"—11 x 1¾")	S-4073							✓	✓	✓	4
		.15	Screw, Hex. Cap (⅝"—11 x 2")	S-3387							✓	✓	✓	4
		.05	Nut, Hex. (⅝"—11)	S-93							✓	✓	✓	8
		.01	Lockwasher, Spring (½")	S-1349							✓	✓	✓	16
		.01	Lockwasher, Spring (⅝")	S-1351							✓	✓	✓	16
		.15	Reinforcement, Gas Tank Bracket Frame	12P2-492A							✓	✓	✓	2
		.20	Anti-Rattle, Gas Tank Bracket	6ZUN0-594C							✓	✓	✓	4
		....	Rivets, Bifurcated (5/32" x ⅞")	S-1015							✓	✓	✓	20
		.40	Strap, Gas Tank	6URK2-15	✓	✓	✓	✓		✓				2
		3.00	Strap, Gas Tank, and Anti-Rattle Assembly	6UN0-610							✓	✓	✓	4
		....	Screw, Hex. Cap	S-9403							✓	✓	✓	4
		.05	Nut, Hex.	S-93							✓	✓	✓	4
		.15	Screw, Hex. Cap	S-1870							✓	✓	✓	4
		.01	Lockwasher, Spring (⅝")	S-1351							✓	✓	✓	12
		.30	Tubing, Gas Tank Vent Pipe ⅜" Bundy	6UVA3-943							✓	✓	✓	1
		....	Elbow, Gas Tank Inlet, Assembly	6UKSA3-916							✓	✓	✓	1
		9.35	Tube, Gas Tank Inlet Titeflex	6UV3-1320							✓	✓	✓	1
		4.70	Valve, Gas Tank ¾" Brass (Walworth #95)	6UV3-22							✓	✓	✓	1
		....	Nipple, Gas Tank Outlet, Assembly	6UKS3-916							✓	✓	✓	1
		3.50	Gauge, Gas Tank	16DFLB3-786-15	✓	✓	✓	✓		✓				1
		3.50	Gauge, Gas Tank, Unit (In Tank)	16DFLB9-786							✓	✓	✓	1
			Gasoline Line and Connections—Tank to Pump											
		.70	Valve, ¼" Gas Shut-Off	6D0-22	✓	✓	✓	✓		✓				1
		.30	Elbow, ⅜" Single	S-5066	✓	✓	✓	✓		✓				1
		.25	Nut, ⅜" Union	S-5005	✓	✓	✓	✓		✓				2
		.20	Tubing, ⅜" Copper—40" Long	S-5146-40	✓									1
		.20	Tubing, ⅜" Copper	S-5146		✓	✓	✓						As
		.20	Tubing, ⅜" Copper—20" Long	S-5146-20						✓				1
		.20	Tubing, ⅜" Copper—16" Long	S-5146-16						✓				1
		2.15	Line, Titeflex (To Pump)—12" Long	6D3-740	✓	✓	✓	✓						1
		.33	Elbow, ¼" Service	S-4617	✓									1
			Gasoline Line and Connections—Pump to Carburetor											
		.06	Union, 5/16" Single	S-5024	✓									1
		.20	Elbow, 5/16" Single	S-5065		✓	✓	✓		✓				1
		.20	Union, 5/16" Single	S-5024A		✓	✓	✓						1
		.20	Nut, 5/16" Union	S-5013	✓	✓	✓	✓						2
		.12	Tubing, 5/16" Copper—54" Long	S-5150-54						✓				1
		.12	Tubing, 5/16" Copper—48" Long	S-5150-48		✓	✓	✓		✓				1
		.06	Loom	16Y0-273	✓	✓	✓	✓		✓				1
0301			**Carburetor, Air Cleaner**											
	457-2	35.00	Carburetor Assembly (Zenith)	6UG4-50	✓									1
	380543	22.50	Carburetor Complete (Stromberg)	6UR4-50		✓								1
	425090	22.50	Carburetor Complete (Stromberg)	6UR4-50A			✓	✓						1
		22.50	Carburetor (Stromberg SF3)	6DA4-50					✓					1
		22.50	Carburetor Complete (Stromberg)	6UDG0-50						✓				1
	457-2	35.00	Carburetor Assembly (Zenith)	6UU3-50							✓	✓	✓	1

THE AUTOCAR COMPANY
ARDMORE, PA.

Master Parts List

03-1 FUEL SYSTEM

CODE	MFR. PART No.	LIST PRICE	NAME OF PART	AUTOCAR PART No.	U-2044	U-4044	U-4144	U-4144-T	C-50	U-5044	U-7144-T	U-8144	U-8144-T	No. REQ.
									MODEL					
0301	......	$0.08	Screw, Hex. Cap ($\frac{7}{16}$"—14 x 1¼")	S-1881							✓	✓	✓	2
	......	.01	Lockwasher, $\frac{7}{16}$" Spring	S-1348							✓	✓	✓	2
	......	1.20	Adapter, Carburetor	6UU3-309A							✓	✓	✓	1
	......	.10	Studs, Carburetor Adapter	6UU2-327							✓	✓	✓	2
	......	.06	Nut, Hex. ($\frac{7}{16}$"—14)	S-431							✓	✓	✓	2
	......	.01	Washer, $\frac{7}{16}$" Spring	S-1348							✓	✓	✓	2
	......	.02	Gasket, Carburetor	6N2-375A							✓	✓	✓	2
	......	.60	Tube, Carburetor Idling By-pass, Assembly	6UU4-913A							✓	✓	✓	1
	B2-52F-1	7.50	Body, Carburetor Throttle, Assembly	6UU0-475							✓	✓	✓	1
	B3-18-E	11.00	Bowl, Carburetor Fuel, Assembly	6UU0-280							✓	✓	✓	1
	C21-105	1.20	Plate, Carburetor Throttle	6UU0-116							✓	✓	✓	1
	C23-26	1.10	Shaft, Carburetor Throttle	6UU0-131							✓	✓	✓	1
	C24-10GX8	1.20	Lever, Carburetor Throttle and Swivel Assembly	6UU0-180							✓	✓	✓	1
	CR28-33 x 5	1.15	Lever, Carburetor Throttle Stop	6UU0-190							✓	✓	✓	1
	C36-12	.95	Piston, Carburetor Pump and Vaccum, Assembly	6UU0-550							✓	✓	✓	1
	C38-17	2.40	Venturi, Carburetor Main (#33)	6UU0-118							✓	✓	✓	1
	C41-9	.25	Valve, Carburetor Check	6UU0-553							✓	✓	✓	1
	C46-6	.35	Screw, Carburetor Idle Adjusting	6UU0-142							✓	✓	✓	1
	C51-3	.75	Jet, Carburetor Accelerating (#25)	6BL0-783							✓	✓	✓	1
	C52-3	.45	Jet, Carburetor Compensator (#32)	6BL0-784							✓	✓	✓	1
	C52-4	.50	Jet, Carburetor Main (#32)	6UU0-449							✓	✓	✓	1
	C54-1	.60	Jet, Carburetor Idling (#16)	6BL0-164							✓	✓	✓	1
	C57-1 x 1	.40	Jet, Carburetor Cap (38-1)	6UU0-785							✓	✓	✓	1
	C66-41	.75	Tube, Carburetor Discharge	6UU0-343							✓	✓	✓	1
	C71-3	2.40	Jet, Carburetor Main, Adjustment Assembly	6UU0-774							✓	✓	✓	1
	C76-21	.85	Well, Carburetor Progressive	6UU0-128							✓	✓	✓	1
	C81-3	.75	Valve, Carburetor Fuel (#55)	6UU0-434							✓	✓	✓	1
	C85-6	.90	Float, Carburetor, Assembly	6UU0-151							✓	✓	✓	1
	C97-10	.80	Valve, Carburetor Power Jet (#13)	6UU0-781							✓	✓	✓	1
	C101-19	2.10	Plate, Carburetor Air Shutter	6UU0-893							✓	✓	✓	1
	C105-130	.60	Shaft, Carburetor Air Shutter	6UU0-167							✓	✓	✓	1
	C106-2	.35	Lever, Carburetor Air Shutter, Assembly	6UU0-168							✓	✓	✓	1
	C109-2	.35	Bracket, Carburetor Air Shutter, Assembly	6BL0-775							✓	✓	✓	1
	C110-1	.05	Clamp, Carburetor Air Shutter Bracket Wire	6UU0-899							✓	✓	✓	1
	C111-17	.10	Spring, Carburetor Idler Adjusting Screw	6UU0-896							✓	✓	✓	1
	C111-62	.10	Spring, Carburetor Throttle Stop Screw	6UU0-442							✓	✓	✓	1
	C120-6	.10	Axle, Carburetor Float	6UU0-346							✓	✓	✓	1
	C120-9	....	Pin, Carburetor Bracket (Part of Item 2)	C120-9							✓	✓	✓	1
	C130-4	.05	Washer, Carburetor Air Shutter Shaft Thrust	6BL0-777							✓	✓	✓	1
	C136-12	.05	Screw, Carburetor Throttle Plate	6UU0-374							✓	✓	✓	1
	C138-61	.05	Screw, Carburetor Fuel Inlet Channel	6UU0-782							✓	✓	✓	1
	C142-28	.25	Gasket,* Carburetor Bowl to Body	C142-28							✓	✓	✓	1
	C148-10	1.10	Body, Carburetor Union	6UU0-485							✓	✓	✓	2
	C149-22	.40	Plug, Carburetor Filter	6UU0-336							✓	✓	✓	1
	C150-1	.20	Screen, Carburetor Filter	6UU0-156							✓	✓	✓	1
	CR9-5	....	Bushing, Carburetor Throttle Shaft (Part of Item 1)	CR9-5							✓	✓	✓	..
	CR26-65	....	Bushing, Carburetor Throttle Lever	CR26-65							✓	✓	✓	1
	CR88-2	.10	Bracket, Carburetor Float (Part of Item 1)	6UU0-218							✓	✓	✓	1
	CR121-10	....	Pin, Carburetor Throttle Stop (Part of Item 1)	CR121-10							✓	✓	✓	1
	CR134-1	.20	Swivel, Carburetor Air Shutter Lever (Part of Item 24)	6UU0-897							✓	✓	✓	1
	CT63-2	....	Pin, Carburetor Lever Bushing Taper	CT63-2							✓	✓	✓	1
	CT63-4	....	Pin, Carburetor Stop Lever Taper	CT63-4							✓	✓	✓	1
	C29-329	2.25	Lever, Carburetor Throttle Shaft and Assembly	6UU0-895							✓	✓	✓	1
	CT62-3	....	Pin, Carburetor Thrust Washer Taper	CT62-3							✓	✓	✓	1
	CT91-1	.10	Plug, Carburetor Governor By-pass ⅛" Pipe	S-4747							✓	✓	✓	1
	CT91-1	.10	Plug, Carburetor Bowl Drain ⅛" Pipe	S-4747							✓	✓	✓	1
	T1S8-6	.05	Screw, Carburetor Air Shutter Swivel (Part of Item 24)	6UU0-894							✓	✓	✓	1
	T1S8-10	.05	Screw, Carburetor Wire Clamp (Part of Item 25)	6UU0-250							✓	✓	✓	1
	T1S10-6	.05	Screw, Carburetor Venturi	6UU0-159							✓	✓	✓	1
	T8S10-9	....	Screw, Carburetor Throttle Lever Clamp	T8S10-9							✓	✓	✓	1
	T8S10-15	.05	Screw, Carburetor Throttle Stop	6UU0-358							✓	✓	✓	1
	T8S31-16	.05	Screw, Carburetor Bowl to Body	6UU0-445							✓	✓	✓	4

Note:—Parts marked (*) are included in 6UU0-901 Gasket Set.

03-2 FUEL SYSTEM

Master Parts List — THE AUTOCAR COMPANY, ARDMORE, PA.

CODE	MFR. PART No.	LIST PRICE	NAME OF PART	AUTOCAR PART No.	U-2044	U-4044	U-4144	U-4144-T	C-50	U-5044	U-7144-T	U-8144	U-8144-T	No. REQ.
0301	T15B6-4	$0.05	Screw, Carburetor Air Shutter Plate	6UU0-117							✓	✓	✓	1
	T21S8	.05	Nut, Carburetor Tube Clamp Screw	6UU0-891							✓	✓	✓	1
	T22S-8	.05	Nut, Carburetor Air Shutter Shaft	6UU0-554							✓	✓	✓	1
	T41-10	....	Lockwasher, Carburetor Venturi Screw	T41-10							✓	✓	✓	..
	T43-6	....	Screw, Carburetor Air Shutter—L.W.	T43-6							✓	✓	✓	..
	CT57-2	.05	Packing, Carburetor Throttle Shaft	6UU0-892							✓	✓	✓	1
	T8S8-7	.05	Screw, Carburetor Throttle Lever Swivel	6UU0-207							✓	✓	✓	1
	CR134-5	.20	Swivel, Carburetor Throttle Lever	6UU0-898							✓	✓	✓	1
	CT52-1	.05	Washer, Carburetor Swivel	6UU0-776							✓	✓	✓	1
	C181-39	.75	Gasket, Carburetor, Set	6UU0-901							✓	✓	✓	1
	T56-48	....	Washer,* Carburetor Power and Accelerator Jet Fibre	T56-48							✓	✓	✓	1
	T56-24	....	Washer,* Carburetor Cap Jet Fibre	T56-24							✓	✓	✓	1
	T56-5	....	Washer,* Carburetor Channel Screw Fibre	T56-5							✓	✓	✓	1
	T56-24	....	Washer,* Carburetor Compensator Jet Fibre	T56-24							✓	✓	✓	1
	T56-2	.05	Washer, Carburetor Discharge Tube Fibre	6UU0-834							✓	✓	✓	1
	T56-15	....	Washer,* Carburetor Filter Plug Fibre	T56-15							✓	✓	✓	1
	T56-23	....	Washer,* Carburetor Fuel Valve Fibre	T56-23							✓	✓	✓	1
	T56-27	....	Washer,* Carburetor Main Jet Fibre	T56-27							✓	✓	✓	1
	T56-13	....	Washer,* Carburetor Main Jet Adjusting Fibre	T56-13							✓	✓	✓	1
	T56-36	....	Washer,* Carburetor Union Body Fibre	T56-36							✓	✓	✓	1
	H90-11260	7.00	Cleaner, Air (United Specialties)	6UKS4-330							✓	✓	✓	1
	......	2.70	Pipe, Air Cleaner and Assembly	6UKSA4-686							✓	✓	✓	1
	......	.05	Screw, Hex. Cap	S-67							✓	✓	✓	2
	......	.01	Screw, Flat Head Machine (¼"—20 x ¾")	S-3843A							✓	✓	✓	2
	......	.01	Nut, Hex. (¼"—20)	S-76							✓	✓	✓	4
	......	.01	Spring, 9/32"—L.W.	S-1345							✓	✓	✓	6
	......	1.80	Elbow, Carburetor	6QT3-401							✓	✓	✓	1
	......	.08	Screw, Hex. Cap (5/16"—18 x 2½")	S-3112							✓	✓	✓	2
	......	.03	Nut, Hex. (5/16"—18)	S-82							✓	✓	✓	2
	......	.01	Washer, 5/16" Spring	S-1346							✓	✓	✓	4
	H80-9471	2.80	Air Cleaner Assembly—Complete (United)	6ZBM0-330	✓									1
	H95-7425	4.00	Air Cleaner Assembly—Complete (United)	6F0-330		✓	✓	✓	✓					1
0302	**Fuel Pump and Filter**													
	1537396	8.00	Pump, Fuel, Assembly (A C Type "D")	6UG4-510	✓									1
	1537203	8.00	Pump, Fuel, Assembly (A C Type "D")	6HF4-510		✓	✓	✓						1
	1537355	8.00	Pump, Fuel, Assembly (A C Type "D")	6N4-510C					✓					1
	1537507	8.00	Pump, Fuel, Assembly (A C Type "D")	6URL4-510						✓				1
	1537722	8.00	Pump, Fuel, Assembly (A C Type "D")	6UU4-510A							✓	✓	✓	1
	855918	.60	Dome, Fuel Pump Air	855918	✓	✓	✓	✓	✓	✓	✓	✓	✓	1
	855493	.01	Cover, Fuel Pump Top Screw	855493	✓	✓	✓	✓	✓	✓	✓	✓	✓	10
	855064	.01	Washer, Fuel Pump Top Cover Screw Lock	855064	✓	✓	✓	✓	✓	✓	✓	✓	✓	10
	855281	.25	Plug, Fuel Pump Valve	855281	✓	✓	✓	✓	✓	✓	✓	✓	✓	1
	855282	.01	Gasket, Fuel Pump Valve Plug	855282	✓	✓	✓	✓	✓	✓	✓	✓	✓	2
	856270	.01	Spring, Fuel Pump Valve	856270	✓	✓	✓	✓	✓	✓	✓	✓	✓	2
	855279	.01	Valve, Fuel Pump	855279	✓	✓	✓	✓	✓	✓	✓	✓	✓	2
	854009	.10	Screen, Fuel Pump	854009	✓	✓	✓	✓	✓	✓				1
	1537710	.35	Bowl, Fuel Pump Metal	1537710							✓	✓	✓	1
	854003	.05	Gasket, Fuel Pump Bowl	854003	✓	✓	✓	✓	✓	✓				1
	854005	.01	Seat, Fuel Pump Bowl	854005	✓	✓	✓	✓	✓	✓				1
	854016	.10	Screw, Fuel Pump Bail and	854016	✓	✓	✓	✓	✓	✓				1
	855763	.05	Nut, Fuel Pump Bail Thumb	855763	✓	✓	✓	✓	✓	✓				1
	1537259	1.55	Body, Fuel Pump	1537259							✓	✓	✓	1
	1537397	2.10	Arm, Fuel Pump Rocker	1537397	✓						✓	✓	✓	1
	1521187	.03	Spring, Fuel Pump Diaphragm	1521187	✓	✓	✓	✓	✓	✓	✓	✓	✓	1
	855253	.02	Spring, Fuel Pump Rocker Arm	855253	✓	✓	✓	✓	✓	✓	✓	✓	✓	1
	855573	.15	Cover, Fuel Pump Bottom	855573	✓	✓	✓	✓	✓	✓	✓	✓	✓	1
	855585	.03	Gasket, Fuel Pump Bottom Cover	855585	✓	✓	✓	✓	✓	✓	✓	✓	✓	1
	132108	.01	Screw, Fuel Pump Bottom Cover	132108	✓	✓	✓	✓	✓	✓	✓	✓	✓	3
	855016	.02	Pin, Fuel Pump Link	855016	✓	✓	✓	✓	✓	✓	✓	✓	✓	2
	855017	.01	Clip, Fuel Pump Link Pin	855017	✓	✓	✓	✓	✓	✓	✓	✓	✓	4

Note:—Parts marked (*) are included in 6UU0-901 Gasket Set.

CODE	MFR. PART No.	LIST PRICE	NAME OF PART	AUTOCAR PART No.	MODEL U-2044	U-4044	U-4144	U-4144-T	C-50	U-5044	U-7144-T	U-8144	U-8144-T	No. REQ.
0302	855574	$0.02	**Link,** Fuel Pump	855574	√	√	√	√	√	√	√	√	√	2
	855532	.01	**Cap,** Fuel Pump Spring	855532	√	√	√	√	√	√	√	√	√	2
	1521972	.20	**Pin,** Fuel Pump Rocker Arm	1521972							√	√	√	1
	1521288	.01	**Washer,** Fuel Pump Rocker Arm Pin	1521288	√	√	√	√	√	√	√	√	√	1
	855389	.37	**Diaphragm,** Fuel Pump, Set (5 Pieces)	855389	√	√	√	√	√	√	√	√	√	1 Set
	855250	.15	**Rod,** Fuel Pump Pull	855250	√	√	√	√	√	√	√	√	√	1
	855213	.05	**Nut,** Fuel Pump Pull Rod	855213	√	√	√	√	√	√	√	√	√	1
	855390	.01	**Lockwasher,** Fuel Pump Pull Rod Nut	855390	√	√	√	√	√	√	√	√	√	1
	855274	.10	**Washer,** Fuel Pump Upper Diaphragm Protector	855274	√	√	√	√	√	√	√	√	√	1
	1521720	.10	**Washer,** Fuel Pump Lower Diaphragm Protector	1521720	√	√	√	√	√	√	√	√	√	1
	855029	.01	**Washer,** Fuel Pump Diaphragm Alignment	855029	√	√	√	√	√	√	√	√	√	1
	855012	.01	**Gasket,** Fuel Pump Pull Rod	855012	√	√	√	√	√	√	√	√	√	1
	855739	2.00	**Seat,** Fuel Pump Top Cover and Valve, Assembly	855739	√	√	√	√	√	√	√	√	√	1
	1522280	.25	**Lever,** Fuel Pump Priming	1522280	√	√	√	√	√	√	√	√	√	1
	854004	....	**Bowl,** Fuel Pump Glass	854004	√	√	√	√		√				1
	1537204	....	**Body,** Fuel Pump	1537204	√	√	√	√		√				1
	1521340	....	**Arm,** Fuel Pump Rocker	1521340		√	√	√		√				1
	1521289	....	**Pin,** Fuel Pump Rocker Arm	1521289	√	√	√	√		√				1
	F328	7.00	**Filter,** Fuel, Assembly (Zenith Model F328)	6URK3-**1020**	√	√	√	√			√	√	√	1
	F1-10	.25	**Gasket,** Fuel Filter Bowl	F1-10	√	√	√	√	√	√	√	√	√	1
	F1-15	.05	**Gasket,** Fuel Filter Bowl Nut	F1-15	√	√	√	√	√	√	√	√	√	..
	F5-7	.50	**Nut,** Fuel Filter Bowl	F5-7	√	√	√	√	√	√	√	√	√	..
	F6-12	6.10	**Element,** Fuel Filter	F6-12	√	√	√	√	√	√	√	√	√	..
	F7-122	4.50	**Head,** Fuel Filter	F7-122	√	√	√	√	√	√	√	√	√	..
	F8-50	2.00	**Bowl,** Fuel Filter	F8-50	√	√	√	√	√	√	√	√	√	..
	F16-27	.10	**Stud,** Fuel Filter Assembly	F16-27	√	√	√	√	√	√	√	√	√	..
	F26-3	.35	**Nut,** Fuel Filter Element	F26-3	√	√	√	√	√	√	√	√	√	..
	F36-9	.15	**Plug,** Fuel Filter Drain	F36-9	√	√	√	√	√	√	√	√	√	..
	F36-9	.15	**Plug,** Fuel Filter ¼″ Pipe	F36-9	√	√	√	√	√	√	√	√	√	..
0303	**Accelerator, Throttle and Choke**													
	......	1.00	**Treadle,** Accelerator	10UU3-**1052**							√	√	√	1
	......	.75	**Bracket,** Accelerator Treadle	10UB3-**336B**							√	√	√	1
	......	.12	**Bushing,** Accelerator Treadle Bracket	10UA1-**15**							√	√	√	2
	......	.04	**Pin,** Accelerator Treadle	10UB2-**729**							√	√	√	1
	......	.01	**Pin,** Rd. St. (⅛″ x 1″)	S-**3253**							√	√	√	1
	......	.05	**Screw,** Fr. Hd. Mach. (¼″—20 x 1¾″)	S-**195**							√	√	√	2
	......	.01	**Nut,** Hex. (¼″—20)	S-**76**							√	√	√	2
	......	.02	**Washer,** $\frac{9}{32}$″ x ¾″ x $\frac{1}{32}$″ Flat	S-**720**							√	√	√	2
	......	.01	**Washer,** $\frac{9}{32}$″ Spring	S-**1345**							√	√	√	2
	......	.25	**Rod,** Accelerator Control	10UU3-**28**							√	√	√	1
	......	.40	**Clevis,** Accelerator Control Rod	10Y1-**136**							√	√	√	1
	......	.06	**Pin,** Accelerator Control Rod Clevis	S-**2228**							√	√	√	1
	......	.01	**Nut,** Hex. (¼″—28)	S-**1416**							√	√	√	1
	......	.01	**Cotter,** $\frac{3}{32}$″ x ½″	S-**2198**							√	√	√	2
	......	.25	**Shaft,** Accelerator Cross	10UU2-**27**							√	√	√	1
	......	.30	**Lever,** Accelerator Cross Shaft	10DFJ2-**72**							√	√	√	1
	......	.01	**Pin,** Rd. St. (⅛″ x 1″)	S-**3253**							√	√	√	1
	......	.80	**Tube,** Accelerator Cross Shaft	10UU2-**361**							√	√	√	1
	......	.12	**Bushing,** Accelerator Cross Shaft Tube	12W1-**79**							√	√	√	2
	......	....	**Bracket,** Accelerator Cross Shaft	10UU3-**336**							√	√	√	1
	......	....	**Clamp,** Accelerator Cross Shaft	10UU2-**198**							√	√	√	2
	......	.06	**Screw,** Hex. Cap (⅜″—16 x 1¼″)	S-**64**							√	√	√	2
	......	.03	**Nut,** Hex. (⅜″—16)	S-**78**							√	√	√	2
	......	.01	**Lockwasher,** $\frac{13}{32}$″ Spring	S-**1347**							√	√	√	4
	......	.90	**Lever,** Carburetor Control Rod	10SA2-**72**							√	√	√	1
	......	.04	**Pin,** Rd. St. (⅛″ x ⅞″)	S-**1546**							√	√	√	1
	......	.05	**Link,** Accelerator Spring	10G2-**79**							√	√	√	1
	......	.12	**Clip,** Accelerator Spring	10SA2-**79**							√	√	√	1
	......	....	**Spring,** Accelerator	10T2-**63**							√	√	√	1
	......	.15	**Rod,** Carburetor Control	10UU3-**194**							√	√	√	1
	......	.40	**Clevis,** Carburetor Control Rod	10Y1-**136**							√	√	√	1

03-4 FUEL SYSTEM

Master Parts List — THE AUTOCAR COMPANY, ARDMORE, PA.

CODE	MFR. PART No.	LIST PRICE	NAME OF PART	AUTOCAR PART No.	U-2044	U-4044	U-4144	U-4144-T	C-50	U-5044	U-7144-T	U-8144	U-8144-T	No. REQ.
0303		$0.06	Pin, Carburetor Control Rod Clevis	S-2228							✓	✓	✓	1
		.01	Nut, Hex. ($\frac{1}{4}''$—28)	S-1416							✓	✓	✓	1
		.01	Cotter, $\frac{3}{32}'' \times \frac{1}{2}''$	S-2198							✓	✓	✓	2
		.75	Wire, Throttle Control, Assembly	10UKS3-1270							✓	✓	✓	1
		.70	Wire, Choke Control, Assembly	10UKS3-750							✓	✓	✓	1
		.05	Bracket, Choke and Throttle Control	10UV2-336							✓	✓	✓	1
		.04	Screw, Hex. Cap ($\frac{5}{16}''$—18 x $\frac{7}{8}''$)	S-15							✓	✓	✓	1
		.03	Nut, Hex. ($\frac{5}{16}''$—18)	S-82							✓	✓	✓	1
		.01	Lockwasher, $\frac{11}{32}''$ Spring	S-1346							✓	✓	✓	2
		.08	Bracket, Choke and Throttle Control	10UVA2-336							✓	✓	✓	1
		.08	Bracket, Choke and Throttle Control	10UVB2-336							✓	✓	✓	1
		.25	Clip, Choke and Throttle Control	10A2-989							✓	✓	✓	3
		.04	Screw, Hex. Cap ($\frac{1}{4}''$—20 x $\frac{3}{4}''$)	S-30							✓	✓	✓	3
		.01	Nut, Hex. ($\frac{1}{4}''$—20)	S-76							✓	✓	✓	3
		.01	Lockwasher, $\frac{9}{32}''$ Spring	S-1345							✓	✓	✓	6
0304	**Fuel Lines**													
		1.65	Line, Tank to Filter Gas (Bundy Tubing)	6UV0-943							✓	✓	✓	1
		.08	Bracket, Tank to Filter Gas Line Clip	10C2-303							✓	✓	✓	3
		.09	Screw, Hex. Cap ($\frac{1}{2}''$—13 x $1\frac{1}{2}''$)	S-1872							✓	✓	✓	3
		.04	Nut, Hex. ($\frac{1}{2}''$—3)	S-77							✓	✓	✓	3
		.01	Lockwasher, $\frac{17}{32}''$ Spring	S-1349							✓	✓	✓	6
		.06	Clip, Gas Line	16S2-44							✓	✓	✓	5
		.04	Screw, Hex. Cap ($\frac{1}{4}''$—20 x $\frac{5}{8}''$)	S-36							✓	✓	✓	5
		.01	Nut, Hex. ($\frac{1}{4}''$—20)	S-76							✓	✓	✓	5
		.01	Washer, $\frac{9}{32}''$ Spring	S-1345							✓	✓	✓	10
		.25	Connector ($\frac{3}{8}'' \times \frac{1}{4}''$) with Comp. Long Nut	S-5865							✓	✓	✓	1
		.70	Cock, Shut Off ($\frac{1}{4}'' \times \frac{1}{4}''$) (45-C-1966)	6D0-22							✓	✓	✓	1
		2.31	Line, Filter to Pump Titeflex Gas	6D3-750							✓	✓	✓	1
		.07	Bracket, Filter to Pump Titeflex Gas Line Clip	6UU2-317							✓	✓	✓	1
	Per Ft.	.06	Loom, Filter to Pump Titeflex Gas Line	16Y-273							✓	✓	✓	$\frac{3}{8}'' \times 2''$
		.20	Ell, $\frac{1}{4}''$ St. (Weatherhead 3400X4)	S-3443							✓	✓	✓	1
		1.15	Loom, Pump to Carb. Gas Line and, Assembly	6UU4-640A							✓	✓	✓	1
		.06	Bracket, Pump to Carb. Gas Line Clip	6UUA2-317							✓	✓	✓	1
		.10	Body, Connector ($\frac{3}{8}'' \times \frac{1}{4}''$)	S-5919							✓	✓	✓	2
0305	**Governor**													
		35.00	Governor (Handy 703—722—138)	2UK0-650	✓									1
		17.25	Governor (Handy V—6—83)	2DB0-650	✓	✓	✓	✓	✓	✓				1
	MA-1377	45.00	Governor (Pierce)	2UU0-650							✓	✓	✓	1
	G-2141	.40	Nut, Governor Adjusting Screw Lock	G-2141							✓	✓	✓	1
	G-2237	.40	Cap, Govenor Throttle Lever Body	G-2237							✓	✓	✓	1
	G-2280	.50	Rod, Governor Throttle, End (Valve Box)	G-2280							✓	✓	✓	1
	G-2281	.60	Rod, Governor Throttle, End	G-2281							✓	✓	✓	1
	G-2282	1.50	Sleeve, Governor Thrust	G-2282							✓	✓	✓	1
	G-2287	.70	Tube, Governor Throttle Rod	G-2287							✓	✓	✓	1
	G-2288	.60	Rod, Governor Throttle	G-2288							✓	✓	✓	1
	G-2289	.20	Screw, Governor Throttle Lever Clevis	G-2289							✓	✓	✓	1
	G-2290	.26	Nut, Governor Throttle Rod Tube Lock	G-2290							✓	✓	✓	1
	G-2291	.80	Screw, Governor Adjusting	G-2291							✓	✓	✓	1
	G-2293	.26	Gasket, Governor Body	G-2293							✓	✓	✓	1
	G-2335	3.25	Gear, Governor	G-2335							✓	✓	✓	1
	G-2373	.26	Crank, Governor Valve Box Bell	G-2373							✓	✓	✓	1
	G-2374	.16	Pin, Governor Bell Crank to Throttle Rod	G-2374							✓	✓	✓	1
	G-2376	.06	Gasket, Governor Valve Box Cover Plate	G-2376							✓	✓	✓	1
	G-5950	.16	Pin, Governor Weight	G-5950							✓	✓	✓	2
	G-2682	.85	Collar, Governor Spring	G-2682							✓	✓	✓	1
	G-2781	.25	Shaft, Governor Rocker	G-2781							✓	✓	✓	1
	G-2784	.16	Spacer, Governor Yoke	G-2784							✓	✓	✓	1
	G-2807	.80	Yoke, Governor	G-2807							✓	✓	✓	1
	G-2811	2.10	Cap, Governor Body	G-2811							✓	✓	✓	1
	G-3770	.55	Bushing, Governor Upper Drive Shaft	G-3770							✓	✓	✓	1

THE AUTOCAR COMPANY, ARDMORE, PA.

Master Parts List

03-5 FUEL SYSTEM

CODE	MFR. PART No.	LIST PRICE	NAME OF PART	AUTOCAR PART No.	MODEL U-2044	U-4044	U-4144	U-4144-T	C-50	U-5044	U-7144-T	U-8144	U-8144-T	No. REQ.
0305	G-3771	$0.55	Bushing, Governor Lower Drive Shaft..................	G-3771							✓	✓	✓	1
	G-6701	18.00	Body, Governor............................	G-6701							✓	✓	✓	1
	G-3978	1.40	Shaft, Governor Drive........................	G-3978							✓	✓	✓	1
	G-3983	1.40	Lever, Governor Throttle......................	G-3983							✓	✓	✓	1
	G-4189	1.40	Weights, Governor...........................	G-4189							✓	✓	✓	2
	A-1870	1.35	Spider, Governor, Assembly.....................	A-1870							✓	✓	✓	1
	SN-1150	.30	Spring, Governor............................	SN-1150							✓	✓	✓	1
	SN-213	.12	Spring, Governor Bumper......................	SN-213							✓	✓	✓	1
	S2A-345	.25	Screw, Governor Bumper Spring Adjusting................	S2A-345							✓	✓	✓	1
	S2B-831	8.00	Box, Governor Valve (Casting Only)................	S2B-831							✓	✓	✓	1
	SC-150	.95	Valve, Governor Butterfly.....................	SC-150							✓	✓	✓	1
	S2D-170	1.00	Shaft, Governor Valve........................	S2D-170							✓	✓	✓	1
	X-11	.05	Pin, Governor Valve Box Bell Crank ($\frac{3}{32}$" x $\frac{3}{4}$")..........	X-11							✓	✓	✓	1
	X-17	.03	Pin, Governor Throttle Rod End (Valve Box) ($\frac{3}{32}$" x $\frac{7}{16}$")....	X-17							✓	✓	✓	1
	X-997	.05	Pin, Governor Spider ($\frac{1}{8}$" x $\frac{3}{4}$")................	X-997							✓	✓	✓	1
	X-74	.07	Pin, Governor Yoke (#1 x $\frac{3}{4}$" Taper)................	X-74							✓	✓	✓	1
	X-198	.06	Screw, Governor Valve Box Cover and Valve Shaft (#6—32)..	X-198							✓	✓	✓	6
	X-234	.12	Nut, Governor Bumper Screw Lock ($\frac{3}{8}$"—24)..............	X-234							✓	✓	✓	1
	X-239	.12	Nut, Governor Drive Shaft Castle ($\frac{3}{8}$"—24).............	X-239							✓	✓	✓	1
	X-317	1.30	Bearing, Governor Thrust Ball.....................	X-317							✓	✓	✓	1
	X-339	.90	Bearing, Governor Drive Shaft Ball...................	X-339							✓	✓	✓	1
	X-461	.12	Key, Governor Gear (Woodruff)....................	X-461							✓	✓	✓	1
	X-513	.02	Key, Governor Throttle Rod End (Valve Box) Cotter........	X-513							✓	✓	✓	1
	X-556	.18	Plate, Governor Valve Box Cover..................	X-556							✓	✓	✓	1
	X-572	.07	Screw, Governor Throttle Lever Set..................	X-572							✓	✓	✓	1
	X-600	.06	Screw, Governor Body Cap (Seal Wire) (#10—24 x $\frac{1}{2}$").....	X-600							✓	✓	✓	2
	X-612	.02	Key, Governor Castle Nut Cotter (Drive Shaft Gear)........	X-612							✓	✓	✓	1
	X-825	.05	Screw, Governor Body Cap (#10—24 x $\frac{5}{8}$").............	X-825							✓	✓	✓	4
	X-454	.06	Plug, Governor Welch ($\frac{9}{16}$") at Rocker Shaft..............	X-454							✓	✓	✓	1

THE AUTOCAR COMPANY,
ARDMORE, PA.

Master Parts List

GROUP
04

ILLUSTRATIONS

GROUP 04

EXHAUST

GROUP 04

Master Parts List

THE AUTOCAR COMPANY, ARDMORE, PA.

As supplied on Autocar Models
U-8144   U-8144-T

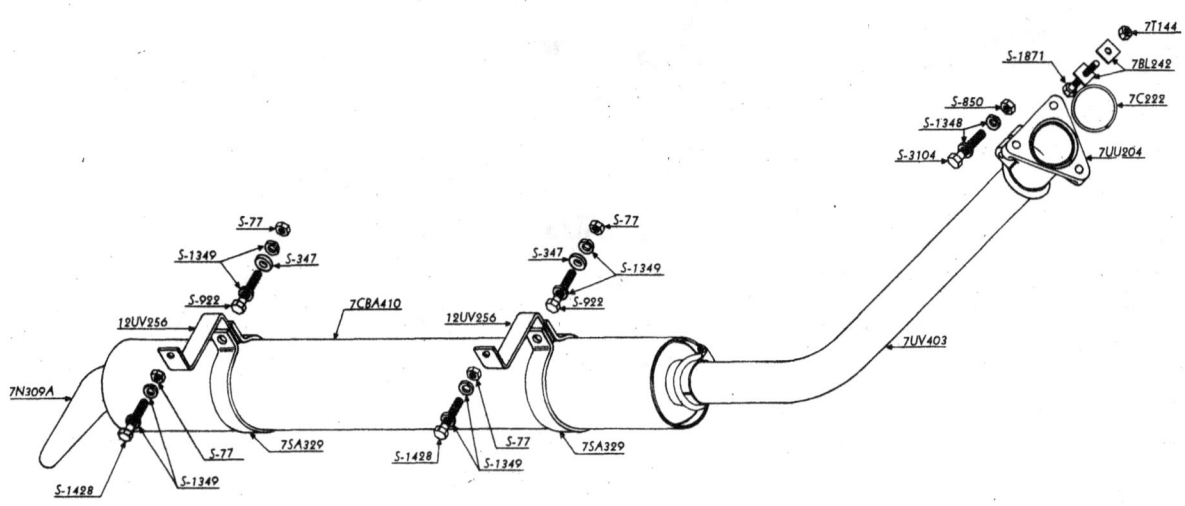

0401—Muffler

THE AUTOCAR COMPANY
ARDMORE, PA.

Master Parts List

04-1 EXHAUST

CODE	MFR. PART No.	LIST PRICE	NAME OF PART	AUTOCAR PART No.	U-2044	U-4044	U-4144	U-4144-T	C-50	U-5044	U-7144-T	U-8144	U-8144-T	No. REQ.
0401			**Muffler**											
	......	$6.00	Muffler Assembly........................	7TE3-10	✓	✓	✓	✓		✓				1
	......	7.80	Muffler Assembly........................	7CBA4-10							✓	✓	✓	1
	......	.15	Bracket, Muffler—Front.................	12UK2-56	✓	✓	✓	✓		✓				1
	......	.50	Bracket, Muffler—Rear..................	12DU2-56	✓	✓	✓	✓		✓				1
	......	.25	Bracket, Muffler........................	12UB2-56							✓	✓	✓	2
	......	.09	Screw, Hex. Cap (½"—13 x 1⅜").......	S-1428							✓	✓	✓	2
	......	.04	Nut, Hex. (½"—13).....................	S-77							✓	✓	✓	2
	......	.01	Lockwasher, ½" Spring..................	S-1349							✓	✓	✓	4
	......	.50	Clip, Muffler...........................	7SA3-29	✓	✓	✓	✓		✓				2
	......	.50	Clip, Muffler...........................	7SA3-29							✓	✓	✓	2
	......	.10	Screw, Hex. Cap (½"—13 x 2")..........	S-922							✓	✓	✓	2
	......	.04	Nut, Hex. (½"—13).....................	S-77							✓	✓	✓	2
	......	.01	Lockwasher, ½" Spring..................	S-1349							✓	✓	✓	4
	......	.06	Washer, $\frac{17}{32}$" x 1⅜" x ¼" Flat St.....	S-347							✓	✓	✓	2
0402			**Exhaust and Tail Pipe**											
	......	2.82	Flange, Exhaust Pipe (Hercules #40314A)...	7BL0-04	✓									1
	......	3.60	Flange, Exhaust Pipe....................	7T2-04		✓	✓	✓		✓				1
	......	2.40	Flange, Exhaust Pipe....................	7UU2-04							✓	✓	✓	1
	......	.12	Gasket, Exhaust Pipe Flange (Hercules #40028A)	7BL0-22	✓									1
	......	.20	Gasket, Exhaust Pipe Flange.............	7M2-22		✓	✓	✓		✓				1
	......	.20	Gasket, Exhaust Pipe Flange.............	7UU2-04							✓	✓	✓	1
	......	.11	Screw, Hex. Cap ($\frac{7}{16}$"—13 x 2½").......	S-1871	✓	✓	✓	✓		✓	✓	✓	✓	3
	......	.30	Nut, Hex. ($\frac{7}{16}$"—13 Brass).............	7T1-44	✓	✓	✓	✓		✓	✓	✓	✓	3
	......	.06	Lockwasher—Sheet Metal...............	7SA2-42	✓	✓	✓	✓		✓				6
	......	.06	Lockwasher, Square.....................	7BL2-42							✓	✓	✓	6
	......	4.10	Pipe, Exhaust...........................	7UK4-03	✓									1
	......	5.65	Pipe, Exhaust...........................	7UG4-03		✓	✓	✓		✓				1
	......	6.50	Pipe, Exhaust...........................	7UV4-03							✓	✓	✓	1
	......	.11	Screw, Hex. Cap ($\frac{7}{16}$"—14 x 3½").......	S-3104							✓	✓	✓	1
	......	.04	Nut, Hex. ($\frac{7}{16}$"—14)...................	S-850							✓	✓	✓	1
	......	.01	Lockwasher, $\frac{15}{32}$" Spring................	S-1348							✓	✓	✓	2
	......	1.35	Pipe, Tail...............................	7UG3-09	✓	✓	✓	✓						1
	......	1.50	Pipe, Tail...............................	7BL3-09						✓				1
	......	1.75	Pipe, Tail...............................	7N3-09A							✓	✓	✓	1
	......	21.20	Arrestor, Flame.........................	7UG0-320	✓	✓	✓	✓						1
	......	.85	Clamp, Flame Arrestor..................	7UG3-29	✓	✓	✓	✓						1
	......	.30	Bracket, Flame Arrestor.................	12UG3-56	✓	✓	✓	✓						1

THE AUTOCAR COMPANY, ARDMORE, PA.

Master Parts List

GROUP 05

ILLUSTRATIONS

GROUP 05

COOLING SYSTEM

GROUP 05

Master Parts List

THE AUTOCAR COMPANY, ARDMORE, PA.

As supplied on Autocar Models
U-8144   U-8144-T

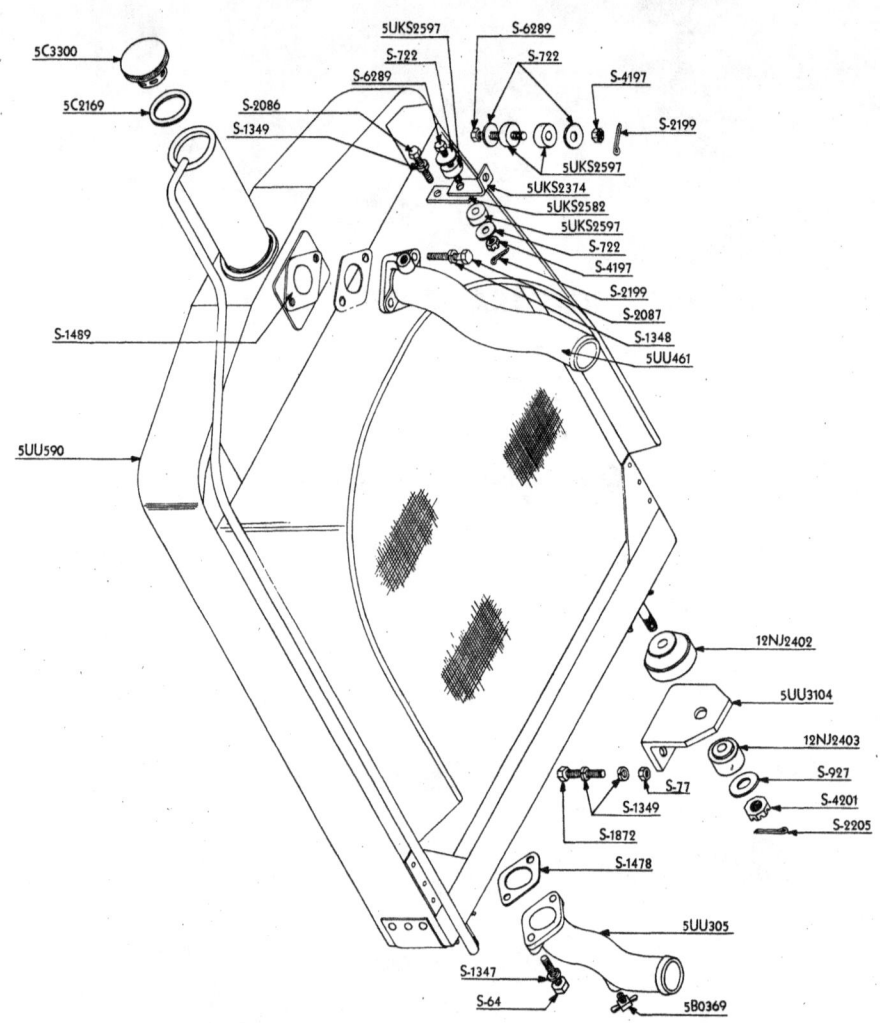

0501—Radiator Shell, Core and Tank

THE AUTOCAR COMPANY,
ARDMORE, PA.

Master Parts List

GROUP 05

As supplied on Autocar Models
U-7144-T   U-8144   U-8144-T

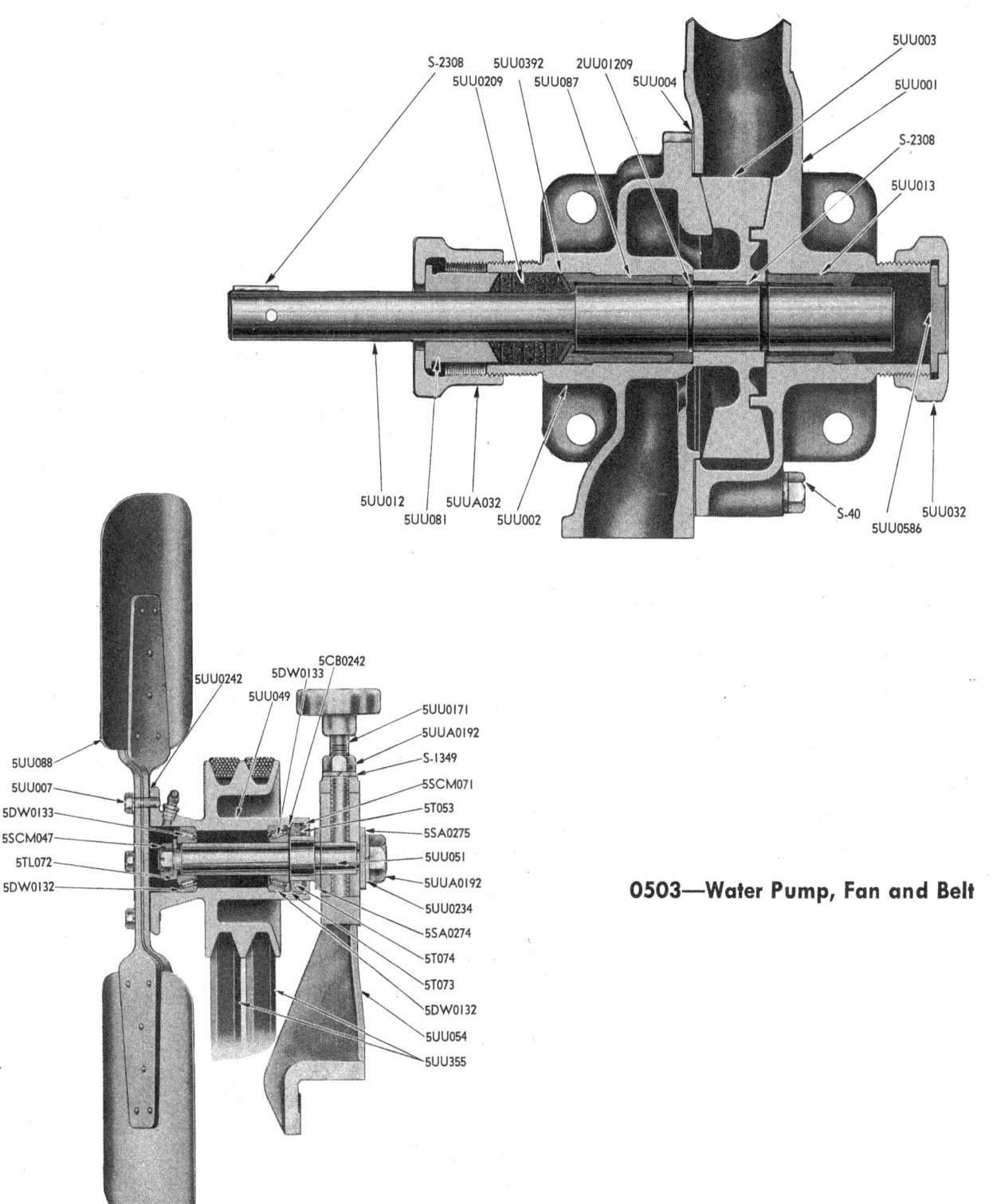

0503—Water Pump, Fan and Belt

GROUP 05

Master Parts List

THE AUTOCAR COMPANY, ARDMORE, PA.

As supplied on Autocar Models U-8144 U-8144-T

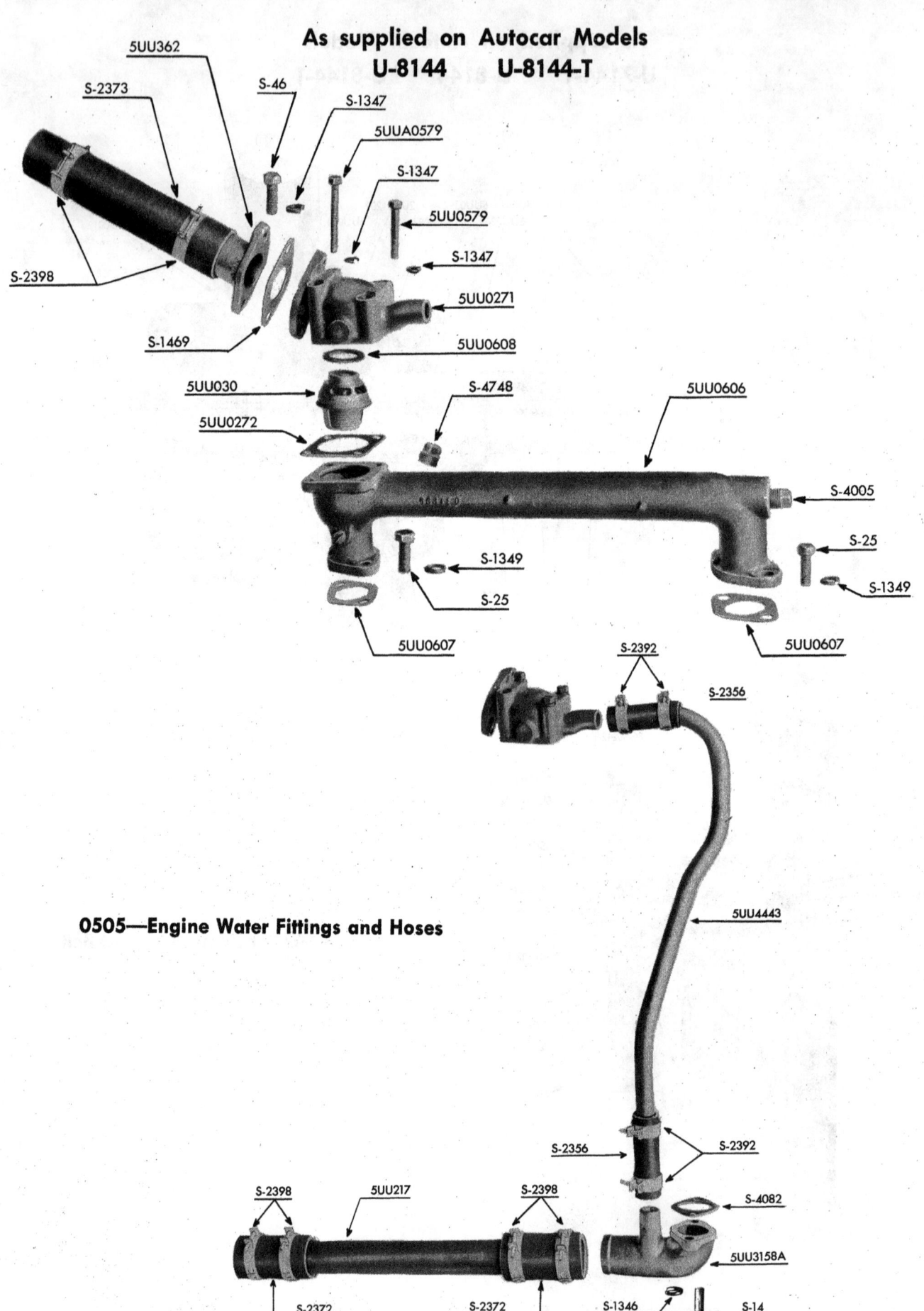

0505—Engine Water Fittings and Hoses

THE AUTOCAR COMPANY,
ARDMORE, PA.

Master Parts List

GROUP 05

As supplied on Autocar Models
U-8144   U-8144-T

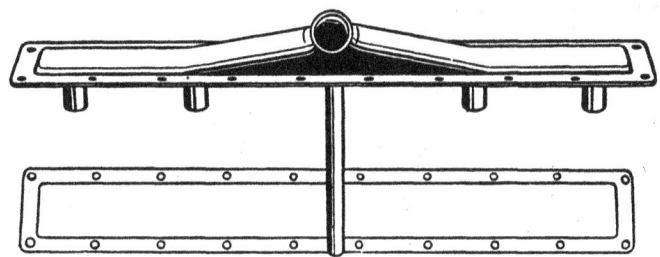

0505—Water Pump Discharge Manifold

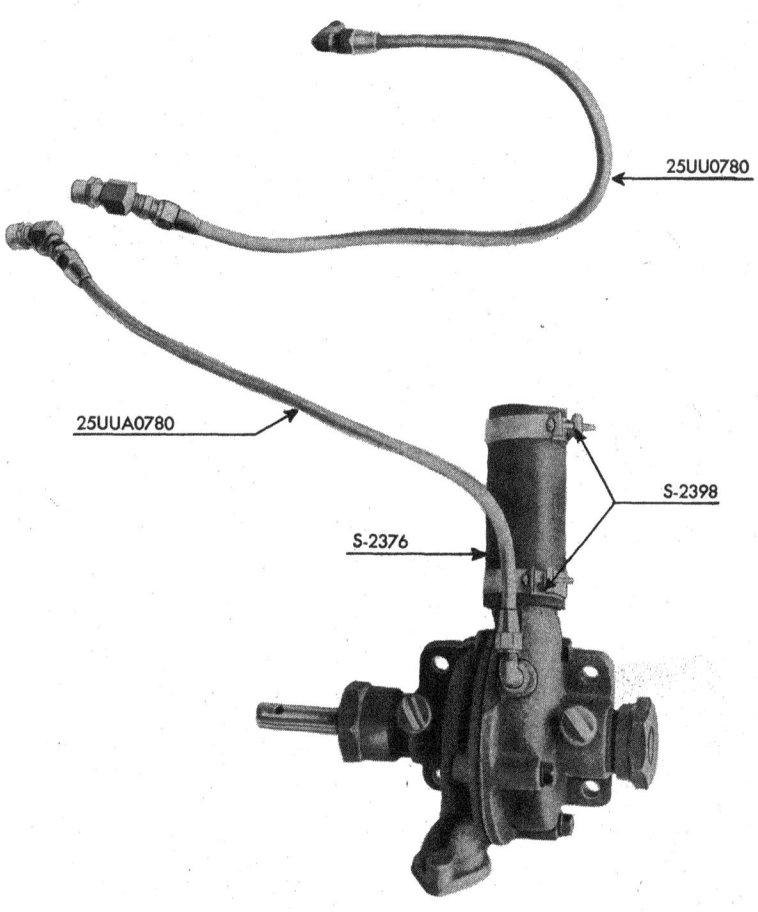

0505—Water Pump Connections

GROUP 05

Master Parts List

THE AUTOCAR COMPANY, ARDMORE, PA.

As supplied on Autocar Models
U-8144   U-8144-T

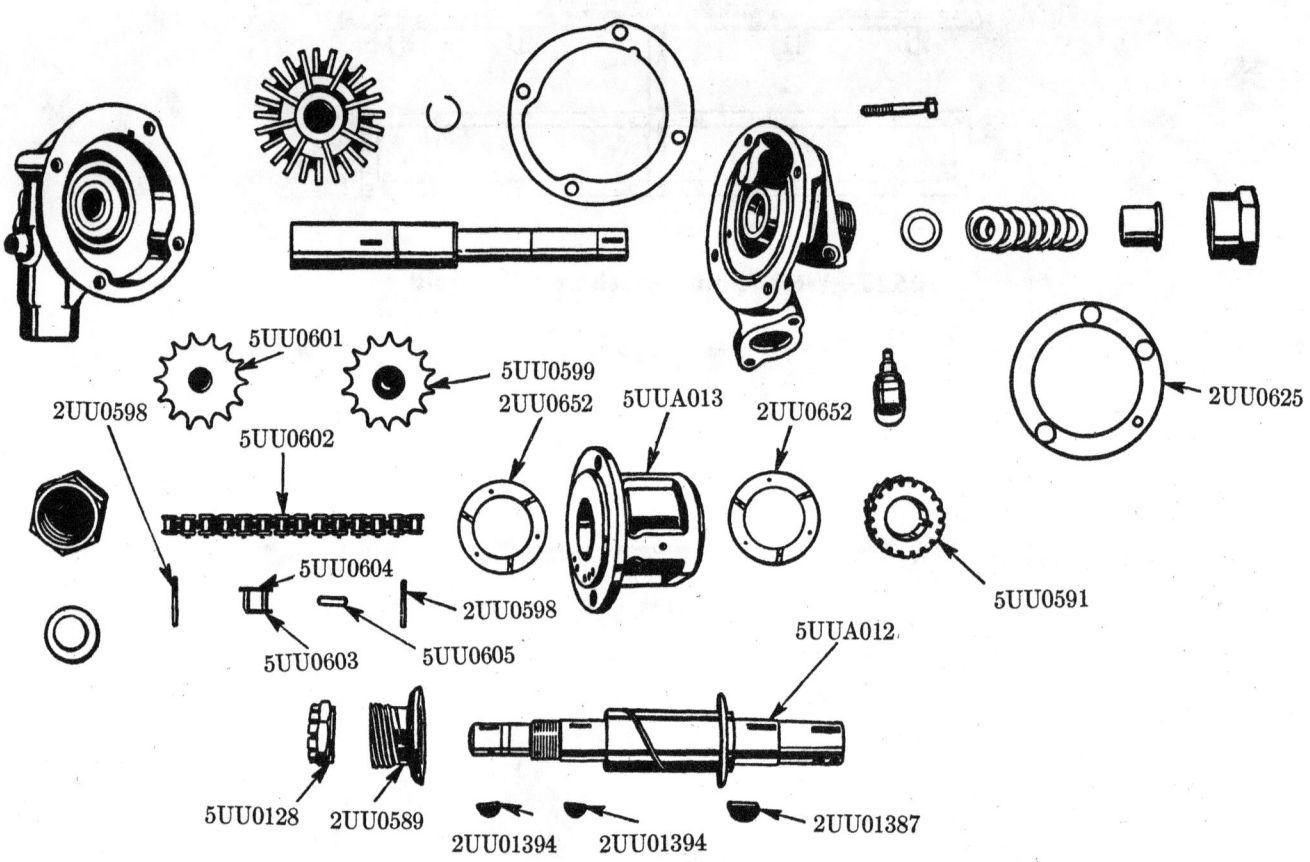

0505—Water Pump Drive Shaft Assembly

CODE	MFR. PART No.	LIST PRICE	NAME OF PART	AUTOCAR PART No.	U-2044	U-4044	U-4144	U-4144-T	C-50	U-5044	U-7144-T	U-8144	U-8144-T	No. REQ.
0501	**Radiator Shell, Core and Tank**													
	......	$56.50	**Core,** Radiator, and Tanks	5UK4-90	√									1
	......	58.40	**Core,** Radiator, and Tanks	5UL4-90		√								1
	......	58.00	**Core,** Radiator, and Tanks	5UL4-90A			√	√						1
	......	77.15	**Core,** Radiator, and Tanks	5UP4-90						√				1
	......	77.25	**Radiator** Assembly	5UU5-90							√	√	√	1
	......	.25	**Cap,** Radiator Filler	5UBL2-300	√	√				√				1
	......	1.50	**Cap,** Radiator Filler	5C0-300			√	√						1
	......	1.50	**Cap,** Radiator Filler	5C3-300							√	√	√	1
	......	.06	**Gasket,** Radiator Filler Cap	5C2-169							√	√	√	1
	......	.15	**Bracket,** Radiator Support	5UK0-104	√									2
	......	1.85	**Bracket,** Radiator Support	5UG4-104		√	√	√		√				2
	......	.50	**Bracket,** Radiator Support	5UU3-104							√	√	√	2
	......	.09	**Screw,** Hex. Cap (½"—13 x 1½")	S-1872							√	√	√	4
	......	.04	**Nut,** Hex. (½"—13)	S-77							√	√	√	4
	......	.01	**Lockwasher,** 17/32" Spring	S-1349							√	√	√	8
	......	.50	**Insulator,** Radiator Support—Top	12NJ2-402							√	√	√	2
	......	.25	**Insulator,** Radiator Support—Bottom	12NJ2-403							√	√	√	2
	......	.04	**Washer,** St. (21/32" x 1⅝" x ⅛")	S-927							√	√	√	2
	......	.05	**Nut,** Hex. (⅝"—11 Slotted)	S-4201							√	√	√	2
	......	.01	**Cotter** (⅛" x 1¼")	S-2205							√	√	√	2
	......	1.50	**Elbow,** Radiator Outlet	5Y3-05	√									1
	......	1.80	**Elbow,** Radiator Outlet	5RM4-05		√	√	√						1
	......	2.25	**Elbow,** Radiator Outlet	5UP4-05						√				1
	......	1.25	**Elbow,** Radiator Outlet	5UU3-05							√	√	√	1
	......	.12	**Gasket,** Permanite	S-1478							√	√	√	1
	......	.06	**Screw,** Hex. Cap (⅜"—16 x 1¼")	S-64							√	√	√	2
	......	.01	**Lockwasher,** 13/32" Spring	S-1347							√	√	√	2
	......	.30	**Cock,** Radiator Drain, ⅜" (Weatherhead #270)	5B0-369	√		√	√		√	√	√	√	1
	......	.30	**Cock,** Radiator Drain	5B01-369		√								1
0501	......	2.00	**Elbow,** Radiator Inlet	5UK4-61	√									1
	......	1.75	**Elbow,** Radiator Inlet	5URM4-61B		√	√	√		√				1
	......	2.80	**Elbow,** Radiator Inlet	5UU4-61							√	√	√	1
	......	.10	**Gasket,** Permanite	S-1489							√	√	√	1
	......	.05	**Screw,** Hex. Cap (7/16"—14 x 1⅜")	S-2087							√	√	√	2
	......	.01	**Lockwasher,** 15/32" Spring	S-1348							√	√	√	2
	......	.25	**Support,** Radiator Top	5UBK2-374	√									1
	......	.10	**Support,** Radiator Top	5URK2-374		√	√	√						1
	......	.15	**Support,** Radiator Top	5URM2-374						√				1
	......	....	**Brace,** Radiator Top	5UKS2-582							√	√	√	1
	......	....	**Screw,** Hex. Cap (½"—13 x ⅞")	S-2086							√	√	√	1
	......	.01	**Lockwasher,** 17/32" Spring	S-1349							√	√	√	1
	......	.12	**Bracket,** Radiator Top Brace	5UKS2-374							√	√	√	1
	......	....	**Washer,** Felt	5UKS2-597							√	√	√	4
	......	....	**Screw,** Hex. Cap (⅜"—16 x 1¾")	S-6289							√	√	√	2
	......	.04	**Nut,** Hex. (⅜"—16)	S-4197							√	√	√	2
	......	.04	**Washer,** St. (13/32" x 1 5/32" x 1/16")	S-722							√	√	√	4
	......	.01	**Cotter,** (3/32" x 1")	S-2199							√	√	√	2
0502	**Thermostat**													
	......	1.80	**Thermostat** (Dole #XT-187)	5A2-30	√		√	√						1
	......	2.75	**Thermostat**	5HF0-30						√				1
	3710-B	2.20	**Thermostat** (Bishop & Babcock)	5UU0-30							√	√	√	1
	11407-A	....	**Gasket,** Thermostat Rubber Seal	5UU0-608							√	√	√	1
0503	**Water Pump, Fan and Belt**													
	40170-CS	16.50	**Pump,** Water, Assembly	5BL0-40	√									1
	104758	15.00	**Pump,** Water, Assembly	5D0-40		√	√	√	√	√				1
	16859-DS	25.80	**Pump,** Water, Assembly	5UU0-40							√	√	√	1
	40165-A	1.00	**Paddle,** Water Pump	5BL0-03	√									1
	B-105095	1.80	**Impeller,** Water Pump	5D0-03		√	√	√	√	√				1

05-2 COOLING

Master Parts List — THE AUTOCAR COMPANY, ARDMORE, PA.

CODE	MFR. PART No.	LIST PRICE	NAME OF PART	AUTOCAR PART No.	U-2044	U-4044	U-4144	U-4144-T	C-50	U-5044	U-7144-T	U-8144	U-8144-T	No. REQ.
0503	15165-A	$4.40	Impeller, Water Pump	5UU0-03							✓	✓	✓	1
	14507-A	.01	Pin, Water Pump Paddle (3/16″)	S-1558	✓									1
	40172-B	3.60	Shaft, Water Pump Paddle	5BL0-12	✓									1
	513-A	.06	Key, Water Pump Paddle (#5 Whitney)	S-2305	✓									1
	2165-A	.02	Key, Water Pump Impeller Shaft	2UU0-1394							✓	✓	✓	2
	C-105098	3.20	Seal, Water Pump Paddle and, Assembly	5D0-378		✓	✓	✓	✓	✓				1
	C-19958	.35	Seal, Water Pump Paddle Carbon	5D0-424		✓	✓	✓	✓	✓				1
	C-19685	.40	Seal, Water Pump Paddle Flexible	5D0-425		✓	✓	✓	✓	✓				1
	C-19686	.15	Guide, Water Pump Paddle Seal Spring	5D0-426		✓	✓	✓	✓	✓				1
	C-106658	.30	Spring, Water Pump Paddle Seal	5D0-427		✓	✓	✓	✓	✓				1
	C-19895	.10	Ring, Water Pump Paddle Seal Clamp	5D0-428		✓	✓	✓	✓	✓				1
	C-19395	.10	Wire, Water Pump Paddle Seal Snap	5D0-429		✓	✓	✓	✓	✓				1
	45735-CS	4.80	Body, Water Pump, and Bushing Assembly	5BL0-01	✓									1
		7.20	Body, Water Pump	5D4-01		✓	✓	✓	✓	✓				1
	15460-C	8.60	Body, Water Pump	5UU0-01							✓	✓	✓	1
	40168-A	.40	Bushing, Water Pump Body	5BL0-13	✓									1
	15465-A	1.58	Bushing, Water Pump Body	5UU0-13							✓	✓	✓	1
	40171-C	5.80	Head, Water Pump, and Bushing Assembly	5BL0-02	✓									1
		3.90	Head, Water Pump	5D3-02		✓	✓	✓	✓	✓				1
	15461-C	9.00	Cover, Water Pump	5UU0-02							✓	✓	✓	1
	22171-B	.52	Bushing, Water Pump Head	5BL0-86	✓									1
	15464-A	1.58	Bushing, Water Pump Cover	5UU0-87							✓	✓	✓	1
	22164-A	.06	Gasket, Water Pump	5BL0-04	✓									1
		.06	Gasket, Water Pump	5SA2-04		✓	✓	✓	✓	✓				1
	15164-A	.06	Gasket, Water Pump Cover	5UU0-04							✓	✓	✓	1
	1864-A	.05	Screw, Water Pump Attaching (3/8″—16 x 1″)	S-23	✓									3
		.04	Screw, Water Pump Head (5/16″—18 x 3/4″)	S-14		✓	✓	✓	✓	✓				4
		.06	Screw, Water Pump Head (5/16″—18 x 2″)	S-3103		✓	✓	✓	✓	✓				3
	2083-A	.04	Screw, Water Pump Cover	5UU0-584							✓	✓	✓	4
	342-A	.01	Lockwasher, Water Pump Cover Screw	S-1347	✓						✓	✓	✓	4
	C-102538	....	Washer, Water Pump Head Screw Copper	C-102538		✓	✓	✓	✓	✓				6
	105097	3.00	Shaft, Water Pump, and Bearing Assembly	5D0-86		✓	✓	✓	✓	✓				1
	16910-B	4.60	Shaft, Water Pump—3/4″ Diameter	5UU0-12							✓	✓	✓	1
	C-102178	.10	Ring, Water Pump Packing Bearing Lock	5D2-264		✓	✓	✓	✓	✓				1
	22177-A	.90	Nut, Water Pump Packing	5BL0-32	✓									1
	15824-A	2.30	Nut, Water Pump Packing—R. H.	5UU0-32							✓	✓	✓	1
	15466-A	1.94	Nut, Water Pump Packing—L. H.	5UUA0-32							✓	✓	✓	1
	22166-A	.35	Packing, Water Pump (4 Pieces per Box)	5BL0-209	✓									6
	16913-A	.06	Packing, Water Pump	5UU0-209							✓	✓	✓	1
	22175-A	.72	Bushing, Water Pump Packing	5BL0-87	✓									1
	1149-A	.10	Washer, Water Pump Packing	5UU0-392							✓	✓	✓	2
	4309-A	.42	Disc, Water Pump Lead	5UU0-586							✓	✓	✓	1
	21265-A	.52	Gland, Water Pump Packing	5BL0-81	✓									1
	16911-A	1.16	Gland, Water Pump	5UU0-81							✓	✓	✓	1
	61-A	.04	Plug, Water Pump Pipe	S-61A							✓	✓	✓	1
	15166-A	.08	Ring, Water Pump Snap	2UU0-1209							✓	✓	✓	1
	749-A	.20	Cup, Water Pump Grease	5TH0-233							✓	✓	✓	2
	2210-A	.06	Dowel, Water Pump	2UU0-1373							✓	✓	✓	2
	17106-A	.20	Lock, Water Pump Packing Nut	5UUB0-32							✓	✓	✓	1
	A-110400	20.50	Fan & Hub Assy. (Schwitzer-Cummins)—21″ with 6 Blades	5UK4-120	✓									1
	A-110634	18.00	Fan Assembly (Schwitzer-Cummins)—21″ with 6 Blades	5UL4-120		✓	✓	✓						1
		18.00	Fan Assembly—20″ with 4 Blades	5SA4-120A						✓				1
	A-110639	14.40	Fan Assembly (Schwitzer-Cummins)—21″ with 6 Blades	5B4-120					✓					1
	A-111545	27.60	Fan Assembly (Schwitzer-Cummins)—Including Bracket	5UU4-80							✓	✓	✓	1
		22.80	Fan Assembly (Schwitzer-Cummins)—Less Bracket	5UU4-120							✓	✓	✓	1
	BF-06954	7.25	Blade, Fan, Assembly—21″ with 6 Blades	5UK0-88	✓									1
	F-06921	4.25	Blade, Fan, Assembly—21″ with 6 Blades	5UL0-88		✓	✓	✓						1
	F-3784	2.90	Blade, Fan, Assembly—20″ with 4 Blades	5SA0-88A						✓				1
	F-06932	5.00	Blade, Fan, Assembly—21″ with 6 Blades	5BA0-88					✓					1
	F-07117	5.70	Blade, Fan, Assembly—22″ with 6 Blades	5UU0-88							✓	✓	✓	1
	C-101572	3.00	Hub, Fan, and Pulley	5ZBLE0-49A	✓									1
	C-6546	2.10	Hub, Fan, and Pulley	5T0-49A		✓	✓	✓	✓	✓				1

05-3 COOLING

THE AUTOCAR COMPANY — ARDMORE, PA.
Master Parts List

CODE	MFR. PART No.	LIST PRICE	NAME OF PART	AUTOCAR PART No.	U-2044	U-4044	U-4144	U-4144-T	C-50	U-5044	U-7144-T	U-8144	U-8144-T	No. REQ.
0503	B-110640	$4.10	Hub, Fan, and Pulley	5B0-49						✓				1
	B-111546	4.60	Hub, Fan, and Pulley	5UU0-49							✓	✓	✓	1
	C-101573	3.00	Spindle, Fan	5ZBLE0-51A	✓									1
	C-8079	2.50	Spindle, Fan	5SA0-51A		✓	✓	✓	✓					1
	C-110641	2.90	Spindle, Fan	5B0-51						✓				1
	C-111844	2.80	Spindle, Fan	5UU0-51							✓	✓	✓	1
	09194	.60	Cup, Fan Spindle Bearing (Timken)	5SCM0-256	✓	✓	✓	✓	✓	✓				2
	07204	1.25	Cup, Fan Spindle Bearing (Timken)	5DW0-132							✓	✓	✓	2
	09074	.20	Cone, Fan Spindle Bearing (Timken)	5SCM0-48	✓	✓	✓	✓	✓	✓				2
	07098	2.05	Cone, Fan Spindle Bearing (Timken)	5DW0-133							✓	✓	✓	2
	C-3806	....	Screw, Fan	5UU0-07	✓									4
	D-624	.04	Screw, Fan	5A0-07		✓	✓	✓	✓					4
	C-16028	.04	Screw, Fan	5UBB0-07						✓				4
	C-3806	.04	Screw, Fan	5UU0-07							✓	✓	✓	4
	C-594	.01	Washer, $\tfrac{5}{16}$" Spring	S-1346	✓	✓	✓	✓		✓	✓	✓	✓	4
	C-2016	.08	Gasket, Fan Hub	5A0-22	✓	✓	✓	✓	✓					1
	C-9095	.08	Gasket, Fan Hub	5UU0-242							✓	✓	✓	1
	C-2750	.08	Nut, Fan Spindle—Front	5SCM0-47	✓	✓	✓	✓	✓	✓	✓	✓	✓	1
	C-17065	.08	Washer, Fan Spindle Nut—Front	5TL0-72	✓						✓	✓	✓	1
	C-4013	.10	Washer, Fan Spindle Nut—Front	5SCM0-72		✓	✓	✓	✓	✓				1
	……	.01	Pin, Cotter ($\tfrac{3}{32}$" x 1")	S-2199	✓	✓	✓	✓	✓					1
	C-7067	.08	Gasket, Fan Bearing Oil	5CB0-242	✓	✓	✓	✓	✓	✓	✓	✓	✓	1
	C-2389	.16	Washer, Fan Bearing Cork Retaining	5SA0-73	✓	✓	✓	✓	✓					1
	C-6061	.12	Washer, Fan Bearing Cork Retaining	5T0-73							✓	✓	✓	1
	C-3814	.10	Washer, Fan Bearing Cork	5SA0-74	✓	✓	✓	✓	✓					1
	C-6062	.10	Washer, Fan Bearing Cork	5T0-74							✓	✓	✓	1
	C-2388	.24	Retainer, Fan Bearing Cork	5SA0-53	✓	✓	✓	✓	✓					1
	C-6060	.24	Retainer, Fan Bearing Cork	5T0-53							✓	✓	✓	1
	C-5098	.04	Wire, Fan Bearing Cork Retainer Lock	5SCM0-71	✓	✓	✓	✓		✓	✓	✓	✓	1
	……	….	Wire, Fan Bearing Cork Retainer Lock	5SA0-113					✓					1
	C-2662	.08	Washer, Fan Bracket Clamp—Front	5SA0-274	✓	✓	✓	✓	✓		✓	✓	✓	1
	C-6063	.12	Washer, Fan Bracket Clamp—Front	5TL0-275						✓				1
	C-2736	.10	Washer, Fan Bracket Clamp—Rear	5SA0-275	✓	✓	✓	✓	✓		✓	✓	✓	1
	……	….	Washer, Fan Bracket Clamp—Rear	5SA0-25						✓				1
	C-2673	.12	Nut, Fan Bracket Clamp	5A0-192	✓									1
	C-1873	.08	Nut, Fan Bracket Clamp	S-2001		✓	✓	✓	✓					1
	C-110377	.12	Nut, Fan Bracket Clamp	5UUA0-192							✓	✓	✓	1
	……	….	Pin, Fan Support Bracket Cotter	……	✓									1
	C-4240	.16	Washer, Fan Bracket Clamp Nut Lock	5UU0-234							✓	✓	✓	1
	A-11239	3.60	Bracket, Fan	5ZBLE0-54A	✓									1
	……	4.80	Bracket, Fan	5RL4-54		✓	✓	✓	✓					1
	A-110934	4.80	Bracket, Fan	5UU0-54							✓	✓	✓	1
	8533-A	….	Screw, Fan Support Bracket (Hercules)	8533-A	✓									2
	……	.09	Screw, Fan Support Bracket	S-1872		✓	✓	✓	✓					2
	312-A	.01	Washer, Spring	S-1349	✓									2
	……	.01	Washer, Spring	S-1349		✓	✓	✓	✓					2
	C-15295	1.50	Screw, Fan Adjusting	5ZBLEQ0-7	✓									1
	……	.70	Screw, Fan Adjuster, Assembly	5T2-170B		✓	✓	✓	✓					1
	C-8500	.70	Screw, Fan Adjusting, Assembly	5UU0-171							✓	✓	✓	1
	C-2692	.06	Nut, Fan Adjusting Screw	5A0-47	✓									1
	C-4913	.16	Nut, Fan Adjusting Screw Lock	5UU0-192							✓	✓	✓	1
	C-2675	.01	Washer, $\tfrac{1}{2}$" Spring	S-1349	✓						✓	✓	✓	1
	42255-B	2.00	Belt, Fan (Hercules)	5UK0-55	✓									2
	……	1.95	Belt, Fan	5T3-55B		✓	✓	✓						1
	……	….	Belt, Fan	5T3-55H					✓					1
	……	1.50	Belt, Fan	5ZTL3-55						✓				1
	77280-B	1.40	Belt, Fan (Hercules)	5UU3-55							✓	✓	✓	1
	C-10825	….	Spacer, Fan Blade ($\tfrac{5}{16}$")	C-10825	✓									1
	……	.95	Spacer, Fan	5UU2-262							✓	✓	✓	1
	……	.04	Gasket, Fan Spacer	5UU2-22							✓	✓	✓	1
	40500-B	….	Pulley, Fan Drive (Hercules)	40500-B	✓									1
	……	4.20	Pulley, Fan Drive	5T3-39A		✓	✓	✓	✓					1

Master Parts List

CODE	MFR. PART No.	LIST PRICE	NAME OF PART	AUTOCAR PART No.	U-2044	U-4044	U-4144	U-4144-T	C-50	U-5044	U-7144-T	U-8144	U-8144-T	No. REQ.
					MODEL									
0503	......	$6.00	**Pulley,** Fan Drive	5UP3-39						✓				1
	4265-A	....	**Key,** Fan Drive Pulley (Hercules)	4265-A	✓									1
	......	.06	**Key,** Fan Drive Pulley	S-2308		✓	✓	✓	✓	✓				1
	......	.08	**Nut,** Fan Drive Pulley Lock—Front	S-2001		✓	✓	✓	✓	✓				1
	......	.06	**Washer,** Fan Drive Pulley Lock Nut	5T2-95		✓	✓	✓	✓	✓				1
	C-7487	.18	**Cap,** Fan Hub	5TL0-341						✓				1
	......	.30	**Spacer,** Fan Bracket	5T2-92						✓				1
	......	.12	**Screw,** Fan Bracket Spacer (½″—13 x 2¾″)	S-4077						✓				1
	......	.01	**Washer,** ½″ Spring	S-1349						✓				1
0505	**Engine Water Fittings and Hoses**													
	17321-C	8.38	**Manifold,** Water Outlet	5UU6-06							✓	✓	✓	1
	60-A	.04	**Plug,** Water Outlet Manifold ¼″ Pipe	S-4748							✓	✓	✓	1
	61-A	.04	**Plug,** Water Outlet Manifold ⅜″ Pipe	S-4004							✓	✓	✓	1
	8179-A	.04	**Gasket,** Water Outlet Manifold	5UU0-607							✓	✓	✓	2
	352-A	.09	**Screw,** Hex. Cap (½″—13 x 1¼″)	S-25							✓	✓	✓	4
	312-A	.01	**Lockwasher,** 17/32″ Spring	S-1349							✓	✓	✓	4
	17904-C	9.50	**Housing,** Thermostat	5UU0-271							✓	✓	✓	1
	80771-A	.10	**Gasket,** Thermostat Housing	5UU0-272							✓	✓	✓	1
	14885-A	.04	**Screw,** Hex. Cap (⅜″—16 x 2⅜″)	5UU0-579							✓	✓	✓	2
	1911-A	.08	**Screw,** Hex. Cap (⅜″—16 x 3⅛″)	5UUA0-579							✓	✓	✓	2
	342-A	.01	**Lockwasher,** 13/32″ Spring	S-1347							✓	✓	✓	4
	......	.75	**Flange,** Water Outlet Hose	5UU3-62							✓	✓	✓	1
	......	.20	**Gasket,** Water Outlet Hose Flange	S-1469							✓	✓	✓	1
	......	.06	**Screw,** Hex. Cap (⅜″—16 x 1⅛″)	S-46							✓	✓	✓	2
	......	.01	**Lockwasher,** 13/32″ Spring	S-1347							✓	✓	✓	2
	......	Ft. .60	**Hose,** Water Outlet (1¾″I.D. x 9½″)	S-2373							✓	✓	✓	1
	......	.10	**Clamp,** Hose (1¾″ I.D. Hose)	S-2398							✓	✓	✓	2
	......	.90	**Pipe,** Radiator Outlet (1¾″ O.D. x 14⅞″)	5UU2-17							✓	✓	✓	1
	......	Ft. .60	**Hose,** Radiator Outlet (1¾″ I.D. x 4″)	S-2372							✓	✓	✓	2
	......	.10	**Clamp,** Radiator Outlet Hose	S-2398							✓	✓	✓	4
	......	.85	**Elbow,** Water Pump Inlet	5UU3-158A							✓	✓	✓	1
	......	.06	**Gasket,** Permanite	S-4082							✓	✓	✓	1
	......	.06	**Screw,** Hex. Cap (⅜″—16 x 1⅛″)	S-46							✓	✓	✓	2
	......	.01	**Lockwasher,** 13/32″ Spring	S-1347							✓	✓	✓	2
	......	3.00	**Pipe,** Water Pump By-Pass	5UU4-443							✓	✓	✓	1
	......	.10	**Hose,** Water Pump By-Pass (1″ I.D. x 3″)	S-2356							✓	✓	✓	2
	......	.15	**Clamp,** Water Pump By-Pass Hose	S-2392							✓	✓	✓	4
	15178-C	7.00	**Manifold,** Water Pump Discharge	5UU0-612							✓	✓	✓	1
	15179-B	.10	**Gasket,** Water Pump Discharge Manifold	5UU0-94							✓	✓	✓	1
	303-A	.03	**Screw,** Hex. Cap (5/16″—18 x ¾″)	S-14							✓	✓	✓	20
	615-A	.01	**Lockwasher,** 11/32″ Spring	S-1346							✓	✓	✓	20
	......	....	**Line,** Compressor Water, Assembly to Manifold	25UU0-780							✓	✓	✓	1
	......	....	**Line,** Compressor Water, Assembly to Pump	25UUA0-780							✓	✓	✓	1
	......	.34	**Hose,** Water Pump Discharge (1¾″ I.D. x 5⅛″)	S-2376							✓	✓	✓	1
	......	.10	**Clamp,** Water Pump Discharge Hose	S-2398							✓	✓	✓	2

THE AUTOCAR COMPANY, ARDMORE, PA.

Master Parts List

GROUP 06

ILLUSTRATIONS

GROUP 06

ELECTRICAL SYSTEM

GROUP 06

Master Parts List

THE AUTOCAR COMPANY, ARDMORE, PA.

As supplied on Autocar Models
U-7144-T U-8144 U-8144-T

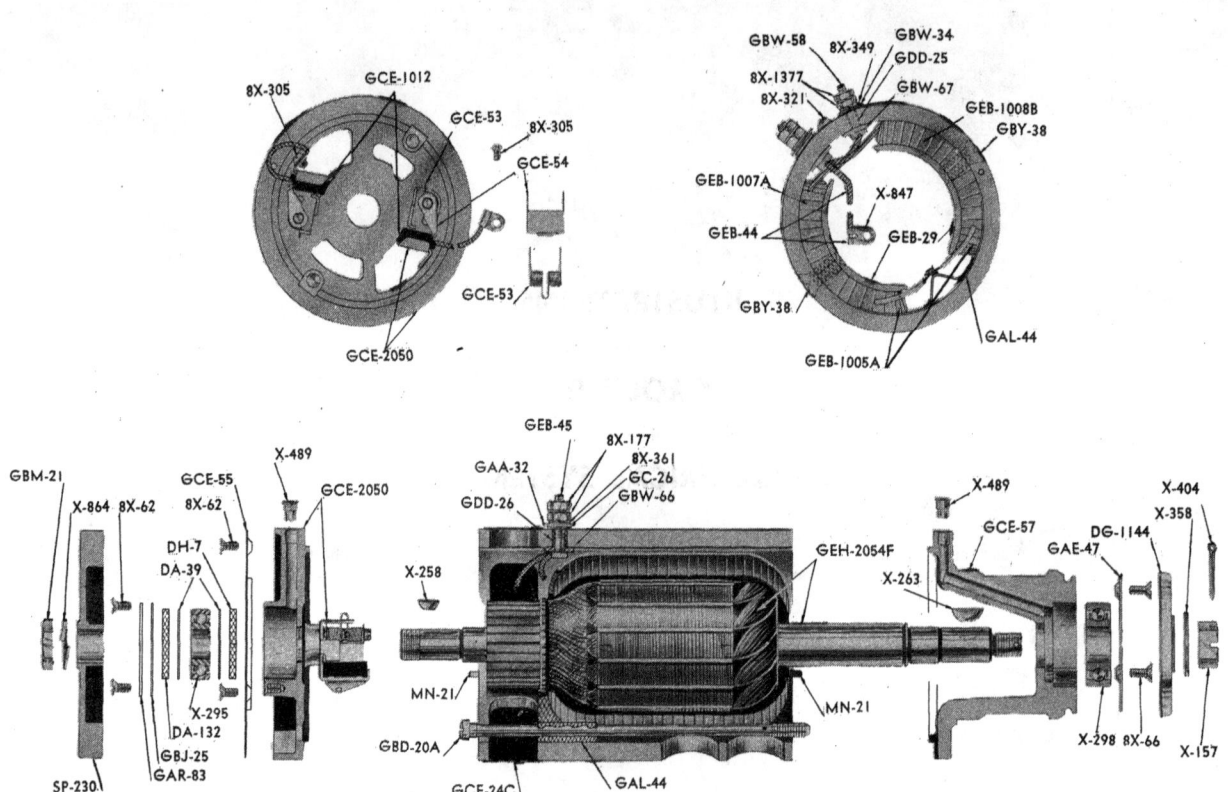

0601—Generator

THE AUTOCAR COMPANY, ARDMORE, PA.

Master Parts List

GROUP 06

As supplied on Autocar Models
U-8144   U-8144-T

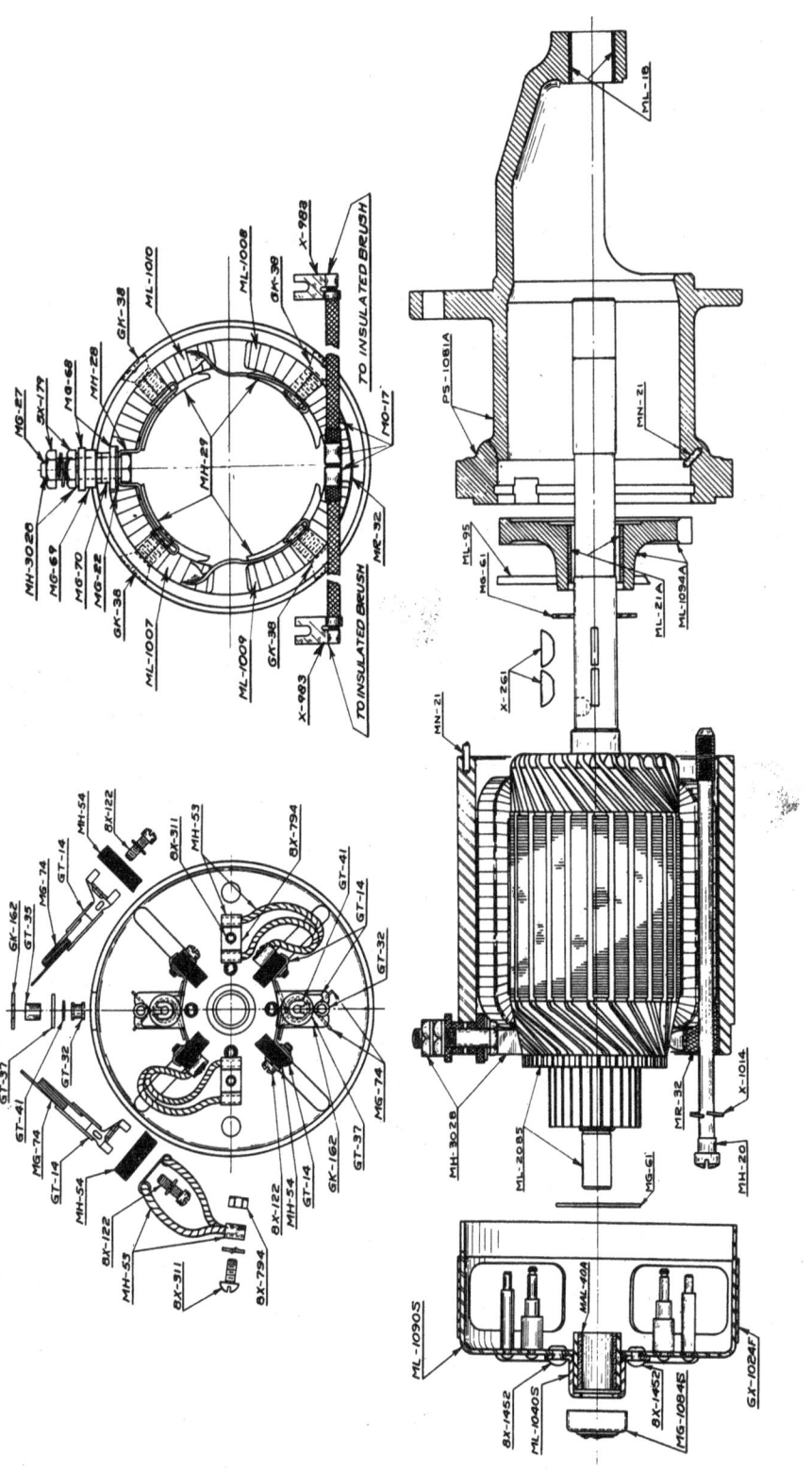

0602—Starting Motor

GROUP 06

Master Parts List

THE AUTOCAR COMPANY, ARDMORE, PA.

As supplied on Autocar Models
U-7144-T U-8144 U-8144-T

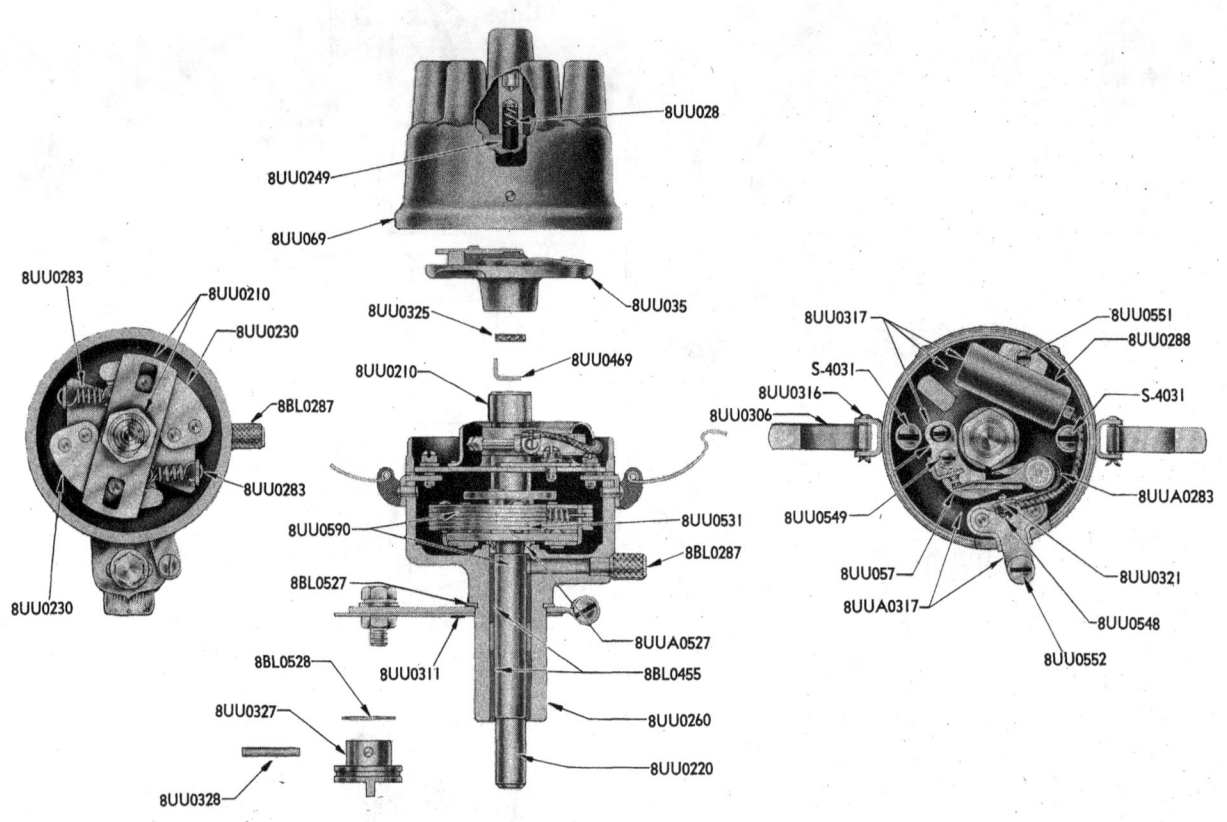

0603—Distributor

THE AUTOCAR COMPANY, ARDMORE, PA.
Master Parts List
GROUP 06

As supplied on Autocar Models U-8144   U-8144-T

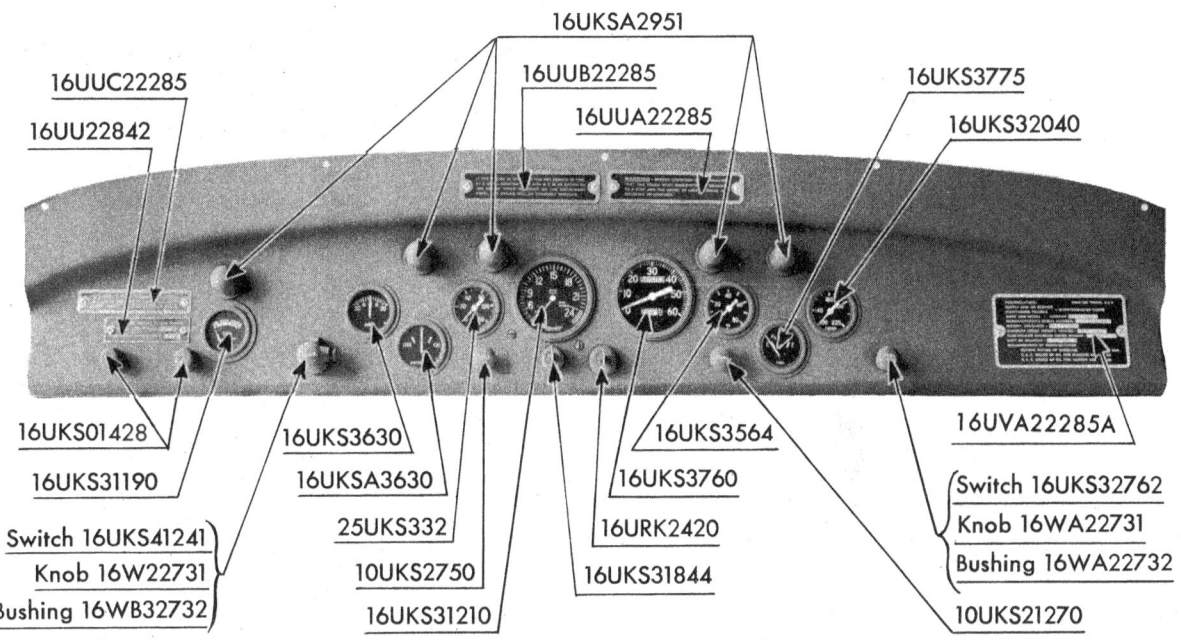

0605—Instruments and Carrier

GROUP 06

Master Parts List

THE AUTOCAR COMPANY, ARDMORE, PA.

As supplied on Autocar Models
U-8144   U-8144-T

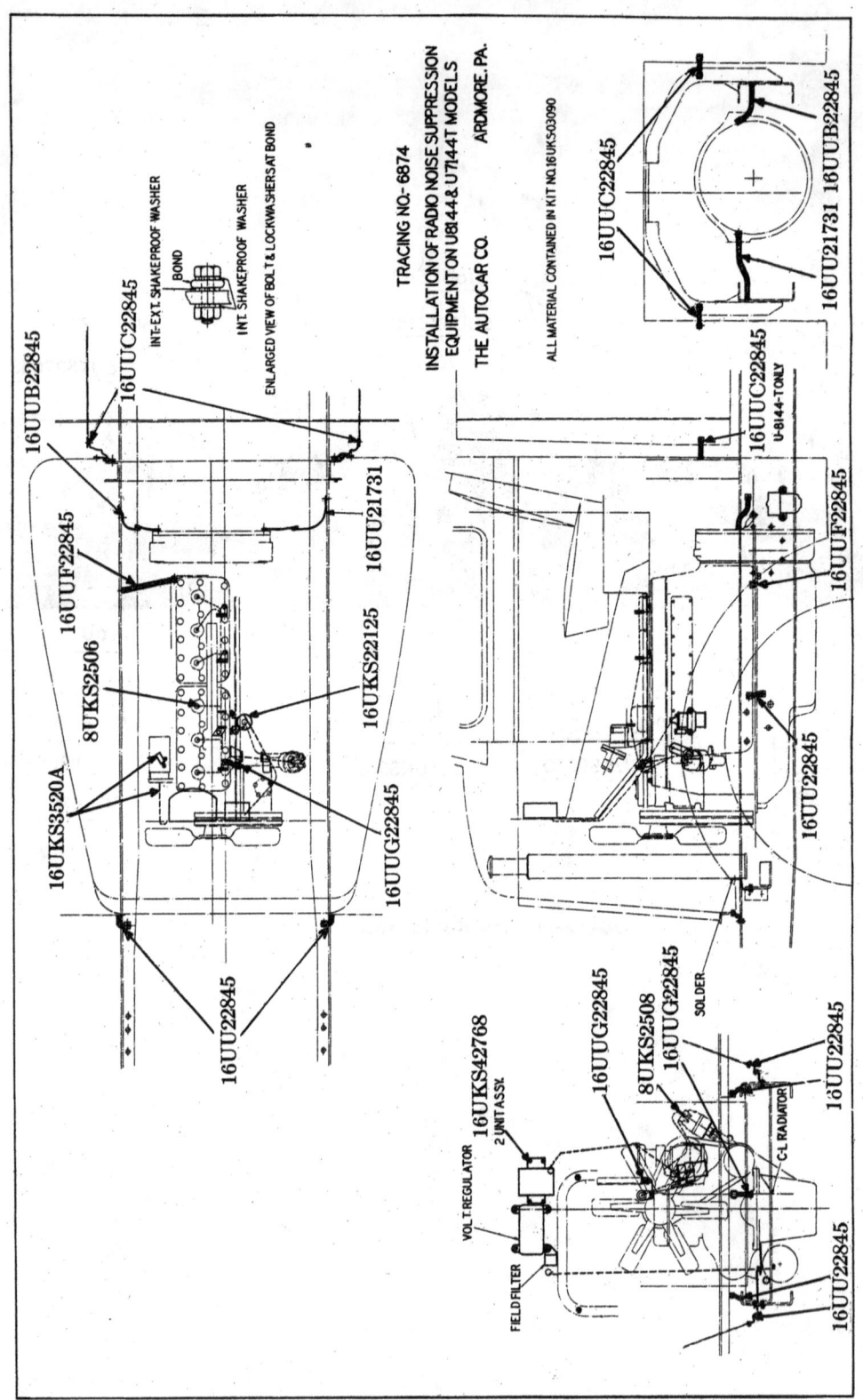

0611—Radio Suppression

THE AUTOCAR COMPANY
ARDMORE, PA.
Master Parts List

06-1 ELECTRICAL

CODE	MFR. PART No.	LIST PRICE	NAME OF PART	AUTOCAR PART No.	U-2044	U-4044	U-4144	U-4144-T	C-50	U-5044	U-7144-T	U-8144	U-8144-T	NO. REQ.
0601	**Generator**													
	GEW-4804	....	Generator Assembly	16BLL0-410	✓									1
	GEW-4801	....	Generator Assembly	16BLK0-410		✓								1
	GEW-4805	....	Generator Assembly	16BLM0-410			✓	✓	✓					1
	GEW-4802	....	Generator Assembly	16BLU0-410							✓	✓	✓	1
	DK-23A	....	Screw, Frame	DK-23A	✓									2
	GBD-20A	$0.11	Screw, Frame	GBD-20A		✓	✓	✓		✓	✓	✓	✓	2
	GAA-32	.01	Insulator, Terminal Post Top	GAA-32	✓	✓	✓	✓	✓	✓	✓	✓	✓	1
	GAL-44	.02	Insulator, Field Connection	GAL-44	✓	✓	✓	✓	✓	✓	✓	✓	✓	1
	GBW-34	.01	Washer, Insulating	GBW-34	✓	✓	✓	✓	✓	✓	✓	✓	✓	1
	GBW-66	.02	Insulator, Terminal Post Bottom	GBW-66	✓	✓	✓	✓	✓	✓	✓	✓	✓	1
	GBW-67	.02	Insulator, Terminal Post Bottom	GBW-67	✓	✓	✓	✓	✓	✓	✓	✓	✓	1
	GBY-38	.01	Screw, Pole Piece	GBY-38	✓	✓	✓	✓	✓	✓	✓	✓	✓	2
	GC-26	.01	Washer, Insulating	GC-26	✓	✓	✓	✓	✓	✓	✓	✓	✓	1
	GDD-25	.04	Bushing, Insulating	GDD-25	✓	✓	✓	✓	✓	✓	✓	✓	✓	1
	GDD-26	.02	Bushing, Insulating	GDD-26	✓	✓	✓	✓	✓	✓	✓	✓	✓	1
	GEB-29	.79	Pole Piece	GEB-29	✓	✓	✓	✓	✓	✓	✓	✓	✓	2
	GEB-1005	....	Coil, Field, Assembly—Complete	GEB-1005	✓									1
	GEB-2005	....	Coil, Field, Assembly—Complete	GEB-2005		✓								1
	GEB-2005A	....	Coil, Field, Assembly—Complete	GEB-2005A			✓	✓	✓					1
	GEB-1005A	4.73	Coil, Field, Assembly—Complete	GEB-1005A							✓	✓	✓	1
	GBW-58	.05	Post, Terminal	GBW-58	✓	✓	✓	✓	✓	✓	✓	✓	✓	1
	GEA-30	....	Lead Assembly	GEA-30	✓	✓								1
	GEB-44	.11	Lead Assembly	GEB-44			✓	✓		✓	✓	✓	✓	1
	X-847	.05	Terminal	X-847	✓	✓	✓	✓	✓	✓	✓	✓	✓	1
	GEB-27	....	Post, Terminal	GEB-27	✓	✓								1
	GEB-45	.05	Post, Terminal	GEB-45			✓	✓	✓	✓	✓	✓	✓	1
	GEB-1007	....	Coil, Field—Left	GEB-1007	✓									1
	GEB-1007B	....	Coil, Field—Left	GEB-1007B			✓	✓	✓					1
	GEB-1007A	2.36	Coil, Field—Left	GEB-1007A		✓					✓	✓	✓	1
	GEB-1008	....	Coil, Field—Right	GEB-1008	✓		✓	✓	✓					1
	GEB-1008A	....	Coil, Field—Right	GEB-1008A		✓								1
	GEB-1008B	2.36	Coil, Field—Right	GEB-1008B							✓	✓	✓	1
	X-959	.05	Terminal	X-959							✓	✓	✓	1
	MN-21	.01	Pin, Dowel	MN-21	✓	✓	✓	✓	✓	✓	✓	✓	✓	2
	5X-177	....	Nut, #14—24 Hex	5X-177	✓	✓								2
	8X-177	.05	Nut, #14—24 Hex	8X-177			✓	✓		✓	✓	✓	✓	2
	X-193	.05	Washer, #14 Lock	X-193	✓	✓	✓	✓	✓	✓	✓	✓	✓	2
	X-196	.05	Washer, #10 Lock	X-196	✓	✓	✓	✓	✓	✓	✓	✓	✓	3
	5X-349	....	Washer, #10 Plain	5X-349		✓								1
	8X-349	.05	Washer, #10 Plain	8X-349	✓		✓	✓		✓	✓	✓	✓	1
	5X-361	....	Washer, ¼ Plain	5X-361		✓								1
	8X-361	.05	Washer, ¼ Plain	8X-361	✓		✓	✓		✓	✓	✓	✓	1
	5X-1377	....	Nut, #10—32 Hex	5X-1377	✓	✓								2
	8X-1377	.05	Nut, #10—32 Hex	8X-1377			✓	✓		✓	✓	✓	✓	2
	8X-140	.05	Nut, #6—32 Hex	8X-140							✓	✓	✓	2
	X-194	.05	Washer, #6 Lock	X-194							✓	✓	✓	1
	8X-1420	.05	Screw, #6—32 x 9/16" Flat Head	8X-1420							✓	✓	✓	1
	8X-177	....	Nut, #14—24 Hex	8X-177		✓								1
	X-193	....	Washer, #14 Lock	X-193			✓	✓	✓	✓				1
	X-196	....	Washer, #10 Lock	X-196	✓	✓	✓	✓	✓					2
	X-203	.05	Washer, 5/16" Lock	X-203	✓	✓	✓	✓	✓	✓	✓	✓	✓	2
	8X-321	.05	Screw, #10—32 x 5/16" Round Head	8X-321	✓	✓	✓	✓	✓	✓	✓	✓	✓	1
	8X-1377	.05	Nut, #10—32 Hex	8X-1377		✓					✓	✓	✓	1
	Commutator End Plate Group													
	DA-39	.01	Guard, Flat Oil	DA-39	✓	✓	✓	✓	✓	✓	✓	✓	✓	2
	DA-132	.02	Washer, Felt	DA-132	✓	✓	✓	✓	✓	✓	✓	✓	✓	1
	DH-7	.04	Washer, Felt	DH-7	✓	✓	✓	✓	✓	✓	✓	✓	✓	1
	GAR-83	.26	Cover, Commutator End Cap	GAR-83	✓	✓	✓	✓	✓	✓	✓	✓	✓	1
	GBJ-25	.11	Gasket, End Cap Cover	GBJ-25	✓	✓	✓	✓	✓	✓	✓	✓	✓	1
	GCE-55	.08	Deflector, Air	GCE-55	✓	✓	✓	✓	✓	✓	✓	✓	✓	1

06-2 ELECTRICAL

Master Parts List

THE AUTOCAR COMPANY, ARDMORE, PA.

CODE	MFR. PART No.	LIST PRICE	NAME OF PART	AUTOCAR PART No.	U-2044	U-4044	U-4144	U-4144-T	C-50	U-5044	U-7144-T	U-8144	U-8144-T	No. REQ.
0601	GCE-3080	....	Plate, Commutator End, Assembly	GCE-3080		✓	✓	✓		✓				1
	GCE-2050	$2.10	Plate, Commutator End, Assembly	GCE-2050	✓						✓	✓	✓	1
	GCE-53	.02	Spring, Brush	GCE-53	✓	✓	✓	✓	✓	✓	✓	✓	✓	2
	GCE-54	.01	Arm, Brush	GCE-54	✓	✓	✓	✓	✓	✓	✓	✓	✓	2
	*GCE-1012	.26	Brush—Main	*GCE-1012	✓	✓	✓	✓	✓	✓	✓	✓	✓	2
	X-195	.05	Washer, #8 Lock	X-195	✓	✓	✓	✓	✓	✓	✓	✓	✓	1
	8X-305	.05	Screw, #8—32 x ¼" Round Head	8X-305	✓	✓	✓	✓	✓	✓	✓	✓	✓	1
	8X-62	.05	Screw, #10—32 x ⅜" Flat Head	8X-62	✓	✓	✓	✓	✓	✓				6
	X-195	.05	Washer, #8 Lock	X-195							✓	✓	✓	1
	X-295	1.21	Bearing, Ball—S.A.E. #203	X-295	✓	✓	✓	✓	✓	✓	✓	✓	✓	1
	8X-305	.05	Screw, #8—32 x ¼" Round Head	8X-305							✓	✓	✓	1
	X-489	.05	Oiler, ¼"—Press in Type	X-489	✓	✓	✓	✓	✓	✓	✓	✓	✓	1
	Armature Assembly Group													
	GAR-1177	....	Thrower, Oil, Assembly	GAR-1177	✓									1
	DG-1144	.26	Thrower, Oil	DG-1144		✓	✓	✓		✓	✓	✓	✓	1
	GBM-21	.05	Nut, Armature Shaft	GBM-21		✓	✓	✓	✓	✓	✓	✓	✓	1
	GEH-2090F	....	Armature Assembly	GEH-2090F	✓									1
	GEH-2054F	16.80	Armature Assembly	GEH-2054F		✓	✓	✓	✓	✓	✓	✓	✓	1
	SP-230	.79	Fan, Ventilating	SP-230	✓	✓	✓	✓	✓	✓	✓	✓	✓	1
	X-156	....	Nut, ½"—20 Slotted	X-156	✓									1
	X-157	.05	Nut, ½"—20 Slotted	X-157		✓	✓	✓	✓	✓	✓	✓	✓	1
	X-258	.05	Key, #3 Woodruff	X-258	✓	✓	✓	✓	✓	✓	✓	✓	✓	1
	GK-174	....	Washer, ½" Plain Steel	GK-174	✓									1
	X-358	.05	Washer, ½" Plain Steel	X-358		✓	✓	✓	✓	✓	✓	✓	✓	1
	X-404	.05	Pin, 3/32" x 1" Cotter	X-404		✓	✓	✓	✓	✓	✓	✓	✓	1
	X-864	.05	Washer, .669 Lock	X-864		✓	✓	✓	✓	✓	✓	✓	✓	1
	X-263	.05	Key, #8 Woodruff	X-263		✓	✓	✓	✓	✓	✓	✓	✓	1
	Drive End Head Assembly													
	GCE-1091	....	Head, Drive End, Assembly	GCE-1091	✓									1
	GCE-1109	....	Head, Drive End, Assembly	GCE-1109		✓	✓	✓		✓				1
	GCE-1057	4.73	Head, Drive End, Assembly	GCE-1057							✓	✓	✓	1
	GAR-95	....	Washer, Felt	GAR-95	✓									1
	GAE-47	.16	Retainer, Bearing	GAE-47		✓	✓	✓		✓	✓	✓	✓	1
	CGE-91	....	Head, Drive End	CGE-91	✓									1
	GX-9	....	Retainer, Drive End Packing	GX-9	✓									1
	GX-10	....	Gasket	GX-10	✓									1
	X-196	....	Washer, #10 Lock	X-196	✓									3
	8X-311	....	Screw, #10—32 x ⅜" Round Head	8X-311	✓									3
	X-378	....	Bearings, Ball—S.A.E. #204	X-378	✓									1
	GCE-109	....	Head, Drive End	GCE-109		✓	✓	✓	✓					1
	GCE-57	1.58	Head, Drive End	GCE-57							✓	✓	✓	1
	Drive End Head Assembly													
	8X-66	.05	Screw, #10—32 x ½" Flat Head	8X-66		✓	✓	✓		✓	✓	✓	✓	3
	X-298	2.73	Bearing, Ball—S.A.E. #204	X-298		✓	✓	✓		✓	✓	✓	✓	1
	X-489	.05	Oiler, ¼"—Press in Type	X-489	✓						✓	✓	✓	1
	Miscellaneous Parts													
	GCE-24C	.11	Band, Head, Assembly	GCE-24C	✓	✓	✓	✓	✓	✓	✓	✓	✓	1
	X-715	.105	Screw, #10—32 x 1¼" Round Head	X-715	✓	✓	✓	✓	✓	✓	✓	✓	✓	1
	8X-794	.05	Nut, #10—32 Square	8X-794	✓	✓	✓	✓	✓	✓	✓	✓	✓	1
	......	1.60	Cable, Generator to Regulator, Assembly	16UKS3-520							✓	✓	✓	1
0602	**Starting Motor**													
	720-T	24.00	Motor, Starting—Assembly (Delco)	16BLA0-560	✓									1
	724-Z	27.50	Motor, Starting—Assembly (Delco)	16RE4-560		✓								1
	1108209	27.50	Motor, Starting—Assembly (Delco)	16REA4-560			✓	✓						1
	721-P	30.00	Motor, Starting—Assembly (Delco)	16ZN4-560						✓				1
	412	27.50	Motor, Starting—Assembly (Delco)	16HM0-560							✓	✓	✓	1
	817790	....	Frame and Field Assembly	817790	✓									1

Note: *Service with Brush Set #GCE-2012-S.

THE AUTOCAR COMPANY
ARDMORE, PA.

Master Parts List

06-3 ELECTRICAL

CODE	MFR. PART No.	LIST PRICE	NAME OF PART	AUTOCAR PART No.	U-2044	U-4044	U-4144	U-4144-T	C-50	U-5044	U-7144-T	U-8144	U-8144-T	No. REQ.
0602	822858	....	Frame and Field Assembly	822858		✓								..
	......	....	Frame and Field Assembly	......			✓	✓						..
	38274	....	Frame and Field Assembly	16HM0-2857							✓	✓	✓	..
	810601	....	Pole Piece	810601	✓	✓	✓	✓		✓				1
	36497	$0.75	Pole Piece	16HM0-2858							✓	✓	✓	1
	815839	....	Frame, End (C.E.)	815839	✓	✓	✓	✓		✓				1
	38290	6.00	Frame, End (C.E.)	16HM0-2859							✓	✓	✓	..
	818002	....	Armature	818002	✓									..
	818134	....	Armature	818134		✓	✓	✓		✓				1
	37895	15.00	Armature	16HM0-2861							✓	✓	✓	..
	16199	.75	Plate, Center Bearing (D.E.)	16HM0-2862							✓	✓	✓	..
	811299	....	Housing, Bendix	811299	✓									..
	828894	....	Housing, Bendix	828894		✓	✓	✓						1
	1842827	....	Housing, Bendix	1842827						✓				1
	16999	7.50	Housing, Bendix	16HM0-2863							✓	✓	✓	..
	817114	....	Band, Commutator Cover	817114	✓	✓	✓	✓		✓				1
	1880355	.60	Band, Cover	16HM0-2864							✓	✓	✓	..
	811553	....	Brush, Motor	811553	✓	✓	✓	✓						..
	828448	....	Brush, Motor	828448						✓				4
	16083	.30	Brush, Motor	16HM0-2865							✓	✓	✓	..
	813521	....	Spring, Brush Arm	813521	✓	✓	✓	✓		✓				4
	34846	.10	Spring, Brush Arm	16HM0-2866							✓	✓	✓	..
	35058	.10	Bearing (D.E.)	16HM0-2867							✓	✓	✓	..
	38296	.10	Bearing (C.E.)	16HM0-2868							✓	✓	✓	..
	810627	....	Coil, Field—Assembly—R. H.	810627	✓									..
	817056	....	Coil, Field—Assembly—R. H.	817056		✓	✓	✓						..
	1838410	....	Coil, Field—Assembly (Upper)	1838410						✓				1
	38276	2.40	Coil, Field—Assembly—R. H.	16HM0-2869							✓	✓	✓	..
	810626	....	Coil, Field—Assembly—L. H.	810626	✓									..
	817055	....	Coil, Field—Assembly—L. H.	817055		✓	✓	✓						..
	1838409	....	Coil, Field—Assembly (Lower)	1838409						✓				1
	38275	2.40	Coil, Field—Assembly—L. H.	16HM0-2871							✓	✓	✓	..
	817077	.05	Washer, Motor Terminal Stud Insulation ($\frac{3}{32}$")	16HM0-2876							✓	✓	✓	..
	809051	....	Bushing, Motor Terminal Stud Insulation ($\frac{1}{16}$")	16HM0-2873							✓	✓	✓	..
	33345	.05	Strip, Field Coil Insulation	16HM0-2874							✓	✓	✓	..
	141552	.01	Screw, Bendix Housing & End Frame L.W. ($\frac{7}{32}$")	S-1344	✓						✓	✓	✓	..
	106497	....	Lockwasher, Brush Field Screw	106497		✓	✓	✓		✓				1
	141551	.01	Lockwasher, Brush Lead Screw (#8)	S-3164	✓						✓	✓	✓	..
	106496	....	Lockwasher, Brush Ground Lead Screw	106496		✓	✓	✓		✓				2
	141553	.01	Lockwasher, Center Bearing Screw ($\frac{9}{32}$")	S-1345	✓						✓	✓	✓	..
	106495	....	Lockwasher, Brush Holder Screw	106495		✓	✓	✓		✓				2
	142248	.01	Lockwasher, Motor Terminal Stud ($\frac{13}{32}$")	S-1347	✓	✓					✓	✓	✓	..
	110730	....	Lockwasher, Field Terminal Stud	110730			✓	✓		✓				2
	805258	.05	Nut, Motor Terminal Stud	16HM0-2875	✓	✓	✓	✓		✓	✓	✓	✓	1
	37872	.05	Screw, Bendix Housing & E. Frame Fasten'g (#10—32 x $\frac{11}{16}$")	16HM0-2934							✓	✓	✓	..
	115607	.01	Screw, Brush to Field Lead (#8—32 x $\frac{5}{16}$")	S-3138							✓	✓	✓	..
	114935	.01	Screw, Brush Ground Lead (#8—32 x $\frac{1}{4}$")	S-208	✓	✓					✓	✓	✓	..
	122159	....	Screw, Brush Ground	122159			✓	✓		✓				2
	828483	.10	Screw, Center Bearing Plate	16HM0-2876							✓	✓	✓	..
	115632	.01	Screw, Center Bearing Plate ($\frac{1}{4}$"—20 x $\frac{5}{8}$")	S-36							✓	✓	✓	..
	826938	....	Stud, Motor Terminal	826938	✓	✓	✓	✓						1
	811601	....	Stud, Motor Terminal	811601						✓				1
	35074	.10	Stud, Motor Terminal	16HM0-2877							✓	✓	✓	..
	828675	.05	Screw, Pole Piece Attaching ($\frac{3}{8}$"—24 x $\frac{5}{8}$")	16HM0-2935	✓	✓				✓	✓	✓	✓	4
	802790	.05	Screw, Pole Piece Attaching ($\frac{5}{16}$"—18 x $\frac{5}{8}$")	16HM0-2936	✓						✓	✓	✓	..
	24754	.05	Washer, Brush Holder Rivet	16HM0-2878							✓	✓	✓	..
	35729	.05	Washer, Brush Holder Rivet Insulating	16HM0-2879							✓	✓	✓	..
	831688	.01	Washer, Motor Terminal Stud ($\frac{3}{8}$" x $\frac{3}{4}$" x $\frac{1}{16}$")	S-334	✓	✓					✓	✓	✓	..
	1838568	.05	Washer, Spacer (C.E.) ($\frac{17}{32}$" x $\frac{7}{8}$" x $\frac{3}{32}$")	S-704						✓				..
	833602	.05	Washer, Spacer (D.E.) (Bendix Shaft)	16HM0-2884	✓	✓	✓	✓		✓	✓	✓	✓	1
	35048	.25	Bushing, Center Plate	16HM0-2884							✓	✓	✓	..
	35586	.05	Clip, Brush Lead	16HM0-2885							✓	✓	✓	..

Note—C.E. Denotes Commutator End. D.E. Denotes Drive End.

06-4 ELECTRICAL — Master Parts List

THE AUTOCAR COMPANY, ARDMORE, PA.

CODE	MFR. PART No.	LIST PRICE	NAME OF PART	AUTOCAR PART No.	U-2044	U-4044	U-4144	U-4144-T	C-50	U-5044	U-7144-T	U-8144	U-8144-T	NO. REQ.
0602	1855434	$0.05	Rivet, Brush Holder.................	16HM0-2886							✓	✓	✓	..
	21428	.05	Bushing, Brush Holder Rivet Insulating............	16HM0-2887							✓	✓	✓	..
	106750	....	Key, Woodruff—No. 6............	106750	✓									..
	124546	.05	Key, Bendix Drive (#8)............	S-2306		✓	✓	✓		✓	✓	✓	✓	..
	810819	.05	Plug, End Welch (C.E.) (¾" Dia.)........	S-4208	✓	✓	✓	✓		✓	✓	✓	✓	1
	36799	....	Clip, Field Lead................	16HM0-2891							✓	✓	✓	..
	134569	....	Nut, Motor Terminal (⅜"—16)........	16HM0-2937							✓	✓	✓	..
	106496	....	Lockwasher, Brush Lead Screw (#8)........	S-3164							✓	✓	✓	..
	107728	....	Screw, Cover Band Bt. Head (#10—32 x 1¼").....	16HM0-2938							✓	✓	✓	..
Bendix Drive Group														
	R11XV	$5.50	Drive, Bendix—Assembly (Eclipse Machine Co.).......	16SA0-880		✓								1
	RCD11FX-10	5.50	Drive, Bendix—Assembly (Eclipse Machine Co.).......	16F0-880			✓	✓						1
	RBCD11FXX-10	7.50	Drive, Bendix—Assembly (Eclipse Machine Co.).......	16ZN0-880						✓				1
	R11X-13	5.50	Drive, Bendix—Assembly (Eclipse Machine Co.).......	16SCM0-880	✓						✓	✓	✓	1
	R11XV-SA	3.90	Sleeve, Bendix—and Shaft Assembly............	16SA0-2004		✓								1
	RCD11F-10SA	3.90	Sleeve, Bendix—and Shaft Assembly............	16ZN0-2004			✓	✓						1
	R11X-13SA	3.90	Sleeve, Bendix—and Shaft Assembly............	16SCM0-2004	✓						✓	✓	✓	1
	SR10-105	.40	Drive, Bendix—Head.............	16T0-595	✓	✓								1
	R11F-5	.40	Drive, Bendix—Head.............	16F0-595			✓	✓						1
	RB11F-5	.65	Drive, Bendix—Head.............	16ZN0-595						✓				1
	R11-21	.07	Clip, Bendix Spring Support..........	16T0-596	✓	✓					✓	✓	✓	1
	R12-8	.01	Lockwasher, Bendix Spring Screw........	16A0-539	✓	✓					✓	✓	✓	2
	R13-8	.01	Lockwasher, Bendix Spring Screw........	16FE0-539			✓	✓		✓				2
	R11-109X	.07	Screw, Bendix Head Spring............	16A0-593	✓	✓					✓	✓	✓	1
	R11F-9X	.08	Screw, Bendix Head Spring............	16F0-593			✓	✓						1
	R11F-109X	.08	Screw, Bendix Head Spring............	16ZN0-593						✓				1
	R11-7X	.06	Screw, Bendix Shaft Spring............	16A0-592	✓	✓					✓	✓	✓	1
	R11F-7X	.07	Screw, Bendix Shaft Spring............	16F0-592			✓	✓						1
	R11F-7XX	.07	Screw, Bendix Shaft Spring............	16ZN0-592						✓				1
	R11-112R	.30	Sleeve, Bendix Service..............	16A0-594	✓	✓					✓	✓	✓	1
	R11F-12R	.30	Sleeve, Bendix Service..............	16ZN0-594			✓	✓						1
	R11-6X	.55	Spring, Bendix Drive...............	16T0-538	✓	✓					✓	✓	✓	1
	......	.55	Spring, Bendix Drive...............	16T0-538			✓	✓						1
	RB11F-6XX	1.50	Spring, Bendix Drive...............	16ZN0-538						✓				1
	F-3120	.01	Ring, Bendix Take-Up..............	16A0-2624			✓	✓						1
Switch and Cable Group														
	SW-4002	.80	Switch, Starting (Auto-Lite)............	16BL2-680B	✓	✓		✓						1
	1881521	27.50	Switch, Starting (Delco-Remy)..........	16UP0-680						✓	✓	✓	✓	1
	......	....	Screw, Hex. Cap (5/16"—18 x ¾").......	S-14						✓	✓	✓	✓	2
	......	.01	Lockwasher, 11/32" Spring............	S-1346						✓	✓	✓	✓	2
	......	.50	Bracket, Starting Switch, Assembly........	16UBL2-443A	✓	✓	✓	✓						1
	......	.25	Bracket, Starting Switch (Part of 16UBL2443A).....	16UBL0-1449C	✓	✓								1
	......	.75	Bracket, Starting Switch, Assembly........	16URK2-443						✓				1
	......	.25	Bracket, Starting Switch (Part of 16UBL2443A).....	16URK2-1449						✓				1
	......	....	Bracket, Starting Switch, and Nut Assembly......	16UKS0-1449							✓	✓	✓	1
	......	....	Screw, Flat Head Machine (⅜"—16 x 1¼").....	S-4261A							✓	✓	✓	2
	......	....	Screw, Hex. Cap (¼"—20 x 1")..........	S-3118							✓	✓	✓	2
	......	....	Nut, Hex. (¼"—20)...............	S-76							✓	✓	✓	2
	......	.01	Lockwasher, 9/32" Spring............	S-1345							✓	✓	✓	4
	......	1.20	Cable, Battery to Starting Switch, Assembly......	16AA9-590	✓	✓	✓	✓						1
	......	2.05	Cable, Battery to Starting Switch, Assembly......	16AJ9-590						✓				1
	......	1.60	Cable, Starting Switch to Battery, Assembly (Positive)...	16UKS3-590							✓	✓	✓	1
	......	1.45	Cable, Starting Switch to Battery, Assembly (Negative)...	16UKSA3-590							✓	✓	✓	1
	......	.95	Breaker, Starting Switch Circuit (Klix-On 60 Amp.)...	16URK2-2649	✓	✓	✓	✓		✓	✓	✓	✓	2
	......	....	Screw, Phillips Flat Head Machine (#10—32 x ½")....	S-1632A	✓	✓	✓	✓	✓	✓	✓	✓	✓	2
	......	....	Nut, Hex. (#10—32)..............	S-3745	✓	✓	✓	✓		✓	✓	✓	✓	2
	......	....	Lockwasher, 9/16" Spring............	S-1344	✓	✓	✓	✓		✓	✓	✓	✓	2
	1455	3.25	Switch, Magnetic Starting (Delco-Remy)........	16UP0-730							✓	✓	✓	1
	......	....	Screw, Phillips Round Head Machine (¼"—20 x ¾")....	S-1614A							✓	✓	✓	2
	......	....	Nut, Hex. (¼"—20)...............	S-76							✓	✓	✓	2

CODE	MFR. PART No.	LIST PRICE	NAME OF PART	AUTOCAR PART No.	U-2044	U-4044	U-4144	U-4144-T	C-50	U-5044	U-7144-T	U-8144	U-8144-T	No. REQ.
0602	......	....	Lockwasher, 9/32" Spring	S-1345							✓	✓	✓	4
	......	$1.35	Cable, Magnetic Switch to Battery, Assembly	16UKSB3-590							✓	✓	✓	1
	......	1.65	Cable, Magnetic Switch to Motor, Assembly	16UKS3-580							✓	✓	✓	1
	Starting Motor													
	811230	....	Bushing, Bendix Housing	811230	✓									..
	810586	....	Lead, Brush & Field Connector	810586	✓									..
	813523	....	Lead, Brush & Field Connector	813523						✓				1
	813554	....	Lead, Ground Brush	813554	✓	✓	✓	✓		✓				2
	811450	....	Lead, Field Brush	811450	✓					✓				1
	810226	....	Spring, Brush	810226	✓	✓	✓	✓		✓				4
	809053	....	Bolt, Through	809053	✓	✓	✓	✓		✓				2
	114503	....	Nut, Field Terminal	114503	✓									..
	802691	....	Wick, Oil (C.E.)	802691	✓	✓	✓	✓		✓				1
	802694	....	Wick, Oil (D.E.)	802694		✓	✓	✓		✓				1
	809591	....	Wick, Oil—and Spring (C.E.)	809591	✓	✓								..
	817313	....	Pin, Brush Holder	817313	✓	✓	✓	✓						2
	817314	....	Pin, Brush Holder Stop	817314	✓	✓	✓	✓						2
	812016	....	Pin, Brush Holder—and Insulation	812016	✓	✓	✓	✓						2
	812015	....	Pin, Brush Holder—and Insulation	812015	✓	✓	✓	✓						
	809593	....	Pin, Dowel (D.E.)	809593	✓	✓	✓	✓		✓				1
	809062	....	Pin, Dowel (1/8—D.E. and C.E.)	809062	✓	✓	✓	✓						1
	115903	....	Screw, Brush to Holder	115903	✓	✓	✓	✓		✓				4
	135616	....	Screw, Brush Lead to Field	135616	✓	✓	✓	✓		✓				1
	809051	....	Washer, Field Terminal Stud	809051	✓	✓	✓	✓		✓				2
	817077	....	Washer, Field Terminal Stud	817077	✓	✓								..
	811388	....	Washer, Armature Shaft Spacer	811388	✓									..
	812496	....	Washer, Bendix Shaft Spacer	812496	✓									..
	812664	....	Washer, Bendix Shaft Spacer	812664	✓									..
	811451	....	Clip, Brush Lead Terminal	811451	✓	✓								..
	816453	....	Clip, Brush Lead Terminal	816453	✓	✓								..
	808933	....	Clip, Brush Lead Terminal	808933	✓	✓								..
	810824	....	Wedge, Field Terminal	810824	✓									..
	809595	....	Oiler	809595	✓									..
	114998	....	Oiler	114998		✓				✓				1
	1880642	....	Oiler	1880642			✓	✓						..
	103884	....	Plug, Oil Well (C.E.)	103884	✓									..
	819362	....	Lead, Brush Connector	819362	✓	✓	✓	✓						..
	1835455	....	Housing, Gear	1835455		✓	✓	✓		✓				1
	809663	....	Shaft, Drive	809663		✓	✓	✓						..
	827518	....	Plate, Drive Housing Cover	827518		✓	✓	✓		✓				1
	817070	....	Gear	817070		✓	✓	✓		✓				1
	810217	....	Bushing, Drive Housing	810217		✓	✓	✓						..
	103865	....	Plug, Oil Hole	103865		✓	✓	✓		✓				1
	103319	....	Lockwasher, Bendix Housing Screw	103319		✓	✓	✓		✓				3
	103319	....	Lockwasher, Gear Case Screw & Through Bolt	103319		✓	✓	✓		✓				2
	1874734	....	Screw, Bendix Housing	1874734		✓	✓	✓		✓				3
	135616	....	Screw, Brush Connector Lead	135616		✓	✓	✓						..
	813134	....	Washer, Felt (D.E.)	813134		✓	✓	✓		✓				1
	817077	....	Washer, Field Terminal Stud	817077		✓	✓	✓						..
	809817	....	Washer, Oil Thrower	809817		✓	✓	✓		✓				1
	1865182	....	Cap, Dust (1 1/8 O.D. & C.E.)	1865182		✓	✓	✓						..
	1881869	....	Screw, Pole Piece	1881869			✓	✓						..
	805790	....	Washer, Field Terminal Stud	805790			✓	✓		✓				1
	1861076	....	Washer, Field Terminal Stud	1861076			✓	✓		✓				2
	805727	....	Washer, Spacer (Between Housing & Gear)	805727			✓	✓		✓				1
	1865182	....	Plug, End (C.E.)	1865182						✓				1
	833602	....	Washer, Spacer (C.E.)	833602						✓				1
	114998	....	Oiler (in Gear Housing)	114998						✓				1
	802691	....	Wick, Oil (in Gear Housing)	802691						✓				1
	826462	....	Bushing, Housing	826462						✓				1
	828846	....	Shaft, Motor Drive	828846						✓				1

06-6 ELECTRICAL — Master Parts List — THE AUTOCAR COMPANY, ARDMORE, PA.

CODE	MFR. PART No.	LIST PRICE	NAME OF PART	AUTOCAR PART No.	U-2044	U-4044	U-4144	U-4144-T	C-50	U-5044	U-7144-T	U-8144	U-8144-T	No. REQ.
0602	809815	....	Washer, Spacer (Between Gear & Motor Drive)	809815						✓				2
	1849774	....	Washer, Spacer (.626 I.D. x 1 1/16" O.D. x 3/32")	1849774						✓				1
0603	**Distributor**													
	1110513	$6.50	Distributor and Gear Assembly (Delco)	8UG0-470	✓									1
	1110097	9.00	Distributor and Gear Assembly (Delco)	8UK0-470		✓	✓	✓						1
	1110043	8.00	Distributor and Gear Assembly	8RLA3-470					✓					1
	1110041	9.00	Distributor and Gear Assembly (Delco)	8URL4-470						✓				1
	1GC-4701-1	8.85	Distributor Assembly (Auto-Lite)	8UU3-470							✓	✓	✓	1
	1G-90	.01	Washer, Thrust	8BL0-528							✓	✓	✓	1
	......	.35	Coupling, Distributor Drive (Grooved)	8URL2-539	✓	✓	✓	✓		✓				1
	1G-644A	.50	Coupling, Distributor Drive	8UU0-327							✓	✓	✓	1
	1G-680	.03	Ring, Lock Spring	8UU0-469							✓	✓	✓	1
	1G-816A	.01	Washer, Thrust	8BL0-527							✓	✓	✓	1
	1838100	.65	Cap, Distributor	8BLA0-69	✓									1
	824987	1.35	Cap, Distributor	8B0-69		✓	✓	✓		✓				1
	824735	.75	Cap, Distributor	8SA0-69A					✓					1
	1GC-1107S	1.15	Cap, Distributor, Assembly	8UU0-69							✓	✓	✓	1
	1G-514	.03	Plunger, Contact Spring	8UU0-249							✓	✓	✓	1
	1G-515	.01	Spring, Contact	8UU0-28							✓	✓	✓	1
	1GC-2130RB	1.25	Drive Shaft and Governor Assembly	8UU0-590							✓	✓	✓	1
	821596	.10	Weight	8BLA0-230	✓									2
	818222	.20	Weight	8SA0-230B		✓	✓	✓	✓	✓				2
	1G-1322	.25	Weight Assembly	8UU0-230							✓	✓	✓	2
	1882978	....	Spring, Weight	1882978	✓									2
	822172	.05	Spring, Weight	8N0-283		✓	✓	✓	✓	✓				2
	811124	.01	Washer, Weight	8SA0-322	✓	✓	✓	✓	✓	✓				2
	1GC-1130R	1.00	Shaft, Drive, Assembly	8UU0-220							✓	✓	✓	1
	1GE-29S	.26	Spring, Weight, Set	8UU0-283							✓	✓	✓	1
	826474	.30	Plate, Breaker	8BLA0-317	✓									1
	821150	.75	Plate, Breaker	8SA0-317		✓	✓	✓	✓	✓				1
	1GC-2148C	1.55	Plate, Breaker, Assembly	8UU0-317							✓	✓	✓	1
	813245	.01	Clip, Breaker Lever Retainer	8SA0-321		✓	✓	✓	✓	✓				1
	1G-676	.02	Clip, Breaker Arm Spring	8UU0-321							✓	✓	✓	1
	1GC-1148	.25	Plate, Breaker, Part Assembly	8UUA0-317							✓	✓	✓	1
	813238	.40	Lever, Breaker, and Point	8SA0-57	✓	✓	✓	✓	✓	✓				1
	1GP-3028FS	.85	Arm, Breaker Point and, Assembly	8UU0-57							✓	✓	✓	1
	1GP-30	.05	Spring, Breaker Arm	8UUA0-283							✓	✓	✓	1
	1869706	.40	Condenser	8BLA0-288	✓									1
	1869704	.45	Condenser	8SA0-288		✓	✓	✓	✓	✓				1
	1GW-3139	.60	Condenser Assembly	8UU0-288							✓	✓	✓	1
	8X-181	.05	Washer, #6 Brass	S-726							✓	✓	✓	1
	8X-302	.05	Screw, #6—32 x 5/16" Hex. Head	8UU0-548							✓	✓	✓	1
	8X-350	.01	Washer, #8 Plain	S-987							✓	✓	✓	1
	8X-884	.05	Screw, #8—32 x 3/16" Fill. Hd.	8UU0-549							✓	✓	✓	1
	8X-1012	.01	Washer, #6 Lock	S-1343							✓	✓	✓	2
	8X-1546	.05	Screw, #6—32 x 5/32" Fill. Hd.	8UU0-551							✓	✓	✓	1
	816803	.05	Washer, Felt	8BLA0-325	✓	✓	✓	✓		✓				1
	IGH-28	.01	Wick, Cam Sleeve Felt	8UU0-325							✓	✓	✓	1
	IGS-99	.02	Washer	8UU0-531							✓	✓	✓	1
	IGS-104	.01	Washer, Thrust	8UUA0-527							✓	✓	✓	1
	816774	.20	Rotor	8BLA0-35	✓									1
	820445	.20	Rotor	8SA0-35		✓	✓	✓	✓	✓				1
	IGS-1016B	.35	Rotor	8UU0-35							✓	✓	✓	1
	820989	1.00	Cam	8BLA0-210	✓									1
	824109	1.00	Cam	824109		✓	✓	✓						1
	822627	1.00	Cam	8SA0-210A					✓					1
	823236	1.35	Cam	8B0-210						✓				1
	IGS-1100RR	1.20	Plate, Cam and Stop	8UU0-210							✓	✓	✓	1
	......	.30	Arm, Advance Control, Assembly	8BL2-311	✓									1
	815877	.35	Arm, Distributor Clamp, and Dial Assembly	8RLA0-311		✓	✓	✓	✓	✓				1
	IGS-2020-1	.70	Arm, Advance Control, Assembly	8UU0-311							✓	✓	✓	1

THE AUTOCAR COMPANY
ARDMORE, PA.

Master Parts List

06-7
ELECTRICAL

CODE	MFR. PART NO.	LIST PRICE	NAME OF PART	AUTOCAR PART No.	MODEL									NO. REQ.
					U-2044	U-4044	U-4144	U-4144-T	C-50	U-5044	U-7144-T	U-8144	U-8144-T	
0603	820668	$1.50	Housing, Distributor	8BLA0-260	✓									1
	1840537	2.00	Housing, Distributor	8SA0-260A		✓	✓	✓	✓	✓				1
	IGS-2062	2.00	Base, Distributor, Assembly	8UU0-260							✓	✓	✓	1
	IG-579A	.16	Bearing, Absorbent Bronze	8BL0-455							✓	✓	✓	2
	816801	.05	Spring, Distributor Cap	8BLA0-28	✓									2
	1871838	.05	Spring, Distributor Cap	8B0-28		✓	✓	✓	✓	✓				2
	IG-694	.03	Spring, Distributor Cap	8UU0-306							✓	✓	✓	2
	805579	.20	Cup, Grease	8SA0-287	✓	✓	✓	✓	✓	✓				1
	X-490	.06	Oiler, Press in Sleeve	8BL0-287							✓	✓	✓	1
	X-1448	.05	Pin, Hinge	8UU0-316							✓	✓	✓	2
	SW-213	.01	Rivet, Distributor Drive	8UU0-328							✓	✓	✓	1
	X-196	.01	Lockwasher, #10	S-1344							✓	✓	✓	3
	8X-870	.01	Screw, #10—32 x 5/16" Fill. Hd.	S-4031							✓	✓	✓	2
	8X-872	.05	Screw, #10—32 x 1/4" Fill. Hd.	8UU0-552							✓	✓	✓	1
	1842541	1.00	Gear, Distributor	2BLA0-855	✓									1
	1110043	1.00	Gear, Distributor	2T2-855					✓					1
	811912	.05	Washer, Distributor Gear Spacer	8SA0-324	✓									1
	1857492	.01	Pin, Distributor Gear	1857492	✓									1
	815003	.05	Pin, Distributor Gear	8SA0-328					✓					1
	528-C	2.60	Coil, Ignition	8SA2-120A	✓	✓	✓	✓		✓				1
	826476	.05	Support, Distributor Cap Spring	8BLA0-329	✓									2
	1847289	.05	Support, Distributor Cap Spring	8B0-329		✓	✓	✓	✓	✓				2
	115607	.01	Screw, Breaker Plate and Cap Spring Support	8SCM0-302A	✓	✓	✓	✓		✓				3
	1836591	1.50	Plate, Mainshaft and Weight	1836591	✓									1
	1842523	1.50	Plate, Mainshaft and Weight	8B0-220					✓					1
	816784	.05	Screw, Contact Support Fastening	8SA0-318	✓	✓	✓	✓		✓				1
	1847341	.25	Support, Contact Point and	8BLA0-09	✓									1
	1848038	.25	Support, Contact Point and	8SA0-09		✓	✓	✓	✓	✓				1
	115417	.01	Screw, Condenser Attaching	8BLA0-302	✓									1
	131951	.01	Screw, Condenser Attaching	8SA0-302		✓	✓	✓	✓	✓				1
	810085	.05	Washer, Spacer (Under Weight Plate)	8BLA0-324	✓									1
	811912	.05	Washer, Spacer (Under Weight Plate)	8SA0-324		✓	✓	✓	✓	✓				1
	820987	.03	Screw, Terminal	6BLA0-472	✓									1
	107715	.01	Screw, Terminal	8B0-303		✓	✓	✓	✓	✓				1
	120622	.02	Nut, Terminal Screw	8BLA0-526	✓									2
	121841	.01	Lockwasher, Terminal Screw	8NFS0-465		✓	✓	✓		✓				1
	821033	.05	Washer, Terminal Screw Bushing	8BLA0-524	✓									1
	821160	.15	Bushing, Terminal Screw	8SA0-326		✓	✓	✓	✓	✓				1
	807716	.02	Washer, Terminal Screw	8BLA0-464	✓									1
	26153	.05	Clamp, Terminal	8SA0-304		✓	✓	✓	✓	✓				1
	816806	.05	Washer, Terminal Screw Insulation	8BLA0-465	✓									1
	820988	.05	Washer, Terminal Screw Insulation	8BLA0-466	✓									1
	820991	.05	Strip, Terminal Screw Insulation	8BLA0-327	✓									1
	811074	.02	Washer, Shim—.005 Thick	8SA0-294	✓									As
	810078	.02	Washer, Shim—.010 Thick	8SA0-295					✓					As
	810074	....	Washer, Shim—.005 Thick	810074					✓					As
	IGB-1171	.25	Arm, Advance Control	8BLA0-284	✓									1
	IG-750	.07	Screw, Advance Control	IG-750	✓									1
	IG-687	.02	Washer, Advance Control Screw Spring	IG-687	✓									1
	IG-688	.01	Washer, Advance Control Screw—Top	IG-688	✓									1
	IG-688A	.01	Washer, Advance Control Screw—Bottom	IG-688A	✓									1
	8X-163	.05	Nut, Advance Control Screw	8X-163	✓									1
	8X-707	.05	Screw, Advance Control Arm Clamp	8X-707	✓									1
	8X-146	.05	Nut, Advance Control Arm Clamp Screw	8X-146	✓									1
	8X-870	.05	Screw, Advance Control Arm	8X-870	✓									1
	809673	.35	Coupling, Distributor	8SA0-327		✓	✓	✓		✓				1
	811912	.05	Washer, Distributor Coupling	8SA0-324		✓	✓	✓		✓				1
	810074	.02	Shim, Distributor Coupling—.005	8SA0-294		✓	✓	✓		✓				As
	810078	.02	Shim, Distributor Coupling—.010	8SA0-295		✓	✓	✓		✓				As
	1857492	....	Pin, Distributor Coupling	1857492		✓	✓	✓						1
	810068	.05	Pin, Distributor Coupling	8BLA0-328						✓				1
	824728	....	Cover, Distributor Cap	824728		✓	✓	✓						1

06-8 ELECTRICAL — Master Parts List

THE AUTOCAR COMPANY, ARDMORE, PA.

CODE	MFR. PART No.	LIST PRICE	NAME OF PART	AUTOCAR PART No.	U-2044	U-4044	U-4144	U-4144-T	C-50	U-5044	U-7144-T	U-8144	U-8144-T	No. REQ.
0603	820133	$0.35	Cover, Distributor Cap	8A0-68						✓				1
	803912	.02	Screw, Distributor Cap Cover	8A0-66		✓	✓	✓		✓				1
	115434	.01	Screw, Distributor Cap Spring Support	8SCM0-315A		✓	✓	✓		✓				2
	107710	....	Screw, Distributor Cap Spring Support	107710					✓					2
	1844542	....	Plate, Mainshaft and Weight	1844542		✓	✓	✓						1
	1842523	....	Plate, Mainshaft and Weight	1842523						✓				1
	......	3.00	Adapter, Distributor	8URL3-541		✓	✓	✓						1
	S-659-V	12.50	Adapter, Tachometer (Stewart-Warner)	16A0-1230						✓				1
	......	.04	Screw, Distributor Adapter (¼"—20 x ⅝")	S-36		✓	✓	✓		✓				1
	......	.01	Washer, ¼" Spring	S-1345		✓	✓	✓		✓				1
	......	.30	Cup, Grease	S-683		✓	✓	✓		✓				1
	......	.01	Pin, Distributor Drive Coupling	S-3253		✓	✓	✓		✓				1
	......	.15	Shaft, Distributor Drive	8URL2-537		✓				✓				1
	......	....	Shaft, Distributor Drive	8UG2-537			✓	✓						1
	......	.05	Shim, Distributor Drive Shaft—.030	8URL2-538		✓	✓	✓		✓				1
	......	1.00	Gear, Distributor Drive	2T2-855		✓	✓	✓		✓				1
	......	....	Pin, Distributor Drive Gear	S-1499		✓	✓	✓		✓				1
	815877	....	Bolt, Distributor Clamp Arm and, Assembly	8A0-62		✓	✓	✓		✓				1
	815095	....	Dial, Distributor Clamp Arm	8A0-83		✓	✓	✓		✓				1
	......	.04	Screw, Distributor Clamp Arm Hold Down	S-30		✓	✓	✓		✓				1
	......	.01	Washer, Spring	S-1345		✓	✓	✓		✓				1
	......	.06	Hook, Ignition Wire	8TE0-18	✓									6
	......	.12	Wire, Ignition	8TE0-449	✓									As
	......	.30	Bracket, Coil Mounting	2UK3-1111	✓									1
	......	.12	Bracket, Coil Mounting	2UG2-1111		✓	✓	✓						1
	......	....	Bracket, Coil Mounting	2UP2-1111						✓				1
	......	.10	Support, Ignition Wires	8UBA2-308		✓	✓	✓						3
	......	.06	Grommet, Ignition Wires Support	8T2-309		✓	✓	✓						3
	813511	....	Spring, Breaker Lever—Fastening	813511						✓				1
	819639	.05	Wick, Felt	8SA0-325						✓				1
	815097	....	Screw, Advance Arm Dial	815097						✓				1
	1847260	....	Screw	1847260						✓				1
	813045	....	Spring	813045						✓				1
	818194	....	Washer	818194						✓				1
0604			**Ignition Coil, Wiring, Plugs and Lock**											
	......	2.90	Cables, Ignition—Assembled	8BN0-610	✓									1
	......	3.20	Wires, Ignition—Assembly	8URL0-610		✓	✓	✓						1
	......	.30	Wire, High Tension—Assembly	8UD0-14A							✓	✓	✓	1
	......	.70	Plug, Spark (Champion H-10)	8D0-23	✓									6
	......	.70	Plug, Spark (Champion #8)	8CB0-23A		✓	✓	✓						6
	......	.70	Plug, Spark (Champion J-10)	8E0-23							✓	✓	✓	6
	IG-4070-H	2.65	Coil (Auto-Lite)	8BLA0-120							✓	✓	✓	1
	......	.06	Nipple, Ignition Wire	8A0-179							✓	✓	✓	6
	......	.25	Wire, Ignition—#1	8-0300-24A							✓	✓	✓	1
	......	.30	Wire, Ignition—#2	8-0300-29A							✓	✓	✓	1
	......	.35	Wire, Ignition—#3	8-0300-34A							✓	✓	✓	1
	......	.40	Wire, Ignition—#4	8-0300-42A							✓	✓	✓	1
	......	.45	Wire, Ignition—#5	8-0300-49A							✓	✓	✓	1
	......	.45	Wire, Ignition—#6	8-0300-50A							✓	✓	✓	1
	......	.10	Plate, Ignition Wire Support	8UBA2-308							✓	✓	✓	3
	......	.40	Plate, Ignition Wire Support	8UU3-308							✓	✓	✓	1
	......	.40	Plate, Ignition Wire Support	8UUA3-308							✓	✓	✓	1
	......	.06	Grommet	8T2-309							✓	✓	✓	7
	......	.15	Wire, Jumper—Coil to Distributor	16BB9-623							✓	✓	✓	1
	......	1.50	Lock, Ignition Switch and	16URK2-420							✓	✓	✓	1
	......	....	Key, Ignition	16UKS0-1113							✓	✓	✓	2

THE AUTOCAR COMPANY
ARDMORE, PA.

Master Parts List

06-9 ELECTRICAL

CODE	MFR. PART No.	LIST PRICE	NAME OF PART	AUTOCAR PART No.	U-2044	U-4044	U-4144	U-4144-T	C-50	U-5044	U-7144-T	U-8144	U-8144-T	No. REQ.
0605	**Instruments**													
	NPN	$40.00	Tachometer (Stewart-Warner)............	16UKS3-1210							✓	✓	✓	1
	90707	1.50	Ammeter................	16RL0-630	✓	✓	✓	✓	✓					1
	400056	1.65	Ammeter (Stewart-Warner)...........	16UKS3-630							✓	✓	✓	1
	......	2.00	Ammeter, Auxiliary (Stewart-Warner).........	16UKSA3-630							✓	✓	✓	1
	93489	2.00	Gauge, Gas.............	16RL0-775	✓	✓	✓	✓	✓					1
	441002	3.60	Gauge, Gas.............	16UKS3-775							✓	✓	✓	1
	93619	4.00	Indicator, Heat..........	16RL0-2040	✓	✓	✓	✓	✓					1
	400059	4.00	Indicator, Heat (Stewart-Warner).........	16UKS3-2040							✓	✓	✓	1
	96808	1.50	Gauge, Oil.............	16RL0-564	✓	✓	✓	✓	✓					1
	400060	4.00	Gauge, Oil (Stewart-Warner)...........	16UKS3-564							✓	✓	✓	1
	584AM	18.00	Speedometer Assembly (Stewart-Warner).........	16UB3-760	✓	✓	✓	✓	✓					1
	590-X	12.00	Speedometer Assembly (Stewart-Warner).........	16UKS3-760							✓	✓	✓	1
	......	5.70	Viscometer (Visco-Meter Corp.).......	16UKS3-1190							✓	✓	✓	1
	......	4.15	Gauge, Air (Stewart-Warner).........	25UKS3-32							✓	✓	✓	1
0606	**Light Switches, Cables, Wiring Harness**													
	7132	1.70	Switch, Blackout (Cole-Hersee).........	16URKB3-1241	✓	✓	✓	✓	✓					1
	1994511	2.75	Switch, Blackout (Delco-Remy).........	16UKS4-1241							✓	✓	✓	1
	1990544	.30	Knob, Blackout Switch........	16W2-2731							✓	✓	✓	1
	1990557	.85	Bushing, Blackout Switch........	16WB3-2732							✓	✓	✓	1
	1997772	.45	Switch, Instrument Panel Light (Delco-Remy).........	16UKS3-2672							✓	✓	✓	1
	1990550	.35	Knob, Instrument Panel Light Switch..........	16WA2-2731							✓	✓	✓	1
	1884374	.15	Bushing, Instrument Panel Light Switch..........	16WA2-2732							✓	✓	✓	1
	5530	.30	Switch, Foot Dimmer (H. A. Douglas).......	16RL2-699A	✓	✓			✓					1
	1997007	.75	Switch, Foot Dimmer (Delco-Remy).........	16W3-699							✓	✓	✓	1
	......	....	Screw, Rd. Hd. Mach. (¼"—20 x ½" Cad.)......	S-6140							✓	✓	✓	2
	......	.01	Washer, 9/32" Spring.........	S-1345							✓	✓	✓	2
	......	.01	Screw, Rd. Hd. Mach. (8"—32 x ⅜" Br.)........	S-3139							✓	✓	✓	3
	......	....	Screw, Wood—#8 L.W.	S-329							✓	✓	✓	3
	215537	3.50	Switch, Stop Light (Westinghouse)......	25NT2-130	✓	✓	✓	✓	✓		✓	✓	✓	1
	......	.04	Screw, Hex. Cap (¼"—20 x 1")........	S-3118							✓	✓	✓	2
	......	.01	Nut, Hex. (¼"—20)........	S-76							✓	✓	✓	2
	......	.01	Lockwasher, 9.32" Spring........	S-1345							✓	✓	✓	4
	1997007	.75	Switch, Foot Dimmer (Delco-Remy).........	16W2-699			✓	✓						1
	SP-2188	.30	Block, Junction—7 Pole (Auto-Lite)......	16UKS3-2676							✓	✓	✓	1
	......	.03	Nut, Rd. Hd. Stove Bolt and (3/16 x 1")......	S-2966A							✓	✓	✓	2
	......	.01	Lockwasher, 7/32" Spring.......	S-1344							✓	✓	✓	4
	......	2.65	Block, Junction—to Inst. Panel Cable Assembly........	16UKS3-740							✓	✓	✓	1
	......	18.25	Cable, Trailer Lighting—Assembly (Warner Elec.).....	16URK3-1570							✓	✓	✓	1
0607	**Headlamps**													
	1004-E	8.00	Headlamp Assembly—Complete 6-V........	16UBK4-460A	✓	✓								2
	1011-E	....	Headlamp Assembly.............	16URKC4-460			✓	✓						2
	1004-E	....	Headlamp Assembly.............	16URKA4-460A						✓				2
	1011-J	9.00	Headlamp Assembly (Guide)........	16UKS4-460							✓	✓	✓	2
	5933524	3.25	Body Assembly—Prime.............	5933524							✓	✓	✓	2
	5931167	1.00	Molding Assembly—Prime...........	5931167							✓	✓	✓	2
	922690	.05	Screw, Molding........	922690							✓	✓	✓	2
	5933522	.85	Connector and Wiring Assembly........	5933522							✓	✓	✓	2
	5932572	.10	Loom—Wire Cover..........	5932572							✓	✓	✓	2
	5930502	.05	Clip—Loom..........	5930502							✓	✓	✓	2
	1878192	.05	Terminal—Douglas........	1878192							✓	✓	✓	4
	1867662	.05	Terminal—Eyelet........	1867662							✓	✓	✓	2
	925000	1.10	Unit, Sealed—6-V........	925000							✓	✓	✓	2
	5931243	.50	Ring, Retaining........	5931243							✓	✓	✓	2
	924552	.05	Screw, Retaining Ring........	924552							✓	✓	✓	6
	5930487	.05	Grommet...........	5930487							✓	✓	✓	2
	......	1.30	Headlamp to Dimmer Switch Cable Assembly......	16UKS3-480							✓	✓	✓	1
	5503-A	3.25	Lamp, Blackout Parking—Assembly........	16UKS3-3050							✓	✓	✓	2
	......	.11	Bulb, Blackout Parking Lamp—#64..........	16GE0-472							✓	✓	✓	2
	5932519	....	Body Assembly...............	16URKB0-2628	✓	✓					✓	✓	✓	2

06-10 ELECTRICAL — Master Parts List
THE AUTOCAR COMPANY, ARDMORE, PA.

CODE	MFR. PART No.	LIST PRICE	NAME OF PART	AUTOCAR PART No.	U-2044	U-4044	U-4144	U-4144-T	C-50	U-5044	U-7144-T	U-8144	U-8144-T	No. REQ.
0607	5932520	....	Body, Headlamp—Assembly	16URKB0-988			✓	✓		✓				2
	5933156	$1.40	Body Assembly	5933156							✓	✓	✓	2
	5932381	....	Door Assembly—Less Lens	16URKB0-2629	✓	✓				✓				2
	5933122	1.60	Door Assembly	5933122							✓	✓	✓	2
	5932419	.05	Screw, Door	5932419							✓	✓	✓	2
	5903375	.15	Gasket, Door	5903375							✓	✓	✓	2
	1878192	.05	Terminal, Wire	1878192							✓	✓	✓	2
	925000	1.10	Lamp, Head—Sealed Beam Unit—6-Volt	16URK0-2648	✓	✓	✓	✓		✓				2
	5932520	....	Body, Headlamp—Assembly	16URKB0-988	✓	✓								2
	5931167	1.00	Molding, Headlamp—Assembly	16URKA0-365	✓	✓	✓	✓						2
	5931243	.35	Ring, Headlamp Retaining	16URKA0-286	✓	✓	✓	✓						2
	5932164	....	Gasket, Door	16URKB0-2632	✓	✓								2
0608	**Tail Lamp**													
	3006-A	3.75	Lamp, Blackout Tail—L. H.	16URKA3-400A	✓	✓				✓				1
	3012-A	3.75	Light, Blackout Tail and Stop—L. H. (Painted Yellow Green)	16URKE4-400			✓	✓						1
	3012-A	3.75	Light, Blackout Tail and Stop—L. H. (Painted Olive Drab)	16URKG4-400			✓	✓						1
	3012-A	3.75	Light, Blackout Tail and Stop, Assembly—L. H.	16URKC4-400							✓	✓	✓	1
	5933078	....	Lamp, Tail, Unit Assembly	5933078							✓	✓	✓	1
	5933104	....	Lamp, Stop, Unit Assembly	5933104							✓	✓	✓	1
	5933231	....	Body Assembly	5933231							✓	✓	✓	1
	5933056	....	Door Assembly	5933056							✓	✓	✓	1
	5933069	....	Screw, Door	5933069							✓	✓	✓	2
	3006-B	3.75	Lamp, Blackout Tail—R. H.	16URK3-400A	✓	✓				✓				1
	3012-B	3.75	Lamp, Blackout Tail & Stop—R. H. (Painted Yellow Green)	16URKD4-400			✓	✓						1
	3012-B	3.75	Light, Blackout Tail & Stop—R. H. (Painted Olive Drab)	16URKF4-400			✓	✓						1
	3012-B	3.75	Lamp, Blackout Tail and Stop, Assembly—R. H.	16URKB4-400							✓	✓	✓	1
	5933078	....	Lamp, Tail, Unit Assembly	5933078							✓	✓	✓	1
	5933121	....	Lamp, Stop, Unit Assembly	5933121							✓	✓	✓	1
	5933231	....	Body Assembly	5933231							✓	✓	✓	1
	5933055	....	Door Assembly	5933055							✓	✓	✓	1
	5933069	....	Screw, Door	5933069							✓	✓	✓	1
	......	....	Cable, Tail Lamp, Assembly	16UBK3-510A	✓									1
	......	3.35	Cable, Tail Light, Assembly	16URK3-510A		✓	✓	✓		✓				1
	......	6.50	Cable, Tail Lamp, Assembly (Chassis)	16UKS4-510							✓	✓	✓	1
	......	....	Reflector, Red (K-D Lamp)	16UKS9-1422							✓	✓	✓	4
	......	.02	Nut, Ph. Rd. Hd. Stove Bolt and (¼" x ¾")	S-2942A							✓	✓	✓	8
	......	1.00	Reflector, Amber (K-D Lamp)	16UKSA9-1422							✓	✓	✓	2
	......	.02	Nut Ph. Rd. Hd. Stove Bolt and (¼" x ¾")	S-2942A							✓	✓	✓	..
0609	**Horn**													
	EA-44	6.25	Horn, Electric, Assembly (Dual) (E. A. Lab.)	16URK3-2740	✓	✓	✓	✓		✓				1
	1880391	10.50	Horn, Dual Electric, Assembly (Delco)	16UKS3-2740							✓	✓	✓	1
	......	.04	Screw, Hex. Cap (¼"—20 x ¾")	S-30							✓	✓	✓	4
	......	.01	Nut, Hex. (¼"—20)	S-76							✓	✓	✓	4
	......	.01	Washer, 9/32" Spring	S-1345							✓	✓	✓	8
	EA-93	....	Relay, Electric Horn (E. A. Lab.)	EA-93	✓	✓	✓	✓						1
	......	....	Relay, Electric Horn (In Horn)	......						✓				1
	1116818	....	Relay, Electric Horn (Delco)	16AW2-2069							✓	✓	✓	1
	......	.01	Screw, Ph. Rd. Hd. Mach. Screw (¼"—20 x ¾")	S-1614A							✓	✓	✓	2
	......	.01	Nut, Hex. (¼"—20)	S-76							✓	✓	✓	2
	......	.01	Lockwasher, 9/32" Spring	S-1345							✓	✓	✓	4
	......	.95	Breaker, Klix-on Automatic Reset Circuit	16URK2-2649	✓	✓	✓	✓		✓				1
	......	4.30	Breaker, Horn Circuit (Klix-on 30-Amp.)	16UKS2-2649							✓	✓	✓	1
0610	**Battery, Starting Cables and Connections**													
	......	24.45	Battery 6-V (Exide)	16TE9-440	✓	✓	✓	✓						1
	XH-13	18.25	Battery 6-V (Exide)	16BL2-440						✓				2
	XH-194	....	Battery (Exide)	16UU0-440							✓	✓	✓	2
	......	4.20	Support, Battery, Assembly	16URK3-1110	✓	✓	✓	✓						1
	......	....	Support, Battery, Assembly	16URK3-1110A						✓				1
	......	4.50	Support, Battery, Assembly	16UKS3-1110							✓	✓	✓	1

THE AUTOCAR COMPANY
ARDMORE, PA.

Master Parts List

06-11
ELECTRICAL

CODE	MFR. PART No.	LIST PRICE	NAME OF PART	AUTOCAR PART No.	U-2044	U-4044	U-4144	U-4144-T	C-50	U-5044	U-7144-T	U-8144	U-8144-T	No. REQ.
0610	......	$0.01	Screw, Hex. Cap (¼″—20 x 1″)................	S-1607A							✓	✓	✓	4
	......	.01	Nut, Hex. (¼″—20)............................	S-76							✓	✓	✓	4
	......	.01	Lockwasher, ¼″ Spring.......................	S-1345							✓	✓	✓	8
	......	....	Frame, Battery Hold Down, Assembly........	16SDL3-1460	✓	✓	✓	✓						1
	......	....	Frame, Battery Hold Down, Assembly........	16URKA3-1460						✓				1
	......	2.00	Frame, Battery Hold Down, Assembly........	16UKS3-1460							✓	✓	✓	1
	......	.03	Nut, Hex. (⅜″—16)............................	S-75							✓	✓	✓	3
	......	.01	Lockwasher, ⅜″ Spring.......................	S-1347							✓	✓	✓	3
	......	....	Side, Battery Shield, Assembly...............	16URK3-1520	✓	✓	✓	✓						1
	......	....	Shield, Battery, Assembly....................	16UKS3-1520							✓	✓	✓	1
	......	.01	Screw, Ph. Rd. Hd. Mach. (¼″—20 x ¾″)......	S-1614A							✓	✓	✓	6
	......	.01	Lockwasher, ¼″ Spring.......................	S-1345							✓	✓	✓	6
	......	....	Lid, Battery...................................	16UKS2-497							✓	✓	✓	1
	......	.01	Washer, Fl. St. (5⁄16″ x ¾″).....................	S-395							✓	✓	✓	2
	......	.01	Lockwasher, ¼″ Spring.......................	S-1345							✓	✓	✓	2
	......	1.65	Cable, Battery Ground, Assembly (Positive)..	16AK9-50	✓	✓								1
	......	1.80	Cable, Battery Ground, Assembly (Positive)..	16AN9-50			✓	✓						1
	......	....	Cable, Battery Ground, Assembly (Positive)..	16UAP9-50						✓				1
	......	1.80	Cable, Battery Ground, Assembly............	16UKS3-50							✓	✓	✓	1
	......	2.50	Battery to Instrument Panel Cable Assembly.	16UKS3-1791							✓	✓	✓	1
0611	Radio Suppression													
	......	....	Suppressors, Spark Plug......................	8UKS2-506							✓	✓	✓	6
	......	....	Suppressor, Distributor.......................	8UKS2-508							✓	✓	✓	1
	......	....	Bracket (Weld to Exhaust Pipe)..............	10E2-303							✓	✓	✓	1
	......	....	Clip Loom (Tinned)...........................	16UC2-44							✓	✓	✓	1
	......	....	Regulator to Generator Cable Assembly.....	16UKS3-520A							✓	✓	✓	1
	......	....	Regulator to Filter Jumper Assembly.........	16UKSS2-623							✓	✓	✓	1
	......	....	Filter to Ammeter Jumper Assembly.........	16UKST2-623							✓	✓	✓	1
	......	....	Switch, Ignition to Filter Jumper Assembly..	16UKSU2-623							✓	✓	✓	1
	......	....	Filter to Coil Jumper Assembly...............	16UKSV2-623							✓	✓	✓	1
	......	....	Grommet (For Regulator Cable).............	16SA2-724							✓	✓	✓	1
	......	....	Condenser, Generator........................	16UKS2-2124							✓	✓	✓	1
	......	....	Condenser, Ignition Coil......................	16UKS2-2125							✓	✓	✓	1
	......	....	Filter, Two Unit, Assembly....................	16UKS3-2768							✓	✓	✓	1
	......	....	Filter, Field, Assembly........................	16UKSA2-2768							✓	✓	✓	1
	......	....	Bond (Cab to Frame).........................	16UU2-2845							✓	✓	✓	2
	......	....	Bond (R. H. Engine Support to Frame)......	16UUB2-2845							✓	✓	✓	1
	......	....	Bond (2—Cab to Cab Support Rear) (2—Body to Cab Sup'rt)	16UUC2-2845							✓	✓	✓	4
	......	....	Bond (Exhaust Pipe to Frame)...............	16UUD2-2845							✓	✓	✓	1
	......	....	Bond (Splash Guard to Frame)..............	16UUE2-2845							✓	✓	✓	2
	......	....	Bond (Crankcase to Frame)..................	16UUF2-2845							✓	✓	✓	1
	......	....	Bond (Radiator to Cross Member)...........	16UUG2-2845							✓	✓	✓	3
	......	....	Bond (Steering Column Bracket).............	16UUH2-2845							✓	✓	✓	1
	......	....	Screw, Hex. Cap (¼″—28 x 1¼″ Cad.)........	S-7437							✓	✓	✓	24
	......	....	Nut (Hex. (¼″—28 Cad.)......................	S-2023							✓	✓	✓	24
	......	....	Screw, Hex. Cap (⅜″—16 x 1¼″ Cad.).......	S-7438							✓	✓	✓	1
	......	....	Screw, Hex. Cap (⅜″—16 x 1¾″ Cad.).......	S-2057							✓	✓	✓	2
	......	....	Nut, Hex. (⅜″—16 Cad.).....................	S-441							✓	✓	✓	2
	......	....	Washer, ¼″ Shakeproof Ext. (Cad.)..........	S-3298							✓	✓	✓	47
	......	....	Washer, ⅜″ Shakeproof Ext. (Cad.)..........	S-3299							✓	✓	✓	3
	......	....	Washer, ½″ Shakeproof Ext. (Cad.)..........	S-3300							✓	✓	✓	5
	......	....	Washer, ¼″ Shakeproof Int. (Cad.)..........	S-3296							✓	✓	✓	30
	......	....	Washer, ½″ Shakeproof Int. (Cad.)..........	S-3295							✓	✓	✓	2

ILLUSTRATIONS

GROUP 07

TRANSMISSION

GROUP 07

Master Parts List

THE AUTOCAR COMPANY, ARDMORE, PA.

As supplied on Autocar Models
U-8144   U-8144-T

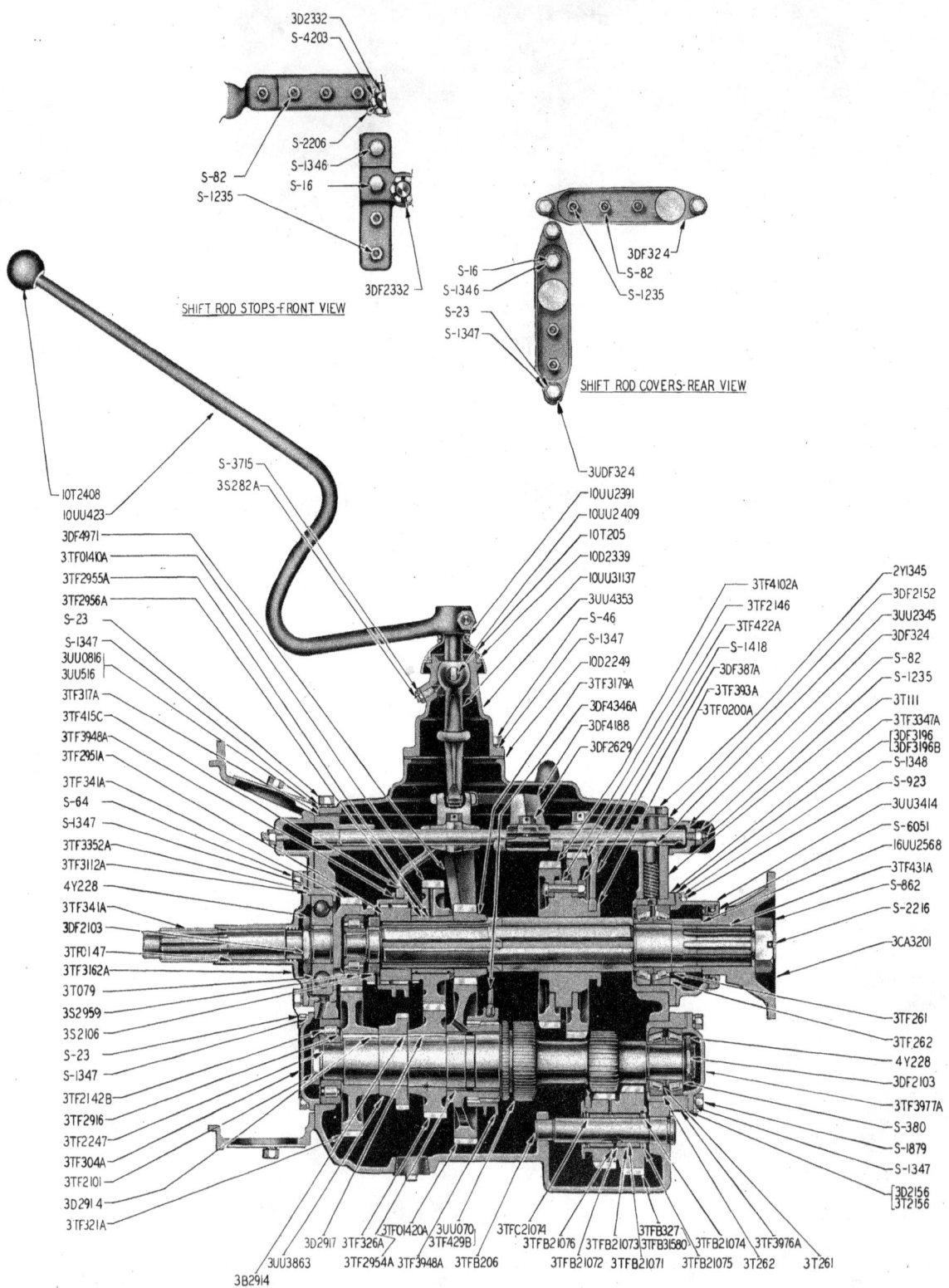

0700—Transmission

Master Parts List

THE AUTOCAR COMPANY, ARDMORE, PA.

GROUP 07

As supplied on Autocar Models
U-8144   U-8144-T

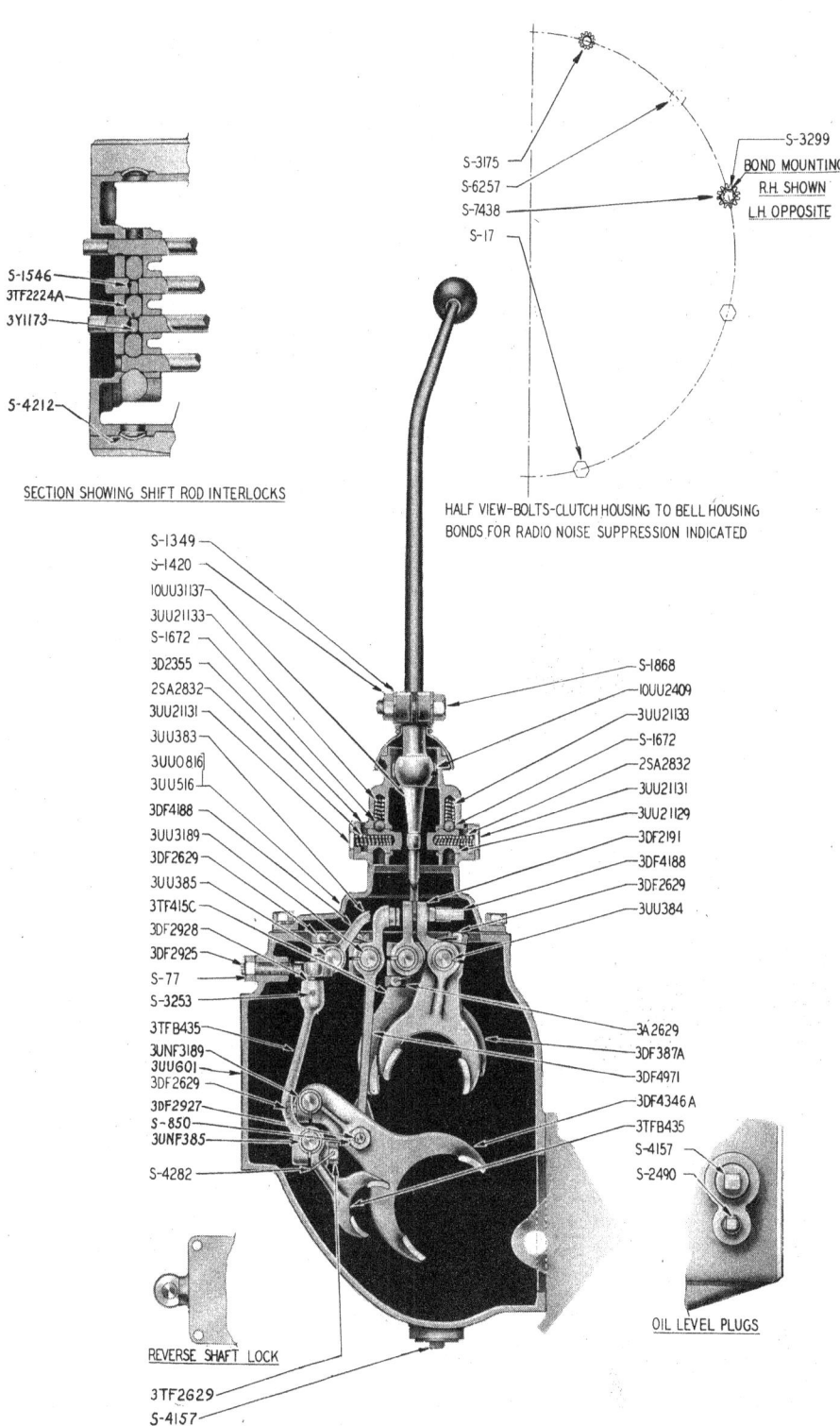

SECTION SHOWING SHIFT ROD INTERLOCKS

HALF VIEW-BOLTS-CLUTCH HOUSING TO BELL HOUSING
BONDS FOR RADIO NOISE SUPPRESSION INDICATED

REVERSE SHAFT LOCK

OIL LEVEL PLUGS

0700—Transmission

Master Parts List

CODE	MFR. PART No.	LIST PRICE	NAME OF PART	AUTOCAR PART No.	MODEL U-2044	U-4044	U-4144	U-4144-T	C-50	U-5044	U-7144-T	U-8144	U-8144-T	No. REQ.
0700	**Transmission**													
	231-F-8	$225.15	**Transmission** Assembly (Clark) (Complete with Clutch Housing and Hand Control)	**3UK6-530**	√									1
	231-F11	185.00	**Transmission** Assembly (with Clutch Housing and **Less** Hand Control) (Part of 3UK6530)	**3UK0-20**	√									1
	52092	....	**Transmission** Gearshift Control Assembly (Complete with Gearshift Lever and Bracket, Connecting Rod, Universal Lever and Top Control Cover) (Part of 3UK6530)	**10UK0-1460**	√									1
	200-V0-268	192.35	**Transmission** Assembly (Clark) (Complete with Clutch Housing and Hand Control)	**3UL6-530**		√	√	√						1
	......	187.90	**Transmission** Assembly (With Clutch Housing and **Less** Hand Control) (Part of 3UL6530)	**3UL0-20**		√	√	√						1
	50987	48.85	**Transmission** Gearshift Control Assembly (Complete with Gearshift Lever and Bracket, Connecting Rod, Universal Lever and Top Control Cover) (Part of 3UL6530)	**10UL0-1460**		√	√	√						1
	......	418.80	**Transmission** Assembly (Complete with Clutch Housing and Hand Control)	**3RL0-530**					√					1
	......	384.00	**Transmission** Assembly (Complete excepting Control Assy.)	**3URL0-20**						√				1
	......	535.80	**Transmission** Assembly (Complete with Clutch Housing and Hand Control)	**3UU0-530**							√	√	√	1
0701	**Transmission Case and Clutch Housing**													
	......	18.00	**Housing,** Clutch (S.A.E. #2)	**3D5-343**	√	√	√	√	√					1
	......	19.50	**Housing,** Clutch—and Bushings Assembly (S.A.E. #2)	**3D0-580**	√	√	√	√	√					1
	......	.09	**Screw,** Clutch Housing (½"—13 x 1¼")	**S-3127**	√	√	√	√	√					10
	......	.01	**Washer,** ½" Spring	**S-1349**	√	√	√	√	√					10
	......	.60	**Bushing,** Clutch Housing	**3UN2-119A**	√	√	√	√	√					2
	......	18.00	**Housing,** Clutch	**3UU5-343**							√	√	√	1
	......	.08	**Screw,** Hex. Cap (⅜"—16 x 1⅛")	**S-6257**							√	√	√	8
	......	.06	**Screw,** Hex. Cap (⅜"—16 x 1½")	**S-17**							√	√	√	2
	......	.05	**Lockwasher,** ⅜" Shakeproof	**S-3174**							√	√	√	10
	......	.04	**Screw,** Headless Set ($\frac{5}{16}$"—18 x 1")	**S-1235**							√	√	√	6
	......	.03	**Nut,** Hex. ($\frac{5}{16}$"—18)	**S-82**							√	√	√	6
	......	.04	**Screw,** Hex. Cap ($\frac{5}{16}$"—18 x ⅝")	**S-16**							√	√	√	2
	......	.01	**Lockwasher,** $\frac{5}{16}$" Spring	**S-1346**							√	√	√	2
	......	.25	**Bushing,** Clutch Control Shaft	**3UN2-119B**							√	√	√	2
	......	.25	**Fitting,** 90 Alemite	**S-4013**							√	√	√	2
	......	.30	**Plate,** Clutch Housing Inspection	**3T3-52A**	√	√	√	√	√					1
	......	.12	**Plate,** Flywheel Sight Hole Cover	**3Y2-113**	√	√	√	√	√					1
	......	.30	**Plate,** Clutch Housing Cover—Top	**3TF3-52**							√	√	√	1
	......	.05	**Screw,** Hex. Cap (⅜"—16 x ¾")	**S-12**	√	√	√	√	√		√	√	√	2
	......	.01	**Lockwasher,** $\frac{13}{32}$" Spring	**S-1347**							√	√	√	2
	......	.30	**Plate,** Clutch Housing Cover—Bottom	**3TF2-469**							√	√	√	1
	......	.05	**Screw,** Hex. Cap (⅜"—16 x ¾")	**S-12**							√	√	√	2
	......	.01	**Lockwasher,** $\frac{13}{32}$" Spring	**S-1347**							√	√	√	2
	3179	19.80	**Case,** Transmission—Bare	**3UK0-01**	√									1
	2490	19.80	**Case,** Transmission—Bare	**3BLA0-01**		√	√	√						1
	......	48.00	**Case,** Transmission—Bare	**3RL6-01**					√	√				1
	......	82.50	**Case,** Transmission—Bare	**3UU6-10**							√	√	√	1
	1079	.12	**Stud,** Transmission Case Clutch Housing	**3BL0-332**	√	√	√	√						4
	......	.20	**Stud,** Transmission Case Clutch Housing	**3D2-332**					√	√				5
	......	.20	**Stud,** Transmission Case Clutch Housing—Short	**3UU2-332**							√	√	√	5
	......	.25	**Stud,** Transmission Case Clutch Housing—Long	**3UUF2-332**							√	√	√	1
	10A-900	.06	**Nut,** Transmission Case Clutch Housing Stud ($\frac{9}{16}$"—18)	**S-84**	√	√	√	√						4
	......	.08	**Nut,** Hex. (¾"—10)	**S-4203**					√	√	√	√	√	5
	......	.01	**Cotter** (⅛" x 1½")	**S-2206**					√	√	√	√	√	6
	......	.03	**Lockwasher,** $\frac{9}{16}$" Spring	**S-1350**	√	√	√	√						4
	......	.02	**Plug,** Drain and Filler	**S-619**	√									1
	......	.20	**Plug,** Drain and Winter Level Pipe (1¼")	**S-4157**	√	√	√	√	√	√	√	√	√	2
	......	.06	**Plug,** Summer Level Pipe (½")	**S-2490**					√	√	√	√	√	1
	......	.04	**Plug,** Welch	**S-4209**							√	√	√	2
	......	.04	**Plug,** Welch (1¼")	**S-4212**					√	√	√	√	√	2

07-1 TRANSMISSION

THE AUTOCAR COMPANY, ARDMORE, PA.

Master Parts List

CODE	MFR. PART No.	LIST PRICE	NAME OF PART	AUTOCAR PART No.	U-2044	U-4044	U-4144	U-4144-T	C-50	U-5044	U-7144-T	U-8144	U-8144-T	NO. REQ.
0701	1398	$0.35	Cover, Power Takeoff Opening	3B3-154	✓	✓	✓	✓						2
		.50	Cover, Power Takeoff Opening	3RL3-154					✓	✓				1
		.50	Cover, Power Takeoff Opening	3UU3-154							✓	✓	✓	1
	1400	.10	Gasket, Power Takeoff Opening Cover	3B3-158	✓	✓	✓	✓						2
		.12	Gasket, Power Takeoff Opening Cover	3RL3-158					✓	✓				1
	7A-600	.05	Screw, Hex. Cap (3/8"—16 x 7/8")	S-50	✓	✓	✓	✓						12
		.05	Screw, Hex. Cap (3/8"—16 x 1")	S-23					✓	✓				6
	1A-600	.01	Lockwasher, 3/8" Spring	S-1347	✓	✓	✓	✓	✓	✓				12
		.08	Gasket, Power Takeoff Opening Cover	3UU3-158							✓	✓	✓	1
		.06	Screw, Hex. Cap (3/8"—16)	S-46							✓	✓	✓	6
		.03	Lockwasher, 1/2" Spring	S-47							✓	✓	✓	6
		6.60	Cover, Transmission—and Dowel Assembly	3RL0-816						✓				1
		6.35	Cover, Transmission—and Dowel Assembly	3UU0-816							✓	✓	✓	1
	2713	8.70	Cover, Transmission Case	3UK0-16	✓									1
	1420	.12	Plug, Gearshift Rod Hole Welch (1 3/16")	S-4235	✓									3
		....	Cover, Transmission—Assembly (Inc. Top Cover Control)			✓	✓	✓						1
	2690	4.57	Cover, Transmission Case	3UBL0-16		✓	✓	✓						1
	1422	.04	Plug, Gearshift Rod Hole Welch (7/8")	S-4209		✓	✓	✓						3
		6.00	Cover, Transmission Case	3RL4-16						✓				1
		6.30	Cover, Transmission Case	3UU5-16							✓	✓	✓	1
		.06	Dowel, Transmission Case Cover	2Y1-345						✓	✓	✓	✓	2
	2733	.10	Gasket, Transmission Case Cover	3BL0-17	✓									1
	1033	.15	Gasket, Transmission Case Cover	3BLA0-17		✓	✓	✓						1
		.15	Gasket, Transmission Case Cover	3D3-17					✓					1
		.35	Gasket, Transmission Case Cover	3TF3-17A							✓	✓	✓	1
	7A-608	.05	Screw, Hex. Cap (3/8"—16 x 1")	S-23	✓	✓	✓	✓	✓	✓	✓	✓	✓	12
	1A-600	.01	Lockwasher, 1/2" Spring	S-1347	✓	✓	✓	✓	✓	✓	✓	✓	✓	12
0702	**Transmission Gears**													
		21.00	Gear, First and Second Speed Slide (40 Teeth)	3RL4-22					✓	✓				1
		43.60	Gear, First and Second Speed Slide—Assembly	3TF0-200A							✓	✓	✓	1
		16.20	Gear, First Speed Slide	3TF4-22A							✓	✓	✓	1
	2687	16.60	Gear, First and Reverse Slide	3UK0-22	✓									1
	3445	11.00	Gear, Mainshaft First and Reverse Slide	3BLA0-22		✓	✓	✓						1
	2685	15.60	Gear, Mainshaft Second Speed	3UK0-102	✓									1
	1019	.75	Washer, Mainshaft Second Speed Gear Locating	3BLA0-1013	✓	✓	✓	✓						1
	1003	.03	Rollers, Mainshaft Second Speed Gear	3BLA0-1016	✓	✓	✓	✓						34
		13.20	Gear, Second Speed Slide	3TF4-102A	✓	✓	✓	✓						1
		8.40	Sleeve, First and Second Speed Slide Gear	3TF3-93A	✓	✓	✓	✓						1
		.30	Bolt, First and Second Speed Slide Gear Sleeve	3TF2-146	✓	✓	✓	✓						8
		.03	Nut, Hex. (3/8"—24)	S-1418	✓	✓	✓	✓						8
	1016	10.50	Gear, Mainshaft Second and Third Speed Slide	3BLA0-1047		✓	✓	✓						1
	3197	11.15	Gear, Mainshaft Third Speed	3UK0-18	✓									1
	1013	9.50	Gear, Mainshaft Third Speed Idler	3BLA0-18		✓	✓	✓						1
	2008	.10	Shim, Mainshaft Third Speed Gear—.003	3BLA0-1044		✓	✓	✓						As
	2009	.10	Shim, Mainshaft Third Speed Gear—.005	3BLA0-1045		✓	✓	✓						As
	1020	.75	Washer, Mainshaft Third Speed Gear Locating	3BLA0-1013		✓	✓	✓						1
	1089	.03	Rollers, Mainshaft Third Speed Gear	3BL0-1016	✓									36
	1051	.10	Ring, Retaining—Mainshaft 3rd Speed Gear Roller Bearing	3BL0-957	✓	✓	✓	✓						1
	1023	.75	Washer, Retainer—Mainshaft 3rd Speed Gear Roller Bearing	3BL0-1014	✓									1
	1064	4.25	Hub, Third and Fourth Speed Shift	3BLA0-951		✓	✓	✓						1
	1073	1.60	Bushing, Mainshaft Third Speed Gear Roller	3BLA0-1017	✓									1
	1052	.10	Lockpin, Mainshaft Third Speed Gear Bushing	3BL0-1015	✓	✓	✓	✓						1
		15.60	Gear, Third and Fourth Speed Slide (28 Teeth)	3RL3-18A					✓	✓				1
		16.65	Gear, Mainshaft Third Speed—and Bushing Assembly	3TF0-1410A							✓	✓	✓	1
		1.20	Bushing, Mainshaft Third Speed Gear	3TF2-955							✓	✓	✓	1
		3.00	Sleeve, Mainshaft Third Speed Gear	3TF2-956A							✓	✓	✓	1
		7.50	Driver, Third and Fourth Speed Clutch	3TF2-951A							✓	✓	✓	1
		9.00	Ring, Third and Fourth Speed Clutch	3TF3-948A							✓	✓	✓	1
	1021	1.50	Bushing, Mainshaft Fourth Gear Bearing	3BLA0-1042		✓	✓	✓						1
	51167	15.00	Gear, Mainshaft Overdrive, and Bushing Assembly	3BLA0-1540		✓	✓	✓						1
	4428-S	1.20	Bushing, Mainshaft Overdrive Gear	3BLA0-1046		✓	✓	✓						1

07-2 Transmission

Master Parts List

THE AUTOCAR COMPANY, ARDMORE, PA.

CODE	MFR. PART No.	LIST PRICE	NAME OF PART	AUTOCAR PART No.	U-2044	U-4044	U-4144	U-4144-T	C-50	U-5044	U-7144-T	U-8144	U-8144-T	No. REQ.
0702	1059	$1.10	Washer, Mainshaft Overdrive Gear Retaining............	3BLA0-1043		✓	✓	✓						1
	......	7.20	Gear, Mainshaft Overspeed Slide (19 Teeth).............	3RL3-179						✓	✓			1
	1058	5.50	Hub, Fourth and Overdrive Shift............................	3BLA0-951		✓	✓	✓						1
	......	7.80	Gear, Mainshaft Overspeed....................................	3TF3-179A							✓	✓	✓	1
	50009	54.00	Countershaft Assembly—Gears and Bearings..............	3BLA0-70		✓	✓	✓						1
	4A-D	.06	Keys, Countershaft Gear (#4 Whitney)....................	S-2302		✓	✓	✓						4
	......	93.50	Countershaft Assembly—Gears and Bearings..............	3RL0-70						✓	✓			1
	......	102.00	Countershaft Assembly—Gears and Bearings..............	3UU0-70							✓	✓	✓	1
	1067	10.50	Gear, Countershaft Drive......................................	3BL0-21	✓									1
	1910	.10	Ring, Countershaft Gear Retaining..........................	3BL0-1012	✓									1
	4A-D	.06	Keys, Countershaft Gear (#3 Whitney)....................	S-2303	✓									2
	1056	11.75	Gear, Countershaft Drive......................................	3BLA0-21		✓	✓	✓						1
	......	12.60	Gear, Countershaft Drive......................................	3RL3-21						✓	✓			1
	......	.30	Key, Countershaft Drive Gear................................	3C2-914						✓	✓			1
	......	14.40	Gear, Countershaft Drive (Use with 3TF2916 Washer)...	3TF3-21A							✓	✓	✓	1
	......	.30	Key, Countershaft Drive Gear................................	3D2-914							✓	✓	✓	1
	......	8.40	Gear, Power Takeoff Driving.................................	3RL3-863						✓	✓			1
	......	9.00	Gear, Power Takeoff Driving.................................	3UU3-863							✓	✓	✓	1
	......	.30	Key, Power Takeoff Driving Gear...........................	3B2-914							✓	✓	✓	1
	1006	4.50	Gear, Second and Reverse Speed Countershaft...........	3BLA0-274		✓	✓	✓						1
	3196	9.00	Gear, Third Speed Countershaft.............................	3UK0-26	✓									1
	1007	6.50	Gear, Third Speed Countershaft.............................	3BLA0-26		✓	✓	✓						1
	1024	.45	Spacer, Second and Third Speed Countershaft Gear.....	3BLA0-108		✓	✓	✓						1
	......	9.00	Gear, Third Speed Countershaft.............................	3RL3-26						✓	✓			1
	......	.50	Key, Third Speed Countershaft Gear.......................	3E2-914						✓	✓			1
	......	10.20	Gear, Third Speed Countershaft.............................	3TF3-26A							✓	✓	✓	1
	......	.30	Key, Third Speed Countershaft Gear.......................	3D2-917							✓	✓	✓	1
	1055	12.25	Gear, Overdrive Countershaft................................	3BLA0-949		✓	✓	✓						1
	......	13.20	Gear, Overdrive Countershaft (37 Teeth)..................	3RL3-949						✓	✓			1
	......	.30	Key, Overdrive Countershaft Gear..........................	3D2-917						✓	✓			1
	......	22.50	Gear, Overdrive Countershaft, and Bushing Assembly..	3TF0-1420A							✓	✓	✓	1
	......	1.20	Bushing, Overdrive Countershaft Gear.....................	3TF2-954							✓	✓	✓	1
	......	9.00	Ring, Overdrive Countershaft Gear Clutch................	3TF3-948A							✓	✓	✓	1
	50320	7.50	Gear, Reverse, Assembly with Bushing....................	3UK0-270	✓									1
	2735	.85	Bushing, Reverse Gear...	3UK0-07	✓									1
	1010	10.00	Gear, Reverse..	3BLA0-27		✓	✓	✓						1
	......	15.00	Gear, Reverse, Assembly with Bushing (14 and 18 Teeth)....	3RL0-270						✓	✓			1
	......	.60	Bushing, Reverse Gear...	3D2-07						✓	✓			1
	......	24.65	Gear, Reverse, Assembly (With Needle Bearings).......	3TFB3-1580							✓	✓	✓	1
	......	14.40	Gear, Reverse..	3TFB3-27							✓	✓	✓	1
	......	5.35	Sleeve, Reverse Gear...	3TFB2-1071							✓	✓	✓	1
	......	.15	Key, Reverse Gear Sleeve.....................................	3TFB2-1072							✓	✓	✓	1
0703	**Transmission Main Drive Pinion and Bearings**													
	11764	13.20	Driveshaft, Transmission, and Gear.........................	3UBT0-41	✓									1
	1177	15.00	Driveshaft, Transmission, and Gear.........................	3BLA0-41		✓	✓	✓						1
	......	21.60	Driveshaft, Transmission, and Gear (29 Teeth)..........	3RL3-41						✓	✓			1
	......	21.00	Gear, Main Drive..	3TF3-41A							✓	✓	✓	1
	47511	8.80	Bearing, Transmission Driveshaft............................	3BLA0-79	✓	✓	✓	✓						1
	6309Z-C-003	8.75	Bearing, Transmission Driveshaft............................	3D2-248A						✓				1
	......	13.70	Bearing, Main Drive Gear (S.K.F. #1749).................	3T0-79							✓	✓	✓	1
	1032	.75	Nut, Transmission Driveshaft Bearing Retaining.........	3BLA0-994	✓	✓	✓	✓						1
	......	.55	Nut, Transmission Driveshaft Lock..........................	3DF2-103						✓	✓	✓	✓	1
	......	.20	Washer, Transmission Driveshaft Lock Nut...............	4Y2-28						✓	✓	✓	✓	1
	......	3.60	Retainer, Main Drive Gear Bearing..........................	3TF3-352A							✓	✓	✓	1
	......	3.60	Retainer, Transmission Driveshaft Bearing................	3D3-352A						✓				1
	2004	2.25	Cap, Transmission Driveshaft Bearing......................	3BLA0-162	✓	✓	✓	✓						1
	1751	.05	Gasket, Transmission Driveshaft Bearing Cap............	3BLA0-452	✓	✓	✓	✓						1
	7A-503	.04	Screw, Trans. Driveshaft Bearing Cap (5/16"—18 x 7/8")..	S-15	✓	✓	✓	✓						2
	1971	.04	Screw, Trans. Driveshaft Bearing Cap—Lower (5/16"—18 x 7/8")	S-1782	✓	✓	✓	✓						2
	1A-600	.01	Lockwasher, 3/8" Spring......................................	S-1347	✓	✓	✓	✓						4
	......	5.10	Cap, Transmission Driveshaft Bearing......................	3D3-162A						✓	✓			1

THE AUTOCAR COMPANY
ARDMORE, PA.

Master Parts List

07-3
TRANSMISSION

CODE	MFR. PART No.	LIST PRICE	NAME OF PART	AUTOCAR PART No.	U-2044	U-4044	U-4144	U-4144-T	C-50	U-5044	U-7144-T	U-8144	U-8144-T	No. REQ.
0703	......	$0.12	Gasket, Transmission Driveshaft Bearing Cap	3D3-112					✓	✓				1
	......	.06	Screw, Transmission Driveshaft Bearing Cap	S-64					✓	✓				1
	......	.06	Screw, Transmission Driveshaft Bearing Cap	S-46					✓	✓				4
	......	.01	Lockwasher, Spring	S-1347					✓	✓				5
	......	4.80	Cap, Main Drive Gear Bearing	3TF3-162A							✓	✓	✓	1
	......	.12	Gasket, Main Drive Gear Bearing Cap	3TF3-112A							✓	✓	✓	1
	......	.06	Screw, Hex. Cap (3/8"—16 x 1¼")	S-64							✓	✓	✓	5
	......	.01	Lockwasher, 13/32" Spring	S-1347							✓	✓	✓	5
	94622 (Hyatt)	1.20	Bearing, Mainshaft Pilot	3BLA0-147	✓	✓	✓	✓						1
	U-1306-TM	7.35	Bearing, Mainshaft Pilot	3RL0-147					✓	✓				1
	......	15.90	Bearing, Main Pilot (S.K.F. #WJM-30)	3TF0-147							✓	✓	✓	1
	43308	7.25	Bearing, Mainshaft Rear	3BLA0-251	✓	✓	✓	✓						1
	332	2.11	Cup, Mainshaft Rear Bearing	3T2-62					✓	✓				2
	336	3.58	Cone, Mainshaft Rear Bearing	3T2-61					✓	✓				2
	......	3.30	Housing, Mainshaft Rear Bearing	3DF3-347					✓	✓				1
	......	.12	Shim, Mainshaft Rear Bearing Retainer	3D3-196					✓	✓				5
	......	.06	Washer, Mainshaft Rear Bearing (5/8" I.D.—2" O.D. x 1/16")	S-380					✓	✓				1
	......	.55	Nut, Mainshaft Rear Bearing Lock	3DF2-103					✓	✓				1
	......	.20	Washer, Mainshaft Rear Bearing Lock Nut	4Y2-28					✓	✓				1
	......	3.41	Cup, Mainshaft Bearing (Timken #372-A)	3TF2-62							✓	✓	✓	2
	......	4.23	Cone, Mainshaft Bearing (Timken #377)	3TF2-61							✓	✓	✓	2
	......	3.30	Housing, Mainshaft Rear Bearing	3TF3-347A							✓	✓	✓	1
	......	.12	Shim, Mainshaft Rear Bearing Retainer—.005"	3DF3-196B							✓	✓	✓	As
	......	.12	Shim, Mainshaft Rear Bearing Retainer—.010"	3DF3-196							✓	✓	✓	5
	1206-TS (Hyatt)	3.60	Bearing, Countershaft Front	3BL0-142	✓									1
	1099	.12	Washer, Countershaft Front Bearing Locating	3BL0-916	✓									1
	1560	.05	Plug, Countershaft Front Bearing Welch	3BL0-1031	✓									1
	1901	.10	Ring, Countershaft Front Bearing Welch Plug Retaining	3BL0-1003	✓									1
	1207-TS (Hyatt)	4.20	Bearing, Countershaft Front	3BLA0-142		✓	✓	✓						1
	1025	.15	Spacer, Countershaft Front Bearing	3BLA0-916		✓	✓	✓						1
	1714	.10	Ring, Countershaft Gear Snap	3BLA0-1012		✓	✓	✓						1
	1561	.12	Plug, Countershaft Front Bearing Welch	3BLA0-1031		✓	✓	✓						1
	1903	.10	Ring, Countershaft Front Bearing Welch Plug Retaining	3BLA0-1003		✓	✓	✓						1
	432	3.09	Cup, Countershaft Front Bearing	4H2-137					✓	✓				1
	449	.64	Cone, Countershaft Front Bearing	3D2-36					✓	✓				1
	......	.25	Washer, Countershaft Front Bearing Thrust	3D2-916					✓	✓				1
	......	.60	Screw, Countershaft Front Bearing Lock	3DF2-923					✓	✓				1
	......	.12	Lockwasher, Countershaft Front Bearing	3DF2-924					✓	✓				1
	......	2.10	Cap, Countershaft Front Bearing	3S3-04B					✓	✓				1
	......	.06	Gasket, Countershaft Front Bearing Cap	3RL2-762					✓	✓				1
	......	.05	Screw, Countershaft Bearing Cap	S-23					✓	✓				4
	......	.01	Lockwasher, Spring	S-1347					✓	✓				4
	......	11.95	Bearing, Countershaft Front (Hyatt #R-1309-TS)	......							✓	✓	✓	1
	......	.30	Washer (Use with 3TF321A Gear)	......							✓	✓	✓	1
	......	.25	Nut, Countershaft Front Bearing Lock	......							✓	✓	✓	1
	......	.09	Washer, Countershaft Front Bearing Lock Nut	......							✓	✓	✓	1
	......	1.20	Cap, Countershaft Front Bearing	......							✓	✓	✓	1
	......	.05	Screw, Hex. Cap (3/8"—16 x 1")	S-23							✓	✓	✓	4
	......	.01	Lockwasher, 13/32" Spring	S-1347							✓	✓	✓	4
	308MFG	7.05	Bearing, Countershaft Rear (Marlin-Rockwell)	3HD0-248	✓									1
	1027	.15	Washer, Countershaft Rear Bearing Retaining	3BLA0-1038	✓									1
	8A-501	.04	Screw, Ctrshft. Rr. Bearing Retaining Washer (5/16"—24 x 3/4")	S-1783	✓									2
	1A-600	.01	Lockwasher, 3/8" Spring	S-1347	✓									2
	2717	1.45	Cap, Countershaft Rear Bearing	3UK0-977	✓									1
	2747	.05	Gasket, Countershaft Rear Bearing Cap	3UK0-156	✓									1
	7A-608	.05	Screw, Countershaft Rear Bearing Cap (3/8"—16 x 1")	S-23	✓									4
	1A-600	.01	Lockwasher, 3/8" Spring	S-1347	✓									4
	41307	6.40	Bearing, Countershaft Rear	3BLA0-248		✓	✓	✓						1
	1166	.10	Shield, Countershaft Rear Bearing	3BLA0-1041		✓	✓	✓						1
	1026	.10	Ring, Countershaft Rear Bearing Spacer	3BLA0-1037		✓	✓	✓						1
	1027	.15	Washer, Countershaft Rear Bearing Retainer	3BLA0-1038		✓	✓	✓						1
	8A-501	.04	Screw, Countershaft Rr. Brg. Ret. Washer (5/16"—24 x 3/4")	S-1783		✓	✓	✓						2

07-4
Transmission

Master Parts List
THE AUTOCAR COMPANY ARDMORE, PA.

CODE	MFR. PART No.	LIST PRICE	NAME OF PART	AUTOCAR PART No.	U-2044	U-4044	U-4144	U-4144-T	C-50	U-5044	U-7144-T	U-8144	U-8144-T	No. REQ.
0703	1A-501	$0.01	Lockwasher, 5/16"........	S-3283		✓	✓	✓						2
	1030	1.45	Cap, Countershaft Rear Bearing........	3BLA0-977		✓	✓	✓						1
	1031	.07	Gasket, Countershaft Rear Bearing Cap........	3BLA0-156		✓	✓	✓						1
	7A-608	.05	Screw, Countershaft Rear Bearing Cap (3/8"—16 x 1").....	S-23		✓	✓	✓						4
	1A-600	.01	Lockwasher, 3/8" Spring........	S-1347		✓	✓	✓						4
	432	3.09	Cup, Countershaft Rear Bearing........	4H2-137					✓	✓				1
	449	.64	Cone, Countershaft Rear Bearing........	3D2-36					✓	✓				1
	......	1.20	Cap, Countershaft Rear Bearing........	3D3-04B					✓	✓				1
	......	.12	Shim, Countershaft Rear Bearing Cap—1/32"........	3D2-156					✓	✓				1
	......	.06	Shim, Countershaft Rear Bearing Cap—.005"........	3T2-156					✓	✓				1
	......	.05	Screw, Countershaft Bearing Cap........	S-23					✓	✓				4
	......	.01	Washer, Spring........	S-1347					✓	✓				4
	......	2.11	Cup, Countershaft Rear Bearing (Timken #332)........	3T2-62							✓	✓	✓	2
	......	3.58	Cone, Countershaft Rear Bearing (Timken #336)........	3T2-61							✓	✓	✓	2
	......	.02	Washer, Plain (1 5/8" I.D. x 2" O.D. x 1/16")........	S-380							✓	✓	✓	1
	......	.55	Nut, Countershaft Rear Bearing Lock........	3DF2-103							✓	✓	✓	1
	......	.20	Washer, Countershaft Rear Bearing Lock Nut........	4Y2-28							✓	✓	✓	1
	......	3.10	Housing, Countershaft Rear Bearing........	3TF3-976A							✓	✓	✓	1
	......	1.00	Cap, Countershaft Rear Bearing........	3TF3-977A							✓	✓	✓	1
	......	.12	Shim, Countershaft Rear Bearing Cap—1/32"........	3D2-156							✓	✓	✓	1
	......	.06	Shim, Countershaft Rear Bearing Cap—.005"........	3T2-156							✓	✓	✓	8
	......	.06	Screw, Hex. Cap (3/8"—16 x 1 3/4")........	S-1879							✓	✓	✓	4
	......	.01	Lockwasher, 3/8" Spring........	S-1347							✓	✓	✓	4
	94622 (Hyatt)	1.20	Bearing, Reverse Gear Shaft........	3BLA0-147		✓	✓	✓						2
	1029	.35	Spacer, Reverse Gear Shaft Bearing........	3BLA0-1039		✓	✓	✓						1
	......	.10	Bearings, Reverse Gear Needle........	3TFB2-1073							✓	✓	✓	106
	......	.65	Washer, Reverse Gear Needle Bearing........	3TFB2-1074							✓	✓	✓	2
	......	1.30	Ring, Reverse Gear Needle Bearing Separator........	3TFB2-1076							✓	✓	✓	1
	......	.50	Ring, Reverse Gear Needle Bearing Snap........	3TFB2-1075							✓	✓	✓	2
0704	**Mainshaft, Counter shaft and Reverse Idler Countershaft**													
	2681	12.85	Mainshaft—Bare........	3UK0-31	✓									1
	......	4.20	Flange, Mainshaft Driving........	3AH0-201	✓	✓								1
	1211	.25	Nut, Mainshaft Driving Flange........	3BLA0-103	✓	✓	✓	✓						1
	1012	16.00	Mainshaft—Bare........	3BLA0-31		✓	✓	✓						1
	......	8.60	Flange, Mainshaft Driving........	3AH3-201			✓	✓						1
	......	26.40	Mainshaft—Bare........	3RL4-31					✓	✓				1
	......	.50	Washer, Mainshaft Drive Thrust........	3RL2-957					✓	✓				1
	......	.25	Key, Mainshaft Driving Flange........	4Y1-163					✓	✓				1
	......	.30	Nut, Mainshaft Driving Flange........	3Y1-103					✓	✓				1
	......	.12	Lockwasher, Mainshaft Driving Flange Nut........	3Y2-106					✓	✓				1
	......	28.20	Mainshaft—Bare........	3TF4-31A							✓	✓	✓	1
	......	.55	Nut, Mainshaft Drive Thrust........	3S2-959							✓	✓	✓	1
	......	8.58	Wrench, Mainshaft Drive Thrust Nut........	X-32556							✓	✓	✓	1
	......	.12	Lockwasher, Mainshaft Drive Thrust Nut........	3S2-106							✓	✓	✓	1
	......	14.40	Flange, Mainshaft Driveshaft........	3CA3-201							✓	✓	✓	1
	......	.60	Nut, Hex. (1 1/2"—16)........	S-862							✓	✓	✓	1
	......	.01	Cotter, 1/8" x 2 1/2"........	S-2216							✓	✓	✓	1
	2686	15.90	Countershaft........	3UK0-29	✓									1
	1005	17.00	Countershaft........	3BLA0-29		✓	✓	✓						1
	......	37.20	Countershaft (16 and 10 Teeth)........	3RL4-29					✓	✓				1
	......	31.50	Countershaft........	3TF4-29B							✓	✓	✓	1
	2719	1.35	Shaft, Reverse Gear........	3UK0-06	✓									1
	1929	.45	Lock, Reverse Gear Shaft........	3BLA0-1018	✓	✓	✓	✓						1
	7A-600	.05	Screw, Reverse Gear Shaft Lock (3/8"—16 x 7/8")........	S-50	✓	✓	✓	✓						1
	1A-600	.01	Lockwasher, 3/8" Spring........	S-1347	✓	✓	✓	✓						1
	1018	2.00	Shaft, Reverse Gear........	3BLA0-06		✓	✓	✓						1
	......	3.30	Shaft, Reverse Gear........	3D2-06A					✓	✓				1
	......	3.30	Shaft, Reverse Gear........	3TFB2-06							✓	✓	✓	1

THE AUTOCAR COMPANY
ARDMORE, PA.

Master Parts List

07-5 TRANSMISSION

CODE	MFR. PART No.	LIST PRICE	NAME OF PART	AUTOCAR PART No.	U-2044	U-4044	U-4144	U-4144-T	C-50	U-5044	U-7144-T	U-8144	U-8144-T	No. REQ.
0705	**Speedometer Drive Gears**													
	90059	$2.50	Gear, Speedometer Drive	16D0-569	✓	✓	✓	✓						1
	93716	2.50	Gear, Speedometer Drive	16B0-569A					✓	✓				1
	......	.06	Key, Speedometer Drive Gear	S-2303					✓	✓				1
	50360	3.15	Case, Speedometer Drive Gear, and Bushing Assembly	3UK0-1370	✓									1
	2716	.05	Gasket, Speedometer Drive Gear Case	3UK0-196	✓									1
	50035	.40	Seal, Speedometer Drive Gear Case Oil	3BLA0-236	✓	✓	✓	✓						1
	7A-608	.05	Screw, Speedometer Drive Gear Case—Short (3/8"—16 x 1")	S-23	✓	✓	✓	✓						3
	7A-606	.08	Screw, Speedometer Drive Gear Case—Long (3/8"—16 x 2¼")	S-40	✓	✓	✓	✓						1
	1A-600	.01	Lockwasher, 3/8" Spring	S-1347	✓	✓	✓	✓						4
	50332	3.15	Case, Speedometer Drive Gear, and Bushing Assembly	3UBT0-1370		✓	✓	✓						1
	......	.20	Plug, Speedometer Hole	3A2-413		✓	✓							1
	2716	.05	Gasket, Speedometer Drive Gear Case	3BL0-196		✓	✓	✓						1
	......	2.90	Case, Speedometer Drive Gear, Assembly	3D0-1370					✓	✓				1
	......	.70	Seal, Speedometer Drive Gear Case Oil	S-6108					✓	✓				1
	......	.08	Screw, Speedometer Drive Gear Case	S-923					✓	✓				4
	......	.01	Lockwasher, 7/16" Spring	S-1348					✓	✓				4
	93171	1.75	Gear, Speedometer Driven (13 Teeth)	16F0-571					✓	✓				1
	93172	1.75	Gear, Speedometer Driven (14 Teeth)	16G0-571					✓	✓				1
	93173	1.75	Gear, Speedometer Driven (15 Teeth)	16H0-571					✓	✓				1
	93174	1.75	Gear, Speedometer Driven (16 Teeth)	16J0-571					✓	✓				1
	93175	1.75	Gear, Speedometer Driven (17 Teeth)	16K0-571					✓	✓				1
	......	.10	Bushing, Speedometer Driven Gear	16T2-572					✓	✓				1
	......	.50	Sleeve, Speedometer Driven Gear	3A0-137B					✓	✓				1
0706	**Shifting Forks and Levers**													
	51094	....	Lever, Gearshift, Assembly	51094	✓									1
	51092	....	Lever, Gearshift, and Bracket Assembly	51092		✓	✓	✓						1
	4196	4.20	Lever, Gearshift—Bare	10UK0-23	✓									1
	4192	4.25	Lever, Gearshift—Bare	10UL0-23		✓	✓	✓						1
	2788	2.65	Bracket, Gearshift Lever Mounting	10UB0-18	✓	✓	✓	✓						1
	2750	.24	Ball, Gearshift Lever	10B0-408	✓	✓	✓	✓						1
	2484	.70	Cover, Gearshift Lever Dust	10UB0-1017	✓	✓	✓	✓						1
	1298	.35	Spring, Gearshift Lever Pivot	10UB0-391	✓	✓	✓	✓						1
	1299	.15	Washer, Gearshift Lever Pivot	10UB0-26	✓	✓	✓	✓						1
	3036	2.50	End, Gearshift Lever	10UB0-767	✓	✓	✓	✓						1
	3061	.08	Pin, Gearshift Lever End (¼")	S-1593	✓	✓	✓	✓						2
	51063	6.35	Rod, Gearshift Connecting, and Yoke Assembly	10UK0-1450	✓									1
	50374	6.50	Rod, Gearshift Connecting, and Yoke Assembly	10UL0-1450		✓	✓	✓						1
	2132	2.90	Yoke, Gearshift Connecting Rod (Lever End)	3UB0-1056	✓	✓	✓	✓						1
	8A-602	.05	Bolt, Gearshift Connecting Rod Yoke (3/8"—24 x 1¾")	S-1457	✓	✓	✓	✓						2
	10A-600	.03	Nut, Gearshift Connecting Rod Yoke Bolt (3/8"—24)	S-1418	✓	✓	✓	✓						2
	1A-600	.01	Lockwasher, 3/8" Spring	S-1347	✓	✓	✓	✓						2
	2129	.05	Pin, Gearshift Connecting Rod Yoke Clevis	3UB0-1055	✓	✓	✓	✓						6
	2785	4.80	Cover, Gearshift Top Control	3UK0-309	✓									1
	2938	1.98	Cover, Gearshift Top Control	3UB0-309		✓	✓	✓						1
	2733	.05	Gasket, Gearshift Top Control Cover	3UB0-961	✓	✓	✓	✓						1
	7A-608	.05	Screw, Gearshift Top Control Cover (3/8"—16 x 1")	S-23	✓	✓	✓	✓						4
	1A-600	.01	Lockwasher, 3/8" Spring	S-1347	✓	✓	✓	✓						4
	2122	5.50	Bracket, Pivot Anchor	3UB0-1052	✓	✓	✓	✓						1
	2080	.12	Ring, Pivot Anchor Bracket Snap	3UB0-1049	✓	✓	✓	✓						1
	2783	2.30	Finger, Shift Lever	3UB0-88	✓	✓	✓	✓						1
	4A-11	.06	Key, Shift Lever Finger (#11 Whitney)	S-2311	✓	✓	✓	✓						1
	1705	.12	Screw, Shift Lever Finger Lock	3UB0-1048	✓	✓	✓	✓						1
	2784	3.00	Shaft, Shift Lever Rocker	3UK0-1058	✓									1
	2935	1.35	Shaft, Shift Lever Rocker	3UB0-1058		✓	✓	✓						1
	3070	.12	Plug, Shift Lever Rocket Shaft Welch (1⅛")	S-4236	✓	✓	✓	✓						1
	2936	4.80	Lever, Gearshift Universal	3UB0-1059	✓	✓	✓	✓						1
	2124	4.20	Bracket, Gearshift Universal Lever	3UB0-1053	✓	✓	✓	✓						1
	2081	.12	Ring, Gearshift Universal Lever Snap	3UB0-1051	✓	✓	✓	✓						1
	1420	.12	Plug, Gearshift Universal Lever Welch (1 3/16")	S-4235	✓	✓	✓	✓						1
	2128	.48	Boot, Gearshift Universal Lever	3UB0-1054	✓	✓	✓	✓						1

For Further Information Refer to Group 0804.

07-6 TRANSMISSION

Master Parts List — The Autocar Company, Ardmore, PA.

CODE	MFR. PART No.	LIST PRICE	NAME OF PART	AUTOCAR PART No.	U-2044	U-4044	U-4144	U-4144-T	C-50	U-5044	U-7144-T	U-8144	U-8144-T	No. REQ.
0706	2934	$3.35	Finger, Gearshift Universal Lever	3UB0-1057	✓	✓	✓	✓						1
	4A-11	.06	Key, Gearshift Universal Lever Finger (#11 Whitney)	S-2311	✓	✓	✓	✓						1
	1705	.12	Screw, Gearshift Universal Lever Finger Lock	3UB0-1048	✓	✓	✓	✓						1
	2722	.85	Rod, First and Second Speed Gearshift	3UK0-84	✓									1
	1085	1.40	Lug, First and Second Speed Gearshift	3BL0-191	✓									1
	2738	1.25	Fork, First and Second Speed Gearshift	3UK0-87	✓									1
	2723	1.15	Rod, Third and Fourth Speed Gearshift	3UK0-83	✓									1
	2732	1.75	Lug, Third and Fourth Speed Gearshift	3UK0-1036	✓									1
	2736	1.80	Fork, Third and Fourth Speed Gearshift	3UK0-15	✓									1
	2721	.85	Rod, Reverse Gearshift	3UK0-85	✓									1
	2949	1.85	Lug, Reverse Gearshift	3UB0-1035	✓									1
	2724	1.80	Fork, Reverse Gearshift	3UK0-35	✓									1
	1566	.10	Screw, Gearshift Fork and Lug	3BL0-629	✓									6
	1042	.50	Plunger, Reverse Gearshift Latch	3BL0-1007	✓									1
	1043	.19	Spring, Reverse Gearshift Latch Plunger	3BL0-1008	✓									1
	10A-500	....	Nut, Reverse Gearshift Latch Adjusting ($\frac{5}{16}''$—24)	S-2024	✓									1
	1844	.05	Spring, Gearshift Rod	3BL0-11	✓									3
	6A-600	.03	Ball, Gearshift Rod Plunger ($\frac{3}{8}''$)	S-1670	✓									3
	1463	.12	Plug, Gearshift Rod Interlock	3BL0-1009	✓									2
	1093	.17	Pin, Gearshift Rod Interlock	3BL0-1011	✓									1
	1423	.03	Plug, Gearshift Rod Interlock Hole Welch ($\frac{5}{8}''$)	S-4207	✓	✓	✓	✓						1
	2702	1.55	Rod, First and Reverse Speed Gearshift	3UBL0-85		✓	✓	✓						1
	1041	1.75	Lug, First and Reverse Speed Gearshift Rod	3BLA0-1035		✓	✓	✓						1
	1038	1.80	Fork, First and Reverse, Second and Third Speed Gearshift	3BLA0-87		✓	✓	✓						2
	1566	.10	Screw, Gearshift Fork and Lug	3BL0-629		✓	✓	✓						6
	1042	.50	Plunger, First and Reverse Gearshift Latch	3BL0-1007		✓	✓	✓						1
	1043	.19	Spring, First and Reverse Gearshift Latch Plunger	3BL0-1008		✓	✓	✓						1
	10A-500	....	Nut, First & Reverse Gearshift Latch Adjusting ($\frac{5}{16}''$—24)	S-2024		✓	✓	✓						1
	2703	1.55	Rod, Second and Third Speed Gearshift	3UBL0-84		✓	✓	✓						1
	1040	1.75	Lug, Second and Third Speed Gearshift	3BLA0-191		✓	✓	✓						1
	2704	1.55	Rod, Fourth and Overdrive Gearshift	3UBL0-83		✓	✓	✓						1
	2705	1.75	Lug, Fourth and Overdrive Gearshift	3UBL0-1036		✓	✓	✓						1
	2727	1.80	Fork, Fourth and Overdrive Gearshift Rod	3UBL0-15		✓	✓	✓						1
	6A-700	.04	Ball, Gearshift Rod Plunger ($\frac{7}{16}''$)	S-1671		✓	✓	✓						3
	1045	.05	Spring, Gearshift Rod	3BLA0-11		✓	✓	✓						3
	6A-700	.04	Ball, Gearshift Rod Interlock	S-1671		✓	✓	✓						4
	1044	.15	Pin, Gearshift Rod Interlock	3BLA0-1011		✓	✓	✓						1
	......	16.80	Lever, Gearshift, Assembly	10DL0-130					✓					1
	......	10.20	Lever, Gearshift—Bare	10DL4-23					✓					1
	......	3.90	Bracket, Gearshift Lever	3D3-353					✓					1
	......	.06	Dowel, Gearshift Lever Bracket	3D2-355					✓					2
	......	.06	Screw, Gearshift Lever Bracket	S-46					✓					4
	......	.01	Lockwasher, Spring	S-1347					✓					4
	......	1.20	Cover, Gearshift Lever Socket	10T2-05					✓					1
	......	.20	Seal, Gearshift Lever Socket Oil	10D2-339					✓					1
	......	.06	Pin, Gearshift Lever Socket	10T1-409					✓					1
	......	.20	Spring, Gearshift Lever Socket	10T2-391A					✓					1
	......	.30	Keeper, Gearshift Lever Socket Spring	10T1-26					✓					2
	......	.01	Pin, Gearshift Lever Socket Spring Keeper	S-1528					✓					1
	......	.35	Gate, Gearshift	10D2-18					✓					1
	......	.10	Gasket, Gearshift Gate	10D2-249					✓					2
	......	2.10	Rod, First and Second Speed Gearshift	3RL3-84					✓	✓				1
	......	3.00	Jaw, First and Second Speed Gearshift Rod	3D2-191					✓	✓				1
	......	4.80	Fork, First and Second Speed Gearshift	3RL3-87					✓	✓				1
	......	.08	Screw, First and Second Speed Gearshift Fork	3D2-629					✓	✓				1
	......	2.10	Rod, Third and Fourth Speed Gearshift	3RL3-83					✓	✓				1
	......	6.00	Fork, Third and Fourth Speed Shift	3RL3-15					✓	✓				1
	......	.35	Screw, Third and Fourth Speed Gearshift Fork	3D2-645					✓	✓				1
	......	2.10	Rod, Overdrive Shift	3RL3-189					✓	✓				1
	......	6.00	Fork, Overdrive Shift	3RL3-346					✓	✓				1
	......	2.10	Rod, Reverse Gearshift	3RL3-85					✓	✓				1
	......	7.20	Arm, Reverse Gearshift Rod	3RL3-188					✓	✓				1

07-7 Transmission

THE AUTOCAR COMPANY, ARDMORE, PA. — Master Parts List

CODE	MFR. PART No.	LIST PRICE	NAME OF PART	AUTOCAR PART No.	U-2044	U-4044	U-4144	U-4144-T	C-50	U-5044	U-7144-T	U-8144	U-8144-T	No. REQ.
0706	......	$4.80	Fork, Reverse Gearshift	3D3-35					✓	✓				1
	......	.40	Screw, Reverse Gearshift Fork	3D2-215					✓	✓				1
	......	.90	Rod, Reverse Gearshift—Lower	3D2-187A					✓	✓				1
	......	.12	Plug, Reverse Gearshift Rod—Lower	3D2-351					✓	✓				1
	......	.35	Screw, Gearshift Arm and Fork	3D2-216					✓	✓				3
	......	.35	Plunger, Gearshift Rod	3T1-345					✓	✓				4
	......	.12	Spring, Gearshift Rod	3T1-11				✓	✓	✓	✓	✓	✓	4
	......	.35	Plunger, Gearshift Rod Interlock	3D2-224					✓	✓				2
	......	.15	Ball, Gearshift Rod Interlock	S-1676					✓	✓				1
	......	.06	Pin, Gearshift Rod Interlock	3Y1-173					✓	✓	✓	✓	✓	2
	......	.04	Stop, Gearshift Rod Interlock Pin (1/8" Dia. x 7/8")	S-1546					✓	✓	✓	✓	✓	2
	......	1.50	Cover, Rear Gearshift Rod—Top	3D3-24					✓	✓				1
	......	.12	Gasket, Rear Gearshift Rod Cover	3D2-152					✓	✓				1
	......	.06	Screw, Rear Gearshift Rod Cover	S-46					✓	✓				2
	......	.01	Lockwasher, Spring	S-1347					✓	✓				2
	......	.06	Screw, Front Gearshift Rod Stop (Overdr., Reverse & Second)	S-1232					✓	✓				3
	......	.10	Screw, Front Gearshift Rod Stop (Direct)	S-1236					✓	✓				1
	......	.06	Screw, Rear Gearshift Rod Stop (Third and Overdrive)	S-1232					✓	✓				2
	......	.10	Screw, Rear Gearshift Rod Stop (Reverse)	S-1233					✓	✓				1
	......	.04	Screw, Gearshift Rod Headless Set (5/16"—18 x 1")	S-1235					✓	✓	✓	✓	✓	5
	......	.03	Nut, Hex. (5/16"—18)	S-82					✓	✓	✓	✓	✓	5
	......	.90	Ball, Gearshift Lever	10T2-408					✓	✓	✓	✓	✓	1
	......	.20	Cover, Reverse Gearshift Rod Plunger	3D0-290					✓	✓				1
	......	.12	Gasket, Reverse Gearshift Rod Plunger Cover	3D2-193					✓	✓				1
	......	.05	Screw, Reverse Gearshift Rod Plunger Cover	S-3377					✓	✓				2
	......	.01	Lockwasher, Spring	S-1346					✓	✓				2
	......	.35	Plunger, Reverse Gearshift Rod—Lower	3UDF2-345					✓	✓				1
	......	.12	Spring, Reverse Gearshift Rod Plunger	3Y1-111A					✓	✓				1
	......	14.50	Housing, Gearshift Finger—Assembly	3URL0-1350						✓				1
	......	10.50	Housing, Gearshift Finger	3URL6-309						✓				1
	......	.15	Gasket, Gearshift Finger Housing	3D3-17						✓				1
	......	1.25	Cap, Gearshift Finger Housing	3UNA3-962						✓				1
	......	.10	Gasket, Gearshift Finger Housing Cap	3URL2-961						✓				1
	......	.50	Plate, Gearshift Finger Housing Cover—and Stop Assembly	3URL0-1630						✓				1
	......	.12	Gasket, Gearshift Finger Housing Cover Plate	3URL2-968						✓				1
	......	4.40	Finger, Gearshift	3URL4-88A						✓				1
	......	1.20	Rod, Gearshift Finger Shift	3URL3-64						✓				1
	......	2.20	Socket, Gearshift Lever—and Bushing Assembly	10U0-1205						✓				1
	......	2.00	Lever, Gearshift	10URL4-23A						✓				1
	......	3.90	Bracket, Gearshift Lever	3D3-353						✓				1
	......	5.80	Support, Gearshift Lever Bracket	10URL4-265A						✓				1
	......	4.75	Stub, Gearshift Lever	10URL3-1137A						✓				1
	......	2.00	Socket, Gearshift Lever	10UN2-1006						✓				1
	......	.06	Key, Gearshift Lever Socket (#5 Whitney)	S-2316						✓				1
	......	.03	Pin, Gearshift Lever Socket (5/16" x 2")	S-3217						✓				1
	......	.35	Bushing, Gearshift Lever Socket	10U2-86						✓				1
	......	.20	Bushing, Gearshift Lever Socket Split	10Y1-84						✓				2
	......	.90	Socket, Gearshift Lever	10URL2-1138						✓				1
	......	1.20	Cover, Gearshift Lever Socket	10BB2-05						✓				1
	......	.25	Pin, Gearshift Lever Socket	10URL2-409						✓				1
	......	.20	Spring, Gearshift Lever Socket	10T2-391A						✓				1
	......	.25	Keeper, Gearshift Lever Socket Spring	10URL2-1166						✓				1
	......	.20	Seal, Gearshift Lever Socket Oil	10D2-339						✓				1
	......	2.20	Ball, Gearshift—Front	10UN2-16A						✓				1
	......	.35	Gate, Gearshift	10D2-18						✓				1
	......	.90	Bushing, Gearshift Lever Ball	10Y1-86						✓				1
	......	.20	Bushing, Gearshift Lever Ball Split	10Y1-84						✓				2
	......	13.25	Tube, Gearshift Connecting—Assembly	10URL4-300A						✓				1
	......	2.00	Tube, Gearshift Connecting	10URL2-64A						✓				1
	......	.40	Stud, Gearshift Connecting Tube	10UD2-1045						✓				1
	......	.12	Nut, Gearshift Connecting Tube Stud (7/8"—14)	S-1409						✓				1
	......	.01	Pin, Gearshift Connecting Tube Stud (3/16" Dia. x 1 5/16")	S-3237						✓				1

07-8 TRANSMISSION
Master Parts List — The Autocar Company, Ardmore, PA.

CODE	MFR. PART No.	LIST PRICE	NAME OF PART	AUTOCAR PART No.	U-2044	U-4044	U-4144	U-4144-T	C-50	U-5044	U-7144-T	U-8144	U-8144-T	No. REQ.
0706	......	$0.04	**Washer,** Gearshift Connecting Tube Stud (7/8" I.D.—1 3/4" O.D. x 1/16")............	S-346						✓				1
	......	.75	**Tube,** Gearshift Connecting—Outer............	10URL2-1014A						✓				1
	......	.12	**Felt,** Gearshift Connecting Tube Bracket............	10UN2-1005						✓				2
	......	.06	**Bracket,** Gearshift Connecting Tube—Retainer, Felt............	10UN2-1015						✓				2
	......	1.80	**Bearing,** Gearshift Connecting Tube—Front............	10BB2-1007						✓				1
	......	1.40	**Bearing,** Gearshift Connecting Tube—Rear............	10UNF2-1008						✓				1
	......	.40	**Line,** Gearshift Front Bearing Oil............	10UA0-1018						✓				1
	......	.25	**Line,** Gearshift Rear Bearing Oil............	10UA0-1019						✓				1
	......	2.40	**Rod,** Third and Fourth Speed Gearshift............	3UU3-83							✓	✓	✓	1
	......	6.00	**Fork,** Third and Fourth Speed Shift............	3TF4-15C							✓	✓	✓	1
	......	.12	**Screw** (7/16"—20 x 1 3/8")............	3A2-629							✓	✓	✓	1
	......	2.10	**Rod,** Overdrive Shift—Upper............	3UU3-189							✓	✓	✓	1
	......	4.50	**Arm,** Overdrive Shift Rod............	3DF4-971							✓	✓	✓	1
	......	5.50	**Fork,** Overdrive Shift............	3DF4-346A							✓	✓	✓	1
	......	2.10	**Rod,** Overdrive Shift—Lower............	3UNF3-189							✓	✓	✓	1
	......	.30	**Pin,** Overdrive Shift Rod Arm and Fork Connecting............	3DF2-927							✓	✓	✓	1
	......	.04	**Nut,** Hex. (7/16"—14)............	S-850							✓	✓	✓	1
	......	1.80	**Rod,** First and Second Speed Gearshift............	3UU3-84							✓	✓	✓	1
	......	3.00	**Jaw,** First and Second Speed Gearshift Rod............	3DF2-191							✓	✓	✓	1
	......	4.80	**Fork,** First and Second Speed Gearshift............	3DF3-87A							✓	✓	✓	1
	......	2.10	**Rod,** Reverse Gearshift—Upper............	3UU3-85							✓	✓	✓	1
	......	6.00	**Arm,** Reverse Gearshift Rod............	3DF4-188							✓	✓	✓	1
	......	.12	**Screw** (7/16"—14 x 1 3/8")............	3DF2-629							✓	✓	✓	5
	......	5.70	**Fork,** Reverse Gearshift............	3TFB4-35							✓	✓	✓	1
	......	.06	**Screw** (7/16"—14 x 1 5/8")............	3TF2-629							✓	✓	✓	1
	......	.12	**Pin,** Reverse Gearshift Rod Arm and Fork Connecting............	3DF2-928							✓	✓	✓	1
	......	.01	**Pin,** Reverse Gearshift Rod Arm & Fork Connecting............	3DF2-928							✓	✓	✓	1
	......	.12	**Screw,** Reverse Gearshift Fork Guide............	3DF2-925							✓	✓	✓	1
	......	.04	**Nut,** Hex. (1/2"—13)............	S-77							✓	✓	✓	1
	......	2.10	**Rod,** Reverse Gearshift—Lower............	3UNF3-85							✓	✓	✓	1
	......	.55	**Plunger,** Gearshift Rod............	3UU2-345							✓	✓	✓	4
	......	.35	**Plunger,** Gearshift Rod Interlock............	3TF2-224A							✓	✓	✓	3
	......	.90	**Cover,** Rear Gearshift Rod—Top............	3DF3-24							✓	✓	✓	1
	......	.70	**Cover,** Rear Gearshift Rod—Side............	3UDF3-24							✓	✓	✓	1
	......	.06	**Gasket,** Rear Gearshift Rod Cover............	3DF2-152							✓	✓	✓	2
	......	.05	**Screw,** Hex. Cap (3/8"—16 x 1")............	S-23							✓	✓	✓	4
	......	.01	**Lockwasher,** 1 3/32" Spring............	S-1347							✓	✓	✓	4
	......	5.35	**Stub,** Gearshift Lever............	10UU3-1137							✓	✓	✓	1
	......	16.80	**Lever,** Gearshift—Only............	10UU4-23							✓	✓	✓	1
	......	.18	**Screw,** Hex. Cap (1/2"—20 x 2 1/4")............	S-1868							✓	✓	✓	1
	......	.04	**Nut,** Hex. (1/2"—20)............	S-1420							✓	✓	✓	1
	......	.01	**Lockwasher,** 1 7/32" Spring............	S-1349							✓	✓	✓	2
	......	4.95	**Bracket,** Gearshift Lever............	3UU4-353							✓	✓	✓	1
	......	.12	**Screw,** Headless Set (5/16"—24 x 1 1/4")............	3S2-82A							✓	✓	✓	1
	......	.06	**Nut,** Hex. (5/16"—24)............	S-2021							✓	✓	✓	1
	......	.06	**Dowel,** Gearshift Lever Bracket............	3D2-355							✓	✓	✓	2
	......	.06	**Screw,** Hex. Cap (3/8"—16 x 1 1/8")............	S-46							✓	✓	✓	4
	......	.01	**Lockwasher,** 1 3/32" Spring............	S-1347							✓	✓	✓	4
	......	1.20	**Cover,** Gearshift Lever Socket............	10T2-05							✓	✓	✓	1
	......	.20	**Seal,** Gearshift Lever Socket Oil............	10D2-339							✓	✓	✓	1
	......	.45	**Pin,** Gearshift Lever Socket Center............	10UU2-409							✓	✓	✓	1
	......	.35	**Spring,** Gearshift Lever Socket............	10UU2-391							✓	✓	✓	1
	......	.10	**Gasket,** Gearshift Gate............	10D2-249							✓	✓	✓	1
	......	.60	**Plunger,** Gearshift Lockout (Reverse)............	3UU2-1129							✓	✓	✓	2
	......	.15	**Spring,** Gearshift Lockout Plunger............	2SA2-832							✓	✓	✓	2
	......	.25	**Nut,** Gearshift Lockout Plunger Spring............	3UU2-1131							✓	✓	✓	2
	......	.06	**Ball,** Gearshift Lockout Plunger............	S-1672							✓	✓	✓	2
	......	.06	**Spring,** Gearshift Lockout Plunger Ball............	3UU2-1133							✓	✓	✓	2

THE AUTOCAR COMPANY, ARDMORE, PA.

Master Parts List

GROUP 08

ILLUSTRATIONS

GROUP 08

TRANSFER CASE

GROUP 08

Master Parts List

THE AUTOCAR COMPANY, ARDMORE, PA.

As supplied on Autocar Models
U-8144   U-8144-T

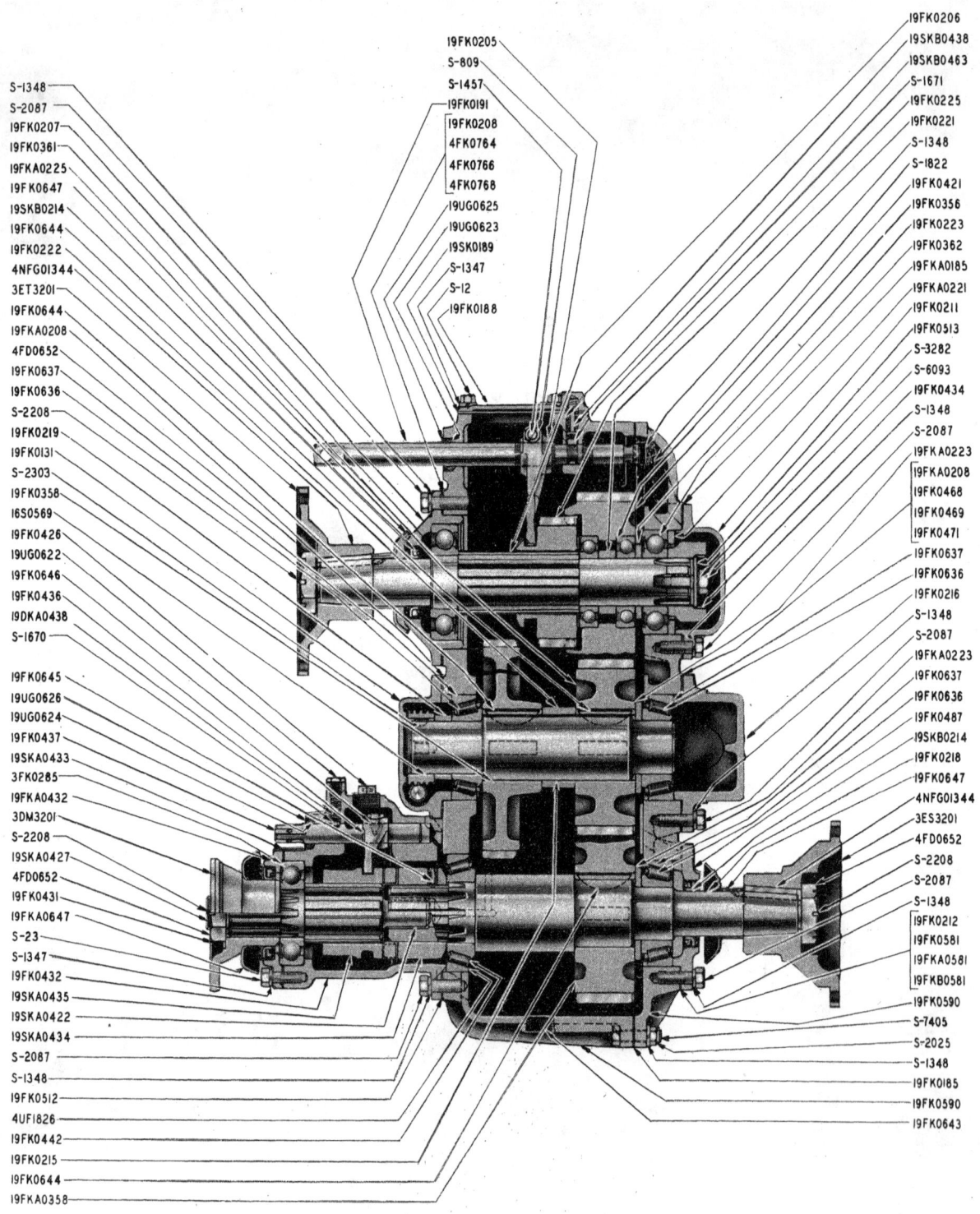

0800—Transmission Transfer Case Assembly

THE AUTOCAR COMPANY, ARDMORE, PA.

Master Parts List

GROUP 08

As supplied on Autocar Models
U-8144   U-8144-T

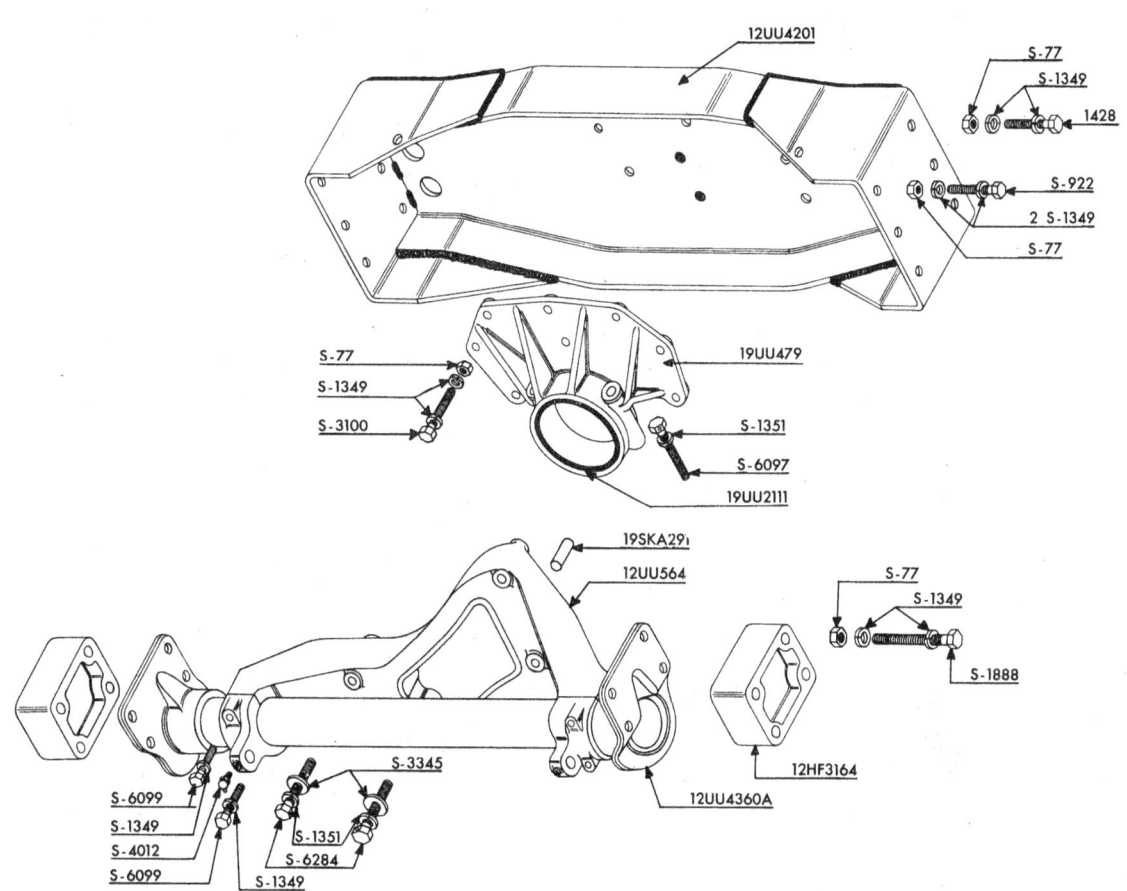

0801—Transfer Case

GROUP 08

Master Parts List

THE AUTOCAR COMPANY, ARDMORE, PA.

As supplied on Autocar Models U-8144   U-8144-T

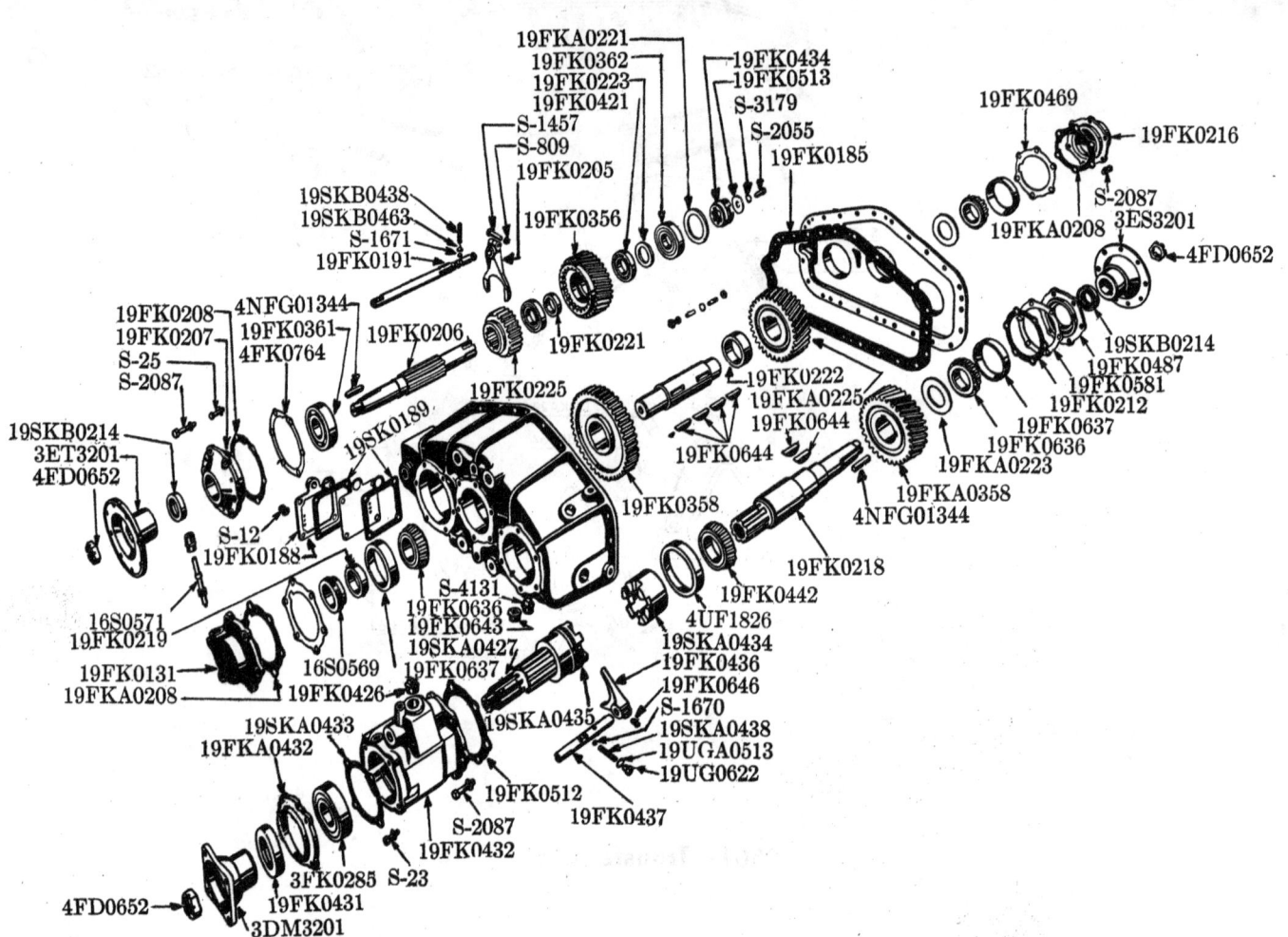

0802—Transfer Case Drive Gear, Mainshaft and Bearings

Master Parts List

THE AUTOCAR COMPANY, ARDMORE, PA.

GROUP 08

As supplied on Autocar Models U-8144   U-8144-T

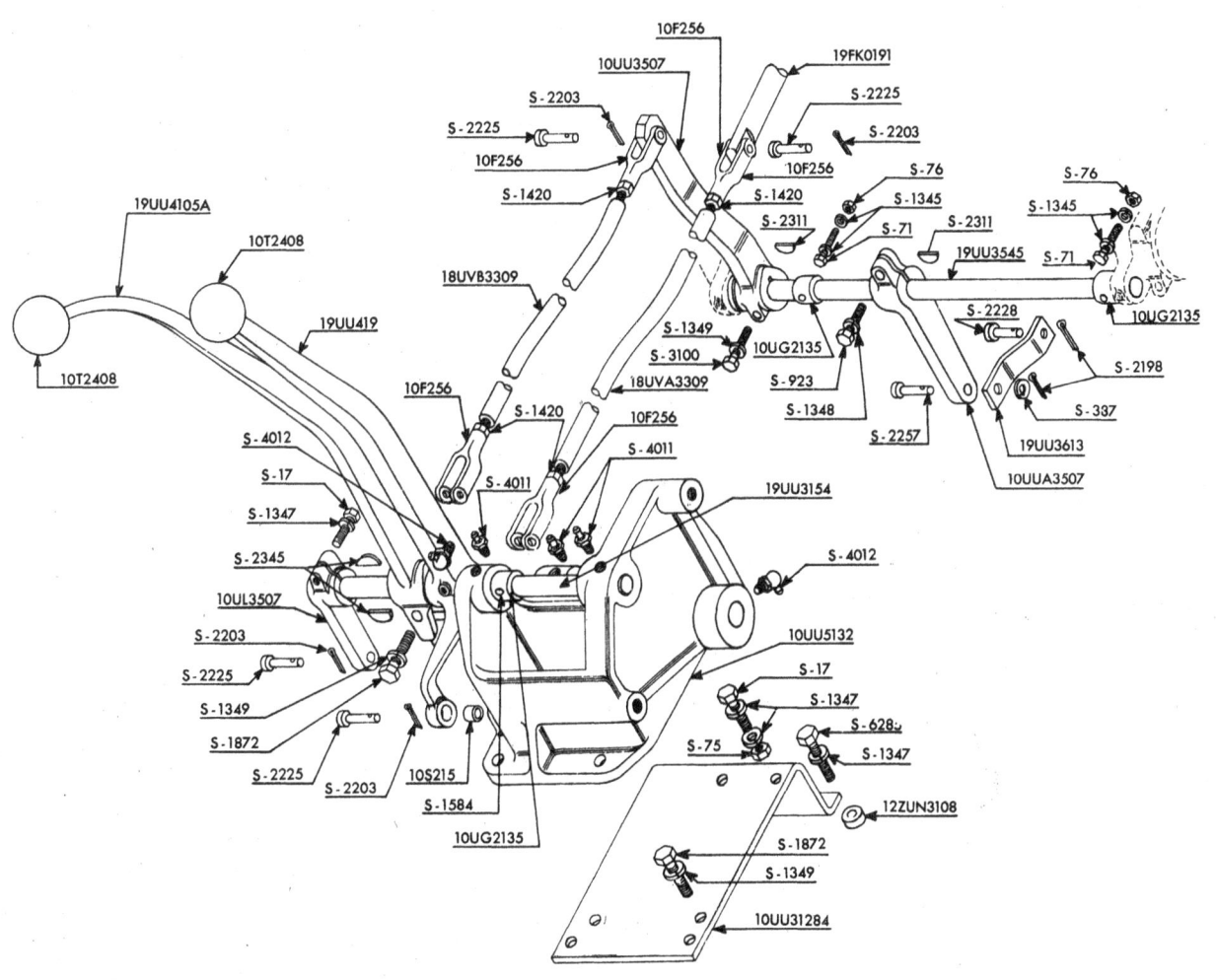

0805—Shifting Shafts, Yokes and Shift Levers

GROUP 08

Master Parts List

THE AUTOCAR COMPANY, ARDMORE, PA.

As supplied on Autocar Models
U-8144   U-8144-T

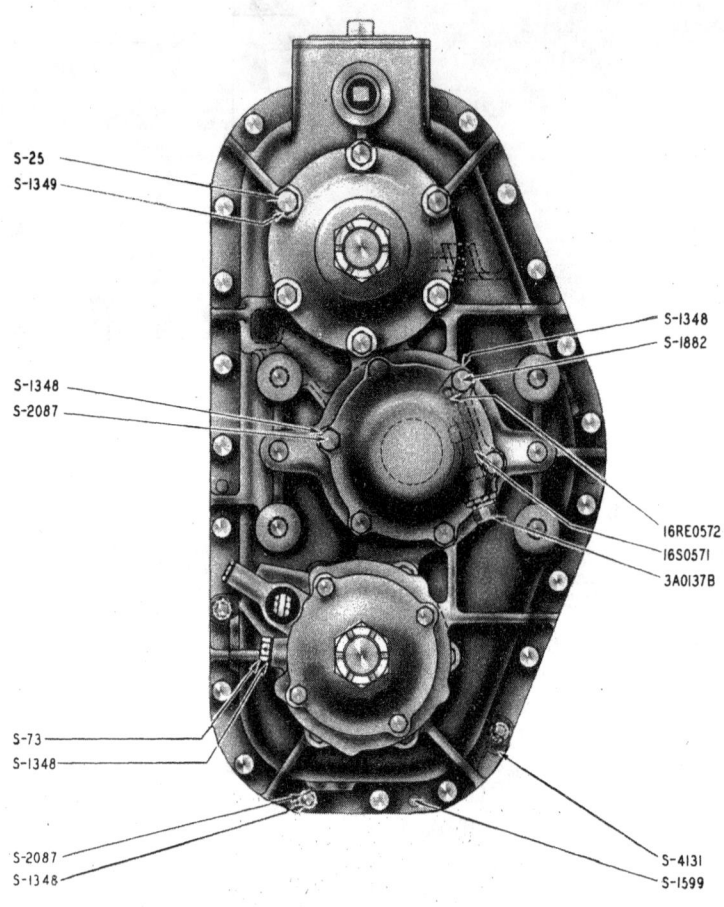

0805—Transfer Case Cover

CODE	MFR. PART No.	LIST PRICE	NAME OF PART	AUTOCAR PART No.	MODEL U-2044	U-4044	U-4144	U-4144-T	C-50	U-5044	U-7144-T	U-8144	U-8144-T	No. REQ.
0800	**Transmission Transfer Case Assembly**													
	T-2-B7-5	$389.50	**Case,** Transfer—Assembly (with Declutching Unit)	19BK0-440	√	√	√	√		√				1
	T-76-2	737.00	**Case,** Transfer—Assembly, Complete (Wisconsin)	19FK2-440							√	√	√	1
0801	**Transfer Case**													
	A-3875-E-5	63.20	**Case,** Transfer—and Cover Assembly	19UG0-590	√	√	√	√		√				1
	A-3875-C-133	117.60	**Case,** Transfer—and Cover Assembly	19FK0-590							√	√	√	1
	2808-G-215	.17	**Gasket,** Transfer Case and Cover	19DKB0-185	√	√	√	√		√				1
	2808-A-365	.17	**Gasket,** Transfer Case and Cover	19FK0-185							√	√	√	1
	#7 x 1¼	.07	**Pin,** Transfer Case and Cover Taper	S-1599	√	√	√	√		√	√	√	√	4
	S-1713	.04	**Screw,** Hex. Cap (7/16″—20 x 1⅝″)	S-7405							√	√	√	18
	N-17	.02	**Nut,** Hex. (7/16″—20)	S-2025							√	√	√	18
	S-2714	.22	**Screw,** Hex. Cap (7/16″—14 x 1⅜″)	S-2087							√	√	√	3
	......	.01	**Lockwasher,** 15/32″ Spring	S-1348							√	√	√	3
	1898-S-201	.32	**Filler,** Transfer Case Oil	19FK0-638							√	√	√	1
	X-300	.03	**Plug,** Transfer Case Oil Filler Pipe (¾″)	S-4131	√	√	√	√		√	√	√	√	1
	P-212	.08	**Plug,** Transfer Case Drain Pipe (1″)	19FK0-643							√	√	√	1
	S-1613	.02	**Screw,** Transfer Case Cover	19UG0-619	√	√	√	√		√				20
	N-16	.03	**Nut,** Transfer Case Cover Screw (⅜″—24)	S-1418	√	√	√	√		√				24
	W-16	.01	**Washer,** ⅜″ Spring	S-1347							√	√	√	46
	......	20.15	**Member,** Transfer Case Cross—Assembly	12UU4-360A							√	√	√	1
	......	.15	**Screw,** Hex. Cap (½″—13 x 3¾″)	S-1888							√	√	√	8
	......	.04	**Nut,** Hex. (½″—13)	S-77							√	√	√	8
	......	.01	**Lockwasher,** 17/32″ Spring	S-1349							√	√	√	16
	......	3.35	**Shim,** Transfer Case Support Tube	12HF3-164							√	√	√	2
	......	18.00	**Support,** Transfer Case—Front	12UU5-64							√	√	√	1
	......	.25	**Insulation,** Transfer Case Bracket	19HF2-111							√	√	√	8
	......	.05	**Pin,** Transfer Case Dowel	19SKA2-91							√	√	√	2
	......	.07	**Screw,** Hex. Cap (½″—13 x 3″)	S-6099							√	√	√	4
	......	.01	**Lockwasher,** 17/32″ Spring	S-1349							√	√	√	4
	......	.20	**Fitting,** #1612 Alemite	S-4012							√	√	√	2
	......	.16	**Screw,** Hex. Cap (⅝″—11 x 4″ Drilled Head)	S-6284							√	√	√	4
	......	.02	**Wire,** #16 B.W.G.—.065″ x 13″ Tie	S-4284							√	√	√	2
	......	.01	**Lockwasher,** 21/32″ Spring	S-1351							√	√	√	4
	......	.04	**Washer,** Flat Steel (6 11/16″ x 1¼″ x ⅛″)	S-3345							√	√	√	4
	......	29.65	**Member,** Transfer Case Support—Rear	12UU4-201							√	√	√	1
	......	.09	**Screw,** Hex. Cap (½″—13 x 1⅜″)	S-1428							√	√	√	10
	......	.10	**Screw,** Hex. Cap (½″—13 x 2″)	S-922							√	√	√	2
	......	.04	**Nut,** Hex. (½″—13)	S-77							√	√	√	12
	......	.01	**Lockwasher,** 17/32″ Spring	S-1349							√	√	√	24
	......	9.10	**Support,** Transfer Case—Rear	19UU4-79							√	√	√	1
	......	.05	**Screw,** Hex. Cap (½″—13 x 1¾″)	S-3100							√	√	√	6
	......	.04	**Nut,** Hex. (½″—13)	S-77							√	√	√	6
	......	.01	**Lockwasher,** 17/32″ Spring	S-1349							√	√	√	12
	......	.75	**Insulator,** Transfer Case	19UU2-111							√	√	√	1
	......	.06	**Screw,** Hex. Cap (⅝″—11 x 3″)	S-6097							√	√	√	2
	......	.01	**Lockwasher,** 21/32″ Spring	S-1351							√	√	√	2
	......	.12	**Wire,** .062″ Soft Iron—15″ Long	S-4107							√	√	√	1
0802	**Transfer Case Drive Gear, Main Shaft and Bearings**													
	3880-H-60	12.15	**Shaft,** Main Drive	19DKB0-206	√	√	√	√		√				1
	2808-B-158	.06	**Gasket,** Main Drive Front and Rear Bearing Cover	19DKA0-212	√	√	√	√		√				2
	3866-C-133	1.25	**Cover,** Main Drive Shaft Rear Bearing	19DKA0-211	√	√	√	√		√				1
	A3866-M-91	3.85	**Cover,** Main Drive Front	19DKB0-570	√	√	√	√		√				2
	A1805-B-80	1.20	**Seal,** Main Drive Front Oil	19DKB0-214	√	√	√	√		√				1
	1829-J-166	.35	**Ring,** Main Drive Shaft Front Bearing	19DKA0-224	√	√	√	√		√				1
	3894-K-167	26.50	**Gear,** Main Drive Shaft Direct Drive (37 Teeth)	19UGA0-225	√	√	√	√		√				1
	3892-V-282	71.60	**Gear,** Main Drive Shaft Slide (21 Teeth)	19UG0-225	√	√	√	√		√				1
	1825-P-16	2.30	**Bushing,** Main Drive Shaft Direct Drive Gear	19DKA0-503	√	√	√	√		√				1
	1846-S-19	.01	**Pin,** Main Drive Shaft Direct Drive Gear Bushing	19DKA0-504	√	√	√	√		√				1
	1309	8.50	**Bearing,** Main Drive Shaft Front (S.K.F.)	3CKA2-285	√	√	√	√		√				1
	1407	6.94	**Bearing,** Main Drive Shaft Rear (S.K.F.)	19DKB0-362	√	√	√	√		√				1

08-2 TRANSFER CASE

Master Parts List

THE AUTOCAR COMPANY, ARDMORE, PA.

CODE	MFR. PART No.	LIST PRICE	NAME OF PART	AUTOCAR PART No.	U-2044	U-4044	U-4144	U-4144-T	C-50	U-5044	U-7144-T	U-8144	U-8144-T	No. REQ.
0802	13399	$0.15	Nut, Main Drive and Driven Shaft	4A0-81	✓	✓	✓	✓	✓	✓				3
	3880-M-325	18.20	Shaft, Main	19FK0-206							✓	✓	✓	1
	......	8.80	Flange, Main Shaft	3ET3-201							✓	✓	✓	1
	16-X-23	.18	Key, Main Shaft Flange	4NFG0-1344							✓	✓	✓	2
	14-X-57	.39	Nut, Main Shaft	4FD0-652							✓	✓	✓	1
	......	.01	Cotter, 1/8" x 2"	S-2208							✓	✓	✓	1
	2848-Q-95	5.10	Clutch, Main Shaft Power Takeoff	19FK0-434							✓	✓	✓	1
	1829-D-394	.15	Washer, Main Shaft Power Takeoff Clutch	19FK0-513							✓	✓	✓	1
	......	.03	Screw, Hex. Cap (1/2"—20 x 1 1/4")	S-2055							✓	✓	✓	1
	......	.04	Lockwasher, 1/2" Shakeproof	S-3179							✓	✓	✓	1
	3892-T-722	27.00	Gear, Main Drive	19FK0-356							✓	✓	✓	1
	7209	3.56	Bearing, Main Drive Gear Idler	19FK0-421							✓	✓	✓	2
	1844-R-174	.86	Spacer, Main Drive Gear Idler Bearing	19FK0-221							✓	✓	✓	1
	1829-C-393	1.15	Washer, Main Drive Gear Idler Bearing	19FK0-223							✓	✓	✓	1
	3892-R-720	13.50	Gear, Main Shaft Low Speed Sliding	19FK0-225							✓	✓	✓	1
	7609	4.68	Bearing, Main Shaft Rear	19FK0-362							✓	✓	✓	1
	1829-A-391	1.75	Washer, Main Shaft Rear Bearing	19FKA0-221							✓	✓	✓	1
	3866-T-384	1.95	Cover, Main Shaft Rear Bearing	19FK0-211							✓	✓	✓	1
	2808-W-361	.05	Gasket, Main Shaft Rear Bearing Cover	19FKA0-185							✓	✓	✓	1
	......	.05	Screw, Hex. Cap (7/16"—14 x 1 3/8")	S-2087							✓	✓	✓	5
	......	.01	Lockwasher, 7/16" Spring	S-1348							✓	✓	✓	5
	7610	5.52	Bearing, Main Shaft Front	19FK0-361							✓	✓	✓	1
	3838-H-86	3.00	Cage, Main Shaft Front Bearing	19FK0-207							✓	✓	✓	1
	A-1805-L-168	.85	Seal, Main Shaft Front Bearing Cage Oil	19SKB0-214							✓	✓	✓	1
	2808-Y-63	.05	Gasket, Main Shaft Front Bearing Cage	19FK0-208							✓	✓	✓	1
	2803-R-616	.07	Shim—Thin	4FK0-764							✓	✓	✓	As
	2803-S-617	.07	Shim—Medium	4FK0-766							✓	✓	✓	As
	2803-T-618	.07	Shim—Thick	4FK0-768							✓	✓	✓	As
	......	.05	Screw, Hex. Cap (7/16"—14 x 1 3/8")	S-2087							✓	✓	✓	5
	......	.01	Lockwasher, 7/16" Spring	S-1348							✓	✓	✓	5
	S-2811	.09	Screw, Hex. Cap (1/2"—13 x 1 1/4")	S-25							✓	✓	✓	1
	W-18	.01	Lockwasher, 1/2" Spring	S-1349							✓	✓	✓	1
0803	**Transfer Case Driven Gears, Shafts and Bearings (Rear Axle Drive)**													
	3892-S-721	19.00	Gear, Driven	19FKA0-358							✓	✓	✓	1
	VX-Woodruff	....	Key, Driven Gear	19FK0-644							✓	✓	✓	2
	3880-N-326	18.20	Shaft, Driven	19FK0-218							✓	✓	✓	1
	3866-R-382	2.10	Cap, Driven Shaft Rear Bearing	19FK0-487							✓	✓	✓	1
	2803-U-619	.07	Shim, Driven Shaft Rear Bearing Cap—Thin	19FK0-581							✓	✓	✓	As
	2803-V-620	.07	Shim, Driven Shaft Rear Bearing Cap—Medium	19FKA0-581							✓	✓	✓	As
	2803-W-621	.07	Shim, Driven Shaft Rear Bearing Cap—Thick	19FKB0-581							✓	✓	✓	As
	2808-Z-362	.05	Gasket, Driven Shaft Rear Bearing Cap	19FK0-212							✓	✓	✓	1
	......	.05	Screw, Hex. Cap (7/16"—14 x 1 3/8")	S-2087							✓	✓	✓	4
	......	.01	Lockwasher, 7/16" Spring	S-1348							✓	✓	✓	4
	A-1805-L-168	.85	Seal, Driven Shaft Rear Bearing Cap Oil	19SKB0-214							✓	✓	✓	1
	49520	3.02	Cup, Driven Shaft Rear Bearing	19FK0-637							✓	✓	✓	1
	49585	5.88	Cone, Driven Shaft Rear Bearing	19FK0-636							✓	✓	✓	1
	3920	3.90	Cup, Driven Shaft Front Bearing	4UF1-826							✓	✓	✓	1
	3979	5.85	Cone, Driven Shaft Front Bearing	19FK0-442							✓	✓	✓	1
	1825-E-57	.35	Bushing, Driven Shaft	19SKA0-422							✓	✓	✓	1
	1829-B-392	1.70	Washer, Driven Shaft	19FKA0-223							✓	✓	✓	1
	14-X-57	.39	Nut, Driven Shaft	4FD0-652							✓	✓	✓	1
	K-2518	.01	Cotter 1/8" x 2"	S-2208							✓	✓	✓	1
	16-X-23	.18	Key, Driven Shaft	4NFG0-1344							✓	✓	✓	1
	5-1-1544	14.40	Flange, Driven Shaft	3ES3-201							✓	✓	✓	1
	Front Axle Drive													
	3880-T-46	11.45	Shaft, Declutch	19SKA0-427							✓	✓	✓	1
	5-X-349	.10	Packing, Declutch Shaft	19UG0-624							✓	✓	✓	1
	1805-W-75	.05	Retainer, Declutch Shaft Packing	19UG0-626							✓	✓	✓	1
	3826-C-185	10.25	Carrier, Declutch Shaft Bearing	19FK0-432							✓	✓	✓	1
	1850-Q-43	.15	Plug, Declutch Shaft Bearing Carrier Oil	19FK0-426							✓	✓	✓	1

08-3 TRANSFER CASE

THE AUTOCAR COMPANY, ARDMORE, PA. — Master Parts List

CODE	MFR. PART No.	LIST PRICE	NAME OF PART	AUTOCAR PART No.	U-2044	U-4044	U-4144	U-4144-T	C-50	U-5044	U-7144-T	U-8144	U-8144-T	No. REQ.
0803	2808-V-360	$0.05	Gasket, Declutch Shaft Bearing Carrier	19FK0-512							✓	✓	✓	1
	......	.05	Screw, Declutch Shaft Bearing Carrier to Case, Hex. Cap ($\frac{7}{16}$"—14 x $1\frac{3}{8}$")	S-2087							✓	✓	✓	6
	......	.01	Lockwasher, $\frac{15}{32}$" Spring	S-1348							✓	✓	✓	6
	3866-P-354	2.40	Cap, Declutch Shaft Bearing	19FKA0-432							✓	✓	✓	1
	2808-P-94	.06	Gasket, Declutch Shaft Bearing Cap	19SKA0-433							✓	✓	✓	1
	S-2610	.05	Screw, Hex. Cap ($\frac{3}{8}$"—16 x 1")	S-23							✓	✓	✓	4
	W-16	.01	Lockwasher, $\frac{13}{32}$" Spring	S-1347							✓	✓	✓	4
	7309	4.82	Bearing, Declutch Shaft	3FK0-285							✓	✓	✓	1
	A-1805-Z-104	1.10	Seal, Declutch Shaft Oil	19FK0-431							✓	✓	✓	1
	2858-S-45	.10	Spring, Declutch Shift Lock	19SKA0-438							✓	✓	✓	1
	3-X-146	.12	Screw, Declutch Shift Lock Spring	19UG0-622							✓	✓	✓	1
	1829-W-257	.04	Lockwasher, Declutch Shift Lock Spring Screw	19UGA0-513							✓	✓	✓	1
	2849-P-94	3.00	Fork, Declutch Shift	19FK0-436							✓	✓	✓	1
	26-X-52	.07	Screw, Declutch Shift Fork	19FK0-646							✓	✓	✓	1
	2848-W-23	10.00	Clutch, Declutch Driving	19SKA0-434							✓	✓	✓	1
	S-664	.12	Lockscrew, Declutch Driving Clutch	19FK0-645							✓	✓	✓	1
	2848-V-22	13.00	Clutch, Declutch Sliding	19SAK0-435							✓	✓	✓	1
	2843-F-110	8.80	Shaft, Declutch Shifting	19FK0-437							✓	✓	✓	1
	1898-R-70	.03	Ball, Declutch Shifting Lock	S-1670							✓	✓	✓	1
	......	5.75	Flange, Declutch Shaft	3DM3-201							✓	✓	✓	1
	14-X-57	.39	Nut, Declutch Shaft	4FD0-652							✓	✓	✓	1
	K-2518	.01	Cotter, $\frac{1}{8}$" x 2"	S-2208							✓	✓	✓	1
	3826-Q-173	11.00	Carrier, Declutching Shaft	19UG0-425	✓	✓	✓	✓		✓				1
	A-3880-N-40	24.00	Shaft, Declutching, and Bushing Assembly	19UG0-427	✓	✓	✓	✓		✓				1
	1825-M-13	1.15	Bushing, Declutching Shaft	19DKA0-502	✓	✓	✓	✓		✓				1
	1308	6.80	Bearing, Declutching Shaft (S.K.F.)	3SA2-285	✓	✓	✓	✓		✓				1
	1829-Y-103	.15	Slinger, Declutching Shaft Bearing Oil	19DKA0-429	✓	✓	✓	✓		✓				1
	A-1805-X-154	1.25	Seal, Declutching Shaft Bearing and Cap Oil	19UG0-431	✓	✓	✓	✓		✓				1
	3866-L-64	2.40	Cap, Declutching Shaft Bearing, Assembly	19UG0-760	✓	✓	✓	✓		✓				1
	2808-J-88	.05	Gasket, Declutching Shaft Bearing Cap	19DKA0-511	✓	✓	✓	✓		✓				1
	2848-C-3	9.00	Clutch, Sliding	19DKA0-435	✓	✓	✓	✓		✓				1
	2849-G-85	3.30	Fork, Declutching Shifter	19UG0-479	✓	✓	✓	✓		✓				1
	2843-P-94	2.55	Shaft, Declutching Shifter	19UG0-437	✓	✓	✓	✓		✓				1
	1898-R-70	.03	Ball, Declutching Shifter Shaft Lock ($\frac{3}{8}$" Dia.)	S-1670	✓	✓	✓	✓		✓				1
	2858-S-45	.10	Spring, Declutching Shifter Shaft Lock	19SKA0-438	✓	✓	✓	✓		✓				1
	3-X-146	.12	Screw, Declutching Shifter Lock Spring	19UG0-622	✓	✓	✓	✓		✓				1
	26-X-52	.07	Screw, Declutching Shifter Lever	19FK0-646	✓	✓	✓	✓		✓				1
	2808-J-88	.05	Retainer, Declutching Shifter Shaft Packing	19UG0-626	✓	✓	✓	✓		✓				1
	5-X-349	.10	Packing, Declutching Shifter Shaft	19UG0-624	✓	✓	✓	✓		✓				1
	1829-K-115	.25	Washer, Declutching Shaft End	19UG2-513	✓	✓	✓	✓		✓				1
	1850-Q-43	.15	Plug, Declutching Carrier Oil Filler	19FK0-426	✓	✓	✓	✓		✓				1
	2808-U-281	.07	Gasket, Declutching Carrier	19DKA0-512	✓	✓	✓	✓		✓				1
	2803-P-224	.14	Shim, Declutching Carrier—Thin	19DKA0-507	✓	✓	✓	✓		✓				As
	2803-Q-225	.14	Shim, Declutching Carrier—Medium	19DKA0-508	✓	✓	✓	✓		✓				As
	2803-R-226	.20	Shim, Declutching Carrier—Thick	19DKA0-509	✓	✓	✓	✓		✓				As
	2808-Q-17	.03	Gasket, Driven Shaft Rear Bearing Cover	19DKA0-477	✓	✓	✓	✓		✓				1
	2803-F-6	.08	Shim, Driven Shaft Rear Bearing Cover—Thin	19DKA0-505	✓	✓	✓	✓		✓				As
	2803-Z-52	.11	Shim, Driven Shaft Rear Bearing Cover—Thick	19DKA0-506	✓	✓	✓	✓		✓				As
	A-3866-D-342	3.85	Cover, Driven Shaft Rear, Assembly	4UG0-1430	✓	✓	✓	✓		✓				1
	3880-P-42	16.95	Shaft, Driven	19DKA0-218	✓	✓	✓	✓		✓				1
	A-1805-M-169	.90	Seal, Driven Shaft Oil	19UG0-214	✓	✓	✓	✓		✓				1
	3894-Q-121	27.00	Gear, Driven Shaft (37 Teeth)	19DKA0-489	✓	✓	✓	✓		✓				1
	432	3.09	Cup, Driven Shaft Bearing (Timken)	4H2-137	✓	✓	✓	✓		✓				2
	438	5.20	Cone, Driven Shaft Bearing (Timken)	4ZGE0-653	✓	✓	✓	✓		✓				2
	16-X-28	.18	Key, Driven Shaft End	4NK0-163	✓	✓	✓	✓		✓				1
0804	Idler Gears, Bearings (Shaft and Caps)													
	3880-G-59	9.35	Shaft, Idler	19DKB0-215	✓	✓	✓	✓		✓				1
	3866-D-108	3.65	Cover, Idler Shaft Front Bearing	19DKA0-216	✓	✓	✓	✓		✓				1
	2808-F-214	.04	Gasket, Idler Shaft Front and Rear Bearing Cover	19DKB0-494	✓	✓	✓	✓		✓				1
	......	3.25	Cover, Idler Shaft Rear Bearing	19DKB3-131	✓	✓	✓	✓		✓				1

08-4 TRANSFER CASE — Master Parts List

THE AUTOCAR COMPANY, ARDMORE, PA.

CODE	MFR. PART No.	LIST PRICE	NAME OF PART	AUTOCAR PART No.	U-2044	U-4044	U-4144	U-4144-T	C-50	U-5044	U-7144-T	U-8144	U-8144-T	No. REQ.
0804	1829-F-6	$1.00	Washer, Idler Shaft Rear Bearing	4BLA0-1178	✓	✓	✓	✓		✓				1
	3892-W-283	20.00	Gear, Idler Shaft Low Speed (41 Teeth)	19UG0-358	✓	✓	✓	✓		✓				1
	3894-R-122	27.00	Gear, Idler Shaft Driven (37 Teeth)	19DKB0-357	✓	✓	✓	✓		✓				1
	1309	8.50	Bearing, Idler Shaft (S.K.F.)	3CKA2-285	✓	✓	✓	✓		✓				2
	1854-S-19	.37	Ring, Idler Shaft Front Bearing Snap	19DKA0-223	✓	✓	✓	✓		✓				1
	3892-U-723	18.75	Gear, Idler	19FKA0-225							✓	✓	✓	1
	VX-Woodruff	....	Key, Idler Gear	19FK0-644							✓	✓	✓	2
	1844-S-175	2.30	Spacer, Idler Gear	19FK0-222							✓	✓	✓	1
	3892-Q-719	19.75	Gear, Low Speed	19FK0-358							✓	✓	✓	1
	VX-Woodruff	....	Key, Low Speed Gear	19FK0-644							✓	✓	✓	2
	1484-T-176	1.05	Spacer, Speedometer Gear	19FK0-219							✓	✓	✓	1
	105946	2.50	Gear, Speedometer Drive (S-W)	16S0-569							✓	✓	✓	1
	3213	.06	Key, Speedometer Drive Gear	S-2303							✓	✓	✓	1
	105944	1.75	Gear, Speedometer Driven (S-W)	16S0-571							✓	✓	✓	1
	78744	.50	Sleeve, Speedometer Driven Gear (S-W)	3A0-137B							✓	✓	✓	1
	T-11012	.10	Bushing, Speedometer Driven Gear (S-W)	16RE0-572							✓	✓	✓	1
	3880-L-324	10.65	Shaft, Idler	19FK0-215							✓	✓	✓	1
	3866-S-383	4.60	Cap, Idler Shaft Front Bearing	19FK0-131							✓	✓	✓	1
	2808-X-362	.05	Gasket, Idler Shaft Front Bearing Cap	19FKA0-208							✓	✓	✓	1
	S-2719	.06	Screw, Hex. Cap ($\frac{7}{16}''$—14 x $2\frac{1}{4}''$)	S-1880							✓	✓	✓	2
	......	.05	Screw, Hex. Cap ($\frac{7}{16}''$—14 x $1\frac{3}{8}''$)	S-2087							✓	✓	✓	4
	......	.01	Lockwasher, $\frac{15}{32}''$ Spring	S-1348							✓	✓	✓	6
	3866-Q-381	3.50	Cap, Idler Shaft Rear Bearing	19FK0-216							✓	✓	✓	1
	2808-X-362	.05	Gasket, Idler Shaft Rear Bearing Cap	19FKA0-208							✓	✓	✓	1
	X-898	.05	Screw, Hex. Cap ($\frac{3}{8}''$—16 x $1''$)	S-23							✓	✓	✓	4
	W-17	.01	Lockwasher, $\frac{13}{32}''$ Spring	S-1347							✓	✓	✓	4
	2803-N-612	.07	Shim, Idler Shaft Front and Rear Bearing Cap—Thin	19FK0-468							✓	✓	✓	2
	2803-P-614	.07	Shim, Idler Shaft Front and Rear Bearing Cap—Medium	19FK0-469							✓	✓	✓	4
	2803-Q-615	.07	Shim, Idler Shaft Front and Rear Bearing Cap—Thick	19FK0-471							✓	✓	✓	6
	49520	3.02	Cup, Idler Shaft Rear Bearing	19FK0-637							✓	✓	✓	1
	49585	5.88	Cone, Idler Shaft Rear Bearing	19FK0-636							✓	✓	✓	1
	49520	3.02	Cup, Idler Shaft Front Bearing	19FK0-637							✓	✓	✓	1
	49585	5.88	Cone, Idler Shaft Front Bearing	19FK0-636							✓	✓	✓	1
	1829-B-392	1.70	Washer, Idler Shaft	19FKA0-223							✓	✓	✓	1
0805	**Shifting Shafts (Yokes and Shift Levers)**													
	3855-M-13	1.35	Cover, Shifter	19UG0-188	✓	✓	✓	✓		✓				1
	3855-H-60	1.40	Cover, Shifting	19FK0-188							✓	✓	✓	1
	S-266	.05	Screw, Hex. Cap ($\frac{3}{8}''$—16 x $\frac{3}{4}''$)	S-12	✓	✓	✓	✓		✓				
	W-16	.01	Washer, $\frac{13}{32}''$ Spring	S-1347							✓	✓	✓	4
	2808-H-2216	.05	Gasket, Shifting Cover	19SK0-189	✓	✓	✓	✓		✓	✓	✓	✓	2
	1898-E-57	.04	Ball, Gear Shift Lock ($\frac{7}{16}''$)	S-1671	✓	✓	✓	✓		✓	✓	✓	✓	1
	2858-A-1	.06	Spring, Gear Shift Lock	19SKB0-438							✓	✓	✓	1
	2858-D-82	.17	Spring, Shifter Lock Plunger	19UG0-202	✓	✓	✓	✓		✓				1
	1846-B-2	.18	Plunger, Gear Shift Lock Spring	19SKB0-463	✓	✓	✓	✓		✓	✓	✓	✓	1
	2849-N-14	3.00	Fork, Shifter	19SK0-205	✓	✓	✓	✓		✓				1
	2849-R-96	2.95	Fork, Gear Shift	19FK0-205							✓	✓	✓	1
	S-1614	.06	Screw, Hex. Cap ($\frac{3}{8}''$—24 x $1\frac{3}{4}''$)	S-1457							✓	✓	✓	1
	N-26	.05	Nut, Hex. ($\frac{3}{8}''$—24 Slotted)	S-809	✓	✓	✓	✓		✓	✓	✓	✓	1
	2843-U-99	6.40	Shaft, Shifter	19UG0-191	✓	✓	✓	✓		✓				1
	2843-R-122	5.30	Shaft, Gear Shift	19FK0-191							✓	✓	✓	1
	5-X-287	.10	Packing, Gear Shift Shaft	19UG0-623	✓	✓	✓	✓		✓	✓	✓	✓	1
	1805-Z-26	.10	Retainer, Gear Shift Shaft Packing	19UG0-625	✓	✓	✓	✓		✓	✓	✓	✓	1
	S-1614-X	.20	Screw, Shifter Fork	19UG0-627	✓	✓	✓	✓		✓				1
	......	.75	Support, Control Bracket	10UU3-1284							✓	✓	✓	1
	......	.30	Spacer, Control Bracket Support	12ZUN3-108							✓	✓	✓	2
	......	.03	Screw, Hex. Cap ($\frac{3}{8}''$—16 x $1\frac{5}{8}''$)	S-6285							✓	✓	✓	2
	......	.01	Lockwasher, $\frac{13}{32}''$ Spring	S-1347							✓	✓	✓	2
	......	12.75	Bracket, Control	10UU5-132							✓	✓	✓	1
	......	.09	Screw, Hex. Cap ($\frac{1}{2}''$—13 x $1\frac{1}{2}''$)	S-1872							✓	✓	✓	2
	......	.01	Lockwasher, $\frac{17}{32}''$ Spring	S-1349							✓	✓	✓	2
	......	.06	Screw, Hex. Cap ($\frac{3}{8}''$—16 x $1\frac{1}{2}''$)	S-17							✓	✓	✓	2

THE AUTOCAR COMPANY
ARDMORE, PA.

Master Parts List

08-5
TRANSFER CASE

CODE	MFR. PART No.	LIST PRICE	NAME OF PART	AUTOCAR PART No.	U-2044	U-4044	U-4144	U-4144-T	C-50	U-5044	U-7144-T	U-8144	U-8144-T	No. REQ.
0805	......	$0.03	Nut, Hex. (3/8"—16)	S-75							✓	✓	✓	2
	......	.01	Lockwasher, 13/32" Spring	S-1347							✓	✓	✓	4
	......	.15	Alemite, Straight (#1610)	S-4011							✓	✓	✓	3
	......	.20	Alemite, 67½° (#1612)	S-4012							✓	✓	✓	1
	......	1.25	Shaft, Control Hand Lever	19UU3-154							✓	✓	✓	1
	......	.20	Collar, Control Hand Lever Shaft	10UG2-135							✓	✓	✓	1
	......	.01	Pin, Steel (1/4" x 1 5/8")	S-1584							✓	✓	✓	1
	......	7.00	Lever, Gear Shift Hand	19UU4-19							✓	✓	✓	1
	......	.90	Ball, Gear Shift Hand Lever	10T2-408							✓	✓	✓	1
	......	.25	Bushing, Gear Shift Hand Lever	10S2-15							✓	✓	✓	1
	......	.20	Alemite, 67½° (#1612)	S-4012							✓	✓	✓	1
	......	4.25	Rod, Gear Shift Control—Assembly	18UVA3-309							✓	✓	✓	1
	......	.04	Nut, Hex. (1/2"—20)	S-1420							✓	✓	✓	2
	......	.50	Clevis, Gear Shift Control Rod	10F2-56							✓	✓	✓	2
	......	.06	Pin, Clevis (1/2" x 1 3/16")	S-2225							✓	✓	✓	2
	......	.01	Pin, Cotter (1/8" x 3/4")	S-2203							✓	✓	✓	2
	......	6.00	Lever, Declutch Shift Hand	19UU4-105A							✓	✓	✓	1
	......	.90	Ball, Declutch Shift Hand Lever	10T2-408							✓	✓	✓	1
	......	.06	Key, #A Woodruff (1/4" x 7/8")	S-2345							✓	✓	✓	1
	......	.09	Screw, Hex. Cap (1/2"—13 x 1 1/2")	S-1872							✓	✓	✓	1
	......	.01	Lockwasher, 17/32" Spring	S-1349							✓	✓	✓	1
	......	.70	Lever, Declutch Shift Control	10UL3-507							✓	✓	✓	1
	......	.06	Key, #A Woodruff (1/4" x 7/8")	S-2345							✓	✓	✓	1
	......	.06	Screw, Hex. Cap (3/8"—16 x 1 1/2")	S-17							✓	✓	✓	1
	......	.01	Lockwasher, 13/32" Spring	S-1347							✓	✓	✓	1
	......	2.00	Rod, Declutch Shift Control—Assembly	18UVB3-309							✓	✓	✓	1
	......	.04	Nut, Hex. (1/2"—20)	S-1420							✓	✓	✓	2
	......	.50	Clevis, Declutch Shift Control Rod	10F2-56							✓	✓	✓	2
	......	.06	Pin, Clevis (1/2" x 1 3/16")	S-2225							✓	✓	✓	2
	......	.01	Pin, Cotter (1/8" x 3/4")	S-2203							✓	✓	✓	2
	......	2.25	Lever, Declutch Shift Transfer	10UU3-507							✓	✓	✓	1
	......	.06	Key, #11 Woodruff (3/16" x 7/8")	S-2311							✓	✓	✓	1
	......	.05	Screw, Hex. Cap (1/2"—13 x 1 3/4")	S-3100							✓	✓	✓	1
	......	.01	Lockwasher, 17/32" Spring	S-1349							✓	✓	✓	1
	......	1.00	Shaft, Declutch Control Cross	19UU3-545							✓	✓	✓	1
	......	.20	Collar, Declutch Control Cross Shaft	10UG2-135							✓	✓	✓	2
	......	.05	Screw, Hex. Cap (1/4"—20 x 2")	S-71							✓	✓	✓	2
	......	.01	Nut, Hex. (1/4"—20)	S-76							✓	✓	✓	2
	......	.01	Lockwasher, 9/32" Spring	S-1345							✓	✓	✓	4
	......	1.80	Lever, Declutch Shift	10UUA3-507							✓	✓	✓	1
	......	....	Key, #11 Woodruff (3/16" x 7/8")	S-2311							✓	✓	✓	1
	......	.08	Screw, Hex. Cap (7/16"—14 x 1 1/2")	S-923							✓	✓	✓	1
	......	.01	Lockwasher, 13/32" Spring	S-1348							✓	✓	✓	1
	......	.20	Link, Declutch Shift Control	19UU3-613							✓	✓	✓	1
	......	.06	Pin, Clevis (3/8" x 1 1/16")	S-2257							✓	✓	✓	1
	......	.06	Pin, Clevis (1/4" x 21/32")	S-2228							✓	✓	✓	1
	......	.03	Washer, Flat Head (3/8")	S-387							✓	✓	✓	1
	......	.01	Pin, Cotter (3/32" x 3/4")	S-2198							✓	✓	✓	2

THE AUTOCAR COMPANY, ARDMORE, PA.

Master Parts List

GROUP 09

ILLUSTRATIONS

GROUP 09

PROPELLER SHAFT and UNIVERSAL JOINTS

GROUP 09

Master Parts List

THE AUTOCAR COMPANY, ARDMORE, PA.

As supplied on Autocar Models
U-8144 U-8144-T

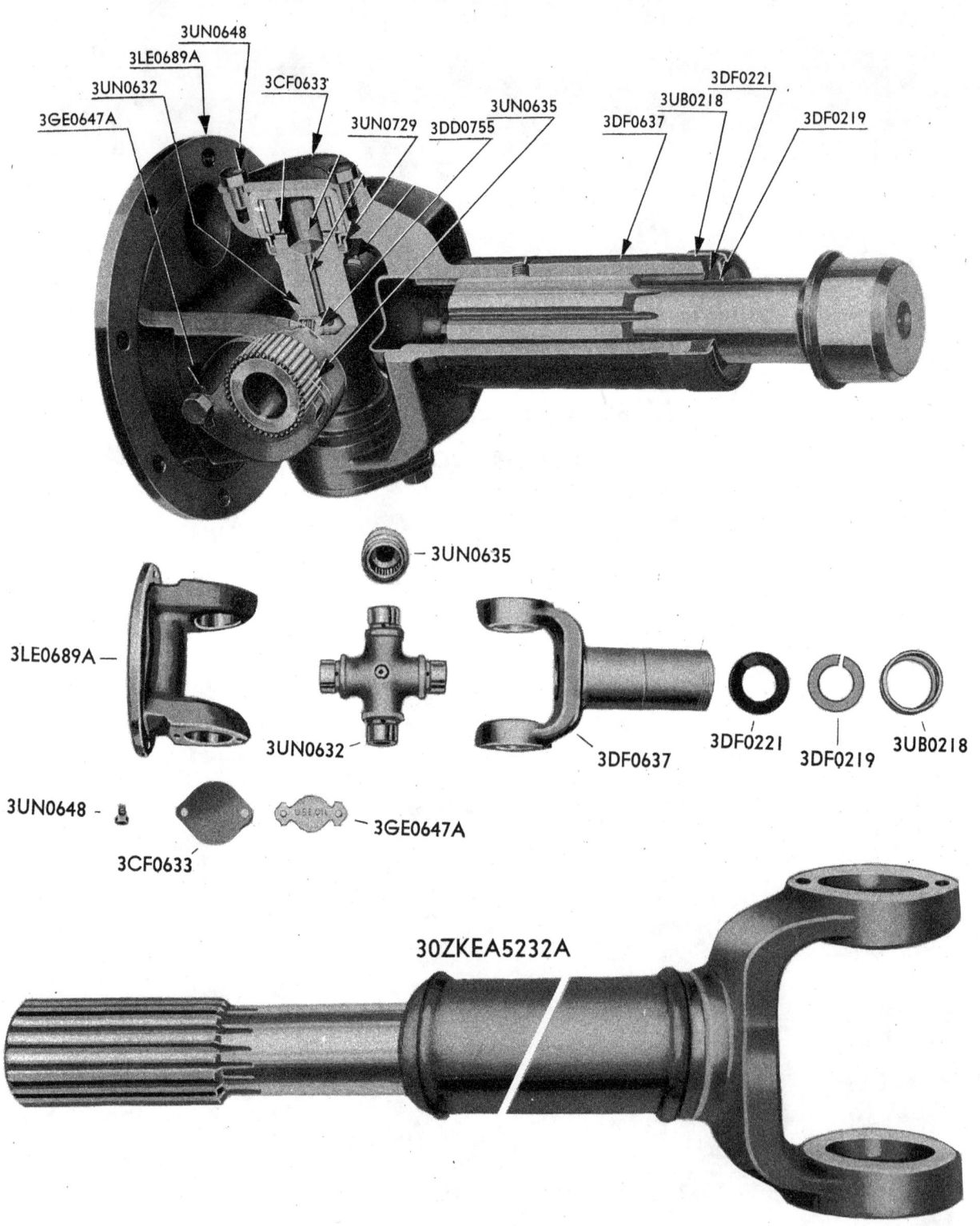

0901—Propeller Shaft Assembly—Spicer

09-1 PROPELLER SHAFT

THE AUTOCAR COMPANY, ARDMORE, PA.

Master Parts List

CODE	MFR. PART No.	LIST PRICE	NAME OF PART	AUTOCAR PART No.	U-2044	U-4044	U-4144	U-4144-T	C-50	U-5044	U-7144-T	U-8144	U-8144-T	No. REQ.
0901	**Propeller Shaft Assembly**													
	7035-SF	$48.00	Driveshaft, Intermediate	30LD4-110	✓	✓	✓	✓		✓				1
	7405-8-SF	89.00	Driveshaft, Intermediate—Series 1600	30LE4-110							✓			1
	.....	67.25	Driveshaft, Intermediate—Series 1600	30ZKE5-232A								✓	✓	1
	.....	44.00	Driveshaft, Front	30ZC5-332B	✓									1
	.....	43.85	Driveshaft, Front	30ZC5-372B		✓	✓	✓						1
	.....	36.50	Driveshaft, Front	30CD5-400					✓					1
	.....	44.65	Driveshaft, Front	30ZC5-433B						✓				1
	.....	56.25	Driveshaft, Front—Series 1500	30ZD5-470J							✓			1
	.....	65.50	Driveshaft, Front—Series 1500	30ZD5-590J								✓	✓	1
	.....	53.75	Driveshaft, Rear	30DD5-340	✓									1
	.....	53.50	Driveshaft, Rear	30DD5-330		✓		✓						1
	.....	53.25	Driveshaft, Rear	30DD5-300			✓							1
	.....	55.25	Driveshaft, Rear	30DD5-530						✓				1
	.....	53.00	Driveshaft, Rear	30DD5-263						✓				1
	.....	67.85	Driveshaft, Rear—Series 1600	30ZDE5-302							✓			1
	.....	68.75	Driveshaft, Rear—Series 1600	30ZDE5-460								✓	✓	1
0902	**Universal Joint**													
	Intermediate Driveshaft													
	.....	28.15	Joint, Slip—Assembly	3ZKE0-140								✓	✓	1
	K4-2-299	4.50	Flange, Joint (5¾″ O.D.)	3ED0-689	✓	✓	✓	✓		✓				2
	K5-2-379	7.50	Flange, Slip Joint (with 6⅝″ Dia. Outside Pilot)	3LE0-689							✓			1
	K5-2-279	6.25	Flange, Fixed Joint (with 6⅝″ Dia. Outside Pilot)	3LE0-689A							✓	✓	✓	2
	K4-3-188X	12.75	Yoke, Slip Joint Sleeve	3LD0-637	✓	✓	✓	✓		✓				1
	K5-3-198	15.15	Yoke, Slip Joint Sleeve	3LE0-637A							✓			1
	K5-3-108	10.57	Yoke, Slip Joint Sleeve	3DF0-637								✓	✓	1
	K4-82-61	17.00	Shaft, Spline Stub	3LD0-753	✓	✓	✓	✓		✓				1
	K5-82-28	17.00	Shaft, Spline Stub	3LE0-753							✓			1
	.....	39.10	Stubshaft Assembly—17 15/16″ Long	30ZKEA5-232A								✓	✓	1
	K4-14-39	.55	Cap, Sleeve Dust	3LD0-218	✓	✓	✓	✓		✓				1
	K5-14-29	.90	Cap, Sleeve Dust	3LE0-218							✓			1
	4-14-19	.55	Cap, Sleeve Dust	3UB0-218								✓	✓	1
	K5-15-23	.11	Washer, Sleeve Dust Cap Steel	3DF0-219	✓	✓	✓	✓		✓				1
	K5-15-23	.11	Washer, Sleeve Dust Cap Steel	3DF0-219								✓	✓	1
	K4-16-93	.11	Washer, Sleeve Dust Cap Felt	3LD0-221	✓	✓	✓	✓		✓				1
	K5-16-33	.20	Washer, Sleeve Dust Cap Felt	3LE0-221							✓			1
	K5-16-23	.11	Washer, Sleeve Dust Cap Felt	3DF0-221								✓	✓	1
	K4-5-78	4.85	Journal Assembly	3LD0-632	✓	✓	✓	✓						2
	K4-5-78	4.85	Journal Assembly	3RG0-632						✓				2
	KR5-5-68	6.35	Journal Assembly	3UN0-632							✓	✓	✓	2
	K4-86-119	.05	Gasket, Journal	3RG0-729	✓	✓	✓	✓		✓				8
	K5-86-79	.05	Gasket, Journal	3UN0-729							✓	✓	✓	8
	K4-76-17	.05	Retainer, Journal Gasket	3RG0-728	✓	✓	✓	✓		✓				8
	K5-76-17	.05	Retainer, Journal Gasket	3UN0-728							✓	✓	✓	8
	980798	.15	Valve, Journal Relief	3DD0-755	✓	✓	✓	✓		✓	✓	✓	✓	1
	K4-6-68X	1.50	Bearing, Needle—Assembly	3RG0-635A	✓	✓	✓	✓		✓				8
	K5-6-78	2.00	Bearing, Needle—Assembly	3UN0-635							✓	✓	✓	8
	K4-70-49	.10	Cap, Needle Bearing	3RG0-633	✓	✓	✓	✓		✓				8
	K5-70-49	.10	Cap, Needle Bearing	3CF0-633							✓	✓	✓	8
	5-73-108	.05	Screw, Needle Bearing Cap	3GE0-648	✓	✓	✓	✓		✓				16
	5-73-109	.05	Screw, Needle Bearing Cap	3UN0-648							✓	✓	✓	16
	98-781	.10	Plate, Needle Bearing Cap Screw Lock	3RG0-647	✓	✓	✓	✓		✓				8
	98-741	.10	Plate, Needle Bearing Cap Screw Lock	3GE0-647A							✓	✓	✓	8
	Front Driveshaft													
	82842-SF	17.25	Joint, Slip, Assembly (Flange Type with Outside Pilot)	3ZC0-140A	✓	✓	✓	✓		✓				1
	.....	21.80	Joint, Slip, Assembly	3ZD0-140D							✓	✓	✓	1
	8128-SF	16.50	Joint, Fixed, Assembly (Flange Type with Outside Pilot)	3DC0-150A	✓	✓	✓	✓		✓				1
	7955-SF	20.95	Joint, Fixed, Assembly (Flange Type without Outside Pilot)	3ED0-150A					✓					1
	K3-2-159	3.50	Flange, Joint (With 2¾″ Dia. Outside Pilot)	3DC0-689	✓	✓	✓	✓		✓				2
	K4-2-299	4.50	Flange, Joint (5¾″ O.D.)	3ED0-689					✓					1

Master Parts List

CODE	MFR. PART No.	LIST PRICE	NAME OF PART	AUTOCAR PART No.	MODEL U-2044	U-4044	U-4144	U-4144-T	C-50	U-5044	U-7144-T	U-8144	U-8144-T	No. REQ.
0902	K4-2-309	$4.50	**Flange,** Joint (With 3¾" Dia. Outside Pilot)	3ZD0-**689**							✓	✓	✓	2
	KL-3-508X	8.00	**Yoke,** Slip Joint Sleeve	3ZC0-**637**	✓	✓	✓	✓		✓				1
	KL4-3-861	10.60	**Yoke,** Slip Joint Sleeve	3ZD0-**637A**							✓	✓	✓	1
	K3-40-341	5.50	**Shaft,** Slip Spline Stub (3" Tube)	3ZC0-**694A**	✓	✓	✓	✓		✓				1
	K3-28-97	4.35	**Yoke,** Fixed Stub (3" Tube)	3DC0-**695**	✓	✓	✓	✓		✓				1
	K4-26-167	5.80	**Yoke,** Fixed Stub (3" Tube)	3RG0-**695A**					✓					1
	......	9.75	**Shaft,** Stub (3" Tube and 1½" Taper)	3CD3-**696**					✓					1
	......	26.25	**Stubshaft** Assembly—$40\frac{11}{16}$" Long	30ZDA5-**470J**							✓			1
	......	35.50	**Stubshaft** Assembly—$52\frac{11}{16}$" Long	30ZDA5-**590J**								✓	✓	1
	3½-14-39	.33	**Cap,** Sleeve Dust	3DC0-**218**	✓	✓	✓	✓		✓				1
	3½-14-19	.33	**Cap,** Sleeve Dust	3RG0-**218**							✓	✓	✓	1
	3½-15-53	.06	**Washer,** Sleeve Dust Cap Steel	3DC0-**219**	✓	✓	✓	✓		✓				1
	4-15-43	.06	**Washer,** Sleeve Dust Cap Steel	3A0-**219**							✓	✓	✓	1
	K3-16-53	.06	**Washer,** Sleeve Dust Cap Felt	3DC0-**221**	✓	✓	✓	✓		✓				1
	K4-16-83	.11	**Washer,** Sleeve Dust Cap Felt	3RG0-**221**							✓	✓	✓	1
	K3-5-108X	3.80	**Journal** Assembly	3DC0-**632A**	✓	✓	✓	✓		✓				2
	K4-5-78	4.85	**Journal** Assembly	3RG0-**632**					✓		✓	✓	✓	2
	K3-86-89	.05	**Gasket,** Journal	3DC0-**729**	✓	✓	✓	✓		✓				8
	K4-86-119	.05	**Gasket,** Journal	3RG0-**729**					✓		✓	✓	✓	8
	K3-76-17	.05	**Retainer,** Journal Gasket	3DC0-**728**	✓	✓	✓	✓		✓				8
	K4-76-17	.05	**Retainer,** Journal Gasket	3RG0-**728**					✓		✓	✓	✓	8
	98-798	.15	**Valve,** Journal Relief	3DD0-**755**	✓	✓	✓	✓	✓		✓	✓	✓	1
	K3-6-68X	1.25	**Bearing,** Needle, Assembly	3DC0-**635A**	✓	✓	✓	✓		✓				8
	K4-6-28	1.50	**Bearing,** Needle, Assembly	3RG0-**635**					✓					4
	K4-6-68	1.50	**Bearing,** Needle, Assembly	3RG0-**635A**							✓	✓	✓	8
	K3-7-39	.05	**Ring,** Needle Bearing Lock	3DC0-**647**	✓	✓	✓	✓		✓				8
	K4-70-49	.10	**Cap,** Needle Bearing	3RG0-**633**					✓		✓	✓	✓	8
	5-73-108	.05	**Screw,** Needle Bearing Cap	3GE0-**648**					✓		✓	✓	✓	8
	98-781	.10	**Plate,** Needle Bearing Cap Screw Lock	3RG0-**647**					✓		✓	✓	✓	8
	......	.25	**Key,** Stub Shaft (1½" Taper Shaft)	4A1-**163**					✓					1
	......	.30	**Nut,** Stub Shaft (1"—20 Threads)	S-**87**					✓					1
	1308	12.00	**Housing,** Steady Bearing, Assembly (Use with 30CD5000—1500 Series Driveshaft Assembly)	3CD4-**970**					✓					1
	Rear Driveshaft													
	6908-SF	21.80	**Joint,** Slip, Assembly (Flange Type without Outside Pilot)	3DW0-**140**	✓	✓	✓	✓	✓	✓				1
	......	14.75	**Joint,** Slip, Assembly	3ZUNC0-**140**							✓	✓	✓	1
	7955-SF	20.95	**Joint,** Slip, Assembly (Flange Type without Outside Pilot)	3ED0-**150A**	✓	✓	✓	✓	✓	✓				1
	K5-2-249	15.30	**Flange,** Fixed Joint	3A0-**689**							✓	✓	✓	1
	K4-2-299	4.50	**Flange,** Joint (5¾" O.D.)	3ED0-**689**	✓	✓	✓	✓	✓	✓				1
	K5-2-279	6.25	**Flange,** Slip Joint (With 6⅝" Dia. Outside Pilot)	3LE0-**689A**							✓	✓	✓	1
	K4-3-88	7.50	**Yoke,** Slip Joint Sleeve	3RG0-**637**	✓	✓	✓	✓	✓	✓				1
	K5-3-108	10.50	**Yoke,** Slip Joint Sleeve	3DF0-**637**							✓	✓	✓	1
	K4-42-591	7.75	**Shaft,** Slip Spline Stub (3" Tube)	3RG0-**694A**	✓	✓	✓	✓	✓	✓				1
	K4-26-167	5.80	**Yoke,** Fixed Stub	3RG0-**695A**	✓	✓	✓	✓	✓	✓				1
	......	19.75	**Stubshaft** Assembly—$24\frac{15}{16}$" Long	30ZDEA5-**302**							✓			1
	......	41.30	**Stubshaft** Assembly—$40\frac{7}{16}$" Long	30ZDEA5-**460**								✓	✓	1
	3½-14-19	.33	**Cap,** Sleeve Dust	3RG0-**218**	✓	✓	✓	✓	✓	✓				1
	4-14-19	.55	**Cap,** Sleeve Dust	3UB0-**218**							✓	✓	✓	1
	4-15-43	.06	**Washer,** Sleeve Dust Cap Steel	3RG0-**219**	✓	✓	✓	✓	✓					1
	4-15-43	.02	**Washer,** Sleeve Dust Cap Steel	3A0-**219**						✓				1
	K5-15-23	.11	**Washer,** Sleeve Dust Cap Steel	3DF0-**219**							✓	✓	✓	1
	K4-16-83	.11	**Washer,** Sleeve Dust Cap Felt	3RG0-**221**	✓	✓	✓	✓	✓	✓				1
	K5-16-23	.11	**Washer,** Sleeve Dust Cap Felt	3DF0-**221**							✓	✓	✓	1
	K4-5-78	4.85	**Journal** Assembly	3RG0-**632**	✓	✓	✓	✓	✓	✓				2
	KR5-5-68	6.35	**Journal** Assembly	3UN0-**632**							✓	✓	✓	2
	K4-86-119	.05	**Gasket,** Journal	3RG0-**729**	✓	✓	✓	✓	✓	✓				8
	K5-86-79	.05	**Gasket,** Journal	3UN0-**729**							✓	✓	✓	8
	K4-76-17	.05	**Retainer,** Journal Gasket	3RG0-**728**	✓	✓	✓	✓	✓	✓				8
	K5-76-17	.05	**Retainer,** Journal Gasket	3UN0-**728**							✓	✓	✓	8
	98-798	.15	**Valve,** Journal Relief	3DD0-**755**	✓	✓	✓	✓	✓	✓	✓	✓	✓	1
	K4-6-68X	1.50	**Bearing,** Needle, Assembly	3RG0-**635A**	✓	✓	✓	✓	✓	✓				8

THE AUTOCAR COMPANY — **Master Parts List** — **09-3 PROPELLER SHAFT**
ARDMORE, PA.

CODE	MFR. PART No.	LIST PRICE	NAME OF PART	AUTOCAR PART No.	MODEL U-2044	U-4044	U-4144	U-4144-T	C-50	U-5044	U-7144-T	U-8144	U-8144-T	No. REQ.
0902	K5-6-78	$2.00	Bearing, Needle, Assembly	3UN0-635							✓	✓	✓	8
	K4-70-49	.10	Cap, Needle Bearing	3RG0-633	✓	✓	✓	✓	✓	✓				8
	K5-70-49	.10	Cap, Needle Bearing	3CF0-633							✓	✓	✓	8
	5-73-108	.05	Screw, Needle Bearing Cap	3GE0-648	✓	✓	✓	✓	✓	✓				16
	5-73-109	.05	Screw, Needle Bearing Cap	3UN0-648							✓	✓	✓	16
	98-781	.10	Plate, Needle Bearing Cap Screw Lock	3RG0-647	✓	✓	✓	✓	✓	✓				8
	98-741	.10	Plate, Needle Bearing Cap Screw Lock	3GE0-647A							✓	✓	✓	8
	Steady Bearing Housing Assembly													
	……	11.10	Housing, Steady Bearing, Assembly	3CD4-970A					✓					1
	1308-F	6.80	Bearing, Steady	3SA2-285					✓					1
	……	6.00	Housing, Steady Bearing	3G4-287					✓					1
	……	.05	Dowel, Steady Bearing Housing	S-1506					✓					4
	A-1784	.30	Cover, Steady Bearing Housing End	3SA2-288					✓					2
	A-1785	.30	Plate, Steady Bearing Housing Felt Seal	3SA2-294					✓					2
	A-1786	.35	Flinger, Steady Bearing Housing	3SA2-293					✓					2
	A-1787	.20	Clamp, Steady Bearing Housing Flinger	3SA2-408					✓					2
	……	.01	Screw, Steady Bearing Housing Flinger Clamp	S-2941					✓					2
	……	.45	Sleeve, Steady Bearing Housing	3CD2-289					✓					1
	……	.35	Nut, Steady Bearing Housing Lock	3CD2-291					✓					1
	……	.05	Washer, Steady Bearing Housing Lock	3CD2-304					✓					1
	A-3280	.20	Ring, Steady Bearing Housing Felt	3SA0-409					✓					2
	X-235-A	.05	Bolt, Steady Bearing Housing	3SA0-418					✓					4
	……	.01	Nut, Steady Bearing Housing Bolt	S-1416					✓					4
	……	.01	Washer, Spring	S-1345					✓					8

THE AUTOCAR COMPANY, ARDMORE, PA.

Master Parts List

GROUP 10

ILLUSTRATIONS

GROUP 10

FRONT AXLE

GROUP 10

Master Parts List

THE AUTOCAR COMPANY, ARDMORE, PA.

As supplied on Autocar Models
U-8144   U-8144-T

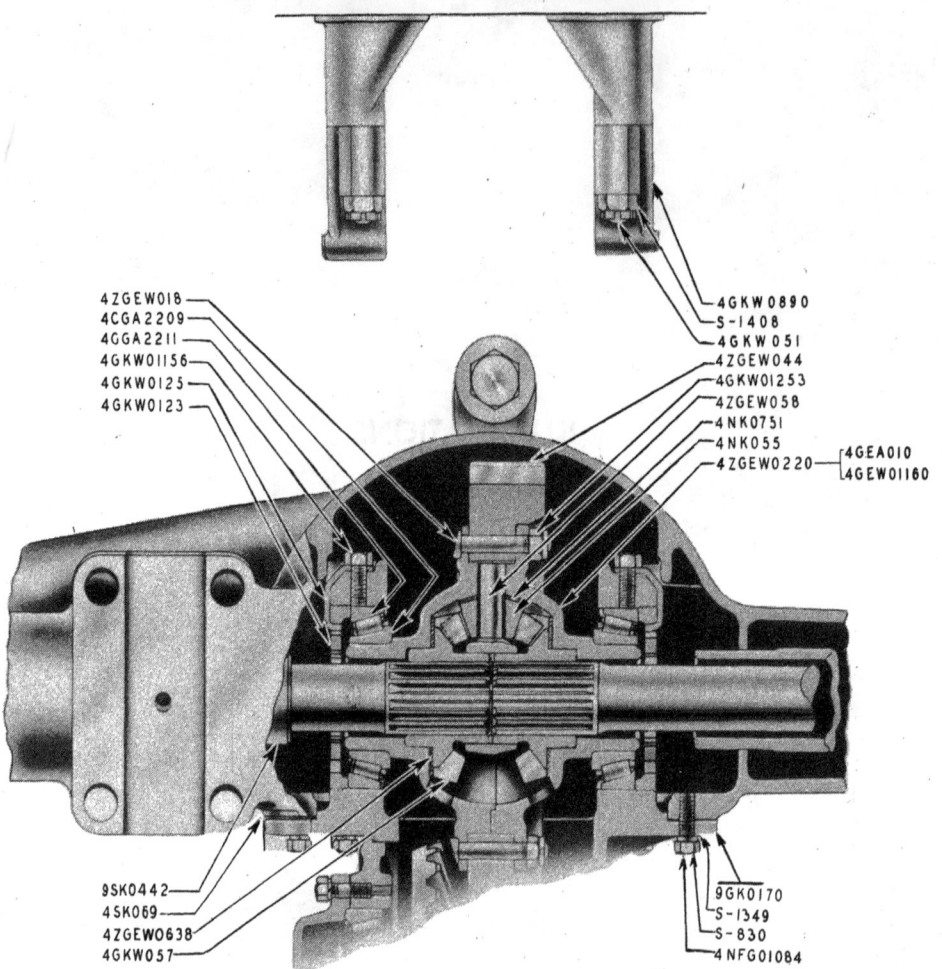

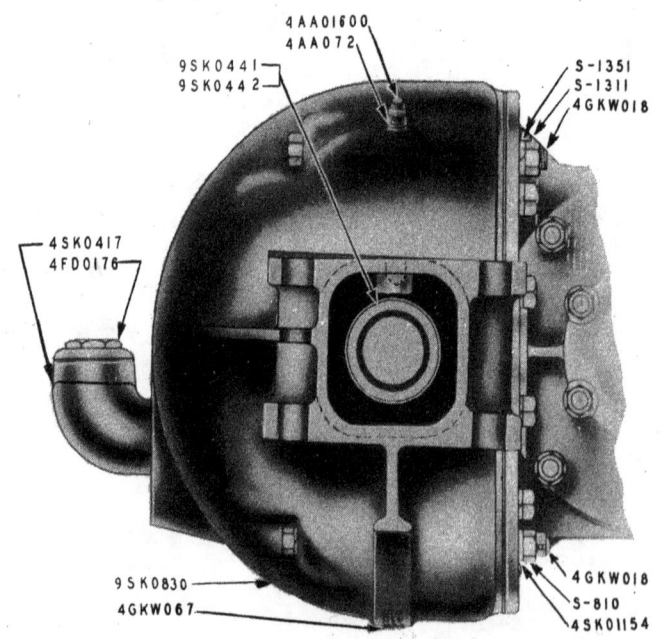

1000—Front Axle Assembly

THE AUTOCAR COMPANY, ARDMORE, PA.

Master Parts List

GROUP 10

As supplied on Autocar Models
U-8144   U-8144-T

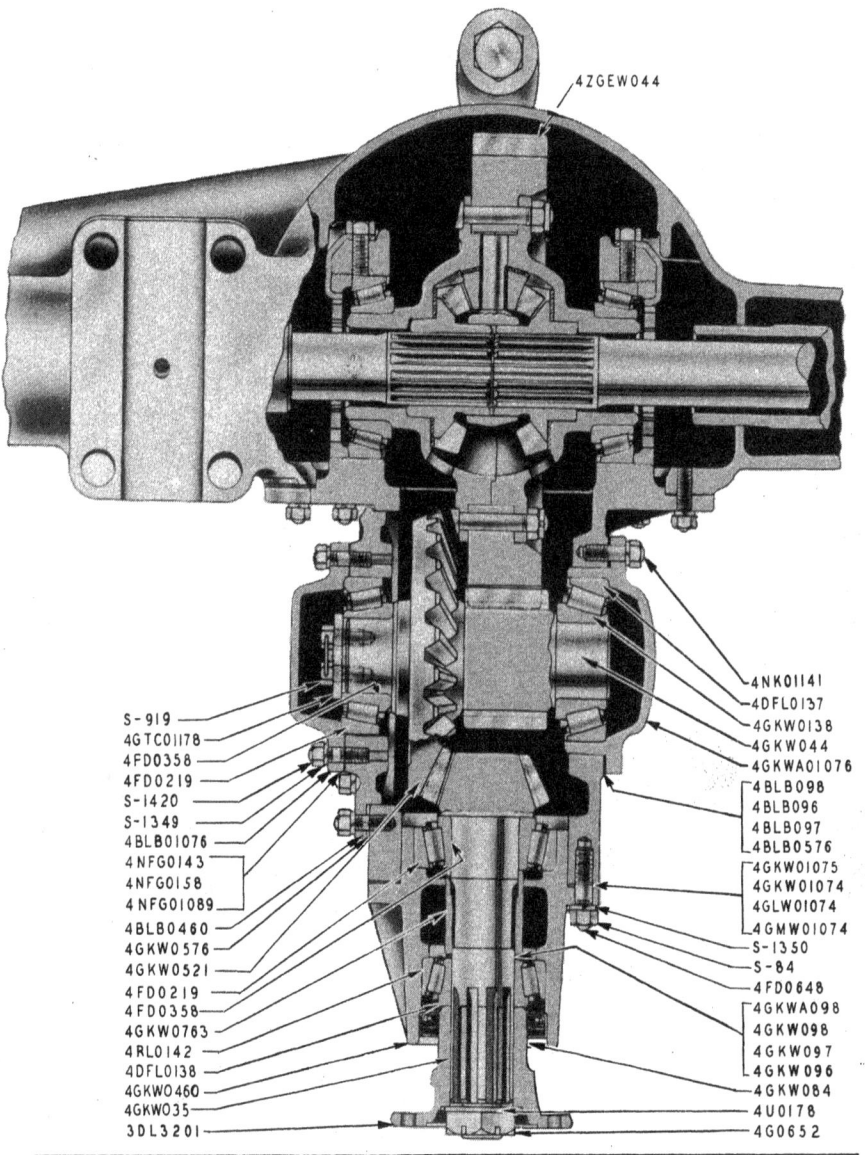

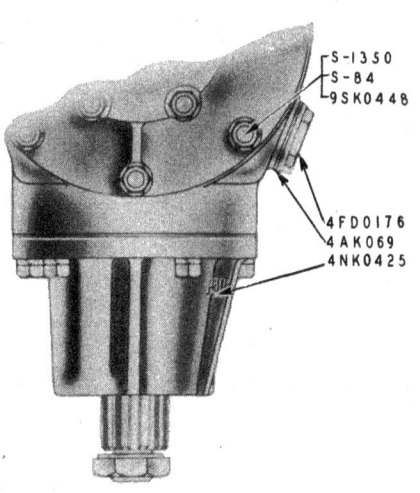

1001—Front Axle Housing

GROUP 10

Master Parts List

THE AUTOCAR COMPANY, ARDMORE, PA.

As supplied on Autocar Models

U-8144   U-8144-T

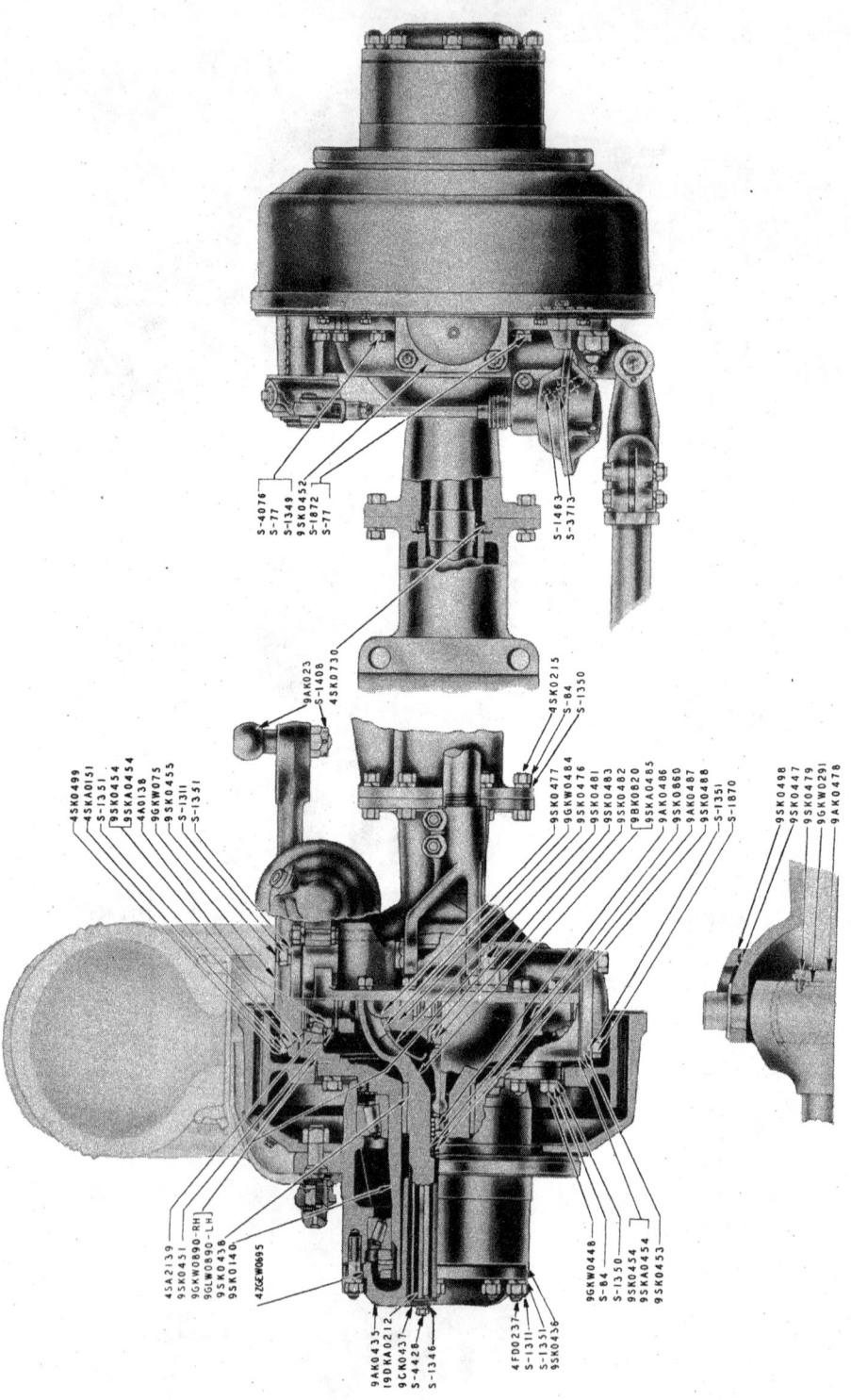

1002—Front Axle Differential and Carrier Assembly

CODE	MFR. PART No.	LIST PRICE	NAME OF PART	AUTOCAR PART No.	MODEL U-2044	U-4044	U-4144	U-4144-T	C-50	U-5044	U-7144-T	U-8144	U-8144-T	No. REQ.
1000	**Front Axle Assembly**													
	......	$335.50	**Axle,** Front—Assembly—Complete	9LL0-**990A**					✓					1
	F-2090-W-73-X-2	1678.00	**Axle,** Front—Assembly	9FKW9-**430**							✓			1
	F-3100-W-X-5	2222.25	**Axle,** Front—Assembly (Ratio 8.148-1) (Timken)	9GKW9-**430**								✓	✓	1
1001	**Front Axle Housing**													
	A3801-X-388	118.70	**Housing** Assembly	9UG0-**830**	✓	✓	✓	✓		✓				1
	A-3801-G-397	180.95	**Housing** Assembly	9FKW0-**830**							✓			1
	A-3801-Y-77	215.40	**Housing** Assembly	9SK0-**830**								✓	✓	1
	P-220	.12	**Plug,** Housing Drain	4UG0-**67**	✓	✓	✓	✓		✓				1
	P-28	.07	**Plug,** Housing Drain	4GKW0-**67**							✓	✓	✓	1
	3885-A-1	1.65	**Filler,** Housing Oil	4SK0-**417**							✓	✓	✓	1
	P-120	.20	**Plug,** Housing Oil Filler ($1\frac{1}{4}$")	S-**4157**	✓	✓	✓	✓		✓				1
	1850-A-1	.42	**Plug,** Housing Oil Filler	4FD0-**176**							✓	✓	✓	1
	2808-N-300	.02	**Gasket,** Housing Oil Filler Plug	4AK0-**69**							✓			1
	......	....	**Gasket,** Housing Oil Filler Plug	....								✓	✓	1
	3816-D-212	4.00	**Tube,** Housing (Short—L. H.)	4UH0-**22**	✓	✓	✓	✓		✓				1
	3816-T-280	10.55	**Sleeve,** Housing—Short	9FKW0-**442**							✓			1
	3816-D-82	6.90	**Sleeve,** Housing—Short	9SK0-**442**								✓	✓	1
	3816-C-211	5.00	**Tube,** Housing (Long—R. H.)	4UGA0-**22**	✓	✓	✓	✓		✓				1
	3816-U-281	14.70	**Sleeve,** Housing—Long	9FKW0-**441**							✓			1
	3816-E-83	16.80	**Sleeve,** Housing—Long	9SK0-**441**								✓	✓	1
	A1199-E-421	.36	**Breather,** Housing Oil	4BL0-**1600**	✓	✓	✓	✓		✓				1
	A1199-C-1043	.14	**Nipple,** Housing Breather	4AA0-**1600**							✓	✓	✓	1
	1898-W-205	.06	**Reducer,** Housing Breather Nipple	4AA0-**72**							✓	✓	✓	1
	4-X-186	.08	**Stud,** Housing to Carrier	4NFG0-**1084**	✓	✓	✓	✓		✓	✓	✓	✓	8
	N-48	.04	**Nut,** Hex. ($\frac{1}{2}$"—20)	S-**830**	✓	✓	✓	✓		✓	✓	✓	✓	8
	W-18	.01	**Lockwasher,** $\frac{17}{32}$" Spring	S-**1349**							✓	✓	✓	8
	W-19	.03	**Washer,** $\frac{9}{16}$" Spring	S-**1350**	✓	✓	✓	✓		✓				16
	4-X-173	.12	**Stud,** Housing and Socket	4FD0-**648**	✓	✓	✓	✓		✓				16
	4-X-173	.12	**Stud,** Housing and Socket	4FD0-**648**							✓			8
	4-X-544	.23	**Stud,** Housing and Socket	4FKW0-**661**							✓			8
	N-19	.01	**Nut,** Housing and Socket	S-**84**							✓			8
	N-112	.06	**Nut,** Housing and Socket	S-**801**							✓			8
	W-19	.01	**Washer,** $\frac{9}{16}$" Spring	S-**1350**							✓			8
	W-112	.02	**Washer,** $\frac{3}{4}$" Spring	S-**1353**							✓			8
	15-X-209	.45	**Bolt,** Housing and Carrier Through	4GKW0-**18**								✓	✓	4
	N-110	.10	**Nut,** Hex. ($\frac{5}{8}$"—18)	S-**1311**								✓	✓	2
	N-210	.10	**Nut,** Hex. ($\frac{5}{8}$"—18 Castle)	S-**810**								✓	✓	2
	W-110	.01	**Lockwasher,** $\frac{21}{32}$" Spring	S-**1351**								✓	✓	2
	1829-P-172	.04	**Lockwasher,** Lug	4SK0-**1154**								✓	✓	4
1002	**Front Axle Differential and Carrier Assembly**													
	G-380-C	445.00	**Carrier,** Differential—Assembly—Complete	9GK0-**170**										
	G-356-H	385.00	**Carrier,** Differential—Assembly—Complete	9FKW0-**170**							✓			1
	G-380-C	445.00	**Carrier,** Differential—Assembly—Complete	9GK0-**170**								✓	✓	1
	2808-E-57	.14	**Carrier,** Differential—to Housing Gasket	4UG0-**483**	✓	✓	✓	✓		✓				1
	2808-U-203	.14	**Gasket,** Housing and Carrier	4NK0-**69**							✓			1
	2808-K-167	.14	**Gasket,** Differential Carrier to Housing	4SK0-**69**								✓	✓	1
	A8-3800-E-57	69.00	**Carrier,** Differential—and Cap Assembly	4FKW0-**890**							✓			1
	A2-3800-U-359	75.95	**Carrier,** Differential—and Cap Assembly	4GKW0-**890**								✓	✓	1
	4-X-153	.21	**Carrier,** Differential—and Cap Stud	4NK0-**122**							✓			1
	4-X-545	.44	**Carrier,** Differential—and Cap Stud	4GKW0-**51**								✓	✓	4
	N-210	.06	**Carrier,** Differential—and Cap Stud Nut ($\frac{5}{8}$"—18)	S-**810**	✓	✓	✓	✓		✓	✓			4
	N-212	.12	**Nut,** Hex. ($\frac{3}{4}$"—16 Castle)	S-**1408**								✓	✓	4
	......	.01	**Cotter** ($\frac{1}{8}$" x $1\frac{1}{2}$")	S-**2206**								✓	✓	4
	482	6.83	**Cone,** Differential Bearing	19SKB0-**443**	✓	✓	✓	✓		✓				2
	1214	10.00	**Bearing,** Differential	4FKW0-**209**							✓			2
	581	10.91	**Cone,** Differential Bearing	4CGA2-**209**								✓	✓	2
	472-A	5.03	**Cup,** Differential Bearing	4D2-**157**	✓	✓	✓	✓		✓				2
	572	6.59	**Cup,** Differential Bearing	4CGA2-**211**								✓	✓	2
	1829-A-27	.25	**Ring,** Differential Bearing Retainer	4NFG0-**1086**							✓			2

10-2 FRONT AXLE

Master Parts List

THE AUTOCAR COMPANY, ARDMORE, PA.

CODE	MFR. PART No.	LIST PRICE	NAME OF PART	AUTOCAR PART No.	U-2044	U-4044	U-4144	U-4144-T	C-50	U-5044	U-7144-T	U-8144	U-8144-T	No. REQ.
1002	2814-E-57	$1.10	Ring, Differential Bearing Adjusting	4GKW0-123								✓	✓	2
	1820-J-10	.15	Lock, Differential Bearing Adjusting Nut	4BMC0-125	✓	✓	✓	✓		✓				2
	1820-N-40	1.10	Lock, Differential Bearing Adjusting Ring	4GKW0-125								✓	✓	2
	S-265-D	.04	Screw, Differential Bearing Adjusting Nut Lock (3/8"—16x5/8")	S-2065	✓	✓	✓	✓		✓				2
	S-288-D	.07	Screw, Differential Bearing Adjusting Ring Lock	4GKW0-1156								✓	✓	2
	A2-3835-Y-51	80.50	Differential Assembly	4UG0-10	✓	✓	✓	✓		✓				1
	A9-3835-N-66	97.00	Differential Assembly	4NKA0-10							✓			1
	A5-3835-L-38	102.00	Differential Assembly	4GEA0-10								✓	✓	1
	A3835-Z-52	43.00	Case, Differential—Assembly	4BMC0-220	✓	✓	✓	✓		✓				1
	A-3835-M-65	34.65	Case, Differential—Assembly	4NK0-220							✓			1
	A-3835-M-39	37.25	Case, Differential—Assembly	4ZGEW0-220								✓	✓	1
	15-X-180	.12	Bolt, Differential Case	4BLA0-18	✓	✓	✓	✓		✓				8
	15-X-112	.27	Bolt, Differential	4NFG0-18							✓			8
	15-X-130	.35	Bolt, Differential	4ZGEW0-18								✓	✓	8
	N-28	.05	Nut, Differential Case Bolt (1/2"—20)	S-811	✓	✓	✓	✓		✓				8
	N-29	.05	Bolt, Differential	4GKW0-1253							✓	✓	✓	8
	K-2410	.01	Cotter	K-2410								✓	✓	8
	2234-K-427	7.35	Gear, Differential Bevel Side	4UG0-57	✓	✓	✓	✓		✓				2
	2834-S-409	13.85	Gear, Differential Side	4NKA0-57							✓			2
	2834-R-408	13.30	Gear, Differential Side	4GKW0-57								✓	✓	2
	1229-Z-728	.29	Washer, Differential Bevel Side Gear Thrust	4BM0-638	✓	✓	✓	✓		✓				2
	1229-X-882	.73	Washer, Differential Side Gear Thrust	4ZGEW0-638							✓	✓	✓	2
	2233-M-429	3.35	Pinion, Differential Bevel Side	4UG0-55	✓	✓	✓	✓		✓				4
	2233-W-413	5.40	Gear, Differential Pinion	4NK0-55							✓	✓	✓	4
	1229-C-861	.13	Washer, Differential Bevel Side Pinion Thrust	4BMA0-751	✓	✓	✓	✓		✓				4
	1829-S-45	.20	Washer, Differential Pinion Gear Thrust	4NK0-751							✓	✓	✓	4
	3878-B-54	4.45	Spider, Differential	4BLA0-58	✓	✓	✓	✓		✓				1
	3878-B-2	8.70	Spider, Differential	4NK0-58							✓			1
	3878-C-3	8.70	Spider, Differential	4ZGEW0-58								✓	✓	1
	A3-3835-Y-51	113.50	Gear, Differential & Ring—Assembly (Specify Ratio)	4UGA0-10	✓	✓	✓	✓		✓				1
	A10-3835-N-66	141.00	Gear, Differential and—Assembly (8.435 Ratio)	4NKA0-80							✓			1
	A10-3835-L-38	156.50	Gear, Differential and—Assembly (8.148 Ratio)	4GEW0-1160								✓	✓	1
1003	**Front Axle Drive Gear and Bearings**													
	3892-S-71	30.00	Gear, Spur (Ring Gear) (8.21 Ratio)	4BMB0-62B	✓	✓	✓	✓		✓				1
	3892-H-34	39.00	Gear, Spur (8.435 Ratio)	4NFG0-62							✓			1
	3892-L-12	50.00	Gear, Spur (8.148 Ratio)	4ZGEW0-44								✓	✓	1
	3891-R-96	23.50	Shaft, Spur Pinion (8.21 Ratio)	4BLA0-44	✓	✓	✓	✓		✓				1
	3891-S-487	40.00	Pinion, Spur (8.435 Ratio)	4FKW0-44							✓			1
	3891-N-482	23.00	Pinion, Spur (8.148 Ratio)	4GKW0-44								✓	✓	1
	615	8.40	Cone, Spur Pinion Bearing	4NK0-136							✓			2
	623	8.40	Cone, Spur Pinion—R. H.	4FD0-358								✓	✓	1
	65225	11.48	Cone, Spur Pinion—L. H.	4GKW0-138								✓	✓	2
	612	5.74	Cup, Spur Pinion—R. H.	4FD0-219								✓	✓	1
	65500	9.30	Cup, Spur Pinion—L. H.	4DFL0-137								✓	✓	1
	A-3838-T-20	9.45	Cage, Spur Pinion Bearing—Assembly	4NKA0-1050							✓			1
	A-3838-G-85	9.45	Cage, Spur Pinion Bearing—Assembly	4FKW0-1050							✓			1
	2808-U-47	.05	Gasket, Spur Pinion Bearing Cage	4NK0-576							✓			2
	2803-X-518	.10	Shim, Spur Pinion Bearing Cage—Thin	4NKB0-553							✓			As
	2803-C-107	.22	Shim, Spur Pinion Bearing Cage—Medium	4NK0-553							✓			As
	2803-D-108	.22	Shim, Spur Pinion Bearing Cage—Thick	4NKA0-553							✓			As
	4-X-186	.08	Stud, Spur Pinion Bearing Cage	4NFG0-1084							✓			6
	4-X-397	.12	Stud, Spur Pinion Bearing Cage	4NK0-1141							✓			6
	N-18	.04	Nut, Spur Pinion Bearing Cage Stud (1/2"—20)	S-1420							✓			12
	W-18	.01	Washer, 1/2" Spring	S-1349							✓			12
	1829-B-54	1.35	Lock, Spur Pinion Bearing	4GTC0-1178								✓	✓	1
	S-278-D	.05	Screw, Spur Pinion Bearing Lock (7/16"—14 x 1")	S-919								✓	✓	2
	3866-G-189	2.85	Cover, Spur Pinion Bearing—R. H.	4BLB0-1076								✓	✓	1
	3866-Z-182	8.85	Cover, Spur Pinion Bearing—L. H.	4GKWA0-1076								✓	✓	1
	A-3838-S-45	10.05	Cage, Spur Pinion Bearing—and Cap Assembly	4BLB0-460								✓	✓	1
	4-X-73	.22	Stud, Spur Pinion Bearing Cover	9SK0-448								✓	✓	1
	4-X-397	.12	Stud, Spur Pinion Bearing Cover	4NK0-1141								✓	✓	18

THE AUTOCAR COMPANY
ARDMORE, PA.

Master Parts List

10-3 FRONT AXLE

CODE	MFR. PART No.	LIST PRICE	NAME OF PART	AUTOCAR PART No.	U-2044	U-4044	U-4144	U-4144-T	C-50	U-5044	U-7144-T	U-8144	U-8144-T	No. REQ.
1003	N-19	$0.06	Nut, Spur Pinion Bearing Cover Stud Hex. (9/16"—18)	S-84								✓	✓	1
	N-18	.04	Nut, Spur Pinion Bearing Cover Stud Hex. (1/2"—20)	S-1420								✓	✓	18
	W-19	.03	Washer, 9/16" Spring	S-1350								✓	✓	1
	W-18	.01	Washer, 1/2" Spring	S-1349								✓	✓	18
	2803-C-55	.12	Shim, Spur Pinion Bearing Cover—R. H.—Thin	4NFG0-143								✓	✓	As
	2803-D-56	.12	Shim, Spur Pinion Bearing Cover—R. H.—Thick	4NFG0-158								✓	✓	As
	2808-V-22	.05	Gasket, Spur Pinion Bearing Cover—R. H.	4NFG0-1089								✓	✓	1
	2803-H-346	.14	Shim, Spur Pinion Bearing Cover—L. H.—Thin	4BLB0-98								✓	✓	As
	2803-K-349	.14	Shim, Spur Pinion Bearing Cover—L. H.—Thick	4BLB0-96								✓	✓	As
	2803-J-348	.14	Shim, Spur Pinion Bearing Cover—Medium	4BLB0-97								✓	✓	As
	2808-L-194	.12	Gasket, Spur Pinion Bearing Cover—L. H.	4BLB0-576								✓	✓	1
	2808-A-183	.09	Gasket, Spur Pinion Bearing Cover—R. H.	4GKW0-576								✓	✓	1
	3889-H-60	25.00	Gear, Bevel (21-T)	4BLA0-521	✓	✓	✓	✓	✓					1
	3889-F-162	38.50	Gear, Bevel	4FKW0-521							✓			1
	3889-W-179	25.00	Gear, Bevel	4GKW0-521								✓	✓	1
	3890-V-74	15.80	Shaft, Bevel Pinion (10-T)	4DK0-35	✓	✓	✓	✓	✓					1
	3890-Z-234	28.15	Pinion, Bevel	4FKW0-35							✓			1
	3890-S-253	26.75	Pinion, Bevel	4GKW0-35								✓	✓	1
	S-265-D	.05	Screw, Bevel Gear (3/8"—16 x 5/8")	S-2065							✓			6
	16-X-23	.18	Key, Bevel Pinion	4NFG0-1344										1
	1227-A-183	.32	Nut, Bevel Pinion	4NKA0-81							✓			1
	16278	.30	Nut, Bevel Pinion	4G0-652								✓	✓	1
	1829-Y-363	.58	Washer, Bevel Pinion Nut	4U0-178								✓	✓	1
	621	8.40	Cone, Bevel Pinion Rear Bearing	4ZGEW0-156							✓			1
	623	8.40	Cone, Bevel Pinion Rear Bearing	4FD0-358								✓	✓	1
	612	5.74	Cup, Bevel Pinion Rear Bearing	4FD0-219								✓	✓	1
	527	6.65	Cone, Bevel Pinion Forward Bearing	4NFG0-138							✓			1
	59200	7.30	Cone, Bevel Pinion Forward Bearing	4DFL0-138								✓	✓	1
	522	4.07	Cup, Bevel Pinion Forward Bearing	3T2-37							✓			1
	59412	4.86	Cup, Bevel Pinion Forward Bearing	4RL0-142								✓	✓	1
	1827-A-105	2.35	Nut, Bevel Pinion Forward Adjusting	4NKA0-29							✓			1
	1827-B-106	2.20	Nut, Bevel Pinion Forward Bearing Jam	4NKB0-81							✓			1
	S-853	.11	Screw, Bevel Pinion Forward Jam Nut	S-422							✓			1
	1829-V-282	.21	Lock, Bevel Pinion Forward Bearing Jam Nut	4NKA0-1204							✓			1
	1829-N-378	.26	Ring, Bevel Pinion Forward Bearing Spinner	4FKW0-762							✓			1
	1844-W-153	1.60	Spacer, Bevel Pinion Bearing	4GKW0-763								✓	✓	1
	2803-A-105	.20	Shim, Bevel Pinion Forward Bearing—Thin	4NK0-98							✓			As
	2803-B-106	.20	Shim, Bevel Pinion Forward Bearing—Medium	4NK0-96							✓			As
	2803-J-582	.43	Shim, Bevel Pinion Forward Bearing—Thick	4FKW0-96							✓			As
	2803-E-239	.11	Shim, Bevel Pinion Bearing	4GKW0-96								✓	✓	As
	2803-F-240	.11	Shim, Bevel Pinion Bearing	4GKW0-97								✓	✓	As
	2803-G-241	.11	Shim, Bevel Pinion Bearing	4GKW0-98								✓	✓	As
	2803-H-242	.11	Shim, Bevel Pinion Bearing	4GKWA0-98								✓	✓	As
	A-3826-T-20	19.50	Cage, Bevel Pinion Forward Bearing, and Cup Assembly	4NK0-460							✓			1
	A-3826-U-151	21.25	Cage, Bevel Pinion Bearing, and Cup Assembly	4GKW0-460								✓	✓	1
	A-3866-M-39	4.60	Cover, Bevel Pinion Forward Bearing, Assembly	4NK0-470							✓			1
	2808-S-45	.06	Gasket, Bevel Pinion Forward Bearing Cover	4NK0-582							✓			1
	S-268	.05	Screw, Bevel Pinion Forward Bearing Cover (3/8"—16 x 1")	S-23							✓			6
	W-16	.01	Washer, 3/8" Spring	S-1347							✓			6
	4-X-542	.13	Stud, Bevel Pinion Forward Bearing Cage	4FKW0-237							✓			6
	N-19	.04	Nut, Bevel Pinion Forward Bearing Cage Stud (9/16"—18)	S-84							✓			6
	W-19	.01	Washer, 9/16" Spring	S-1350							✓			6
	2803-K-115	.55	Shim, Bevel Pinion Forward Bearing	4NK0-1023							✓			As
	2803-L-116	.55	Shim, Bevel Pinion Forward Bearing	4NKA0-1023							✓			As
	2803-M-117	.55	Shim, Bevel Pinion Forward Bearing	4NKB0-1023							✓			As
	2803-N-118	.55	Shim, Bevel Pinion Forward Bearing	4NKC0-1023							✓			As
	2803-P-120	.65	Shim, Bevel Pinion Forward Bearing	4NKD0-1023							✓			As
	2803-D-472	.56	Shim, Bevel Pinion Forward Bearing	4NKE0-1023							✓			As
	2803-E-473	.56	Shim, Bevel Pinion Forward Bearing	4NKF0-1023							✓			As
	2803-F-474	.56	Shim, Bevel Pinion Forward Bearing	4NKG0-1023							✓			As
	2803-G-475	.56	Shim, Bevel Pinion Forward Bearing	4NKH0-1023							✓			As
	2803-H-476	.56	Shim, Bevel Pinion Forward Bearing	4NKJ0-1023							✓			As

10-4 FRONT AXLE — Master Parts List

THE AUTOCAR COMPANY, ARDMORE, PA.

CODE	MFR. PART NO.	LIST PRICE	NAME OF PART	AUTOCAR PART NO.	U-2044	U-4044	U-4144	U-4144-T	C-50	U-5044	U-7144-T	U-8144	U-8144-T	NO. REQ.
1003	2803-Y-519	$0.17	Shim, Bevel Pinion Bearing Cage—Thin	4GKW0-1074								✓	✓	As
	2803-Z-520	.17	Shim, Bevel Pinion Bearing Cage—Medium	4GLW0-1074								✓	✓	As
	2803-A-521	.17	Shim, Bevel Pinion Bearing Cage—Thick	4GMW0-1074								✓	✓	As
	2808-T-46	.07	Gasket, Bevel Pinion Forward Bearing Cage	4NK0-143							✓			1
	2808-K-323	.08	Gasket, Bevel Pinion Bearing Cage	4GKW0-1075								✓	✓	1
	4-X-173	.12	Stud, Bevel Pinion Bearing Cage	4FD0-648								✓	✓	6
	N-19	.06	Nut, Bevel Pinion Bearing Cage Stud Hex. (9/16″—18)	S-84								✓	✓	6
	W-19	.03	Washer, 9/16″ Spring	S-1350								✓	✓	6
	1850-B-28	.06	Plug, Bevel Pinion Bearing Cage	4NK0-425							✓	✓	✓	1
	A-1805-Y-155	1.20	Seal, Bevel Pinion Forward Bearing Oil	4FKW0-84							✓			1
	A-1805-Y-155	1.60	Seal, Bevel Pinion Bearing Cage Oil	4GKW0-84								✓	✓	1
	1850-A-1	.42	Plug, Gear Carrier Inspection	4FD0-176							✓	✓	✓	1
	2808-N-300	.02	Gasket, Gear Carrier Inspection Plug	4AK0-69							✓	✓	✓	1
	G-316	310.00	Mounting Assembly (Gear Carrier Assembly)	4BA0-170	✓	✓	✓	✓		✓				1
	A2-3800-K-63	47.50	Mounting with Caps (Gear Carrier)	4UG0-920	✓	✓	✓	✓		✓				1
	4-X-177	.21	Stud, Differential Support Cap	4NFG0-122	✓	✓	✓	✓		✓				4
	A3838-Y-25	7.35	Cap, Mounting Side, and Cup Assembly	4BLA0-460	✓	✓	✓	✓		✓				2
	2803-T-150	.17	Shim, Mounting Side Cap—Medium	4BLA0-97	✓	✓	✓	✓		✓				As
	2803-V-152	.17	Shim, Mounting Side Cap—Thick	4BLA0-96	✓	✓	✓	✓		✓				As
	2803-U-151	.17	Shim, Mounting Side Cap—Thin	4BLA0-98	✓	✓	✓	✓		✓				As
	2808-F-58	.05	Gasket, Mounting Side Cap—R. H.	4BLA0-143	✓	✓	✓	✓		✓				2
	4-X-186	.08	Stud, Mounting Side Cap, and Cage	4NFG0-1084	✓	✓	✓	✓		✓				14
	N-18	.04	Nut, Mounting Side Cap and Cage Stud (1/2″—20)	S-1420	✓	✓	✓	✓		✓				14
	P-120	.20	Plug, Mounting Inspection	S-4157	✓	✓	✓	✓		✓				1
	A3826-Z-26	14.00	Cage, Pinion Bearing, and Cap Assembly	4UG0-40	✓	✓	✓	✓		✓				1
	2803-Q-147	.14	Shim, Pinion Bearing Cage—Thin	4BLA0-98A	✓	✓	✓	✓		✓				As
	2803-R-148	.14	Shim, Pinion Bearing Cage—Medium	4BLA0-97A	✓	✓	✓	✓		✓				As
	2803-S-149	.14	Shim, Pinion Bearing Cage—Thick	4BLA0-96A	✓	✓	✓	✓		✓				As
	A1805-F-162	1.60	Seal, Pinion Bearing Cage Oil	4UG0-84	✓	✓	✓	✓		✓				1
	49175	5.65	Cone, Pinion Bearing—Front	4UG0-1232	✓	✓	✓	✓		✓				1
	49368	3.35	Cup, Pinion Bearing—Front	4RL0-137	✓	✓	✓	✓		✓				1
	59200	7.30	Cone, Pinion Bearing—Rear	4DFL0-138	✓	✓	✓	✓		✓				1
	59412	4.86	Cup, Pinion Bearing—Rear	4RL0-142	✓	✓	✓	✓		✓				1
	1829-U-21	.50	Ring, Pinion Bearing Spinner	4UG0-762	✓	✓	✓	✓		✓				1
	1844-W-23	3.00	Spacer, Pinion Shaft Bearing	4UG0-584	✓	✓	✓	✓		✓				1
	2803-L-142	.12	Shim, Pinion Shaft Bearing	4UG0-96	✓	✓	✓	✓		✓				As
	2803-M-143	.12	Shim, Pinion Shaft Bearing	4UGA0-96	✓	✓	✓	✓		✓				As
	2803-N-144	.12	Shim, Pinion Shaft Bearing	4UH0-96	✓	✓	✓	✓		✓				As
	2803-P-146	.12	Shim, Pinion Shaft Bearing	4UHA0-96	✓	✓	✓	✓		✓				As
	16-X-28	.18	Key, Pinion Shaft	4NK0-163	✓	✓	✓	✓		✓				1
	59175	7.30	Cone, Bevel Gear Bearing	4RL0-141	✓	✓	✓	✓		✓				2
	59412	4.86	Cup, Bevel Gear Bearing	4RL0-142	✓	✓	✓	✓		✓				2
	13399	.15	Nut, Bevel Pinion Shaft	4A0-81	✓	✓	✓	✓		✓				1
	4-X-322	.11	Stud, Gear Shaft Bearing Washer	4BLA0-648	✓	✓	✓	✓		✓				2
	1820-K-11	.05	Lock, Gear Shaft Bearing Stud	4BLA0-28	✓	✓	✓	✓		✓				1
	1827-G-7	.75	Lock, Pinion Bearing	4UG0-28	✓	✓	✓	✓		✓				1
	1827-N-14	2.00	Nut, Pinion Bearing Adjusting	4UG0-936	✓	✓	✓	✓		✓				1
	1829-E-5	.25	Washer, Pinion Bearing Lock	4NK0-1204	✓	✓	✓	✓		✓				1
	1850-B-28	.06	Plug, Carrier	4NK0-425	✓	✓	✓	✓		✓				1
	1829-F-6	1.00	Washer, Gear Shaft Bearing Lock	4BLA0-1178	✓	✓	✓	✓		✓				1
	2214-R-18	1.50	Nut, Differential Bearing Adj.	4BMC0-119	✓	✓	✓	✓		✓				2
	W-18	.01	Washer, 1/2″ Spring	S-1349	✓	✓	✓	✓		✓				14
1006	Steering Knuckle, Flange and Arm													
	A-3811-G-7	34.00	Knuckle, Steering—Assembly	9ZDK0-140	✓	✓	✓	✓		✓	✓			2
	A-3111-Z-520	28.10	Knuckle, Steering, & Bushing Assy.—R. H. (1½″—12 Thds.)	9NA0-140					✓					1
	A-3111-A-521	28.10	Knuckle, Steering, & Bushing Assy.—L. H. (1½″—12 Thds.)	9NA0-150					✓					1
	A-3897-V-100	72.10	Knuckle, Steering, and Bushing Assembly	9SK0-140								✓	✓	2
	1225-R-252	.32	Bushing, Steering Knuckle (With Oil Groove)	9N0-16A					✓					1
	3101-C-55	2.25	Pin, Steering Knuckle (Without Oil Groove)	9N0-14A					✓					1
	1825-G-7	.80	Bushing, Steering Knuckle	9ZDK0-438							✓			2
	1825-E-31	3.05	Bushing, Steering Knuckle	9SK0-438								✓	✓	2

THE AUTOCAR COMPANY
ARDMORE, PA.

Master Parts List

10-5
FRONT AXLE

CODE	MFR. PART No.	LIST PRICE	NAME OF PART	AUTOCAR PART No.	U-2044	U-4044	U-4144	U-4144-T	C-50	U-5044	U-7144-T	U-8144	U-8144-T	No. REQ.
1006	5-X-290	$1.20	Washer, Steering Knuckle Felt....................	9ZDK0-447	✓	✓	✓	✓		✓				2
	5-X-322	2.15	Felt, Steering Knuckle...........................	9SK0-447								✓	✓	2
	1805-M-39	.25	Retainer, Steering Knuckle Felt..................	9SK0-451								✓	✓	2
	1818-E-5	.15	Spring, Trunnion Socket Felt.....................	9ZDK0-498	✓	✓	✓	✓		✓				2
	1818-B-2	.15	Spring, Steering Knuckle Felt Pressure............	9SK0-498								✓	✓	2
	3-X-145	.07	Screw, Steering Knuckle Stop....................	9FKW0-501							✓			2
	S-2814	.05	Screw, Steering Knuckle Stop Hex. Cap (½"—20 x 1¾")...	S-1430								✓	✓	2
	N-37	.04	Nut, Steering Knuckle Stop Screw ($\frac{7}{16}$"—14)...........	S-3920							✓			2
	N-38	.04	Nut, Steering Knuckle Stop Screw................	N-38								✓	✓	2
	41125	4.39	Cone, Steering Knuckle Bearing..................	4ZDK0-138	✓	✓	✓	✓	✓	✓				4
	53176	6.27	Cone, Steering Knuckle Bearing..................	4A0-138								✓	✓	4
	41286	1.97	Cup, Steering Knuckle Bearing...................	4ZDK0-139	✓	✓	✓	✓	✓	✓				4
	53387	3.11	Cup, Steering Knuckle Bearing...................	4SA2-139								✓	✓	4
	3866-J-62	2.25	Cap, Steering Knuckle Bearing—Upper............	9ZDK0-452	✓	✓	✓	✓		✓	✓			1
	3866-F-136	3.90	Cap, Steering Knuckle Bearing—Upper............	9SK0-452								✓	✓	1
	4-X-389	.18	Stud—Long.....................................	9SK0-455								✓	✓	4
	4-X-526	.17	Stud—Short....................................	4SKA0-151								✓	✓	4
	N-110	.10	Nut, Hex. (⅝"—18).............................	S-1311								✓	✓	4
	13-X-14	.06	Nut, Hex......................................	4SK0-499								✓	✓	4
	W-110	.01	Washer, $\frac{21}{32}$" Spring............................	S-1351								✓	✓	8
	3866-X-63	3.45	Cap, Steering Knuckle Bearing—Lower...........	9UG0-453	✓	✓	✓	✓		✓				2
	3866-J-62	2.25	Cap, Steering Knuckle Bearing—Lower...........	9ZDK0-453							✓			2
	3866-G-137	3.85	Cap, Steering Knuckle Bearing—Lower...........	9SK0-453								✓	✓	2
	S-21012	.15	Screw, Hex. Cap (⅝"—11 x 1½").................	S-1870								✓	✓	8
	W-110	.01	Lockwasher, $\frac{21}{32}$" Spring........................	S-1351								✓	✓	8
	2803-H-138	.07	Shim, Steering Knuckle Bearing Cap—Thin.......	9ZDK0-454	✓	✓	✓	✓		✓	✓			As
	2803-W-309	.17	Shim, Steering Knuckle Bearing Cap—Thin.......	9SK0-454								✓	✓	20
	2803-J-140	.07	Shim, Steering Knuckle Bearing Cap—Thick......	9ZDKA0-454	✓	✓	✓	✓		✓				As
	S-2812	.05	Screw, Steering Knuckle Bearing Cap (½"—13 x 1⅞")...	S-4329							✓			8
	W-18	.01	Washer, ½" Spring.............................	S-1349							✓			8
	40-X-661	.07	Stud, Steering Knuckle Bearing Cap..............	9FKW0-455							✓			8
	N-48	.04	Nut, Steering Knuckle Bearing Cap Stud (½"—20)........	S-3747							✓			8
	W-18	.10	Washer, ½" Spring.............................	S-1349							✓			8
	2803-W-309	.17	Shim, Steering Knuckle Bearing Cap—Thick......	9SKA0-454								✓	✓	4
	A2-3897-N-274	32.50	Flange, Steering Knuckle, Assembly—R. H.......	9UG0-890	✓	✓	✓	✓		✓				1
	A2-3897-U-307	43.00	Flange, Steering Knuckle Companion, Assembly.....	9FKW0-890							✓			1
	A2-3897-Q-277	46.00	Flange, Steering Knuckle Companion, Assembly.....	9GKW0-890								✓	✓	1
	A3-3897-N-274	32.50	Flange, Steering Knuckle, Assembly—L. H.......	9UGA0-890	✓	✓	✓	✓		✓				1
	A3-3897-U-307	43.00	Flange, Steering Knuckle Companion, Assembly.....	9FKWA0-890							✓			1
	A3-3897-Q-277	46.00	Flange, Steering Knuckle Companion, Assembly.....	9GLW0-890								✓	✓	1
	4-X-607	.18	Stud ($\frac{9}{16}$"—18)................................	9GKW0-448								✓	✓	24
	N-49	.06	Nut, Stud ($\frac{9}{16}$"—18)..........................	S-84								✓	✓	24
	W-19	.03	Lockwasher, $\frac{19}{32}$" Spring.......................	S-1350								✓	✓	24
	S-2810	.03	Screw, Hex. Cap (½"—20 x 1¼")..................	S-2055								✓	✓	4
	S-2815	.05	Screw, Hex. Cap (½"—20 x ⅞")..................	S-2815								✓	✓	16
	N-48	.04	Nut, Hex. (½"—20).............................	S-830								✓	✓	16
	W-18	.01	Lockwasher, $\frac{17}{32}$" Spring.......................	S-1349								✓	✓	20
	4-X-317	.13	Stud, Steering Knuckle Companion Flange.........	9ZDK0-448							✓			24
	N-46	.02	Nut, Steering Knuckle Companion Flange (⅜"—24)........	S-814							✓			24
	W-16	.01	Washer, ⅜" Spring.............................	S-1347							✓			24
	S-1716	.25	Screw, Steering Knuckle Companion Flange ($\frac{7}{16}$"—20 x 2⅛")..	S-4332							✓			4
	S-1712	.05	Screw, Steering Knuckle Companion Flange ($\frac{7}{16}$"—20 x 1½")..	S-6226							✓			20
	N-47	.02	Nut, Steering Knuckle Companion Flange Screw ($\frac{7}{16}$"—20)....	S-3756							✓			24
	W-17	.06	Washer, $\frac{7}{16}$" Spring............................	S-1348							✓			24
	A3897-S-201	37.80	Socket, Trunnion, Assembly (Spindle Pin Pocket).....	9ZDK0-860	✓	✓	✓	✓		✓				2
	A-3897-X-310	37.45	Socket, Trunnion, Assembly......................	9FKW0-860							✓			2
	A-3897-R-96	98.20	Socket, Trunnion, Assembly......................	9SK0-860								✓	✓	2
	3-X-135	.17	Screw, Hex. Cap ($\frac{9}{16}$"—18 x).................	4SK0-215								✓	✓	16
	N-49	.06	Nut, Hex. ($\frac{9}{16}$"—18)...........................	S-84								✓	✓	16
	2847-D-56	2.64	Pin, Trunnion Socket Bearing (Spindle Pin)........	9ZDK0-444	✓	✓	✓	✓		✓				4
	5-X-290	1.20	Felt, Trunnion Socket...........................	9ZDK0-447							✓			2
	1818-E-5	.15	Spring, Trunnion Socket.........................	9ZDK0-498							✓			2

Master Parts List

CODE	MFR. PART No.	LIST PRICE	NAME OF PART	AUTOCAR PART No.	MODEL									No. REQ.
					U-2044	U-4044	U-4144	U-4144-T	C-50	U-5044	U-7144-T	U-8144	U-8144-T	
1006	3933-E-83	$11.30	**Arm,** Steering	10UG0-**93**	✓	✓	✓	✓		✓				1
	3933-Z-78	10.10	**Arm,** Steering	9FKW0-**75**							✓			1
	3933-E-31	18.80	**Arm,** Steering	9GKW0-**75**								✓	✓	1
	2910-C-3	1.74	**Stud,** Steering Arm Ball (1½″ Dia.)	9UGA0-**23**	✓	✓	✓	✓		✓	✓			1
	2910-R-18	4.50	**Stud,** Steering Arm Ball	9AK0-**23**								✓	✓	1
	13-X-36	.12	**Nut,** Steering Arm Ball Stud	9UG02-**94**	✓	✓	✓	✓		✓	✓			1
	N-212	.12	**Nut,** Steering Arm Ball (¾″—16)	S-**1408**								✓	✓	1
1007	**Axle Shaft and Universal Joint**													
	A-3802-F-370	24.25	**Axle,** Driving, and Pilot Seat Assembly—R. H. (Inner)—Long	9UK0-**1030**	✓	✓	✓	✓		✓				1
	A-3802-E-551	21.65	**Axle,** Front Driving—Long	9FKW0-**433**							✓			1
	A-3802-B-496	30.05	**Axle,** Front Driving—Long	9GKW0-**433**								✓	✓	1
	A-3802-G-371	21.00	**Axle,** Driving, and Pilot Assembly—L. H. (Inner)—Short	9UG0-**1030**	✓	✓	✓	✓		✓				1
	2203-K-271	.07	**Shim,** Steering Knuckle	9N0-**68**					✓					2
	7-X-9	.13	**Key,** Steering Knuckle Pin Draw	9N0-**288**					✓					As
	7-X-10	.13	**Key,** Steering Knuckle Pin Draw	9SA0-**109**					✓					As
	7-X-8	.15	**Key,** Steering Knuckle Pin Draw	9N0-**507**					✓					As
	......	.03	**Nut,** Steering Knuckle Pin Draw Key	S-**75**					✓					2
	......	.01	**Washer,** Spring	S-**1347**					✓					2
	5-X-66	.05	**Felt,** Steering Knuckle Pin	9N0-**173**					✓					2
	1205-E-187	.06	**Retainer,** Steering Knuckle Pin Felt	9N0-**302**					✓					2
	1218-V-48	.08	**Spring,** Steering Knuckle Pin Felt Retainer	9N0-**506**					✓					2
	T-138	2.90	**Bearing,** Steering Knuckle Thrust	9N0-**51**					✓					2
	1205-F-188	.18	**Retainer,** Steering Knuckle Thrust Bearing	9N0-**508**					✓					2
	1250-R-44	.05	**Plug,** Steering Knuckle Bottom	9N0-**168**					✓					2
	3133-P-354	7.55	**Arm,** Steering—R. H. (Single)	9N0-**12**					✓					1
	......	11.95	**Arm,** Steering—L. H. (Double)	9N0-**13**					✓					1
	26-X-15	.09	**Screw,** Steering Arm Stop	9N0-**501**					✓					2
	14-X-42	.06	**Nut,** Steering Arm Stop Screw	9N0-**502**					✓					2
	16-X-1	.13	**Key,** Steering Arm	9TA0-**77**					✓					2
	14-X-3	.18	**Nut,** Steering Arm	4G0-**624**					✓					2
	......	3.00	**Pin,** Steering Arm Ball (1½″)	9TE2-**23**					✓					1
	......	.12	**Nut,** Steering Arm Ball Pin (1½″ Ball)	S-**4221**					✓					1
	......	.10	**Nut,** Steering Arm Ball Pin (1¼″ Ball)	S-**810**					✓					1
	......	.90	**Adapter,** Steering Arm Nut Pin (1¼″ Ball)	10ZDT2-**596**					✓					1
	......	1.15	**Pin,** Steering Arm Ball	10A2-**266**					✓					1
	......	45.00	**Center,** Front Axle—Only	9NFD0-**07**					✓					1
	A-3802-E-551	19.35	**Axle,** Front Driving—Short	9FKW0-**434**							✓			1
	A-3802-A-495	23.65	**Axle,** Front Driving—Short	9GKW0-**434**								✓	✓	1
	3870-D-4	13.50	**Flange,** Driving	9ZDK0-**435**	✓	✓	✓	✓		✓	✓			2
	3870-E-57	26.95	**Flange,** Front Driving Axle	9AK0-**435**								✓	✓	2
	2808-W-75	.08	**Gasket,** Driving Flange	9ZDK0-**436**	✓	✓	✓	✓		✓	✓			2
	2808-H-164	.08	**Gasket,** Front Driving Axle Flange	9SK0-**436**								✓	✓	2
	S-266	.05	**Screw,** Front Driving Axle Drive Flange Puller (⅜″—16 x ¾″)	S-**12**	✓	✓	✓	✓		✓	✓			4
	S-2610	.06	**Screw,** Front Driving Axle Flange Puller (⅜″—16 x 1¼″)	S-**64**								✓	✓	4
	N-36	.03	**Nut,** Front Driving Axle Flange Puller Screw (⅜″—16)	S-**75**	✓	✓	✓	✓		✓	✓	✓	✓	4
	4-X-509	.10	**Stud,** Front Driving Axle Drive Flange	4ZDK0-**237**	✓	✓	✓	✓		✓	✓			16
	4-X-180	.18	**Stud,** Front Driving Axle Flange and Hub	4FD0-**237**								✓	✓	16
	N-48	.04	**Nut,** Front Driving Axle Flange Stud (½″—20)	S-**3747**							✓			16
	N-110	.10	**Nut,** Front Driving Axle Flange and Hub Stud (⅝″—18)	S-**1311**								✓	✓	16
	W-18	.01	**Washer,** ½″ Spring	S-**1349**							✓			16
	W-110	.01	**Lockwasher,** 21/32″ Spring	S-**1351**								✓	✓	16
	1246-Y-233	.07	**Dowel,** Front Driving Axle Flange	4ZGEW0-**695**								✓	✓	8
	A-1805-A-53	.70	**Seal,** Front Driving Axle Oil	4SK0-**730**								✓	✓	2
	1829-T-332	.47	**Retainer,** Front Driving Axle Oil Seal	9CK0-**437**								✓	✓	2
	S-157	.02	**Screw,** Front Driving Axle Oil Seal Retainer	S-157								✓	✓	4
	W-15	.01	**Lockwasher,** 11/32″ Spring	S-**1346**								✓	✓	4
	2808-H-138	.04	**Gasket,** Front Driving Axle Oil Seal Retainer	19DKA0-**212**								✓	✓	2
	A-3897-B-236	144.29	**Drive,** Universal, Assembly	9ZDK0-**820**	✓	✓	✓	✓		✓	✓			2
	A1-3897-Z-234	217.30	**Drive,** Universal, Assembly	9BK0-**820**								✓	✓	2
	3897-Z-234	122.10	**Drive,** Universal only	9SKA0-**485**								✓	✓	2
	1898-N-66	.99	**Plunger,** Universal Drive Pilot Pin	9AK0-**486**	✓	✓	✓	✓		✓	✓	✓	✓	2

THE AUTOCAR COMPANY
ARDMORE, PA.

Master Parts List

10-7 FRONT AXLE

CODE	MFR. PART No.	LIST PRICE	NAME OF PART	AUTOCAR PART No.	MODEL U-2044	U-4044	U-4144	U-4144-T	C-50	U-5044	U-7144-T	U-8144	U-8144-T	No. REQ.
1007	1898-P-68	$0.37	Spring, Universal Drive Pilot Pin Plunger	9AK0-487	✓	✓	✓	✓		✓	✓	✓	✓	2
	1898-Q-69	.26	Spring, Universal Drive Buffer	9SK0-488	✓	✓	✓	✓		✓	✓	✓	✓	2
	1898-G-59	1.00	Ball, Universal Joint Cage	9ZDK0-477	✓	✓	✓	✓		✓	✓			12
	1898-T-72	2.40	Ball, Universal Drive	9SK0-477								✓	✓	12
	1898-F-58	15.40	Cage, Universal Joint	9ZDK0-476	✓	✓	✓	✓		✓	✓			2
	1898-S-71	19.15	Cage, Universal Drive Ball	9SK0-476								✓	✓	2
	1898-L-64	6.30	Pilot, Universal Joint	9ZDK0-481	✓	✓	✓	✓		✓	✓			2
	1898-X-76	12.60	Pilot, Universal Drive	9SK0-481								✓	✓	2
	1898-M-65	5.43	Pin, Universal Joint Pilot	9ZDK0-482	✓	✓	✓	✓		✓	✓			2
	1898-Y-77	5.80	Pin, Universal Drive Pilot	9SK0-482								✓	✓	2
	1898-A-53	1.70	Seat, Universal Joint Pilot Pin	9ZDK0-483	✓	✓	✓	✓		✓	✓			2
	1898-B-80	4.05	Seat, Universal Drive Pilot Pin	9SK0-483								✓	✓	2
	1898-H-60	20.30	Race, Universal Joint Cage Ball, Inner	9ZDK0-484	✓	✓	✓	✓		✓	✓			2
	A-1898-J-114	25.90	Race, Universal Drive Inner	9GKW0-484								✓	✓	2
	1898-L-194	.39	Spacer, Universal Drive Inner Race	9GKW0-291								✓	✓	2
	1898-J-62	6.50	Retainer, Universal Joint and Shaft	9ZDK0-478	✓	✓	✓	✓		✓	✓			2
	1898-K-193	5.55	Retainer, Universal Drive and Shaft	9AK0-478								✓	✓	2
	1199-A-651	.21	Screw, Universal Joint Retainer	9ZDK0-479	✓	✓	✓	✓		✓	✓			6
	1898-W-75	.37	Screw, Universal Drive and Shaft Retainer	9SK0-479								✓	✓	12
	S-2812	.12	Screw, Carrier to Housing (½″—13 x 1⅛″)	S-19	✓	✓	✓	✓		✓				8
	N-49	.06	Nut, Socket to Housing Stud (⅝″—18)	S-84	✓	✓	✓	✓		✓				16
	A1805-B-54	.70	Seal, Driving Axle Oil, Assembly	4ZDK0-395	✓	✓	✓	✓		✓				2
	1854-H-8	.18	Ring, Universal Joint Retainer Snap	9ZDK0-504	✓	✓	✓	✓		✓	✓			2
	2858-C-3	.25	Spring, Universal Joint	9ZDK0-511	✓	✓	✓	✓		✓	✓			2
	1825-G-7	.80	Bushing, Universal Drive	9ZDK0-438	✓	✓	✓	✓		✓				2
	2808-H-138	.04	Gasket, Drive Shaft Retainer	9NK0-473	✓	✓	✓	✓		✓				2
	5-X-346	.15	Packing, Universal Joint	9ZDK0-489	✓	✓	✓	✓		✓	✓			2
	1844-D-30	.88	Spacer, Universal Joint	9ZDK0-499	✓	✓	✓	✓		✓	✓			2
	3897-B-236	75.30	Cage, Universal Drive—Outer	9ZDK0-485							✓			2
	A-1898-U-203	.56	Fitting, Universal Drive Relief	4HDA0-1363							✓			4
	4-X-317	.13	Stud, Steering Knuckle	9ZDK0-448	✓	✓	✓	✓		✓				24
	1825-S-71	.45	Bushing, Tie Rod Bolt	9SK0-449	✓	✓	✓	✓		✓				2
	4-X-186	.08	Stud, Steering Knuckle Bearing Cap	4NFG0-1084	✓	✓	✓	✓		✓				8

ILLUSTRATIONS

GROUP 11

REAR AXLE

GROUP 11

Master Parts List

THE AUTOCAR COMPANY, ARDMORE, PA.

As supplied on Autocar Models
U-8144 U-8144-T

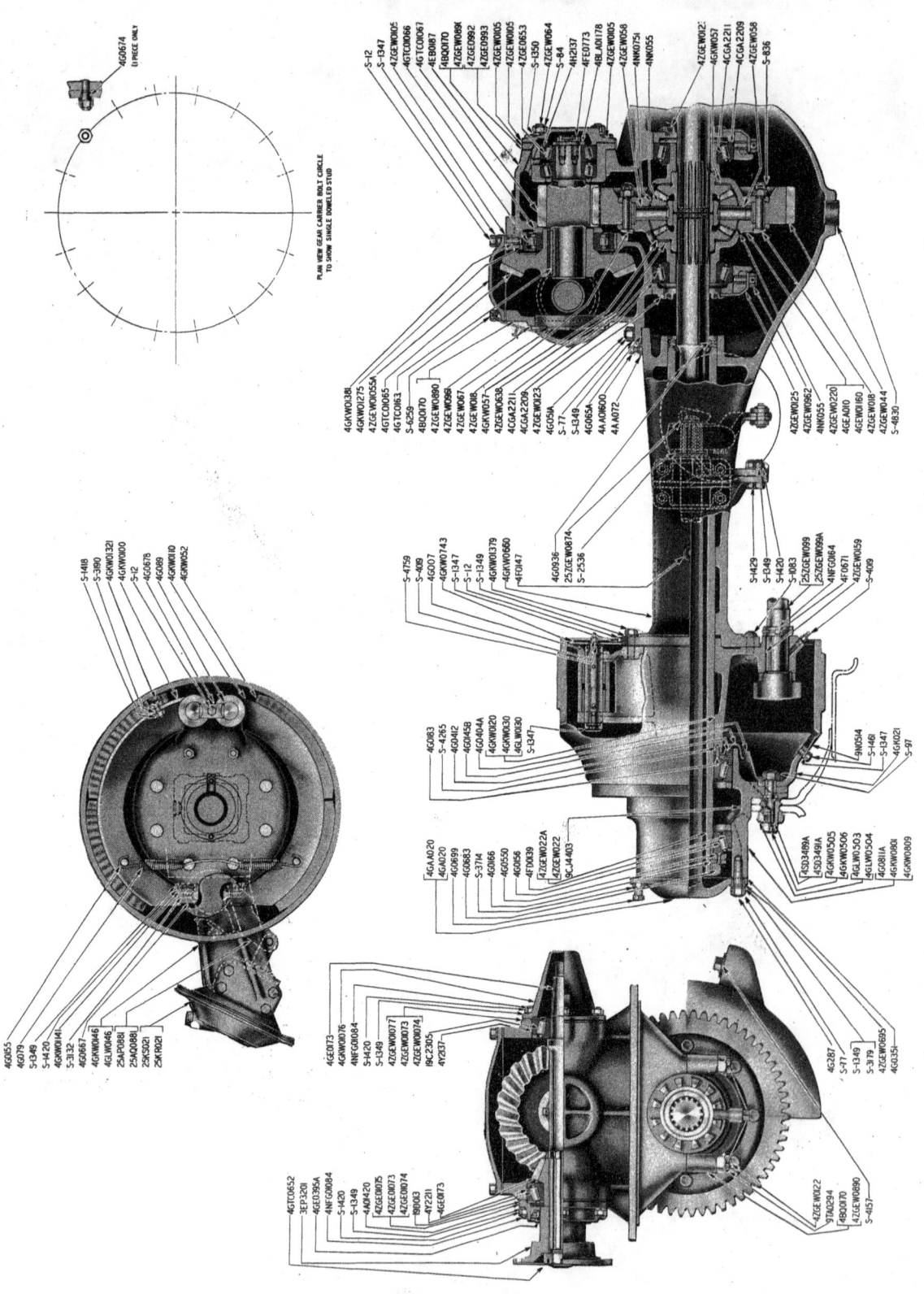

1100—Rear Axle Assembly

THE AUTOCAR COMPANY, ARDMORE, PA.

Master Parts List

GROUP 11

As supplied on Autocar Models
U-8144   U-8144-T

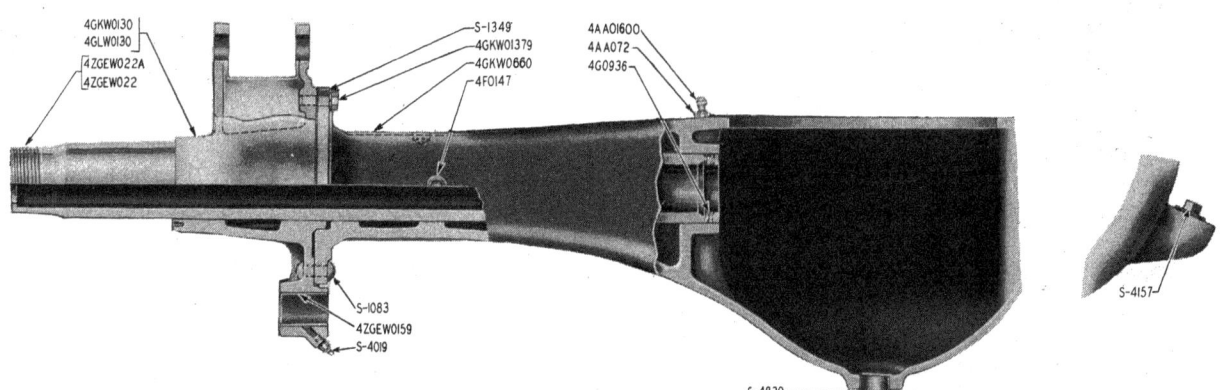

1101—Housing Assembly

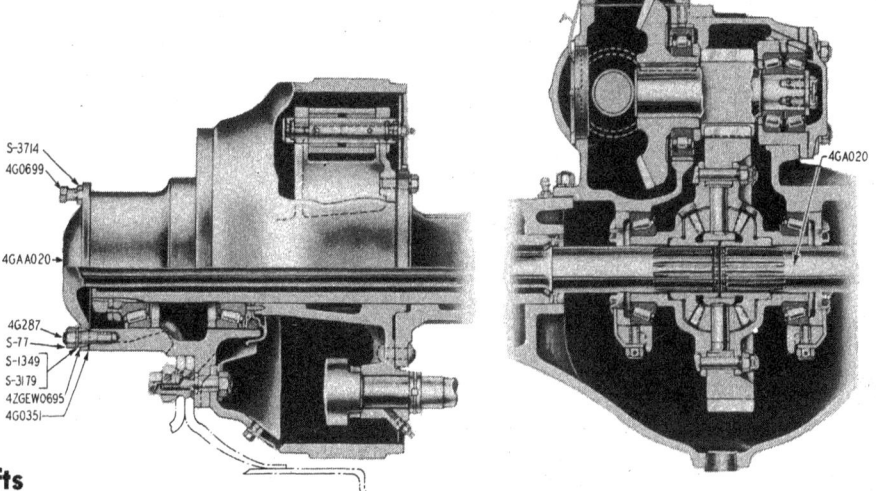

1102—Rear Axle Drive Shafts

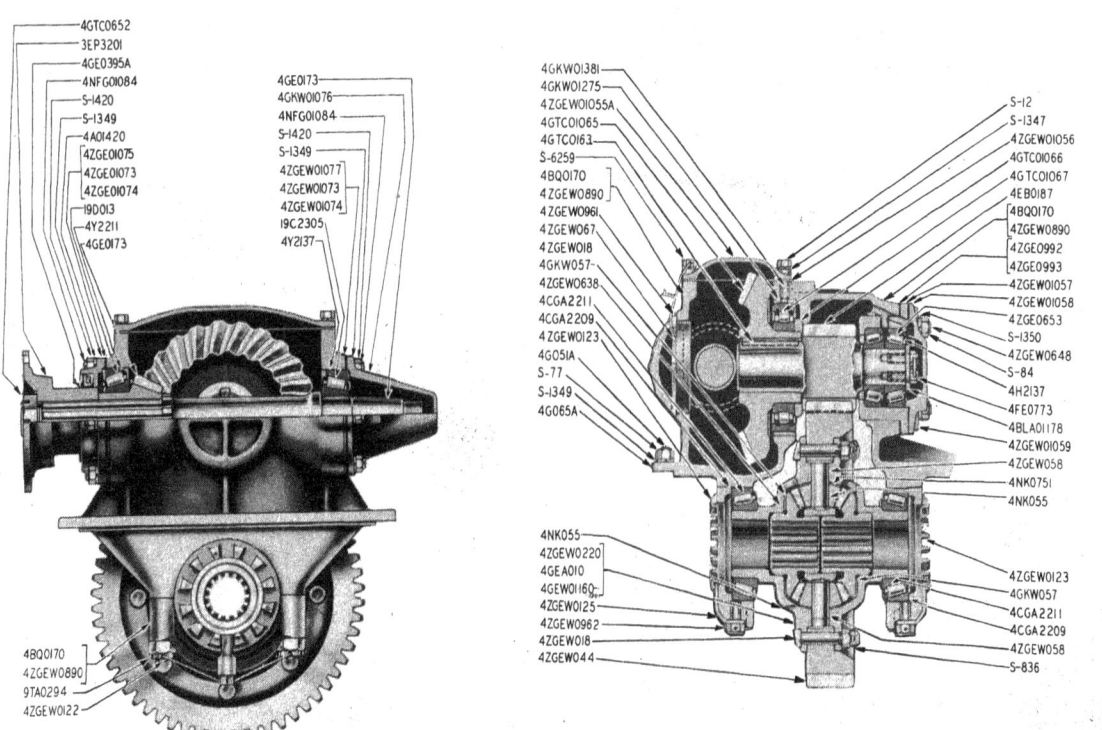

1103—Rear Axle Differential and Carrier Assembly

Master Parts List

CODE	MFR. PART No.	LIST PRICE	NAME OF PART	AUTOCAR PART No.	MODEL U-2044	U-4044	U-4144	U-4144-T	C-50	U-5044	U-7144-T	U-8144	U-8144-T	No. REQ.
1100	**Rear Axle Assembly**													
	R-2090-W-X-1	1086.00	**Axle,** Rear, Assembly (Timken)	4FKW9-**970**							✓			1
	R-3100-W-X-2	845.25	**Axle,** Rear, Assembly, Ratio 8.148, (Timken)	4GKW9-**970**								✓	✓	1
1101	A-3801-Z-286	129.75	**Housing** and Studs Assembly (Without Tubes)	4UG0-**660**	✓	✓	✓	✓		✓				1
	......	197.50	**Housing** and Tubes Assembly	4TGA0-**660**					✓					1
	A-2-3201-E-1045	231.85	**Housing** Assembly	4FKW0-**270**							✓			1
	A-1-3201-G-1047	357.60	**Housing** Assembly	4GKW0-**660**								✓	✓	1
	P-28	.07	**Plug,** Housing Drain	4GKW0-**67**	✓	✓	✓	✓		✓				1
	X-300	.09	**Plug,** Housing Drain (¾"—Square Head)	S-**4131**							✓			1
	X-1278	.20	**Plug,** Housing Drain (1¼"—Hollow Head)	S-**4830**								✓	✓	1
	3885-A-1	1.65	**Neck,** Housing Filler	4SK0-**417**	✓	✓	✓	✓		✓				1
	1850-A-1	.42	**Plug,** Housing Filler Neck	4FD0-**176**	✓	✓	✓	✓		✓				1
	X-1182	.32	**Elbow,** Housing Filler	4D0-**417**							✓			1
	X-136	.06	**Plug,** Housing Filler (1"—Square Head)	S-**4156**							✓			1
	X-1011	.20	**Plug,** Housing Filler (1¼"—Square Head)	S-**4157**								✓	✓	1
	A-1199-E-421	.36	**Breather,** Housing Oil	4BL0-**1600**	✓	✓	✓	✓		✓				1
	A-1199-C-1043	.14	**Nipple,** Housing Breather	4AA0-**1600**							✓	✓	✓	1
	1898-W-205	.06	**Bushing,** Housing Breather Nipple	4AA0-**72**								✓	✓	1
	2808-E-57	.14	**Gasket,** Housing	4UG0-**483**	✓	✓	✓	✓		✓				1
	......	178.00	**Axle,** Rear, Housing—Bare	4TG7-**01C**					✓					1
	A-3211-Q-953	26.50	**Brake,** Rear, Spider and Bushing Assembly	4FKW0-**120**							✓			2
	A-1-3211-M-949	78.10	**Brake,** Rear, Spider and Sleeve Assembly—Long	4GKW0-**130**								✓	✓	1
	A-2-3211-M-949	73.70	**Brake,** Rear, Spider and Sleeve Assembly—Short	4GLW0-**130**								✓	✓	1
	1225-G-85	.32	**Bushing,** Rear Brake Spider	4G0-**159**							✓			2
	1225-R-200	.21	**Bushing,** Rear Brake Spider	4ZGEW0-**159**								✓	✓	4
	3-X-111	.14	**Screw,** Rear Brake Spider and Housing	4GKW0-**1379**								✓	✓	4
	X-531	.01	**Lockwasher,** $\frac{17}{32}$" Spring	S-**1349**								✓	✓	4
	X-1273	.05	**Rivet,** Rear Brake Spider and Housing Screw (½" x 2")	S-**1083**								✓	✓	12
	3280-G-1515	6.85	**Spacer,** Rear Brake Spider and Brake Flange	4FKW0-**1022**							✓			2
	X-1372	.07	**Rivet,** Rear Brake Spider and Brake Flange Spacer	S-**1084**							✓			16
	A-1898-U-203	.56	**Fitting,** Rear Brake Spider Grease Relief, Assembly	4HDA0-**1363**							✓			4
	26-X-16	.15	**Screw,** Housing Sleeve Retaining	4DG0-**147**							✓			2
	26-X-19	.26	**Screw,** Housing Sleeve Retaining	4F0-**147**								✓	✓	2
	......	1.40	**Seal,** Axle Shaft Oil, Assembly	4F0-**730**					✓					2
	......	.70	**Seal,** Axle Shaft Oil	S-**6106**					✓					2
	......	.65	**Retainer,** Axle Shaft Oil Seal	4F2-**145**					✓					2
	......	.12	**Screw,** Axle Shaft Oil Seal Retainer Set	4TF2-**147**					✓					2
	A-1205-Y-181	.93	**Seal,** Axle Shaft Oil	4FKW0-**395**							✓			2
	A-1205-P-146	.86	**Seal,** Axle Shaft Oil	4G0-**936**								✓	✓	2
	......	.30	**Fitting,** Alemite Measuring (⅛")	S-**4019**								✓	✓	2
	3816-J-36	21.65	**Tube,** Housing	4UG0-**22**	✓	✓	✓	✓		✓				2
	......	22.20	**Tube,** Rear Axle	4TF4-**22A**					✓					2
	......	.12	**Screw,** Rear Axle Tube Set	4Y1-**147**					✓					2
	......	.04	**Nut,** Hex. (½"—13)	S-**77**					✓					2
	3216-T-1034	22.00	**Sleeve,** Housing	4FKW0-**22**							✓			2
	3216-J-114	40.65	**Sleeve,** Housing—Long	4ZGEW0-**22A**								✓	✓	1
	3216-H-112	36.25	**Sleeve,** Housing—Short	4ZGEW0-**22**								✓	✓	1
	......	14.00	**Cover,** Rear Axle Housing Rear	4F4-**116**					✓					1
	......	.12	**Cover,** Rear Axle Housing Rear, and Carrier Gasket	4N3-**483**					✓					2
	......	.06	**Dowel,** Rear Axle Housing Rear Cover	4SA2-**111A**					✓					2
	......	.12	**Screw,** Rear Cover Cap	4TG2-**1035**					✓					2
	......	.04	**Lockwasher,** $\frac{25}{32}$" Spring	S-**1353**					✓					2
	......	.12	**Plug,** Rear Cover Drain (1"—Square Head)	S-**4156**					✓					3
	......	.12	**Screw,** Hex. Cap (⅝"—11 x 1")	S-**4321**					✓					6
	......	.01	**Lockwasher,** $\frac{21}{32}$" Spring	S-**1351**					✓					8
1102	**Rear Axle Drive Shafts**													
	3802-A-53	25.15	**Axle,** Driving	4DK0-**38**	✓	✓	✓	✓		✓				2
	2808-G-59	.07	**Gasket,** Driving Axle Flange	4UG0-**1143**	✓	✓	✓	✓		✓				2
	4-X-522	.17	**Stud,** Driving Axle Flange	4UG0-**87**	✓	✓	✓	✓		✓				24
	A-1805-Y-103	.90	**Seal,** Driving Axle Oil	4UG0-**730**	✓	✓	✓	✓		✓				2

THE AUTOCAR COMPANY
ARDMORE, PA.

Master Parts List

11-1 Rear Axle

CODE	MFR. PART No.	LIST PRICE	NAME OF PART	AUTOCAR PART No.	U-2044	U-4044	U-4144	U-4144-T	C-50	U-5044	U-7144-T	U-8144	U-8144-T	No. REQ.
1102	2853-N-14	.90	Sleeve, Driving Axle Oil Seal	4UG0-1165	✓	✓	✓	✓		✓				2
	......	25.20	Shaft, Axle, and Flange Assembly	4TF0-20					✓					2
	......	14.40	Shaft, Axle	4TF3-38					✓					2
	......	9.60	Flange, Axle	4TF4-11					✓					2
	......	.12	Screw, Axle Flange Jack	4TF2-174					✓					4
	......	.03	Nut, Hex. (3/8"—16)	S-75					✓					4
	3202-K-1779	19.00	Shaft, Axle, Assembly	4FKW0-38							✓			2
	2208-A-105	.07	Gasket	4NFG0-531							✓			2
	1246-B-28	.26	Dowel	4G0-674							✓			6
	1229-V-516	.03	Lockwasher, 1/2" Shakeproof Ext.	S-3179							✓			6
	A-3202-P-1394	27.00	Shaft, Axle, Assembly—Long	4GAA0-20								✓	✓	1
	A-3202-N-1392	24.80	Shaft, Axle, Assembly—Short	4GA0-20								✓	✓	1
	X-1680	.06	Screw, Puller	4G0-699								✓	✓	4
	X-768	.05	Nut, Hex. (1/2"—16 L. H.)	S-3714								✓	✓	4
	2208-S-19	.05	Gasket	4G0-351								✓	✓	2
	T-7308	.19	Stud (1/2"—13)	4G2-87								✓	✓	14
	X-370	.04	Nut, Hex. (1/2"—13)	S-77								✓	✓	14
	X-1363	.01	Lockwasher, 1 1/32" Spring	S-1349								✓	✓	8
	1229-V-516	.04	Lockwasher, 1/2" Shakeproof Ext.	S-3179								✓	✓	6
	1246-Y-233	.07	Dowel	4XGEW0-695								✓	✓	6
1103	**Rear Axle Differential and Carrier Assembly**													
	G-316	310.00	Mounting Assembly (Gear Carrier Assembly)	4BA0-170	✓	✓	✓	✓		✓				1
	......	437.00	Carrier, Gear, Assembly	4FC0-170					✓					1
	A-9-3800-L-64	412.00	Carrier, Differential, Assembly—Complete (Ratio 8.43)	4FKW0-170							✓			1
	A-23-3800-X-24	532.00	Carrier, Differential, Assembly—Complete (Ratio 8.148)	4BQ0-170								✓	✓	1
	A-2-3800-K-63	47.50	Mounting with Caps (Gear Carrier)	4UG0-920	✓	✓	✓	✓		✓				1
	......	75.00	Carrier, Gear, with Caps	4FA0-920					✓					1
	A-5-3800-L-64	83.50	Carrier, Differential, and Cap Assembly	4FKWA0-890							✓			1
	A-2-3800-X-24	110.00	Carrier, Differential, and Cap Assembly	4ZGEW0-890								✓	✓	1
	4-X-177	.21	Stud, Differential Support Cap	4NFG0-122	✓	✓	✓	✓		✓	✓			4
	......	.10	Nut, Hex. (5/8"—18)	S-810	✓	✓	✓	✓		✓	✓			4
	......	1.00	Bolt, Differential Cap	4SA2-122					✓					4
	4-X-175	.44	Stud, Differential Carrier and Cap	4ZGEW0-122								✓	✓	4
	13453	.16	Nut, Differential Carrier and Cap Stud	9TA0-294								✓	✓	4
	......	.02	Lockwire (.065")	S-4284								✓	✓	1
	A-3838-Y-25	7.35	Cap, Mounting Side, and Cup Assembly	4BLA0-460	✓	✓	✓	✓		✓				2
	2803-T-150	.17	Shim, Mounting Side Cap—Medium	4BLA0-97	✓	✓	✓	✓		✓				As
	2803-V-152	.17	Shim, Mounting Side Cap—Thick	4BLA0-96	✓	✓	✓	✓		✓				As
	2803-U-151	.17	Shim, Mounting Side Cap—Thin	4BLA0-98	✓	✓	✓	✓		✓				As
	4-X-186	.08	Cap, Mounting Side, and Cage Stud	4NFG0-1084	✓	✓	✓	✓		✓				14
	N-18	.04	Nut, Hex. (1/2"—20)	S-1420	✓	✓	✓	✓		✓				14
	P-120	.20	Plug, Mounting Inspection	S-4157	✓	✓	✓	✓		✓				1
	2808-F-58	.05	Gasket, Mounting Side Cap—R. H.	4BLA0-143	✓	✓	✓	✓		✓				2
	......	.12	Gasket, Gear Carrier and Housing	4N3-483					✓					2
	......	.12	Plug, Pipe (1" Square Head)	S-4156					✓					3
	......	.05	Washer, Differential Cap Screw Lock	4FC2-175					✓					4
	......	.06	Pin, Housing Front Cover Dowel	4F2-111					✓					2
	......	.65	Cover, Hand Hole, and Baffle Assembly	4F0-1040					✓					1
	......	.05	Screw, Hex. Cap (3/8"—16 x 3/4")	S-12					✓					2
	......	.01	Lockwasher, 1 3/32" Spring	S-1347					✓					2
	......	.85	Trough, Oil	4F4-1273					✓					1
	......	.03	Washer, Oil Trough	4F2-1334					✓					2
	......	.12	Screw, Hex. Cap (7/16"—14 x 1 1/8")	S-6085					✓					2
	......	.03	Washer, Flat Lead (15/32" x 23/32" x 1/8")	S-3329					✓					2
	......	.10	Gasket, Drive Gear Case Inspection Cover	4DF2-06					✓					1
	......	.30	Dowel, Gear Carrier Cap	4F2-1199					✓					2
	1850-C-3	1.40	Plug, Differential Carrier Inspection	4NFG0-32							✓			1
	1850-D-4	1.95	Plug, Differential Carrier Inspection	4ZGEW0-67								✓	✓	1
	2808-F-6	.12	Gasket, Differential Carrier Inspection Plug	4NFG0-06							✓			1
	2808-D-4	.06	Gasket, Differential Carrier Inspection Plug	4ZGEW0-961								✓	✓	1
	......	.02	Lockwire (.065")	S-4284								✓	✓	1

Master Parts List

CODE	MFR. PART No.	LIST PRICE	NAME OF PART	AUTOCAR PART No.	MODEL U-2044	U-4044	U-4144	U-4144-T	C-50	U-5044	U-7144-T	U-8144	U-8144-T	No. REQ.
1103	3855-B-2	3.30	**Cover,** Differential Carrier Top	4NFG0-1098							✓			1
	3855-A-1	6.05	**Cover,** Differential Carrier Top	4ZGEW0-1055A								✓	✓	1
	2808-E-5	.07	**Gasket,** Differential Carrier Top Cover	4NFG0-1099							✓			1
	2 X-36	.03	**Screw,** Hex. Cap ($\frac{3}{8}$″—16 x 1$\frac{1}{4}$″)	S-64							✓			8
	X-529	.01	**Lockwasher,** $\frac{13}{32}$″ Spring	S-1347							✓			1
	3855-E-57	.08	**Gasket,** Differential Carrier Top Cover	4ZGEW0-1056								✓	✓	8
	2-X-33	.05	**Screw,** Hex. Cap ($\frac{3}{8}$″—16 x $\frac{3}{4}$″)	S-12								✓	✓	9
	10 X-75	.06	**Screw,** Hex. Cap ($\frac{3}{8}$″—16 x $\frac{3}{4}$″ Dr. Hd.)	S-6259								✓	✓	1
	X-529	.01	**Lockwasher,** $\frac{13}{32}$″ Spring	S-1347								✓	✓	10
	2208-K-193	.18	**Gasket,** Differential Carrier to Housing	4ZDK0-65							✓			1
	T-6551	.20	**Gasket,** Differential Carrier to Housing	4G0-65A								✓	✓	1
	4-X-186	.08	**Stud,** Carrier to Housing	4NFG0-1084	✓	✓	✓	✓		✓				4
	N-48	.02	**Nut,** Hex. ($\frac{1}{2}$″—20)	S-830	✓	✓	✓	✓		✓				5
	S-2812	.12	**Screw,** Hex. ($\frac{1}{2}$″—13 x 1$\frac{1}{8}$″)	S-19	✓	✓	✓	✓		✓				8
	W-18	.01	**Lockwasher,** $\frac{17}{32}$″ Spring	S-1349	✓	✓	✓	✓		✓				12
	4X-101	.18	**Stud,** Differential Carrier to Housing	4ZDK0-661							✓			2
	4X3	.09	**Stud,** Differential Carrier to Housing	4A2-87							✓			8
	13X7	.06	**Nut,** Hex. ($\frac{5}{8}$″—18)	S-1311							✓			2
	13X4	.03	**Nut,** Hex. ($\frac{7}{16}$″—20)	S-1419							✓			8
	X-530	.01	**Lockwasher,** $\frac{15}{32}$″ Spring	S-1348							✓			8
	X-533	.01	**Lockwasher,** $\frac{21}{32}$″ Spring	S-1351							✓			2
	4X-186	.17	**Stud,** Differential Carrier to Housing ($\frac{1}{2}$″—13)	4G0-51A								✓	✓	16
	13X5	.04	**Nut,** Hex. ($\frac{1}{2}$″—13)	S-77								✓	✓	16
	1246-B-28	.26	**Dowel**	4G0-674								✓	✓	1
	X-1363	.01	**Lockwasher,** $\frac{17}{32}$″ Spring	S-1349								✓	✓	16
	A-3-3855-Y-51	113.50	**Differential** and Ring Gear Assembly	4UGA0-10	✓	✓	✓	✓		✓				1
	A-2-3835-Y-51	80.50	**Differential** Assembly	4UG0-10	✓	✓	✓	✓		✓				1
	......	96.50	**Differential** Assembly	4TF4-10C					✓					1
	A-9-3835-N-66	97.00	**Differential** Assembly	4NKA0-10							✓			1
	A-5-3835-L-38	102.00	**Differential** Assembly	4GEA0-10								✓	✓	1
	A-3835-Z-52	43.00	**Case,** Differential Assembly	4BMC0-220	✓	✓	✓	✓		✓				1
	......	26.25	**Case,** Differential—R. H.	4TF4-161A					✓					1
	......	26.25	**Case,** Differential—L. H.	4TF4-162A					✓					1
	A-3835-M-65	34.65	**Case,** Differential Assembly	4NK0-220							✓			1
	A-3835-M-39	37.25	**Case,** Differential Assembly	4ZGEW0-220								✓	✓	2
	15-X-180	.12	**Bolt,** Differential Case	4BLA0-18	✓	✓	✓	✓		✓				8
	N-28	.05	**Nut,** Hex. ($\frac{1}{2}$″—20)	S-811	✓	✓	✓	✓		✓				8
	......	.30	**Bolt,** Differential and Spur Gear	4Y2-18					✓					12
	......	.06	**Nut,** Hex. ($\frac{1}{2}$″—13)	S-4199					✓					12
	15-X-112	.27	**Bolt,** Differential Case	4NFG0-18							✓			8
	13-X-12	.07	**Nut,** Hex. ($\frac{9}{16}$″—18)—Castle	S-836							✓			8
	15-X-130	.35	**Bolt,** Differential Case	4ZGEW0-18								✓	✓	2
	13-X-12	.06	**Nut,** Hex. ($\frac{9}{16}$″—18)—Castle	S-836								✓	✓	2
	......	.01	**Pin,** Cotter ($\frac{1}{8}$″ x 1$\frac{1}{4}$″)	S-2205								✓	✓	2
	3878-B-54	4.45	**Spider,** Differential	4BLA0-58	✓	✓	✓	✓		✓				1
	3878-B-2	8.70	**Spdier,** Differential	4NK0-58							✓			1
	3878-C-3	8.70	**Spider,** Differential	4ZGEW0-58								✓	✓	1
	2234-K-427	7.35	**Gear,** Differential Side	4UG0-57	✓	✓	✓	✓		✓				2
	......	14.00	**Gear,** Differential Side	4M4-57					✓					2
	2834-S-409	13.85	**Gear,** Differential Side	4NKA0-57							✓			2
	2834-R-408	13.30	**Gear,** Differential Side	4GKW0-57								✓	✓	2
	1229-Z-728	.29	**Washer,** Differential Side Gear Thrust	4BM0-638	✓	✓	✓	✓		✓				2
	......	1.90	**Washer,** Differential Side Gear Thrust	4TF2-597					✓					2
	1229-X-822	.75	**Washer,** Differential Side Gear Thrust	4ZGEW0-638							✓	✓	✓	2
	2233-M-429	3.35	**Pinion,** Differential Bevel Side	4UG0-55	✓	✓	✓	✓		✓				4
	......	9.00	**Pinion,** Differential	4C3-55A					✓					4
	2233-W-413	5.40	**Gear,** Differential Pinion	4NK0-55							✓	✓	✓	4
	1229-C-861	.13	**Washer,** Differential Bevel Side Pinion Thrust	4BMA0-751	✓	✓	✓	✓		✓				4
	......	.60	**Washer,** Differential Pinion Thrust	4TF2-751					✓					4
	1829-S-45	.65	**Washer,** Differential Pinion Gear Thrust	4NK0-751							✓	✓	✓	4
	......	8.75	**Cross,** Differential	4M4-58					✓					1
	3891-Q-121	23.50	**Shaft,** Spur Pinion (6.62 Ratio)	4BMC0-1061	✓									1

THE AUTOCAR COMPANY
ARDMORE, PA.

Master Parts List

11-3
REAR AXLE

CODE	MFR. PART No.	LIST PRICE	NAME OF PART	AUTOCAR PART No.	U-2044	U-4044	U-4144	U-4144-T	C-50	U-5044	U-7144-T	U-8144	U-8144-T	NO. REQ.
1103	3891-R-96	23.50	**Shaft,** Spur Pinion (8.21 Ratio)................	4BLA0-44		✓	✓	✓		✓				1
	3892-C-133	30.00	**Gear,** Spur (6.62 Ratio).........................	4BMC0-62	✓									1
	3892-S-71	30.00	**Gear,** Spur (8.21 Ratio).........................	4BMB0-62B		✓	✓	✓		✓				1
	A-10-3835-N-66	141.00	**Differential** and Gear Assembly (8.43 Ratio) (Consists of Diff. Assy. 4NKA010 and following 3 items)..............	4NKA0-80							✓			1
	3892-H-34	39.00	**Gear** Spur (8.43 Ratio).........................	4NFG0-62							✓			1
	15-X-112	.27	**Bolt,** Differential Case.........................	4NFG0-18							✓			8
	13-X-12	.07	**Nut,** Hex. ($\frac{9}{16}''$—18)—Castle................	S-836							✓			8
	A-10-3835-L-38	156.50	**Differential** and Gear Assembly (Consists of Diff. Assy. 4GEA010 and following 4 items)................	4GEW0-1160								✓	✓	1
	3892-L-12	50.00	**Gear,** Spur (8.148 Ratio).......................	4ZGEW0-44								✓	✓	1
	15-X-130	.35	**Bolt,** Differential Case.........................	4ZGEW0-18								✓	✓	6
	13-X-12	.06	**Nut,** Hex. ($\frac{9}{16}''$—18)—Castle................	S-836								✓	✓	6
	......	.01	**Pin,** Cotter ($\frac{1}{8}''$ x $1\frac{1}{4}''$)................	S-2205								✓	✓	6
	2214-R-18	1.50	**Nut,** Differential Bearing Adjusting..............	4BMC0-119	✓	✓	✓	✓		✓				2
	S-265-D	.04	**Screw,** Hex. Cap ($\frac{3}{8}''$—16 x $\frac{5}{8}''$)............	S-2065	✓	✓	✓	✓		✓				2
	W-18	.01	**Lockwasher,** $\frac{1}{32}''$ Spring...................	S-1349	✓	✓	✓	✓		✓				14
1104	**Rear Axle Side Gears and Pinions.**													
	(Also Listed in Assemblies in Group 1103).													
	2234-K-427	7.35	**Gear,** Differential Side.........................	4UG0-57	✓	✓	✓	✓		✓				2
	......	14.00	**Gear,** Differential Side.........................	4M4-57					✓					2
	2834-S-409	13.85	**Gear,** Differential Side.........................	4NKA0-57							✓			2
	2834-R-408	13.30	**Gear,** Differential Side.........................	4GKW0-57								✓	✓	2
	1229-Z-728	.29	**Washer,** Differential Side Gear Thrust...........	4BM0-638	✓	✓	✓	✓		✓				2
	......	1.90	**Washer,** Differential Side Gear Thrust...........	4TF2-597					✓					2
	1229-X-822	.75	**Washer,** Differential Side Gear Thrust...........	4ZGEW0-638								✓	✓	2
	2233-M-429	3.35	**Pinion,** Differential Bevel Side...................	4UG0-55	✓	✓	✓	✓		✓				4
	......	9.00	**Pinion,** Differential............................	4C3-55A					✓					4
	2233-W-413	5.40	**Gear,** Differential Pinion.......................	4NK0-55								✓	✓	4
	1229-C-861	.13	**Washer,** Differential Bevel Side Pinion Thrust.....	4BMA0-751	✓	✓	✓	✓		✓				4
	......	.60	**Washer,** Differential Pinion Thrust..............	4TF2-751					✓					4
	1829-S-45	.65	**Washer,** Differential Pinion Gear Thrust.........	4NK0-751								✓	✓	4
	3878-B-54	4.45	**Spider,** Differential............................	4BLA0-58	✓	✓	✓	✓		✓				1
	3878-B-2	8.70	**Spider,** Differential............................	4NK0-58							✓			1
	3878-C-3	8.70	**Spider,** Differential............................	4ZGEW0-58								✓	✓	1
1105	**Rear Axle Drive Gears and Bearings**													
	472-A	5.03	**Cup,** Differential Bearing......................	4D2-157	✓	✓	✓	✓		✓				2
	......	6.18	**Cup,** Differential Bearing......................	4Y2-211					✓					2
	581	10.91	**Cup,** Differential Bearing......................	4CGA2-209								✓	✓	2
	482	6.83	**Cone,** Differential Bearing.....................	19SKB0-443	✓	✓	✓	✓		✓				2
	......	2.40	**Cone,** Differential Bearing.....................	4TF2-123					✓					2
	572	6.59	**Cone,** Differential Bearing.....................	4CGA2-211								✓	✓	2
	1828-E-5	8.40	**Bearing,** Differential..........................	4NFG0-567							✓			2
	1829-A-27	.25	**Ring,** Differential Bearing Retaining.............	4NFG0-1086							✓			2
	......	2.40	**Adjuster,** Differential Bearing..................	4TF2-123					✓					2
	2814-B-2	1.95	**Ring,** Differential Bearing Adjusting..............	4ZGEW0-123								✓	✓	2
	1820-J-10	.15	**Lock,** Differential Bearing Adjusting Nut.........	4BMC0-125	✓	✓	✓	✓		✓				2
	......	.20	**Bar,** Differential Bearing Adjuster Lock..........	4TF2-125					✓					2
	......	.06	**Pin,** Clevis ($\frac{3}{8}''$—$1\frac{1}{16}''$)...................	S-2257					✓					2
	1820-B-2	.09	**Lock,** Differential Bearing Adjusting.............	4ZGEW0-125								✓	✓	2
	2X-30	.16	**Screw,** Differential Bearing Adjusting Lock.......	4ZGEW0-962								✓	✓	2
	......	46.20	**Gear,** Spur Driving (6.4 Ratio—46 Teeth)........	4F4-1272					✓					1
	......	46.20	**Gear,** Spur Driving (6.94 Ratio—47 Teeth).......	4E4-1272					✓					1
	......	46.20	**Gear,** Spur Driving (7.56 Ratio—48 Teeth).......	4D4-1272					✓					1
	......	46.20	**Gear,** Spur Driving (8.27 Ratio—49 Teeth).......	4C4-1272					✓					1
	......	46.20	**Gear,** Spur Driving (9.00 Ratio—50 Teeth).......	4B4-1272					✓					1
	3892-H-34	39.00	**Gear,** Spur (8.43 Ratio) (See also Group 1103)....	4NFG0-62							✓			1
	3892-L-12	50.00	**Gear,** Spur (8.148 Ratio) (See also Group 1103)...	4ZGEW0-44								✓	✓	1
	3891-U-437	23.50	**Pinion,** Spur (8.43 Ratio).......................	4A0-723							✓			1
	3891-D-420	27.00	**Pinion,** Spur (8.148 Ratio)......................	4EB0-187								✓	✓	1

11-4 REAR AXLE

Master Parts List — THE AUTOCAR COMPANY, ARDMORE, PA.

CODE	MFR. PART No.	LIST PRICE	NAME OF PART	AUTOCAR PART No.	U-2044	U-4044	U-4144	U-4144-T	C-50	U-5044	U-7144-T	U-8144	U-8144-T	No. REQ.
1105	3780	5.11	Cone, Spur Pinion Bearing................	4A0-568							✓			2
	438	5.20	Cone, Spur Pinion Bearing................	4ZGEW0-653								✓	✓	2
	3720	2.92	Cup, Spur Pinion Bearing.................	9UF0-825							✓			2
	432	3.09	Cup, Spur Pinion Bearing.................	4H2-137								✓	✓	2
	3826-R-148	6.25	Cage, Spur Pinion Bearing................	4A0-1087							✓			1
	3826-N-14	5.80	Cage, Spur Pinion Bearing................	4ZGEW0-1057								✓	✓	1
	2803-C-55	.12	Shim, Spur Pinion Bearing—Thin.........	4NFG0-143							✓			As
	2803-W-23	.15	Shim, Spur Pinion Bearing—Thin.........	4ZGE0-992								✓	✓	As
	2803-D-56	.12	Shim, Spur Pinion Bearing—Thick........	4NFG0-158							✓			As
	2803-X-24	.15	Shim, Spur Pinion Bearing—Thick........	4ZGE0-993								✓	✓	As
	2208-V-22	.05	Gasket, Spur Pinion Bearing Cover.......	4FKW0-351							✓			1
	2808-K-11	.06	Gasket, Spur Pinion Bearing Cage and Cover..	4ZGEW0-1058								✓	✓	2
	3866-J-322	2.45	Cover, Spur Pinion Bearing...............	4A0-1088							✓			1
	3866-R-18	3.20	Cover, Spur Pinion Bearing Cage.........	4ZGEW0-1059								✓	✓	1
	4-X-220	.18	Stud, Spur Pinion Bearing ($\tfrac{9}{16}''$ Dia.)...	9SKA0-448							✓			1
	4-X-362	.13	Stud, Spur Pinion Bearing ($\tfrac{1}{2}''$ Dia.)....	4A0-648							✓			5
	13-X-6	.05	Nut, Hex. ($\tfrac{9}{16}''$—18)...............	S-84							✓			1
	13-X-5	.03	Nut, Hex. ($\tfrac{1}{2}''$—20)................	S-1420							✓			5
	X-532	.01	Lockwasher, $\tfrac{19}{32}''$ Spring...........	S-1350							✓			1
	X-531	.01	Lockwasher, $\tfrac{17}{32}''$ Spring...........	S-1349							✓			5
	4-X-250	.13	Stud, Spur Pinion Bearing Cage...........	4ZGEW0-648								✓	✓	6
	13-X-6	.06	Nut, Hex. ($\tfrac{9}{16}''$—18)...............	S-84								✓	✓	6
	X-532	.01	Lockwasher, $\tfrac{19}{32}''$ Spring...........	S-1350								✓	✓	6
	1829-U-359	1.25	Washer, Spur Pinion Bearing.............	4A0-178							✓			1
	3-X-109	.08	Screw, Hex. Cap ($\tfrac{1}{2}''$—20 x $1\tfrac{1}{4}''$).....	S-2055							✓			2
	1829-F-6	1.00	Washer, Spur Pinion Bearing.............	4BLA0-1178								✓	✓	1
	2-X-32	.15	Screw, Spur Pinion Bearing Washer.......	4FE0-773								✓	✓	2
	A-3826-Z-26	14.00	Cage, Pinion Bearing, and Cup Assembly...	4UG0-40	✓	✓	✓	✓		✓				1
	......	25.90	Cage, Drive Pinion, and Cup Assembly....	4FA0-460					✓					1
	......	59.50	Cage, Drive Pinion, Assembly.............	4FA0-40					✓					1
	2803-Q-147	.14	Shim, Pinion Bearing Cage—Thin.........	4BLA0-98A	✓	✓	✓	✓		✓				As
	2803-R-148	.14	Shim, Pinion Bearing Cage—Medium.....	4BLA0-97A	✓	✓	✓	✓		✓				As
	2803-S-149	.14	Shim, Pinion Bearing Cage—Thick........	4BLA0-96A	✓	✓	✓	✓		✓				As
	......	.11	Shim, Drive Pinion Cage (.005)...........	4F9-143					✓					As
	......	.20	Shim, Drive Pinion Cage (.007)...........	4FA9-143					✓					As
	A-1805-F-162	1.60	Seal, Pinion Bearing Cage Oil.............	4UG0-84	✓	✓	✓	✓		✓				1
	......	.85	Seal, Drive Pinion Cage Oil...............	S-7214					✓					1
	49175	5.65	Cone, Pinion Bearing—Front.............	4UG0-1232	✓	✓	✓	✓		✓				1
	......	6.60	Cone, Drive Pinion Bearing—Front.......	4FA0-141					✓					1
	49368	3.35	Cup, Pinion Bearing—Front..............	4RL0-137	✓	✓	✓	✓		✓				1
	......	7.30	Cup, Drive Pinion Bearing—Front........	4FA0-142					✓					1
	59200	7.30	Cone, Pinion Bearing—Rear..............	4DFL0-138	✓	✓	✓	✓		✓				1
	......	10.71	Cone, Drive Pinion Bearing—Rear........	4FB0-141					✓					1
	59412	4.86	Cup, Pinion Bearing—Rear...............	4RL0-142	✓	✓	✓	✓	✓	✓				1
	1829-U-21	.50	Ring, Pinion Bearing Spinner.............	4UG0-762	✓	✓	✓	✓		✓				1
	1844-W-23	3.00	Spacer, Pinion Shaft Bearing..............	4UG0-584	✓	✓	✓	✓		✓				1
		1.85	Spacer, Drive Pinion Bearing.............	4F2-1022					✓					1
	2803-L-142	.12	Shim, Pinion Shaft Bearing...............	4UG0-96	✓	✓	✓	✓		✓				As
	2803-M-143	.12	Shim, Pinion Shaft Bearing...............	4UGA0-96	✓	✓	✓	✓		✓				As
	2803-N-144	.12	Shim, Pinion Shaft Bearing...............	4UH0-96	✓	✓	✓	✓		✓				As
	2803-P-146	.12	Shim, Pinion Shaft Bearing...............	4UHA0-96	✓	✓	✓	✓		✓				As
	......	.06	Shim, Drive Pinion Bearing Spacer—Thick...	4N2-1023					✓					As
	......	.06	Shim, Drive Pinion Bearing Spacer—Thin..	4N2-1028					✓					As
	......	.05	Screw, Hex. Cap ($\tfrac{1}{2}''$—13 x $1\tfrac{3}{4}''$).....	S-3100					✓					6
	......	.01	Lockwasher, $\tfrac{17}{32}''$ Spring...........	S-1349					✓					6
	3889-H-60	25.00	Gear, Bevel (21 Teeth)....................	4BLA0-521	✓	✓	✓	✓		✓				1
	3889-N-170	29.00	Gear, Bevel..............................	4A0-521							✓			1
	3889-Z-156	32.00	Gear, Bevel..............................	4GTC0-1065								✓	✓	1
	59175	7.30	Cone, Bevel Gear Bearing................	4RL0-141	✓	✓	✓	✓		✓				2
	59412	4.86	Cup, Bevel Gear Bearing.................	4RL0-142	✓	✓	✓	✓		✓				2
	U-1217-TAM	32.60	Bearing, Bevel Gear.....................	4A0-1091							✓			1
	1844-Q-147	.72	Spacer, Bevel Gear Bearing...............	4A0-1067							✓			1

THE AUTOCAR COMPANY
ARDMORE, PA.
Master Parts List

11-5 Rear Axle

CODE	MFR. PART No.	LIST PRICE	NAME OF PART	AUTOCAR PART No.	U-2044	U-4044	U-4144	U-4144-T	C-50	U-5044	U-7144-T	U-8144	U-8144-T	NO. REQ.
1105	16-X-51	.27	Key, Bevel Gear	4A0-1251							✓			1
	U-1218-TAM	49.55	Bearing, Bevel Gear	4GTC0-1066								✓	✓	1
	1821-F-6	5.15	Sleeve, Bevel Gear Bearing	4GKW0-1275								✓	✓	1
	1844-L-116	1.25	Spacer, Bevel Gear Bearing	4GTC0-1067								✓	✓	1
	26-X-61	.18	Screw, Bevel Gear	4GKW0-1381								✓	✓	1
	16-X-48	.32	Key, Bevel Gear	4GTC0-163								✓	✓	1
	3890-V-74	15.80	Shaft, Bevel Pinion (10 Teeth)	4DK0-35	✓	✓	✓	✓		✓				1
	16-X-28	.18	Key, Pinion Shaft	4NK0-163	✓	✓	✓	✓		✓				1
	13399	.15	Nut, Bevel Pinion Shaft	4A0-81	✓	✓	✓	✓		✓				1
	4-X-322	.11	Stud, Gear Shaft Bearing Washer	4BLA0-648	✓	✓	✓	✓		✓				2
	1820-K-11	.05	Lock, Gear Shaft Bearing Stud	4BLA0-28	✓	✓	✓	✓		✓				1
	1829-F-6	1.00	Lockwasher, Gear Shaft Bearing	4BLA0-1178	✓	✓	✓	✓		✓				1
	1827-G-7	.75	Lock, Pinion Bearing	4UG0-28	✓	✓	✓	✓		✓				1
	1827-N-14	2.00	Nut, Pinion Bearing Adjusting	4UG0-936	✓	✓	✓	✓		✓				1
	1829-E-5	.25	Lockwasher, Pinion Bearing	4NK0-1204	✓	✓	✓	✓		✓				1
	1850-B-28	.06	Plug, Carrier	4NK0-425	✓	✓	✓	✓		✓				1
	......	29.25	Pinion, Drive, and Shaft (11 Teeth)	4FA4-35					✓					1
	......	.35	Nut, Drive Pinion Shaft Flange	3B2-103					✓					1
	......	.12	Lockwasher, Drive Pinion Shaft Flange Nut	3B2-262					✓					1
	......	.30	Key, Drive Pinion Shaft Flange	3B2-261					✓					1
	......	.35	Nut, Drive Pinion Bearing Lock	3B2-101					✓					2
	......	.20	Washer, Drive Pinion Bearing Lock Nut	4Y2-28A					✓					1
	......	3.00	Cap, Drive Pinion Cage	4F3-583A					✓					1
	......	.03	Gasket, Drive Pinion Cage Cap	4FA3-582					✓					1
	3890-U-21	14.25	Pinion, Bevel	4NFG0-35							✓			1
	......	98.75	Gear, Jackshaft Spur and Bevel, Assembly	4FA0-960					✓					1
	......	38.40	Gear, Jackshaft Spur (6.4 Ratio—17 Teeth)	4F4-1271					✓					1
	......	38.40	Gear, Jackshaft Spur (6.94 Ratio—16 Teeth)	4E4-1271					✓					1
	......	38.40	Gear, Jackshaft Spur (7.56 Ratio—15 Teeth)	4D4-1271					✓					1
	......	38.40	Gear, Jackshaft Spur (8.27 Ratio—14 Teeth)	4C4-1271					✓					1
	......	38.40	Gear, Jackshaft Spur (9.00 Ratio—13 Teeth)	4B4-1271					✓					1
	......	36.95	Gear, Jackshaft Bevel (26 Teeth)	4FA4-521					✓					1
	......	10.71	Cone, Jackshaft Bearing—R. H.	4SK0-136					✓					1
	......	11.48	Cone, Jackshaft Bearing—L. H.	4FA0-139					✓					1
	......	7.30	Cup, Jackshaft Bearing—R. H.	4FA0-142					✓					1
	......	9.30	Cup, Jackshaft Bearing—L. H.	4DFL0-137					✓					1
	......	14.50	Retainer, Jackshaft Bearing, and Cup Assembly—R. H.	4FA0-1050					✓					1
	......	5.00	Retainer, Jackshaft Bearing—R. H.	4F4-113A					✓					1
	......	.04	Plug, Welch (1¼″)	S-4212					✓					2
	......	12.50	Retainer, Jackshaft Bearing, and Cup Assembly—L. H.	4FA0-1100					✓					1
	......	5.00	Retainer, Jackshaft Bearing—L. H.	4F4-114A					✓					1
	......	.45	Clamp, Jackshaft Bearing	4E2-1265					✓					1
	......	.11	Shim, Jackshaft Bearing Cap (.005)	4F9-143					✓					As
	......	.20	Shim, Jackshaft Bearing Cap (.007)	4FA9-143					✓					As
	A-1-3880-N-274	34.32	Pinion, Bevel, and Thru Shaft Assembly	4GE0-173								✓	✓	1
	......	15.75	Flange, Drive	3EP3-201								✓	✓	1
	......	.07	Screw, Hex. Cap (⅜″-24 x 1 5⁄16″)	3Y2-222A								✓	✓	8
	......	.03	Nut, Hex. (⅜″-24)	S-1418								✓	✓	8
	......	.01	Lockwasher, 13⁄32″ Spring	S-1347								✓	✓	16
	3880-C-3	18.50	Shaft, Thru	4NFG0-1092							✓			1
	13399	.15	Nut, Thru Shaft	4A0-81							✓			1
	16-X-29	.25	Key, Thru Shaft	4UU0-1251							✓			1
	14-X-58	.53	Nut, Thru Shaft—Forward	4GTC0-652								✓	✓	1
	......	.01	Cotter (⅛″ x 2″)	S-2208								✓	✓	1
	560	8.29	Cone, Thru Shaft Forward Bearing	4Y2-141							✓			1
	570	8.66	Cone, Thru Shaft Forward Bearing	19D0-13								✓	✓	1
	552-A	6.17	Cup, Thru Shaft Forward Bearing	4NFG0-142							✓			1
	563	6.18	Cup, Thru Shaft Forward Bearing	4Y2-211								✓	✓	1
	A-3866-D-4	3.70	Cover, Thru Shaft Forward Bearing, Assembly	4NFG0-1420							✓			1
	A-3866-B-28	7.15	Cover, Thru Shaft Forward Bearing, Assembly	4A0-1420								✓	✓	1
	A-1805-A-79	1.05	Seal, Thru Shaft Forward Bearing Cover Oil	4FKWA0-84							✓			1
	A-1205-M-325	2.50	Seal, Thru Shaft Forward Bearing Cover Oil	4GE0-395A								✓	✓	1

11-6 REAR AXLE

Master Parts List

THE AUTOCAR COMPANY, ARDMORE, PA.

CODE	MFR. PART No.	LIST PRICE	NAME OF PART	AUTOCAR PART No.	U-2044	U-4044	U-4144	U-4144-T	C-50	U-5044	U-7144-T	U-8144	U-8144-T	No. REQ.
1105	4-X-186	.08	**Stud,** Thru Shaft Forward Bearing Cover...................	4NFG0-1084							✓	✓	✓	6
	13-X-5	.03	**Nut,** Hex. ($\frac{1}{2}''$—20).................................	S-1420							✓	✓	✓	6
	X-531	.01	**Lockwasher,** $\frac{17}{32}''$ Spring....................	S-1349							✓	✓	✓	6
	2808-N-14	.05	**Gasket,** Thru Shaft Forward Bearing Cover.................	4ZGE0-1075								✓	✓	1
	2803-R-44	.15	**Shim,** Thru Shaft Forward Bearing Cover—Thin...........	4NFG0-1096							✓			As
	2803-G-33	.21	**Shim,** Thru Shaft Forward Bearing Cover—Thin...........	4ZGE0-1073								✓	✓	As
	2803-S-45	.15	**Shim,** Thru Shaft Forward Bearing Cover—Thick..........	4NFG0-1097							✓			As
	2803-H-34	.21	**Shim,** Thru Shaft Forward Bearing Cover—Thick..........	4ZGE0-1074								✓	✓	As
	527	6.65	**Cone,** Thru Shaft Rear Bearing.........................	4NFG0-138							✓			1
	536	6.99	**Cone,** Thru Shaft Rear Bearing.........................	19C2-305								✓	✓	1
	522	4.07	**Cup,** Thru Shaft Rear Bearing..........................	3T2-37							✓			1
	532-A	5.04	**Cup,** Thru Shaft Rear Bearing..........................	4Y2-137								✓	✓	1
	3866-H-8	3.00	**Cover,** Thru Shaft Rear Bearing........................	4NFG0-1093							✓			1
	3866-L-12	3.20	**Cover,** Thru Shaft Rear Bearing........................	4GKW0-1076								✓	✓	1
	2808-P-16	.05	**Gasket,** Thru Shaft Rear Bearing.......................	4ZGEW0-1077								✓	✓	1
	2803-P-42	.12	**Shim,** Thru Shaft Rear Bearing Cover—Thin.............	4NFG0-1094							✓			As
	2803-J-36	.18	**Shim,** Thru Shaft Rear Bearing Cover—Thin.............	4ZGEW0-1073								✓	✓	As
	2803-Q-43	.12	**Shim,** Thru Shaft Rear Bearing Cover—Thick............	4NFG0-1095							✓			As
	2803-K-37	.18	**Shim,** Thru Shaft Rear Bearing Cover—Thick............	4ZGEW0-1074								✓	✓	As
	4-X-186	.08	**Stud,** Thru Shaft Rear Bearing Cover....................	4NFG0-1084							✓	✓	✓	6
	13-X-5	.04	**Nut,** Hex. ($\frac{1}{2}''$—20).................................	S-1420							✓	✓	✓	6
	X-529	.01	**Lockwasher,** $\frac{17}{32}''$ Spring....................	S-1349							✓	✓	✓	6

ILLUSTRATIONS

GROUP 12

BRAKES

GROUP 12

Master Parts List

THE AUTOCAR COMPANY, ARDMORE, PA.

As supplied on Autocar Models
U-8144   U-8144-T

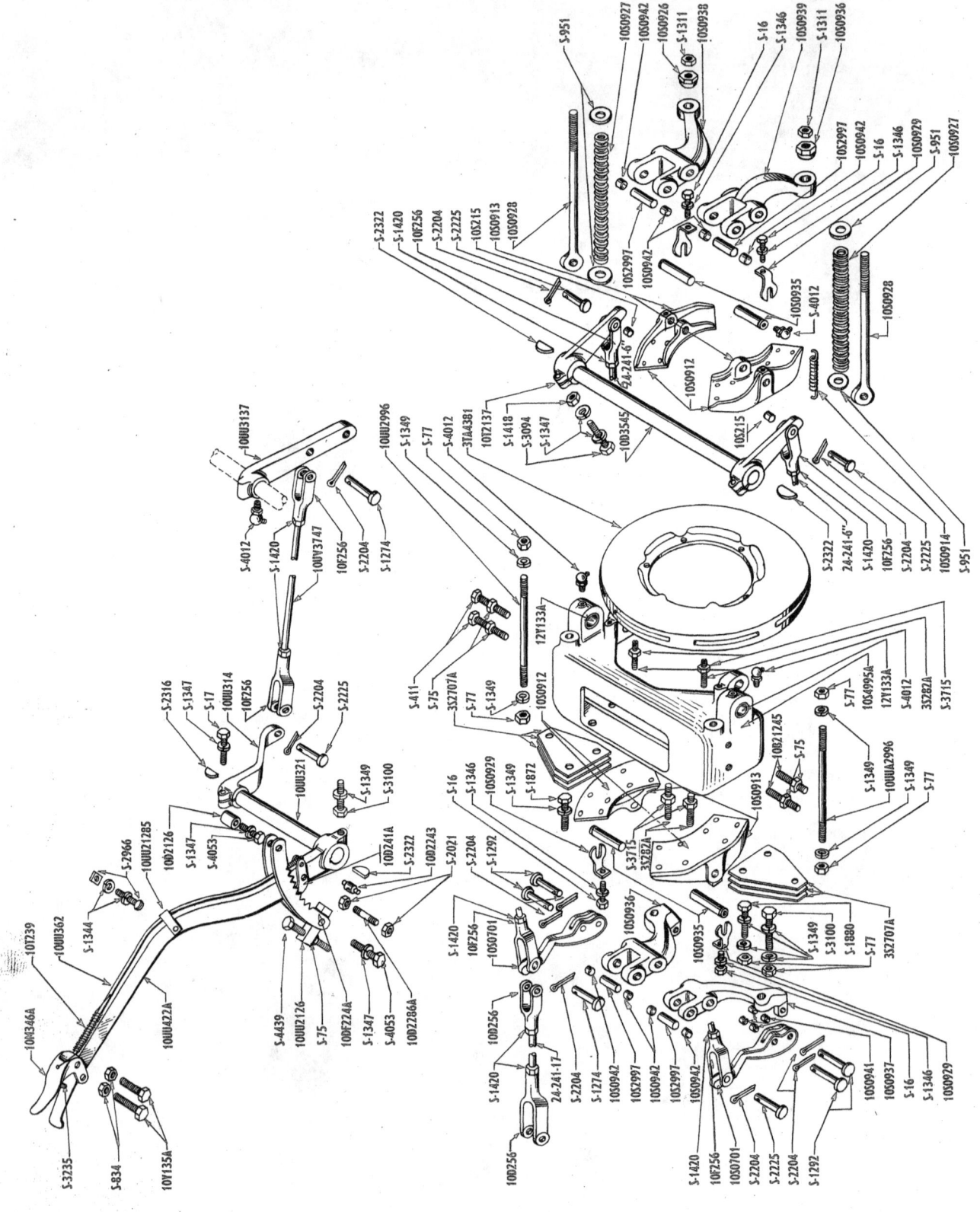

1201—Hand Brake

THE AUTOCAR COMPANY, ARDMORE, PA.

Master Parts List

GROUP 12

As supplied on Autocar Models
U-8144 U-8144-T

FRONT

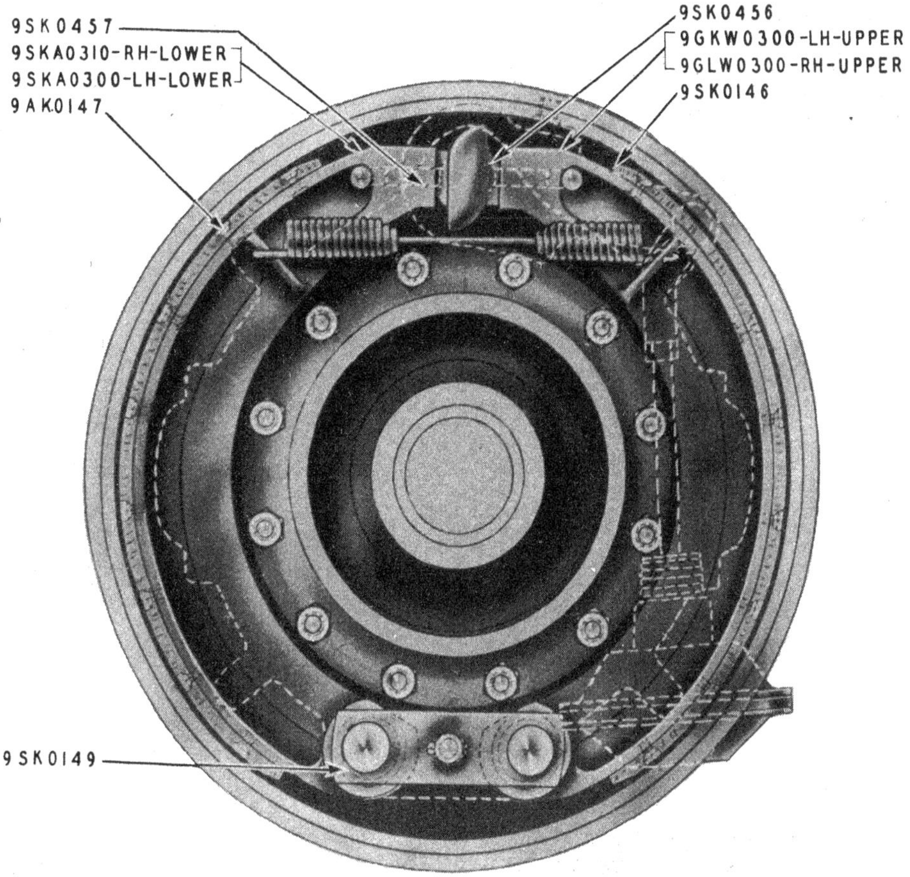

- 9SK0457
- 9SKA0310-RH-LOWER
- 9SKA0300-LH-LOWER
- 9AK0147
- 9SK0456
- 9GKW0300-LH-UPPER
- 9GLW0300-RH-UPPER
- 9SK0146
- 9SK0149

1202—

Brake Shoes and Facings

REAR

- S-3132
- 4G0667
- S-1418
- S-3190
- 4GKW01321
- 4GKW0100
- 4G089
- 4GKW0110
- 4GKW052

GROUP 12

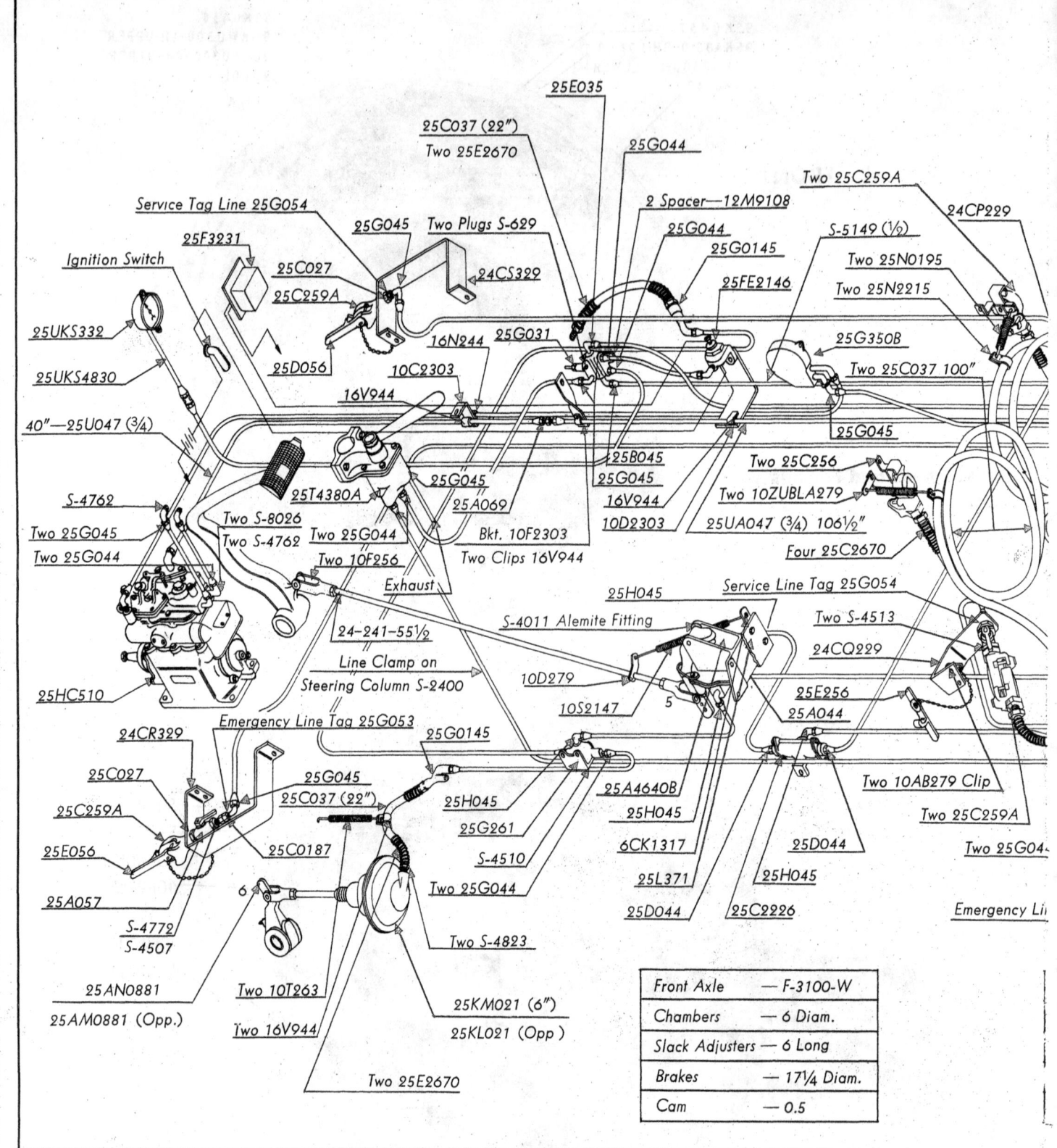

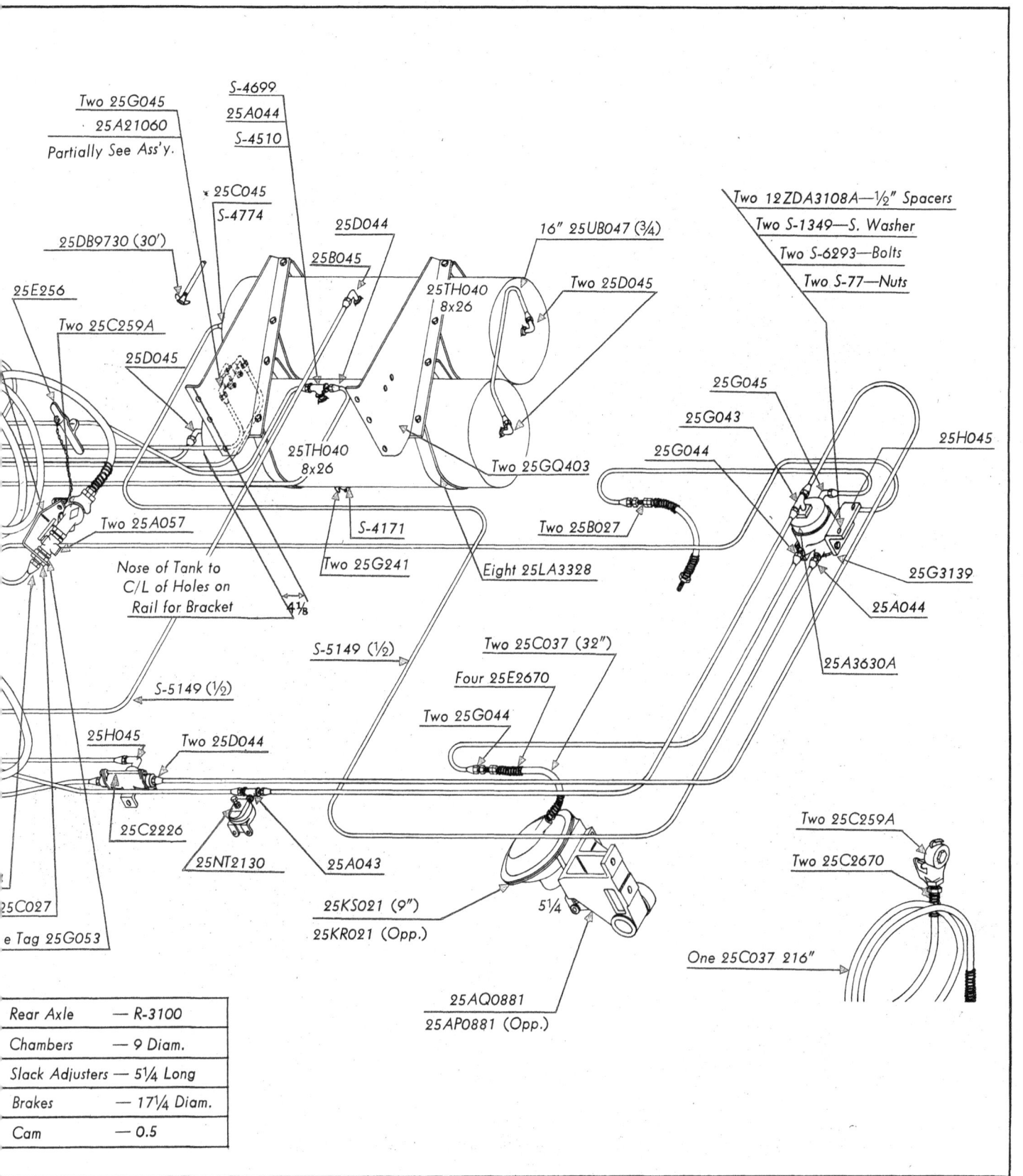

1205—Air Brake

GROUP 12

Master Parts List

THE AUTOCAR COMPANY, ARDMORE, PA.

As supplied on Autocar Models
U-8144   U-8144-T

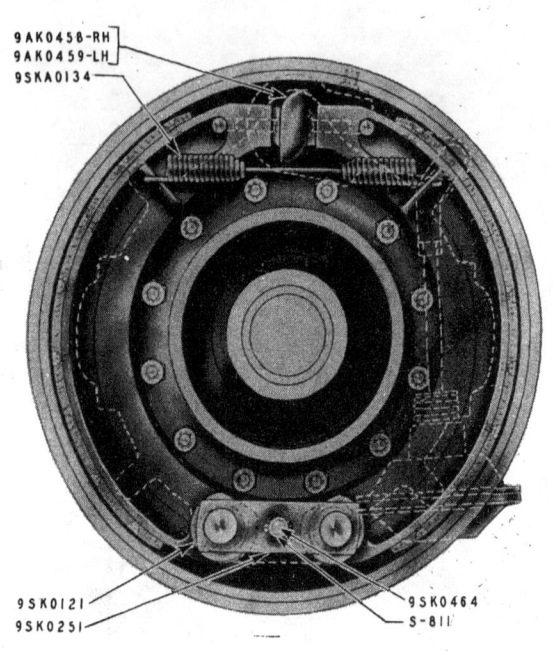

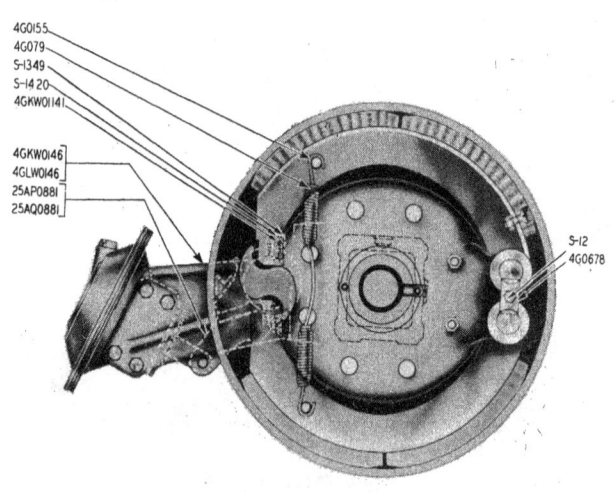
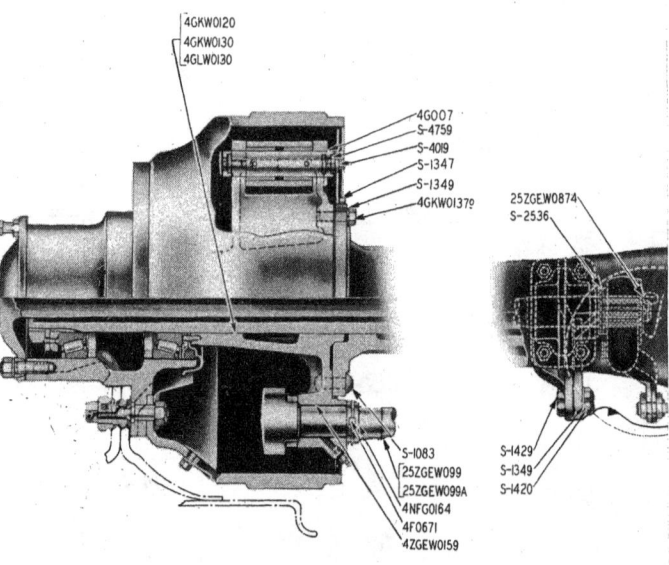

1203—Brake Shoe Support, Guide Springs, Pins, Anchor Plate

12-1 HAND BRAKES

THE AUTOCAR COMPANY, ARDMORE, PA. — Master Parts List

CODE	MFR. PART NO.	LIST PRICE	NAME OF PART	AUTOCAR PART NO.	U-2044	U-4044	U-4144	U-4144-T	C-50	U-5044	U-7144-T	U-8144	U-8144-T	NO. REQ.
1201	American Cable—		**Hand Brakes**											
	C-653	4.84	Arm, Lever—Front L. H.	C-653	✓	✓	✓	✓	✓					1
	C-655	2.10	Arm, Lever—Rear L. H.	C-655	✓	✓	✓	✓	✓					1
	C-598-2	.08	Bushing	C-598-2	✓	✓	✓	✓	✓					4
	C-506	2.50	Shoe, Brake	C-506	✓	✓	✓	✓	✓					2
	C-134	.58	Lining, Brake Shoe	C-134	✓	✓	✓	✓	✓					2
	C-241	.01	Rivet, Brake Shoe	C-241	✓	✓	✓	✓	✓					16
	C-598-4	.08	Bushing	C-598-4	✓	✓	✓	✓	✓					4
	C-509	.40	Pin, Brake Shoe	C-509	✓	✓	✓	✓	✓					2
	C-542	.04	Retainer, Brake Shoe Pin	C-542	✓	✓	✓	✓	✓					2
	C-535	.02	Screw, Hex. Head Cap ($\frac{5}{16}$"—18 x $\frac{5}{8}$")	C-535	✓	✓	✓	✓	✓					2
	C-536	.01	Lockwasher ($\frac{5}{16}$")	C-536	✓	✓	✓	✓	✓					2
	C-479	.45	Rod, Tie	C-479	✓	✓	✓	✓	✓					1
	C-513	.14	Release, Spring Lever Arm	C-513	✓	✓	✓	✓	✓					1
	C-483-A	.18	Nut, Spherical	C-483-A	✓	✓	✓	✓	✓					1
	C-493	.01	Washer, Plain	C-493	✓	✓	✓	✓	✓					2
	C-270	.02	Nut ($\frac{1}{2}$"—20)	C-270	✓	✓	✓	✓	✓					2
	C-557	2.16	Lever, Brake Operating (L. H. offset)	C-557	✓	✓	✓	✓	✓					1
	RA-12-6	.02	Pin, Clevis ($\frac{1}{2}$")	RA-12-6	✓	✓	✓	✓	✓					2
	C-514	.05	Spring, Brake Shoe	C-514	✓	✓	✓	✓	✓					1
	C-604	.07	Screw, Set (Shoe Adjusting)	C-604	✓	✓	✓	✓	✓					2
	C-541	.02	Nut, Lock ($\frac{3}{8}$"—16)	C-541	✓	✓	✓	✓	✓					2
	40-S-5-D	22.00	Disc, Brake (14")	40-S-5-D	✓	✓	✓	✓	✓					..
	C-507	.26	Pin, Anchor	C-507	✓	✓	✓	✓	✓					..
	C-598-2	.08	Bushing	C-598-2	✓	✓	✓	✓	✓					..
	.....	4.75	Bracket, Anchor	10NC4-995	✓	✓	✓	✓	✓					..
	.....	.12	Screw, Set	3S2-82A	✓	✓	✓	✓	✓					..
	.....	.01	Nut, Lock	S-3715	✓	✓	✓	✓	✓					..
	.....	....	Brake, Disc, Assembly (American Cable)	10NR7-990							✓			1
	.....	....	Brake, Disc, Assembly (American Cable)	10SB7-990								✓	✓	1
	.....	9.60	Lever, Hand Brake and Bushing Assembly—Bare	10DF0-872						✓				1
	.....	.25	Bushing, Hand Brake Lever	10S2-15						✓				1
	.....	12.40	Lever, Hand Brake, Assembly—with Pawl Rod, Etc.	10DF0-100						✓				1
	.....	12.40	Lever, Hand Brake, Assembly	10UU0-100							✓	✓	✓	1
	.....	6.75	Lever, Hand Brake—Only	10UU4-22A							✓	✓	✓	1
	.....	.05	Screw, Hex. Cap ($\frac{1}{2}$"—13 x $1\frac{1}{4}$")	S-3100							✓	✓	✓	1
	.....	.01	Lockwasher, $\frac{11}{32}$" Spring	S-1349							✓	✓	✓	1
	.....	.06	Pin, Hand Brake Lever	S-3235						✓	✓	✓	✓	1
	.....	2.70	Spoon, Hand Brake Lever	10M3-46						✓				1
	.....	1.45	Spoon, Hand Brake Lever	10M3-46A							✓	✓	✓	1
	.....	.10	Bolt, Spoon ($\frac{1}{4}$"—20 x $1\frac{33}{64}$")	10Y1-35A						✓	✓	✓	✓	2
	.....	.01	Nut, Hex. ($\frac{1}{4}$"—20)	S-834						✓	✓	✓	✓	2
	.....	.20	Pawl, Hand Brake Lever	10D2-41A						✓	✓	✓	✓	2
	.....	.12	Pin, Hand Brake Lever Pawl	10D2-286A						✓	✓	✓	✓	1
	.....	.06	Nut, Hex. ($\frac{5}{16}$"—24)	S-2021						✓	✓	✓	✓	2
	.....	.40	Rod, Hand Brake Lever Pawl	10DF2-62						✓				1
	.....	.01	Nut, Pawl Rod End	S-1416						✓				1
	.....	.75	Rod, Hand Brake Lever Pawl	10UU3-62							✓	✓	✓	1
	.....	.12	End, Pawl Rod	10D2-243						✓	✓	✓	✓	1
	.....	.25	Spring, Pawl Rod	10T2-39						✓				1
	.....	.06	Guide, Hand Brake Lever Pawl Rod	10UU2-1285							✓	✓	✓	1
	.....	.03	Bolt, Pawl Rod Guide ($\frac{3}{16}$" x 1")	S-2966							✓	✓	✓	1
	.....	.01	Lockwasher, $\frac{7}{32}$" Spring	S-1344							✓	✓	✓	2
	.....	.60	Sector, Hand Brake Lever	10DF2-24A						✓	✓	✓	✓	2
	.....	.06	Screw, Hex. Cap ($\frac{3}{8}$"—16 x $1\frac{5}{8}$")	S-4053							✓	✓	✓	2
	.....	.01	Lockwasher, $\frac{13}{32}$" Spring	S-1347							✓	✓	✓	2
	.....	.08	Bolt, Hand Brake Lever Sector	S-40						✓				2
	.....	.03	Nut, Hand Brake Lever Sector Bolt	S-75						✓				2
	.....	.01	Lockwasher, $\frac{13}{32}$" Spring	S-1347						✓				2
	.....	.12	Washer, Sector Spacer	10D2-126						✓	✓	✓	✓	1
	.....	.60	Spacer, Hand Brake Lever Sector—Front	10A2-126						✓				1
	.....	.01	Screw, Hand Brake Lever Stop	S-4051						✓				1

12-2 HAND BRAKES

Master Parts List — The Autocar Company, Ardmore, PA.

CODE	MFR. PART No.	LIST PRICE	NAME OF PART	AUTOCAR PART No.	U-2044	U-4044	U-4144	U-4144-T	C-50	U-5044	U-7144-T	U-8144	U-8144-T	NO. REQ.
1201	......	.03	Nut, Hand Brake Lever Stop Screw	S-75					✓					1
	......	.90	Spacer, Hand Brake Lever Sector—Top	10UU2-126							✓	✓	✓	1
	......	.05	Screw, Hex. Cap (3/8"—16 x 3 1/4")	S-4439							✓	✓	✓	1
	......	.03	Nut, Hex. (3/8"—16)	S-75							✓	✓	✓	1
	......	.80	Shaft, Hand Brake	10UU3-21							✓	✓	✓	1
	......	.06	Key, #C Woodruff (5/16" x 1 1/8")	S-2322							✓	✓	✓	1
	......	.06	Key, #15 Woodruff (1/4" x 1")	S-2316							✓	✓	✓	1
	......	1.65	Lever, Hand Brake Shaft	10UU3-14							✓	✓	✓	1
	......	.06	Screw, Hex. Cap (3/8"—16 x 1 1/2")	S-17							✓	✓	✓	1
	......	.01	Lockwasher, 13/32" Spring	S-1347							✓	✓	✓	1
	......	.50	Clevis, Hand Brake Shaft Lever	10F2-56							✓	✓	✓	1
	......	.04	Nut, Hex. (1/2"—20)	S-1420							✓	✓	✓	1
	......	.06	Pin, Clevis (1/2" x 1 9/16")	S-2225							✓	✓	✓	1
	......	.01	Pin, Cotter (1/8" x 1")	S-2204							✓	✓	✓	1
	......	.45	Rod, Hand Brake—Forward	10UU4-747							✓			1
	......	.65	Rod, Hand Brake—Forward	10UV3-747								✓	✓	1
	......	.50	Clevis, Transfer Lever	10F2-56								✓	✓	1
	......	.04	Nut, Hex. (1/2"—20)	S-1420								✓	✓	1
	......	1.80	Lever, Hand Brake Transfer	10UU3-137								✓	✓	1
	......	.20	Fitting, 67 1/2° Alemite (1/8")	S-4012								✓	✓	1
	......	.50	Clevis	10D2-56								✓	✓	1
	......	.04	Nut, Hex. (1/2"—20)	S-1420								✓	✓	1
	......	.06	Pin, Clevis (1/2" x 2 1/4")	S-1274								✓	✓	1
	......	.01	Pin, Cotter (1/8"—1")	S-2204								✓	✓	1
	......	.30	Rod, Hand Brake—Rear	24-241-17							✓	✓	✓	1
	......	.50	Clevis	10D2-56							✓	✓	✓	2
	......	.04	Nut, Hex. (1/2"—20)	S-1420							✓	✓	✓	2
	......	.06	Pin, Clevis (1/2" x 2 1/4")	S-1274							✓	✓	✓	1
	......	.01	Pin, Cotter (1/8" x 1")	S-2204							✓	✓	✓	1
	......	2.16	Lever, Disc Brake Operating (R. H. Off Set)	10NA0-701							✓			1
	......	2.16	Lever, Disc Brake Operating (L. H. Off Set)	10NB0-701							✓			1
	......	.08	Bushing, Lever Arm	10NA0-941							✓			8
	......	.07	Pin, Clevis (.496 x 1 17/64")	S-1285							✓			4
	......	.01	Pin, Cotter (1/8" x 1")	S-2204							✓			4
	......	2.65	Lever, Disc Brake Link Brake Operating	10S0-701								✓	✓	2
	......	.09	Bushing, Lever Arm	10S0-941								✓	✓	8
	C-479	.35	Rod, Disc Brake Lever Arm Tie	10NA9-28							✓			2
	......	.01	Washer, Disc Brake Lever Arm Tie Rod	S-978							✓			4
	C-513	.14	Spring, Disc Brake Lever Arm Release	10NA0-927							✓			2
	C-483-A	.18	Nut, Spherical	10NA0-926							✓			2
	......	.04	Nut, Hex. (1/2"—20)	S-1420							✓			2
	C-521	.65	Rod, Disc Brake Link Tie (Lever Arm)	10S0-928								✓	✓	2
	......	.06	Pin, Clevis (.605" x 1 29/32")	S-1292								✓	✓	2
	......	.01	Pin, Cotter (1/8" x 1")	S-2204								✓	✓	2
	......	.03	Washer, Flat (7/8" x 1 1/2" x 1/8")	S-951								✓	✓	4
	C-523	.18	Spring, Lever Arm Release	10S0-927								✓	✓	2
	C-539	.22	Nut, Spherical	10S0-926								✓	✓	2
	......	.10	Nut, Hex. (5/8"—18)	S-1311								✓	✓	2
	......	3.50	Arm, Disc Brake Link Lever—Front R. H.	10NB0-936							✓			1
	C-498-4	.08	Bushing, Brake Shoe	10NA0-942							✓			2
	......	.12	Pin, Anchor	10NA0-997							✓			1
	......	.12	Screw, Anchor Pin Set	3S3-82A							✓			1
	......	.01	Nut, Hex. (5/16"—24)	S-3715							✓			1
	C-532	6.10	Arm, Disc Brake Link Lever—Front R. H.	10S0-936								✓	✓	1
	C-598-3	.09	Bushing, Brake Shoe	10S0-942								✓	✓	2
	......	.30	Pin, Anchor	10S2-997								✓	✓	1
	......	.06	Pin, Clevis (.605" x 1 29/32")	S-1292								✓	✓	1
	......	.01	Pin, Cotter (1/8" x 1")	S-2204								✓	✓	1
	C-506	2.50	Shoe, Disc Brake—Front	10NA0-913							✓			2
	559	.89	Lining, Disc Brake Shoe	10N3-912							✓			2
	C-241	.01	Rivet, Disc Brake Shoe Lining	10N0-911							✓			16
	C-509	.40	Pin, Disc Brake Shoe	10NA0-935							✓			2

THE AUTOCAR COMPANY, ARDMORE, PA.

Master Parts List

12-3 HAND BRAKES

CODE	MFR. PART NO.	LIST PRICE	NAME OF PART	AUTOCAR PART NO.	U-2044	U-4044	U-4144	U-4144-T	C-50	U-5044	U-7144-T	U-8144	U-8144-T	NO. REQ.
1201	C-542	.04	Retainer, Disc Brake Shoe Pin	10NA0-929							✓			2
		.04	Screw, Hex. Cap (5/16″—18 x 7/8″)	S-16							✓			2
		.01	Lockwasher, 17/32″ Spring	S-1346							✓			2
		3.26	Shoe, Disc Brake—Front	10S0-913								✓	✓	2
		1.08	Lining, Disc Brake Shoe	10S0-912								✓	✓	2
		.01	Rivets, Disc Brake Shoe Lining	10TA0-911								✓	✓	16
		.50	Pin, Brake Shoe Link	10S0-935								✓	✓	2
		.20	Fitting, 67½° Alemite (1/8″)	S-4012								✓	✓	2
		.05	Retainer, Brake Shoe Pin	10S0-929								✓	✓	2
		.04	Screw, Hex. Cap (5/16″—18 x 5/8″)	S-16								✓	✓	2
		.01	Lockwasher, 17/32″ Spring	S-1346								✓	✓	2
		4.84	Arm, Disc Brake Link Lever—Front—L. H.	16NC0-936							✓			1
	C-498-4	.08	Bushing, Brake Shoe	10NA0-942							✓			2
		.12	Pin, Anchor	10NA0-997							✓			1
		.12	Screw, Anchor Pin Set	3S3-82A							✓			1
		.01	Nut, Hex. (5/16″—24)	S-3715							✓			1
		6.10	Arm, Disc Brake Link Lever—Front—L. H.	10S0-937								✓	✓	1
		.09	Bushing, Brake Shoe	10S0-942								✓	✓	2
		.30	Pin, Anchor	10S2-997								✓	✓	1
		.06	Pin, Clevis (.605″ x 1 23/32″)	S-1292								✓	✓	1
		.01	Pin, Cotter (1/8″ x 1″)	S-2204								✓	✓	1
		14.40	Bracket, Disc Brake Anchor	10NA5-995A							✓			1
		.25	Bushing, Anchor Bracket	12Y1-33A							✓			2
		8.60	Support, Anchor Bracket End	12R4-101							✓			2
		.03	Washer, Flat (17/32″ x 1 1/8″ x 1/16″)	S-709							✓			4
		.01	Lockwasher, 17/32″ Spring	S-1349							✓			8
		.04	Nut, Hex. (1/2″—20)	S-1420							✓			4
		14.40	Bracket, Shaft Brake Anchor	10S4-995A								✓	✓	1
		.25	Bushing, Anchor Bracket	12Y1-33A								✓	✓	2
		.20	Fitting, 67½° Alemite (1/8″)	S-4012								✓	✓	2
		.12	Screw, Set (5/16″—24 x 1 1/4″)	3S2-82A								✓	✓	4
		.01	Nut, Hex. (5/16″—24)	S-3715								✓	✓	4
		.10	Screw, Set (3/8″—16 x 2 1/2″)	S-411								✓	✓	2
		.10	Screw, Adjusting	10B2-1245								✓	✓	2
		.03	Nut, Hex. (3/8″—16)	S-75								✓	✓	4
		.25	Shim	3S2-707A							✓	✓	✓	6
		.15	Bolt, Through (1/2″—13 x 7 9/16″)	10UU2-996							✓	✓	✓	1
		.04	Nut, Hex. (1/2″—13)	S-77								✓	✓	2
		.01	Lockwasher, 17/32″ Spring	S-1349								✓	✓	2
		.09	Screw, Hex. Cap (1/2″—13 x 1 1/2″)	S-1872								✓	✓	2
		.01	Lockwasher, 17/32″ Spring	S-1349								✓	✓	4
		.11	Screw, Hex. Cap (1/2″—13 x 2 1/4″)	S-1880								✓	✓	1
		.05	Screw, Hex. Cap (1/2″—13 x 1 3/4″)	S-3100								✓	✓	1
		.01	Lockwasher, 17/32″ Spring	S-1349								✓	✓	4
		.04	Nut, Hex. (1/2″—13)	S-77								✓	✓	2
		.15	Bolt, Through (1/2″—13 x 7 3/16″)	10UUA2-996							✓	✓	✓	1
		.01	Lockwasher, 17/32″ Spring	S-1349								✓	✓	2
		.04	Nut, Hex. (1/2″—13)	S-77								✓	✓	2
		24.00	Disc, Driveshaft Brake	3TA4-381								✓	✓	1
		1.60	Shaft, Disc Brake Cross	10D3-545							✓	✓	✓	1
		.06	Key, #C Woodruff (5/16″ x 1 1/8″)	S-2322							✓	✓	✓	2
		1.50	Lever, Brake	10T2-137							✓			3
		.08	Screw, Hex. Cap (3/8″—16 x 2″)	S-3383							✓			2
		.03	Nut, Hex. (3/8″—16)	S-75							✓			2
		.01	Lockwasher, 13/32″ Spring	S-1347							✓			2
		.25	Bushing	10S2-15							✓			2
		1.50	Lever, Brake	10T2-137								✓	✓	2
		.06	Screw, Hex. Cap (3/8″—24 x 2″)	S-3094								✓	✓	2
		.03	Nut, Hex. (3/8″—24)	S-1418								✓	✓	2
		.01	Lockwasher, 13/32″ Spring	S-1347								✓	✓	2
		.25	Bushing	10S2-15								✓	✓	2
		.18	Rod, Disc Brake Push	24-241-6							✓	✓	✓	2

12-4 HAND BRAKES

Master Parts List — The Autocar Company, Ardmore, PA.

CODE	MFR. PART No.	LIST PRICE	NAME OF PART	AUTOCAR PART No.	U-2044	U-4044	U-4144	U-4144-T	C-50	U-5044	U-7144-T	U-8144	U-8144-T	No. REQ.
1201	......	.50	Clevis	10F2-56							✓	✓	✓	4
	......	.04	Nut, Hex. ($\frac{1}{2}"$—20)	S-1420							✓	✓	✓	4
	......	.06	Pin, Clevis ($\frac{1}{2}" \times 1\frac{9}{16}"$)	S-2225							✓	✓	✓	3
	......	.01	Pin, Cotter ($\frac{1}{8}" \times 1"$)	S-2204							✓	✓	✓	3
	C-506	2.50	Shoe, Disc Brake—Rear	10NA0-913						✓				2
	559	.89	Lining, Disc Brake Shoe	10N3-912						✓				2
	C-241	.01	Rivet, Disc Brake Shoe Lining	10N0-911						✓				16
	C-514	.05	Spring, Disc Brake Shoe	10S0-914						✓				2
	......	.10	Screw, Disc Brake Shoe Adjusting	10B2-1245						✓				4
	......	.03	Nut, Hex. ($\frac{5}{16}"$—18)	S-82						✓				4
	......	3.26	Shoe, Disc Brake—Rear	10S0-913								✓	✓	2
	......	1.08	Lining, Brake Shoe	10S0-912								✓	✓	2
	......	.01	Rivets, Brake Shoe Lining	10TA0-911								✓	✓	16
	......	.05	Spring, Disc Brake Shoe	10S0-914								✓	✓	2
	......	3.50	Arm, Disc Brake Lever—Rear—R. H.	10NB0-938						✓				1
	......	.08	Bushing, Brake Shoe	10NA0-942						✓				2
	......	.12	Pin, Anchor	10NA0-997						✓				1
	......	.40	Pin, Disc Brake Shoe	10NA0-935						✓				2
	......	.04	Retainer, Disc Brake Shoe Pin	10NA0-929						✓				2
	......	.04	Screw, Hex. Cap ($\frac{5}{16}"$—18 x $\frac{5}{8}"$)	S-16						✓				2
	......	.01	Lockwasher, $\frac{11}{32}"$ Spring	S-1346						✓				2
	......	5.15	Arm, Disc Brake Lever—Rear—R. H.	10S0-938								✓	✓	1
	......	.09	Bushing, Brake Shoe	10S0-942								✓	✓	2
	......	.30	Pin, Anchor	10S2-997								✓	✓	1
	......	.50	Pin, Brake Shoe Link	10S0-935								✓	✓	1
	......	.05	Retainer, Brake Shoe Pin	10S0-929								✓	✓	1
	......	.04	Screw, Hex. Cap ($\frac{5}{16}"$—18 x $\frac{5}{8}"$)	S-16								✓	✓	1
	......	.01	Lockwasher, $\frac{11}{32}"$ Spring	S-1346								✓	✓	1
	......	4.36	Arm, Disc Brake Lever—Rear—L. H.	10NC0-938						✓				1
	......	.08	Bushing, Brake Shoe	10NA0-942						✓				2
	......	.12	Pin, Anchor	10NA0-997						✓				1
	......	.40	Pin, Brake Shoe Link	10NA0-935						✓				2
	......	.04	Retainer, Brake Shoe Pin	10NA0-929						✓				2
	......	.04	Screw, Hex. Cap ($\frac{5}{16}"$—18 x $\frac{5}{8}"$)	S-16						✓				2
	......	.01	Lockwasher, $\frac{11}{32}"$ Spring	S-1346						✓				2
	......	5.15	Arm, Disc Brake Lever—Rear—L. H.	10S0-939								✓	✓	1
	......	.09	Bushing, Brake Shoe	10S0-942								✓	✓	2
	......	.30	Pin, Anchor	10S2-997								✓	✓	1
	......	.50	Pin, Brake Shoe Link	10S0-935								✓	✓	1
	......	.05	Retainer, Brake Shoe Pin	10S0-929								✓	✓	1
	......	.04	Screw, Hex. Cap ($\frac{5}{16}"$—18 x $\frac{5}{8}"$)	S-16								✓	✓	1
	......	.01	Lockwasher, $\frac{11}{32}"$ Spring	S-1346								✓	✓	1
	......	1.80	Bracket, Hand Brake Lever Fulcrum	10D2-132					✓					1
	......	.06	Gasket	10D2-244					✓					1
	......	.04	Nut, Hex. ($\frac{1}{2}"$—20)	S-3747					✓					1
	......	.01	Washer, Flat ($\frac{17}{32}" \times 1" \times \frac{3}{32}"$)	S-374					✓					1
	......	.01	Pin, Cotter ($\frac{3}{32}" \times 1"$)	S-2199					✓					1
	......	.06	Screw, Hex. Cap ($\frac{3}{8}"$—16 x $\frac{1}{8}"$)	S-46					✓					2
	......	.01	Lockwasher, $\frac{13}{32}"$ Spring	S-1347					✓					2
	......	.12	Wick, Hand Brake Lever Oiler	10D2-311					✓					1
	......	.01	Washer, Hand Brake Lever Oiler Wick	S-3184					✓					1
1202	**Brake Shoes and Facings**													
	A-3-3822-T-306	14.00	Shoe, Front Brake, Assembly	4HDA0-100	✓	✓	✓	✓		✓				4
	A-2-3822-E-5	33.50	Shoe, Front Brake, Assembly	9GKW0-300							✓			2
	A-2-3822-F-6	33.50	Shoe, Front Brake, Assembly	9GLW0-300							✓			2
	A-2-3822-F-6	33.50	Shoe, Front Brake, Assembly—R. H.—Upper	9GLW0-300								✓	✓	1
	A-1-3833-E-5	33.50	Shoe, Front Brake, Assembly—R. H.—Lower	9SKA0-310								✓	✓	1
	A-2-3822-E-5	33.50	Shoe, Front Brake, Assembly—L. H.—Upper	9GKW0-300								✓	✓	1
	A-1-3822-F-6	33.50	Shoe, Front Brake, Assembly—L. H.—Lower	9SKA0-300								✓	✓	1
	2240-J-530	1.95	Lining, Front Brake Shoe (Undrilled)	4UG0-52	✓	✓	✓	✓		✓				4
	X-1838	.01	Rivet, Front Brake Shoe Lining ($\frac{3}{16}" \times \frac{7}{16}"$)	S-1020	✓	✓	✓	✓		✓				56

THE AUTOCAR COMPANY
ARDMORE, PA.

Master Parts List

12-5 HAND BRAKES

CODE	MFR. PART No.	LIST PRICE	NAME OF PART	AUTOCAR PART No.	U-2044	U-4044	U-4144	U-4144-T	C-50	U-5044	U-7144-T	U-8144	U-8144-T	No. REQ.
1202	2840-H-8	4.75	Lining, Front Brake Shoe	9SK0-146							✓	✓	✓	4
	17-X-40	.02	Rivet, Front Brake Shoe Lining	9AK0-147							✓	✓	✓	56
	1225-U-99	.18	Bushing, Front Brake Shoe	4UG0-89	✓	✓	✓	✓		✓				4
	1825-K-37	.60	Bushing, Front Brake Shoe	4SK0-149							✓	✓	✓	4
	2217-Z-26	1.15	Plate, Front Brake Shoe Wear	25SD0-235	✓	✓	✓	✓		✓				4
	X-1678	.02	Screw, Flat Head Machine ($\frac{5}{16}''$—18 x $\frac{3}{4}''$)	S-4280	✓	✓	✓	✓		✓				4
	2817-A-1	1.50	Plate, Front Brake Shoe Wear	9SK0-456							✓	✓	✓	4
	1846-M-65	.18	Pin, Front Brake Shoe Wear Plate	9SK0-457							✓	✓	✓	4
	A-3-3222-T-306	14.00	Shoe, Rear Brake, Assembly	4HDA0-100	✓	✓	✓	✓		✓				4
	A-27-3222-B-210	30.00	Shoe, Rear Brake, Assembly—Upper	4FKW0-100							✓			2
	A-28-3222-S-253	31.50	Shoe, Rear Brake, Assembly—Upper	4GKW0-100								✓	✓	2
	A-28-3222-B-210	30.00	Shoe, Rear Brake, Assembly—Lower	4FKW0-110							✓			2
	A-29-3222-S-253	31.50	Shoe, Rear Brake, Assembly—Lower	4GKW0-110								✓	✓	2
	1225-U-99	.35	Bushing, Rear Brake Shoe	4UG0-89	✓	✓	✓	✓		✓				4
	15015	.35	Bushing, Rear Brake Shoe	4G0-89							✓	✓	✓	8
	2240-J-530	5.05	Lining, Rear Brake Shoe (Undrilled)	4UG0-52	✓	✓	✓	✓		✓				8
	X-1838	.01	Rivet, Rear Brake Shoe Lining ($\frac{3}{16}''$ x $\frac{7}{16}''$)	S-1020	✓	✓	✓	✓		✓				56
	2240-M-585	3.45	Lining, Rear Brake Shoe	4FKW0-52							✓			8
	X-1915	.07	Bolt, Rear Brake Shoe Lining	4G0-666							✓			32
	X-541	.02	Nut, Hex. ($\frac{3}{8}''$—16)	S-78							✓			32
	X-529	.01	Lockwasher, $\frac{13}{32}''$ Spring	S-1347							✓			32
	2240-U-983	4.35	Lining, Rear Brake Shoe	4GKW0-52								✓	✓	8
	10-X-180	.10	Bolt, Rear Brake Shoe Lining	4GKW0-1321								✓	✓	32
	13-X-3	.03	Nut, Hex. ($\frac{3}{8}''$—24)	S-1418								✓	✓	32
	1229-E-421	.01	Washer, $\frac{3}{8}''$ Shakeproof Int.	S-3190								✓	✓	32
	2217-C-26	1.15	Plate, Rear Brake Show Wear	25SD0-235	✓	✓	✓	✓		✓				4
	1846-X-1678	.02	Screw, Flat Head Machine ($\frac{5}{16}''$—18 x $\frac{3}{4}''$)	S-4280	✓	✓	✓	✓		✓				4
	2297-L-168	1.30	Plate, Brake Shoe Wear	4G0-667							✓	✓	✓	4
	X-1916	.06	Screw, Flat Head Machine ($\frac{5}{16}''$—18 x $1''\frac{3}{16}$)	S-3132							✓	✓	✓	4
1203	Brake Shoe Support, Guide, Springs, Pins, Anchor Plate													
	2258-B-184	.26	Spring, Front Brake Shoe	4HDA0-79	✓	✓	✓	✓		✓				2
	2858-S-97	.65	Spring, Front Brake Shoe	9SKA0-134							✓	✓	✓	2
	1259-M-39	1.70	Pin, Front Brake Shoe Anchor	4HDA0-1317	✓	✓	✓	✓		✓				4
	N-112	.03	Nut, Hex. ($\frac{3}{4}''$—16)	S-3753	✓	✓	✓	✓		✓				4
	W-112	.04	Lockwasher, $\frac{25}{32}''$ Spring	S-1353	✓	✓	✓	✓		✓				4
	1859-A-27	3.30	Pin, Front Brake Shoe Anchor	9FKW0-121							✓			4
	N-116	.12	Nut, Hex. ($1''$—14)	S-3736							✓			4
	W-116	.04	Lockwasher, $1''$ Spring	S-1357							✓			4
	1859-B-2	4.75	Pin, Front Brake Shoe Anchor	9SK0-121								✓	✓	4
	N-116	.12	Nut, Hex. ($1''$—14)	S-3736								✓	✓	4
	W-116	.04	Lockwasher, $1''$ Spring	S-1357								✓	✓	4
	......	.56	Fitting, Excess Grease Relief	4HDA0-1363								✓	✓	4
	R-158	.01	Rivet, Front Brake Shoe Anchor Pin Bracket	4UG0-1326	✓	✓	✓	✓		✓				12
	1229-M-221	.03	Washer, Front Brake Shoe "C"	25G0-103	✓	✓	✓	✓		✓				4
	1845-A-1	.95	Link, Front Brake Shoe	9SK0-251							✓	✓	✓	2
	10-X-258	.20	Bolt, Front Brake Shoe Link	9FKW0-464							✓			2
	N-28	.05	Nut, Hex. ($\frac{1}{2}''$—20)	S-811							✓			2
	10-X-189	.15	Bolt, Front Brake Shoe Link	9SK0-464								✓	✓	2
	N-28	.05	Nut, Hex. ($\frac{1}{2}''$—20)	S-811								✓	✓	2
	2810-J-244	6.00	Camshaft, Front Brake—R. H.	9BKT0-458	✓	✓				✓				1
	2610-N-248	9.50	Camshaft, Front Brake—R. H.	9UG0-458			✓	✓						1
	2810-T-254	10.30	Camshaft, Front Brake—R. H.	9FKW0-458							✓			1
	2810-E-239	13.00	Camshaft, Front Brake—R. H.	9AK0-458								✓	✓	1
	2810-K-245	6.00	Camshaft, Front Brake—L. H.	9BKT0-459	✓	✓				✓				1
	2810-P-250	9.50	Camshaft, Front Brake—L. H.	9UGA0-459			✓	✓						1
	2810-S-253	10.30	Camshaft, Front Brake—L. H.	9FKW0-459							✓			1
	2810-F-240	13.00	Camshaft, Front Brake—L. H.	9AK0-459								✓	✓	1
	R-5706	.10	Washer, Front Brake Shaft	25G0-106A	✓	✓	✓	✓		✓				2
	1854-B-80	.05	Ring, Front Brake Shaft Snap	4UG0-1353	✓	✓	✓	✓		✓	✓	✓	✓	2
	1825-G-59	.60	Bushing, Front Brake Camshaft	9AK0-461							✓	✓	✓	4
	......	.56	Fitting, Excess Grease Relief	4HDA0-1363							✓	✓	✓	2

Master Parts List

CODE	MFR. PART No.	LIST PRICE	NAME OF PART	AUTOCAR PART No.	MODEL U-2044	U-4044	U-4144	U-4144-T	C-50	U-5044	U-7144-T	U-8144	U-8144-T	No. REQ.
1203	A-2-3275-A-53	7.70	**Adjuster,** Front Brake Slack—R. H.	25BC0-310	✓	✓				✓				1
	A-2-3275-A-53	7.70	**Adjuster,** Front Brake Slack—L. H.	25BD0-310	✓	✓				✓				1
	3-X-2	.24	**Bolt,** Front Brake Slack Adjuster Clamp	25G0-244	✓	✓				✓				2
	13-X-5	.04	**Nut,** Hex. (½"—20)	S-1420	✓	✓				✓				2
	X-531	.01	**Lockwasher,** $\frac{17}{32}$" Spring	S-1349	✓	✓				✓				2
	215091	14.00	**Adjuster,** Front Brake Slack (Bendix-Westinghouse)	25KH0-881		✓	✓	✓						2
	220688	10.00	**Adjuster,** Front Brake Slack (Bendix-Westinghouse)	25AR0-881							✓			2
	216781	18.00	**Adjuster,** Front Brake Slack—L. H. (Bendix-Westinghouse)	25AM0-881								✓	✓	1
	216782	18.00	**Adjuster,** Front Brake Slack—L. H. (Bendix-Westinghouse)	25AN0-881								✓	✓	1
	A-3899-J-166	7.50	**Bracket,** Front Brake Shaft, and Bushing Assembly	4UG0-900	✓	✓	✓	✓		✓				2
	1825-H-8	.50	**Bushing,** Front Brake Shaft Bracket—Long	4UG0-149	✓	✓	✓	✓		✓				2
	1825-J-10	.35	**Bushing,** Front Brake Shaft Bracket—Short	4UGA0-149	✓	✓	✓	✓		✓				2
	S-11012	.30	**Screw,** Hex. Cap (⅝"—18 x 1½")	S-2085	✓	✓	✓	✓		✓				4
	N-110	.10	**Nut,** Hex. (⅝"—18)	S-1311	✓	✓	✓	✓		✓				4
	2858-B-184	.26	**Spring,** Rear Brake Shoe Return	4HDA0-79	✓	✓	✓	✓		✓				2
	2258-A-53	.80	**Spring,** Rear Brake Shoe	4G0-79							✓	✓	✓	4
	1246-H-60	.11	**Pin,** Rear Brake Shoe Spring	4G0-155							✓	✓	✓	4
	1259-M-39	1.70	**Pin,** Rear Brake Shoe Anchor	4HDA0-1317	✓	✓	✓	✓		✓				4
	1259-B-28	2.75	**Pin,** Rear Brake Shoe Anchor	4G0-07							✓	✓	✓	4
	13-X-22	.07	**Nut,** Rear Brake Shoe Anchor Pin	4HDA0-1078	✓	✓	✓	✓		✓				4
	X-240	.01	**Rivet,** Rear Brake Shoe Anchor Bracket ($\frac{5}{16}$" x 1")	S-563	✓	✓	✓	✓		✓				12
	1229-D-108	.06	**Lock,** Rear Brake Shoe Anchor Pin	4G0-678							✓			2
	X-702	.03	**Screw,** Hex. Cap ($\frac{5}{16}$"—18 x ⅞")	S-15							✓			2
	X-528	.01	**Lockwasher,** $\frac{11}{32}$" Spring	S-1346							✓			2
	1229-D-108	.06	**Lock,** Rear Brake Shoe Anchor Pin	4G0-678								✓	✓	2
	X-725	.05	**Screw,** Hex. Cap (⅜"—16 x ¾")	S-12								✓	✓	2
	X-529	.01	**Lockwasher,** $\frac{13}{32}$" Spring	S-1347								✓	✓	2
	......	.35	**Fitting,** Alemite Measuring (⅛")	S-4019								✓	✓	4
	......	.08	**Bushing,** ¼" x ⅛" Galv. Pipe	S-4759								✓	✓	4
	2810-R-252	9.65	**Camshaft,** Rear Brake—R. H.	4UG0-251	✓	✓	✓	✓		✓				1
	2210-C-1641	12.35	**Camshaft,** Rear Brake—R. H.	4FKW0-153							✓			1
	2210-F-786	11.95	**Camshaft,** Rear Brake—R. H.	25ZGEW0-99								✓	✓	1
	2810-Q-251	9.65	**Camshaft,** Rear Brake—L. H.	4UG0-252	✓	✓	✓	✓		✓				1
	2210-B-1640	12.35	**Camshaft,** Rear Brake—L. H.	4FKW0-154							✓			1
	2210-E-785	11.95	**Camshaft,** Rear Brake—L. H.	25ZGEW0-99A								✓	✓	1
	1199-R-434	.69	**Collar,** Rear Brake Camshaft	4NFG0-164							✓	✓	✓	2
	26-X-4	.15	**Screw,** Rear Brake Camshaft Collar	4F0-671							✓	✓	✓	2
	1229-R-122	.05	**Washer,** Spacing (1$\frac{17}{32}$" I.D.—2¼" O.D. x $\frac{1}{16}$")	S-2536							✓	✓	✓	4
	A-3299-C-289	2.85	**Bracket,** Rear Brake Camshaft Support, and Bushing Assy	4HDA0-900	✓	✓	✓	✓		✓				2
	1225-H-86	.23	**Bushing,** Rear Brake Camshaft Support Bracket	25G0-233A	✓	✓	✓	✓		✓				2
	1229-M-221	.03	**Washer,** Rear Brake Shoe Anchor Pin "C"	25G0-103	✓	✓	✓	✓		✓				4
	3-X-75	.10	**Bolt,** Rear Brake Camshaft Support Bracket	25N0-112	✓	✓	✓	✓		✓				4
	13-X-7	.10	**Nut,** Hex. (⅝"—18)	S-1311	✓	✓	✓	✓		✓				4
	X-533	.01	**Lockwasher,** $\frac{21}{32}$" Spring	S-1351	✓	✓	✓	✓		✓				4
	1229-X-518	.03	**Washer,** Shakeproof Ext.	S-3173	✓	✓	✓	✓		✓				4
	A-3899-P-172	8.90	**Bracket,** Rear Brake Shaft, and Bushing Assembly—R. H.	25KC0-65	✓	✓	✓	✓		✓				1
	A-3899-Q-173	8.90	**Bracket,** Rear Brake Shaft, and Bushing Assembly—L. H.	25KD0-65	✓	✓	✓	✓		✓				1
	1825-P-42	.35	**Bushing,** Rear Brake Shaft Bracket—Thin	4AKB0-149	✓	✓	✓	✓		✓				1
	R-5706	.10	**Washer,** Rear Brake Shaft	25G0-106A	✓	✓	✓	✓		✓				2
	3299-C-523	5.50	**Bracket,** Rear Brake Camshaft and Diaphragm—R. H.	4HDAA0-146							✓			1
	3299-F-916	5.55	**Bracket,** Rear Brake Camshaft and Diaphragm—R. H.	4GKW0-146								✓	✓	1
	3299-B-522	5.50	**Bracket,** Rear Brake Camshaft and Diaphragm—L. H.	4HDA0-146							✓			1
	3299-E-915	5.55	**Bracket,** Rear Brake Camshaft and Diaphragm—L. H.	4GLW0-146								✓	✓	1
	4-X-2	.24	**Stud,** Rear Brake Camshaft and Diaphragm Bracket	4HDA0-957							✓			4
	13-X-7	.06	**Nut,** Hex. (⅝"—18)	S-1311							✓			4
	X-533	.01	**Lockwasher,** $\frac{21}{32}$" Spring	S-1351							✓			4
	4-X-335	.18	**Stud,** Rear Axle Housing and Camshaft Bracket	4GKW0-1141								✓	✓	8
	......	.04	**Nut,** Hex. (½"—20)	S-1420								✓	✓	8
	X-531	.01	**Lockwasher,** $\frac{17}{32}$" Spring	S-1349								✓	✓	8
	215530	14.25	**Adjuster,** Rear Brake Shaft Slack	25AK0-881	✓	✓	✓	✓		✓				2
	217919	10.25	**Adjuster,** Rear Brake Slack—R. H. (Bendix-Westinghouse)	25AS0-881							✓			1
	220484	15.00	**Adjuster,** Rear Brake Slack—R. H. (Bendix-Westinghouse)	25AP0-881								✓	✓	1

CODE	MFR. PART No.	LIST PRICE	NAME OF PART	AUTOCAR PART No.	U-2044	U-4044	U-4144	U-4144-T	C-50	U-5044	U-7144-T	U-8144	U-8144-T	No. REQ.
					MODEL									
1203	217920	10.25	**Adjuster,** Rear Slack—L. H. (Bendix-Westinghouse)	25AT0-**881**							√			1
	220483	15.00	**Adjuster,** Rear Brake Slack—L. H. (Bendix-Westinghouse)	25AQ0-**881**								√	√	1
	1229-R-122	.04	**Washer,** Rear Brake Slack Adjuster Spacer	S-**2536**	√	√	√	√		√				4
	A-1229-J-166	.26	**Washer,** Rear Brake Slack Adjuster Retaining, Assembly	25SD0-**173**							√			2
	1229-C-107	.10	**Washer,** Rear Brake Slack Adjuster	25ZGEW0-**874**								√	√	2
	1854-B-80	.05	**Ring,** Rear Brake Slack Adjuster Retainer Snap	4UG0-**1353**	√	√	√	√		√				2
	3911-U-73	10.35	**Spider,** Rear Brake	4UG0-**928**	√	√	√	√		√				2
	2808-L-168	.07	**Gasket,** Rear Brake Spider to Housing	4NKA0-**1205**	√	√	√	√		√				2
	S-1816	.10	**Bolt,** Rear Brake Spider to Housing (½″—20 x 2″)	S-**1438**	√	√	√	√		√				8
	3-X-135	.17	**Bolt,** Rear Brake Spider to Housing	4SK0-**215**	√	√	√	√		√				8
	A-3211-M-949	37.00	**Spider,** Rear Brake, and Bushing Assembly	4GKW0-**120**								√	√	2
	A-3211-Q-953	26.50	**Spider,** Rear Brake, and Bushing Assembly	4FKW0-**120**							√			2
	A1-3211-M-949	78.10	**Spider,** Rear Brake, and Sleeve Assembly—Long	4GKW0-**130**								√	√	1
	A2-3211-M-949	73.70	**Spider,** Rear Brake, and Sleeve Assembly—Short	4GLW0-**130**								√	√	1
	1225-G-85	.32	**Bushing,** Rear Brake Spider	4G0-**159**							√			2
	1225-R-200	.21	**Bushing,** Rear Brake Spider	4ZGEW0-**159**								√	√	4
	3-X-111	.14	**Screw,** Rear Brake Spider and Housing	4GKW0-**1379**								√	√	4
	X-531	.01	**Lockwasher,** 17/32″ Spring	S-**1349**								√	√	4
	X-1273	.05	**Rivet,** Rear Brake Spider and Housing Screw (½″ x 2″)	S-**1083**								√	√	12
1204	**Brake Pedal (See Group 0204 for Support)**													
	......	.55	**Pad,** Brake Pedal	10FE2-**292**	√	√	√	√		√	√			1
	......	.55	**Pad,** Brake Pedal	10FE2-**292**								√	√	1
	......	.01	**Bolt,** Flat Head Stove, and Nut (¼″ x 1⅛″)	S-**3053**								√	√	2
	......	.01	**Lockwasher,** 9/32″ Spring	S-**1345**								√	√	2
	......	2.20	**Lever,** Brake Pedal	10UB5-**54**	√	√	√	√		√	√			1
	......	.25	**Ring,** Lock	10D2-**988**	√	√	√	√		√	√			2
	......	.25	**Bushing**	10T1-**25**	√	√	√	√		√	√			2
	......	2.20	**Lever,** Brake Pedal	10UB5-**54**								√	√	1
	......	.01	**Washer,** Flat St. (1 1/32″ x 1¾″ x 3/32″)	S-**3316**								√	√	1
	......	.25	**Ring,** Lock	10D2-**988**								√	√	1
	......	.20	**Fitting,** 67½° Alemite (⅛″)	S-**4012**								√	√	1
	......	.25	**Bushing**	10T1-**25**								√	√	2
	......	.30	**Rod,** Brake	**24-241-20**	√									1
	......	.25	**Rod,** Brake	**24-241-15**	√					√				1
	......	.80	**Rod,** Brake	**24-241-57**			√	√						1
	......	.80	**Rod,** Brake Control	**24-241-59**							√			1
	......	.80	**Rod,** Brake Control	**24-241-55½**								√	√	1
	......	.50	**Clevis,** Brake Control Rod	10F2-**56**	√	√	√	√		√				11
	......	.50	**Clevis,** Brake Control Rod	10F2-**56**							√	√	√	2
	......	.12	**Pin,** Brake Control Rod Clevis	S-**2225**							√	√	√	2
	......	.04	**Nut,** Hex. (½″—20)	S-**1420**							√	√	√	2
	......	.01	**Cotter,** ⅛″ x 1″	S-**2204**							√	√	√	2
	......	.40	**Spring,** Brake Control Rod Return	10S2-**147**	√	√	√	√		√	√	√	√	1
	......	.12	**Clip,** Brake Control Rod Return Spring	10D2-**79**			√	√			√	√	√	1
	......	.12	**Clip,** Brake Control Rod Return Spring	10Y2-**79**	√	√				√				1
	......	.25	**Link,** Brake Control Rod Return	6CK1-**317**								√	√	1
	......	.45	**Pin,** Off-Set Clevis	10B2-**1104A**	√	√				√				2
1205	**Air Brake Tank, Compressor, Governor, Valves**													
	......	18.75	**Tank,** Air	25TH4-**40**	√	√	√	√		√				2
	217321	12.95	**Tank,** Air Reservoir	25TH0-**40**							√	√	√	2
	215310	1.00	**Cock,** Drain	25G2-**41**	√	√	√	√		√	√	√	√	2
	......	.20	**Strap,** Air Reservoir Tank	25LA3-**328**	√	√	√	√		√	√			8
	......	.20	**Strap,** Air Reservoir Tank	25LA3-**328**								√	√	8
	......	.10	**Screw,** Hex. Cap (⅜″—16 x 3½″)	S-**1436**								√	√	4
	......	.08	**Screw,** Hex. Cap (⅜″—16 x 1¼″)	S-**2042**								√	√	8
	......	.03	**Nut,** Hex. (⅜″—16)	S-**75**								√	√	12
	......	.01	**Lockwasher,** 13/32″ Spring	S-**1347**								√	√	24
	......	.25	**Bracket,** Air Tank	25GC2-**03**	√									2
	......	.45	**Bracket,** Air Tank	25GG3-**03**	√	√	√	√		√				2
	......	.35	**Bracket,** Air Tank	25LE3-**03**		√	√	√		√				2

Master Parts List

CODE	MFR. PART No.	LIST PRICE	NAME OF PART	AUTOCAR PART No.	MODEL U-2044	U-4044	U-4144	U-4144-T	C-50	U-5044	U-7144-T	U-8144	U-8144-T	No. REQ.
1205	......	.50	**Bracket,** Air Tank	25GH3-**03**							✓			2
	......	.50	**Bracket,** Air Tank	25LA3-**03**							✓			2
	......	1.00	**Bracket,** Air Reservoir Tank	25GQ4-**03**								✓	✓	2
	......	.09	**Screw,** Hex. Cap (½"—13 x 1⅜")	S-**1428**								✓	✓	6
	......	.04	**Nut,** Hex. (½"—13)	S-**77**								✓	✓	6
	......	.01	**Lockwasher,** $\frac{17}{32}$" Spring	S-**1349**								✓	✓	6
	......	.04	**Lockwasher,** $\frac{17}{32}$" Shakeproof Ext.	S-**3179**								✓	✓	6
	215186	10.00	**Indicator,** Low Pressure	25FE2-**146**	✓	✓	✓	✓		✓	✓			1
	215186	10.00	**Indicator,** Low Pressure	25FE2-**146**								✓	✓	1
	......	.04	**Screw,** Hex. Cap (¼"—20 x 1")	S-**3118**								✓	✓	2
	......	.01	**Nut,** Hex. (¼"—20)	S-**76**								✓	✓	2
	......	.01	**Lockwasher,** $\frac{9}{32}$" Spring	S-**1345**								✓	✓	4
	11120	7.85	**Buzzer,** Low Pressure Indicator	25E2-**231**	✓	✓	✓	✓						1
	11120	9.00	**Buzzer,** Low Pressure Indicator	25B2-**231**						✓				1
	......	11.90	**Buzzer,** Low Pressure Indicator	25F3-**231**							✓			1
	......	11.90	**Buzzer,** Low Pressure Indicator	25F3-**231**								✓	✓	1
	......	.03	**Bolt,** Round Head Stove, and Nut ($\frac{3}{16}$" x 1")	S-**2966A**								✓	✓	2
	......	.01	**Lockwasher,** $\frac{7}{32}$" Spring	S-**1344**								✓	✓	4
	205105	2.25	**Valve,** Safety	25G0-**31**	✓	✓	✓	✓		✓	✓	✓	✓	1
	......	3.00	**Valve,** Tire Inflating, Assembly	25A0-**110**	✓	✓	✓	✓		✓				1
	......	3.00	**Valve,** Air Supply, Assembly	25A2-**110**							✓			1
	......	3.00	**Valve,** Air Supply, Assembly	25A2-**1060**								✓	✓	1
	......	.04	**Screw,** Hex. Cap (¼"—20 x 1")	S-**3118**								✓	✓	2
	......	.01	**Nut,** Hex. (¼"—20)	S-**76**								✓	✓	2
	......	.01	**Lockwasher,** $\frac{9}{32}$" Spring	S-**1345**								✓	✓	4
	215995	17.80	**Governor**	25G3-**50B**	✓	✓	✓	✓		✓	✓			1
	215995	17.80	**Governor**	25G3-**50B**								✓	✓	1
	......	.04	**Screw,** Round Head Machine (¼"—20 x 1")	S-**6156**								✓	✓	2
	......	.01	**Nut,** Hex. (¼"—20)	S-**76**								✓	✓	2
	......	.01	**Lockwasher,** $\frac{9}{32}$" Spring	S-**1345**								✓	✓	4
	......	158.50	**Compressor,** Air	25HB5-**10**	✓									1
	......	158.50	**Compressor,** Air	25BB4-**10**		✓	✓	✓		✓				1
	......	220.00	**Compressor,** Air	25HC5-**10**							✓			1
	2UE7¼VW	220.00	**Compressor**	25HC4-**10**								✓	✓	1
	1173-A	.08	**Screw,** Hex. Cap ($\frac{7}{16}$"—14 x 1¼")	S-**1881**								✓	✓	4
	335-A	.01	**Lockwasher,** $\frac{7}{16}$" Spring	S-**1348**								✓	✓	4
	17909-D	10.08	**Bracket,** Compressor Mounting	25UU0-**417**							✓	✓	✓	1
	35703-A	.04	**Gasket,** Compressor Mounting Bracket	25UU0-**219**							✓	✓	✓	1
	4625-A	.04	**Screw,** Compressor Mounting Bracket	2UU0-**319**							✓	✓	✓	1
	312-A	.01	**Lockwasher,** $\frac{17}{32}$" Spring	S-**1349**							✓	✓	✓	2
	3136-A	.03	**Screw,** Flat Head Machine (½"—13 x 1¼")	S-**3826**							✓	✓	✓	2
	4594-A	.01	**Lockwasher,** ½" Shakeproof Csnk. Int.	S-**3294**							✓	✓	✓	2
	8552-A	.45	**Screw,** Compressor Mounting Bracket	2UUA0-**319**							✓	✓	✓	1
	17134-B	8.50	**Pulley,** Compressor	25UU0-**317**							✓	✓	✓	1
	215748	41.75	**Valve,** Hand Control	25T4-**380A**		✓		✓		✓	✓	✓	✓	1
	217698	5.25	**Valve,** Double Check	25C2-**226**	✓	✓	✓	✓		✓	✓			2
	217698	5.25	**Valve,** Double Check	25C2-**226**								✓	✓	2
	......	.05	**Screw,** Hex. Cap ($\frac{5}{16}$"—18 x 1¼")	S-**32**								✓	✓	2
	......	.03	**Nut,** Hex. ($\frac{5}{16}$"—18)	S-**82**								✓	✓	2
	......	.01	**Lockwasher,** $\frac{11}{32}$" Spring	S-**1346**								✓	✓	2
	215537	3.50	**Switch,** Stop Light	25NT2-**130**	✓	✓	✓	✓		✓	✓			1
	215537	3.50	**Switch,** Stop Light	25NT2-**130**								✓	✓	1
	......	.04	**Screw,** Hex. Cap (¼"—20 x 1")	S-**3118**								✓	✓	2
	......	.01	**Nut,** Hex. (¼"—20)	S-**76**								✓	✓	2
	......	.01	**Lockwasher,** $\frac{9}{32}$" Spring	S-**1345**								✓	✓	4
	......	56.80	**Valve,** Air Application, and Stop Assembly	25A4-**650B**	✓									1
	......	58.80	**Valve,** Air Application, and Stop Assembly	25A0-**650B**		✓	✓	✓		✓				1
	216213	56.00	**Valve,** Brake Application	25A4-**640B**							✓			1
	216213	56.00	**Valve,** Brake Application	25A4-**640B**								✓	✓	1
	......	.05	**Screw,** Hex. Cap ($\frac{5}{16}$"—18 x 1½")	S-**31**								✓	✓	4
	......	.03	**Nut,** Hex. ($\frac{5}{16}$"—18)	S-**82**								✓	✓	4
	......	.01	**Lockwasher,** $\frac{11}{32}$" Spring	S-**1346**								✓	✓	8

THE AUTOCAR COMPANY
ARDMORE, PA.

Master Parts List

12-9
HAND BRAKES

CODE	MFR. PART No.	LIST PRICE	NAME OF PART	AUTOCAR PART No.	U-2044	U-4044	U-4144	U-4144-T	C-50	U-5044	U-7144-T	U-8144	U-8144-T	No. REQ.
1205	.....	.10	Ell, 1/8" P. T. Alemite M & F................	S-4021								✓	✓	1
	.....	.20	Oiler, 67½° Alemite (1/8")...................	S-4012								✓	✓	1
	.....	.90	Bracket, Brake Application Valve...........	25AE3-71	✓									1
	.....	.45	Bracket, Brake Application Valve...........	25AD4-71		✓			✓	✓				1
	.....	.90	Bracket, Brake Application Valve...........	25L3-71			✓	✓			✓			1
	.....	.90	Bracket, Brake Application Valve...........	25L3-71								✓	✓	1
	.....	.09	Screw, Hex. Cap (½"—13 x 1½")...........	S-1872								✓	✓	2
	.....	.04	Nut, Hex. (½"—13)........................	S-77								✓	✓	2
	.....	.01	Lockwasher, 17/32" Spring................	S-1349								✓	✓	4
	205000	5.00	Valve, Quick Release.....................	25G2-61	✓	✓	✓	✓	✓	✓	✓			1
	205000	5.00	Valve, Quick Release.....................	25G2-61								✓	✓	1
	.....	.05	Screw, Hex. Cap (5/16"—18 x 1⅛")..........	S-27								✓	✓	2
	.....	.03	Nut, Hex. (5/16"—18).....................	S-82								✓	✓	2
	.....	.01	Lockwasher, 11/32" Spring................	S-1346								✓	✓	4
	217383	23.75	Valve, Relay............................	25A3-630A	✓	✓	✓	✓	✓	✓	✓			1
	217383	23.75	Valve, Relay............................	25A3-630A								✓	✓	1
	.....	.06	Screw, Hex. Cap (3/8"—16 x 1½")..........	S-17								✓	✓	2
	.....	.03	Nut, Hex. (3/8"—16).....................	S-75								✓	✓	2
	.....	.01	Lockwasher, 13/32" Spring................	S-1347								✓	✓	4
	.....	.85	Bracket, Relay Valve.....................	25G3-139								✓	✓	1
	.....	.20	Spacers, ½"............................	12ZDA3-108A								✓	✓	2
	.....	.10	Screw, Hex. Cap (½"—13 x 2")..........	S-6293								✓	✓	2
	.....	.04	Nut, Hex. (½"—13)......................	S-77								✓	✓	2
	.....	.01	Lockwasher, 17/32" Spring...............	S-1349								✓	✓	2
1206	**Brake Tubes, Clips, Brackets, Springs**													
	.....	.40	Bracket, Hose Coupling Frame—L. H.—Front..	24CR3-29								✓	✓	1
	.....	.40	Bracket, Hose Coupling Frame—R. H.—Front..	24CS3-29								✓	✓	1
	.....	.05	Screw, Hex. Cap (5/16"—18 x 1¼")..........	S-32								✓	✓	4
	.....	.03	Nut, Hex (5/16"—18).....................	S-82								✓	✓	4
	.....	.01	Lockwasher, 11/32" Spring................	S-1346								✓	✓	8
	.....	.25	Bracket, Hose Coupling Frame—R. H.—Rear..	24CP2-29								✓	✓	1
	.....	.25	Bracket, Hose Coupling Frame—L. H.—Rear..	24CQ2-29								✓	✓	1
	.....	.06	Screw, Hex. Cap (3/8"—16 x 1½")..........	S-17								✓	✓	4
	.....	.03	Nut, Hex. (3/8"—16).....................	S-75								✓	✓	4
	.....	.01	Lockwasher, 13/32" Spring................	S-1347								✓	✓	8
	.....	.60	Bracket, Hose Spring.....................	25N0-194		✓	✓	✓		✓				1
	200661	.50	Spring, Hose............................	25N0-195		✓	✓	✓		✓	✓			2
	.....	.12	Clip, Hose..............................	25N2-215		✓	✓	✓		✓	✓			2
	.....	.25	Clip, Hose Spring........................	10ZUBLA2-79							✓	✓	✓	2
	220129	1.25	Coupling, Dummy.......................	25A0-56	✓	✓	✓	✓	✓					1
	212227	.75	Coupling, Dummy Hose (Rear of Body)......	25C2-56							✓	✓	✓	2
	.....	.05	Screw, Hex. Cap (5/16"—18 x 1")..........	S-3120							✓	✓	✓	4
	.....	.03	Nut, Hex. (5/16"—18)....................	S-82							✓	✓	✓	4
	.....	.01	Lockwasher, 11/32" Spring...............	S-1346							✓	✓	✓	8
	.....	.35	Bracket, Air Line Rubber Mounting........	25ZT2-287	✓									1
	.....	.10	Plate, Air Line Rubber Mounting Bracket....	25ZT2-288	✓									1
	.....	.10	Rubber, Air Line Mounting...............	25ZT2-289	✓									1
	.....	.45	Tube, Frame............................	25C2-27	✓	✓	✓	✓		✓				2
1207	**Brake Wheel Chamber**													
	.....	17.00	Chamber, Front Brake—R. H..............	25KE0-21	✓	✓	✓	✓		✓				1
	.....	17.00	Chamber, Front Brake—L. H..............	25FK0-21	✓	✓	✓	✓		✓				2
	220686	16.50	Diaphragm, Front Brake—R. H............	25KP0-21							✓			1
	220687	16.50	Diaphragm, Front Brake—L. H............	25KN0-21							✓			1
	217362	16.50	Diaphragm, Front Brake, Assembly—R. H....	25KL0-21								✓	✓	1
	217361	16.50	Diaphragm, Front Brake, Assembly—L. H....	25KM0-21								✓	✓	1
	4-X-663	.36	Stud, Front Brake Diaphragm.............	4GKW0-361								✓	✓	2
	N-18	.04	Nut, Hex. (½"—20)......................	S-1420								✓	✓	2
	W-18	.01	Lockwasher, 17/32" Spring...............	S-1349								✓	✓	2
	4-X-362	.13	Stud, Front Brake Diaphragm.............	4A0-648							✓			2
	N-18	.04	Nut, Hex. (½"—20)......................	S-1420							✓			2

12-10 HAND BRAKES

Master Parts List
THE AUTOCAR COMPANY, ARDMORE, PA.

CODE	MFR. PART No.	LIST PRICE	NAME OF PART	AUTOCAR PART No.	U-2044	U-4044	U-4144	U-4144-T	C-50	U-5044	U-7144-T	U-8144	U-8144-T	No. REQ.
1207	W-18	.01	Lockwasher, 17/32″ Spring	S-1349								✓	✓	2
	......	18.50	Chamber, Rear Brake—R. H.	25KC0-21	✓	✓	✓	✓		✓				1
	......	18.50	Chamber, Rear Brake—L. H.	25KD0-21	✓	✓	✓	✓		✓				1
	220371	18.50	Diaphragm, Rear Brake	25KQ0-21							✓			2
	216814	18.50	Diaphragm, Rear Brake—R. H.	25KS0-21								✓	✓	1
	216813	18.50	Diaphragm, Rear Brake—L. H.	25KR0-21								✓	✓	1
	15-X-79	.09	Screw, Hex. Cap (1/2″—20 x 1 1/2″)	S-1429								✓	✓	8
	13-X-5	.04	Nut, Hex. (1/2″—20)	S-1420								✓	✓	8
	X-531	.01	Lockwasher, 17/32″ Spring	S-1349								✓	✓	8
1208	**Brake Dust Shields**													
	A-3836-F-136	5.70	Shield, Front Brake Dust—R. H.	4UG0-120	✓	✓	✓	✓		✓				1
	A1-3836-F-136	5.70	Shield, Front Brake Dust—L. H.	4UG0-447	✓	✓	✓	✓		✓				1
	3836-L-142	1.85	Shield, Front Brake Dust	4FKW0-744							✓			2
	S-265	.02	Screw, Hex. Cap (3/8″—16 x 5/8″)	S-53							✓			12
	W-16	.01	Lockwasher, 13/32″ Spring	S-1347							✓			12
	3836-N-144	2.30	Shield, Front Brake Dust	4GKW0-744								✓	✓	2
	S-264	.04	Screw, Hex. Cap (3/8″—16 x 1/2″)	S-1461								✓	✓	16
	W-16	.01	Lockwasher, 13/32″ Spring	S-1347								✓	✓	16
	1898-E-213	.12	Plate, Front Brake Dust Shield Mounting	4GKW0-254								✓	✓	8
	S-267	.05	Screw, Hex. Cap (3/8″—16 x 7/8″)	S-50								✓	✓	8
	W-16	.01	Lockwasher, 13/32″ Spring	S-1347								✓	✓	8
	3880-H-34	.42	Slinger, Front Brake Dust Shield Oil	9UG0-750	✓	✓	✓	✓		✓				2
	A-3236-H-502	12.05	Shield, Rear Brake Dust, Assembly	4HDA0-254	✓	✓	✓	✓		✓				2
	3236-N-690	.95	Shield, Rear Brake Dust	4GKW0-743								✓	✓	4
	X-725	.05	Screw, Hex. Cap (3/8″—16 x 3/4″)	S-12								✓	✓	12
	X-529	.01	Lockwasher, 13/32″ Spring	S-1347							✓	✓	✓	12
1209	**Brake Lines, Pipes, Hoses**													
	......	Ft. .20	Tubing, 3/8″ O.D. Copper, #19 Ga.	S-5146	✓									As
	......	Ft. .20	Tubing, 3/8″ O.D. Copper, #19 Ga.	S-5146							✓			88 Ft.
	......	Ft. .20	Tubing, 3/8″ O.D. Copper, #19 Ga.	S-5146								✓	✓	176 Ft.
	......	Ft. .30	Tubing, 1/2″ O.D. Copper, #18 Ga.	S-5149	✓	✓	✓	✓		✓				As
	......	Ft. .30	Tubing, 1/2″ O.D. Copper, #18 Ga.	S-5149							✓			24 Ft.
	......	Ft. .30	Tubing, 1/2″ O.D. Copper, #18 Ga.	S-5149								✓	✓	32 Ft.
	......	Ft. .37	Tubing, 3/4″ O.D. Copper, #20 Ga.	S-5147	✓	✓	✓	✓		✓				As
	......	Ft. .37	Tubing, 3/4″ O.D. Copper, #20 Ga.	S-5147							✓			16 Ft.
	......	1.25	Tubing, 3/4″ O.D. Copper, #20 Ga. (40″ long)	25U0-47								✓	✓	1
	......	3.35	Tubing, 3/4″ O.D. Copper, #20 Ga. (106 1/2″ long)	25UA0-47								✓	✓	1
	......	.50	Tubing, 3/4″ O.D. Copper, # Ga. (16″ long)	25UB0-47								✓	✓	1
	......	Ft. .37	Tubing, 3/4″ O.D. Copper, #20 Ga.	S-5147								✓	✓	16 Ft.
	......	Ft. .06	Loom (7/16″ I.D.)	16B0-273							✓			84 Ft.
	......	Ft. .06	Loom (7/16″ I.D.)	16B0-273								✓	✓	155 Ft.
	......	Ft. .06	Loom (9/16″ I.D.)	16C0-273							✓			20 Ft.
	......	Ft. .06	Loom (9/16″ I.D.)	16C0-273								✓	✓	30 Ft.
	......	Ft. .20	Loom (1 3/8″ I.D.)	16E0-273								✓	✓	12 Ft.
	215766	3.11	Hose Assembly	25D0-660	✓	✓	✓	✓		✓				2
	215768	3.53	Hose Assembly	25E0-660	✓	✓	✓	✓		✓				2
	215901	13.30	Hose, Air, Assembly (162 1/2″)	25A0-350	✓									1
	220570	14.95	Hose Assembly (216″) for Emerg. Tow—Incl. next 3 items	25B0-350							✓	✓	✓	1
	BW-101-M	Ft. .55	Hose (3/8″ I.D.) (208″)	25C0-37							✓	✓	✓	1
	220165	4.15	Coupling, Hose	25C2-59A							✓	✓	✓	2
	......	.95	Connector, Hose	25C2-670							✓	✓	✓	2
	101-M	Ft. .55	Hose Assembly (100″ Each)	25C0-37		✓	✓	✓		✓				2
	101-M	Ft. .55	Hose Assembly (216″ Each)	25C0-37		✓	✓	✓		✓				1
	215604	12.00	Hose Assembly (112 1/2″) for Semi-Trailer—Incl. next 3 items	25C0-350							✓	✓	✓	2
	BW-101-M	Ft. .55	Hose (3/8″ I.D.) (105″)	25C0-37							✓	✓	✓	2
	220165	4.15	Coupling, Hose	25C2-59A							✓	✓	✓	4
	......	.95	Connector, Hose	25C2-670							✓	✓	✓	4
	217766	3.15	Coupling, Hose	25C2-59	✓	✓	✓	✓		✓				2

12-11 HAND BRAKES

THE AUTOCAR COMPANY, ARDMORE, PA.

Master Parts List

CODE	MFR. PART No.	LIST PRICE	NAME OF PART	AUTOCAR PART No.	U-2044	U-4044	U-4144	U-4144-T	C-50	U-5044	U-7144-T	U-8144	U-8144-T	No. REQ.
1209	220129	1.25	Hose, Dummy	25A0-56	✓	✓	✓	✓	✓					3
	216581	1.50	Coupling, Dummy Hose	25GF2-56	✓	✓	✓	✓	✓					1
	212227	.75	Coupling, Dummy Hose	25C2-56		✓		✓	✓					2
	220635	2.15	Coupling, Dummy Hose (With Bleeder Vent)	25D2-56							✓	✓	✓	1
	220636	1.00	Coupling, Dummy Hose (No Vent)	25E2-56							✓	✓	✓	3
		2.45	Cocks, Shut-Off	25A0-57	✓	✓	✓	✓	✓			✓	✓	3
	205086	4.00	Cocks, Cut-Off	25GA0-57							✓			3
	220306	3.50	Valve, Single Check	25C0-187	✓	✓	✓	✓		✓	✓	✓	✓	1
		.80	Stud, Clamping	25B0-27							✓	✓	✓	2
	205730	1.50	Stud, Clamping	25C0-27							✓	✓	✓	4
	201499	.15	Tag, Emergency Line	25G0-53	✓	✓	✓	✓		✓	✓	✓	✓	2
	201500	.15	Tag, Service Line	25G0-54	✓	✓	✓	✓		✓	✓	✓	✓	2
	BW-101-M	Ft. .55	Hose, Chamber (3/8" I.D.)	25C0-37							✓	✓	✓	9 Ft.
		.30	Spring, Chamber Hose	10T2-63							✓	✓	✓	2
	215536	1.05	Connectors, Chamber Hose	25E2-670							✓	✓	✓	8
	205420	.30	Connector, Frame Tube and Hose	25G0-145	✓	✓	✓	✓		✓	✓	✓	✓	2
		9.75	Hose, Tire Inflation, Assembly	25C9-730	✓	✓	✓	✓		✓				1
		9.75	Hose, Tire Inflation, Assembly (30 Ft.)	25DB9-730							✓	✓	✓	1
	215535	.95	Connector, Flexible Hose	25C2-670		✓	✓	✓		✓				6
	100248	5.00	Gauge, Air Pressure, Assembly	25DFL3-32	✓	✓	✓	✓		✓				1
		2.90	Hose, Air Gauge, Assembly	25UKS4-830							✓	✓	✓	1
	212322	.40	Tee, Tubing (Manifold)	25E2-35	✓	✓	✓	✓		✓	✓	✓	✓	1
	217051	.50	Tee, 3/8"	25A0-43	✓						✓	✓	✓	1
	205103	.50	Tee, 3/8"	25G0-43		✓	✓	✓		✓				1
	217525	.40	Connector, 1/2" Tube	25A0-44	✓	✓	✓	✓		✓	✓	✓	✓	3
	205824	.25	Connector, 3/8" Tube	25D0-44	✓	✓	✓	✓		✓	✓	✓	✓	5
	205053	.25	Connector, 3/8" Tube	25G0-44	✓	✓	✓	✓		✓	✓	✓	✓	13
	215762	1.10	Connector, 3/4" Tube	25C0-44	✓	✓	✓	✓		✓				2
	215709	.45	Elbow, 1/2"	25B0-45	✓	✓	✓	✓		✓	✓	✓	✓	2
	216310	.50	Elbow, 1/2"	25C0-45		✓		✓		✓		✓	✓	1
	216520	.55	Elbow, 3/4"	25D0-45	✓	✓					✓	✓	✓	3
	205102	.30	Elbow, 3/8"	25G0-45	✓	✓	✓	✓		✓	✓	✓	✓	12
	205522	.30	Elbow, 3/8"	25E0-45		✓								1
	205829	.25	Elbow, 3/8"	25H0-45	✓	✓	✓	✓		✓	✓	✓	✓	6
	217504	1.05	Elbow, 3/8"	25J0-45	✓	✓								1
		.12	Clip, Tubing	10D2-79							✓	✓	✓	1
		.20	Clip, Tubing	10AB2-79							✓	✓	✓	2
		.05	Clip, Tubing	16N2-44								✓	✓	1
		.05	Clip, Tubing	16V9-44							✓	✓	✓	6
		.07	Bracket, Tubing Clip	10C2-303								✓		1
		.01	Screw, Hex. Cap (1/4"—20 x 3/4")	S-1614A								✓		1
		.01	Nut, Hex. (1/4"—20)	S-76								✓		1
		.01	Lockwasher, 9/32" Spring	S-1345								✓		2
		.07	Bracket, Tubing Clip	10D2-303								✓		1
		.01	Screw, Hex. Cap (1/4"—20 x 3/4")	S-1614A								✓		1
		.01	Nut, Hex. (1/4"—20)	S-76								✓		1
		.01	Lockwasher, 9/32" Spring	S-1345								✓		2
		.07	Bracket, Tubing Clip	10F2-303								✓		1
		.05	Spacer, Tubing	12M9-108								✓		2
		.05	Plug, Square Head (1/4")	S-4737	✓	✓	✓	✓		✓				2
		.02	Plug, 1/4" Pipe	S-629							✓	✓	✓	2
		.07	Nipple, 1/4" Close	S-4507	✓	✓	✓	✓		✓	✓	✓	✓	1
		.10	Nipple, 3/8" Close	S-4510	✓	✓	✓	✓		✓	✓	✓	✓	2
		.10	Nipple, 1/2" x 3"	S-4570	✓	✓	✓	✓		✓				1
		.30	Nipple, 1/2" x 3" (Brass)	S-4571		✓								1
		.28	Tee, 3/8"	S-4699	✓	✓	✓	✓		✓	✓	✓	✓	1
		.20	Bushing, 1/4" x 3/8" Reducing	S-4762	✓	✓	✓	✓		✓	✓	✓	✓	3
		.12	Bushing, 1/4" x 1/2" Reducing	S-4772	✓	✓	✓	✓		✓	✓	✓	✓	1
		.20	Bushing, 1/2" x 3/4" Reducing	S-4774	✓	✓					✓	✓	✓	1
		.12	Ell, 1/4"—90°	S-4171	✓	✓	✓	✓		✓				1
		.30	Ell, 1/2"—45°	S-4610	✓	✓								1
		.40	Ell, 1/4"—45° (Brass)	S-4821	✓	✓	✓	✓		✓	✓	✓	✓	4

12-12
HAND BRAKES
Master Parts List
THE AUTOCAR COMPANY
ARDMORE, PA.

CODE	MFR. PART No.	LIST PRICE	NAME OF PART	AUTOCAR PART No.	MODEL									No. REQ.
					U-2044	U-4044	U-4144	U-4144-T	C-50	U-5044	U-7144-T	U-8144	U-8144-T	
1209	......	.45	Ell, ½″—90° (Brass).................	S-4822	✓	✓	✓	✓		✓	✓			1
	......	.20	Ell, ¼″—45° Service.................	S-4823							✓	✓	✓	2
	......	.25	Ell, ½″—90° Service.................	S-4620							✓	✓	✓	1
	......	.10	Nipple, ½″ Galvanized Close........	S-4513	✓	✓	✓	✓		✓	✓	✓	✓	2
	......	.32	Ell, ¼″—45° Street.................	S-8026	✓	✓	✓	✓		✓	✓	✓	✓	1
	......	.06	Screw, Set (⅜″—16 x ⅜″)............	S-408		✓	✓	✓	✓					1

THE AUTOCAR COMPANY,
ARDMORE, PA.

Master Parts List

GROUP 13

ILLUSTRATIONS

GROUP 13

WHEELS, HUBS and DRUMS

GROUP 13 — Master Parts List
The Autocar Company, Ardmore, PA.

As supplied on Autocar Models
U-8144 U-8144-T

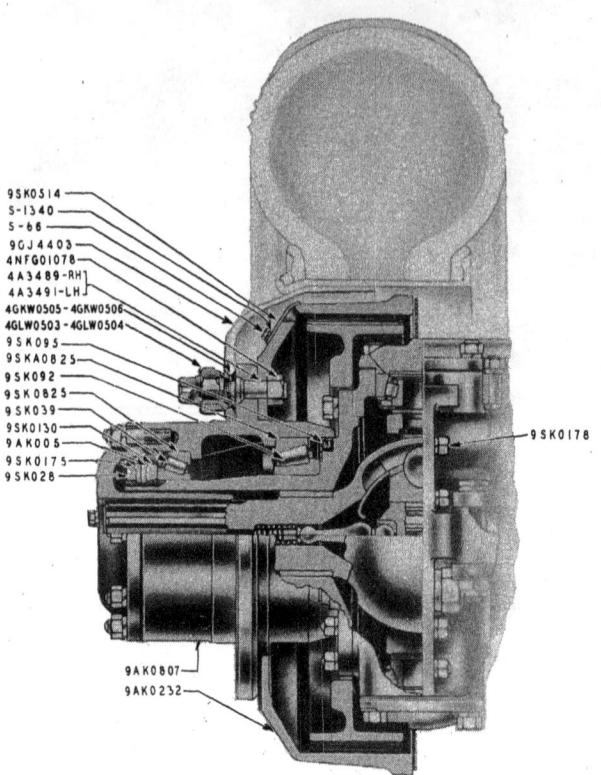

FRONT AXLE

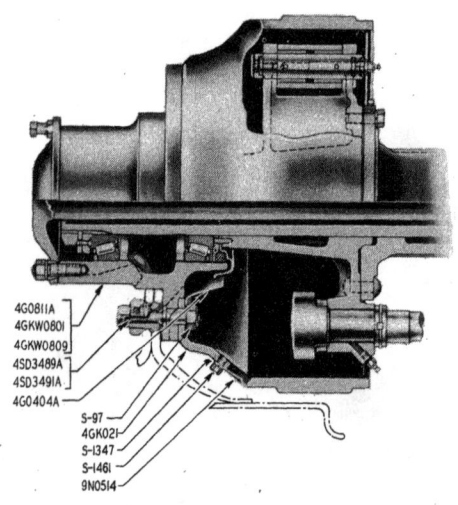
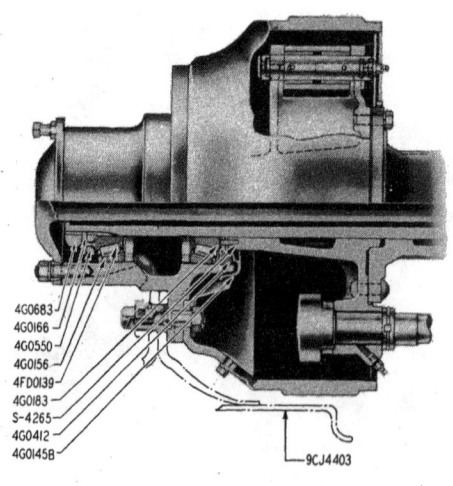

REAR AXLE

1301—Wheels Assembly

THE AUTOCAR COMPANY
ARDMORE, PA.

Master Parts List

13-1 WHEEL ASSEMBLY

CODE	MFR. PART No.	LIST PRICE	NAME OF PART	AUTOCAR PART No.	U-2044	U-4044	U-4144	U-4144-T	C-50	U-5044	U-7144-T	U-8144	U-8144-T	No. REQ.
1301			**Wheel Assembly, Bearings, Retainers and Wrenches**											
	33287	26.00	Wheel, Front and Rear (Budd)..................	9BJ0-403	✓	✓	✓	✓		✓				6
	44470	26.00	Wheel, Front and Rear (Budd-L-Rim)................	9KA4-403							✓			6
	B-45520	37.00	Wheel, Front and Rear (Budd)..................	9CJ4-403								✓	✓	7
	594	11.05	Cone, Front Wheel Bearing—Inner..................	4FD0-138	✓	✓	✓	✓		✓	✓			2
	71450	28.85	Cone, Front Wheel Bearing—Inner..................	9SK0-92								✓	✓	2
	498	9.53	Cone, Front Wheel Bearing—Outer..................	19SKB0-445	✓	✓	✓	✓		✓	✓			2
	52400	17.00	Cone, Front Wheel Bearing—Outer..................	9SK0-39								✓	✓	2
	592-A	9.26	Cup, Front Wheel Bearing—Inner..................	4FD0-139	✓	✓	✓	✓		✓	✓			2
	71750	22.65	Cup, Front Wheel Bearing—Inner..................	9SKA0-825								✓	✓	2
	493	4.77	Cup, Front Wheel Bearing—Outer..................	19SKB0-444	✓	✓	✓	✓		✓	✓			2
	52637	8.50	Cup, Front Wheel Bearing—Outer..................	9SK0-825								✓	✓	2
	1827-H-34	1.65	Nut, Front Wheel Bearing Adjusting..................	9ZDK0-28	✓	✓	✓	✓		✓	✓			2
	A-1827-W-49	2.00	Nut, Front Wheel Bearing Adjusting..................	9SK0-130								✓	✓	2
	1829-G-85	.40	Washer, Front Wheel Bearing Lock..................	9ZDK0-175	✓	✓	✓	✓		✓	✓			2
	1829-W-153	.20	Washer, Front Wheel Bearing Adjusting Nut..................	9SK0-175								✓	✓	2
	1827-H-34	1.65	Nut, Front Wheel Bearing Lock..................	9ZDK0-28	✓	✓	✓	✓		✓				2
	3856-B-2	1.35	Wrench, Front Wheel Bearing Nut..................	4FKW0-696							✓			1
	5-X-285	.60	Washer, Front Wheel Bearing Felt..................	9ZDK0-95	✓	✓	✓	✓		✓	✓			2
	1805-X-24	.25	Retainer, Front Wheel Bearing Felt Washer..................	9ZDK0-178	✓	✓	✓	✓		✓				2
	1827-X-50	1.00	Nut, Front Wheel Bearing Adjusting Jam..................	9SK0-28								✓	✓	2
	1829-V-152	.88	Lock, Front Wheel Bearing Adjusting Jam Nut..................	9AK0-05								✓	✓	2
	5-X-328	1.10	Felt, Front Wheel Bearing..................	9SK0-95								✓	✓	2
	1805-X-24	.25	Retainer, Front Wheel Bearing Felt..................	9ZDK0-178							✓			2
	1829-B-158	.40	Retainer, Front Wheel Bearing Felt..................	9SK0-178								✓	✓	2
	563	6.18	Cup, Rear Wheel Bearing—Inner..................	4Y2-211	✓	✓	✓	✓		✓				2
	552-A	6.17	Cup, Rear Wheel Bearing—Outer..................	4NFG0-142	✓	✓	✓	✓		✓				2
	5520	6.34	Cup, Rear Wheel Bearing..................	4NFG0-157							✓			2
	592-A	9.26	Cup, Rear Wheel Bearing..................	4FD0-139								✓	✓	4
	566	8.66	Cone, Rear Wheel Bearing—Inner..................	4UG0-141	✓	✓	✓	✓		✓				2
	560	8.29	Cone, Rear Wheel Bearing—Outer..................	4Y2-141	✓	✓	✓	✓		✓				2
	5557	10.56	Cone, Rear Wheel Bearing..................	4NFG0-156							✓			4
	596	11.05	Cone, Rear Wheel Bearing..................	4G0-156								✓	✓	4
	5-X-139	.39	Felt, Rear Wheel Bearing..................	4DG0-412							✓			2
	5-X-18	.79	Felt, Rear Wheel Bearing..................	4G0-412								✓	✓	2
	1205-M-117	.23	Retainer, Rear Wheel Bearing Felt—Inner..................	4DG0-145							✓			2
	3105-L-64	.54	Retainer, Rear Wheel Bearing Felt—Inner..................	4G0-145B								✓	✓	2
	1205-L-116	.11	Retainer, Rear Wheel Bearing Felt—Outer..................	4DG0-183							✓			2
	3105-N-14	.85	Retainer, Rear Wheel Bearing Felt—Outer..................	4G0-183								✓	✓	2
	X-1677	.01	Screw, Flat Head Machine (¼"—20 x ⅝")..................	S-4265								✓	✓	4
	1229-W-283	.10	Washer, Rear Wheel Bearing Felt Retainer..................	4DG0-107							✓			2
	AT-6880	1.55	Nut, Rear Wheel Bearing Adjusting..................	4D0-550	✓	✓	✓	✓		✓	✓			2
	A-14X-15	1.50	Nut, Rear Wheel Bearing Adjusting..................	4G0-550								✓	✓	2
	T-3564	.84	Nut, Rear Wheel Bearing Jam..................	4D0-25	✓	✓	✓	✓		✓	✓			2
	14X-14	1.70	Nut, Rear Wheel Bearing Jam..................	4G0-683								✓	✓	2
	T-3840	.35	Washer, Rear Wheel Bearing Lock..................	4D0-166	✓	✓	✓	✓		✓	✓			2
	1229-V-48	.66	Washer, Rear Wheel Bearing Nut..................	4G0-166								✓	✓	2
	5-X-282	.60	Washer, Rear Wheel Bearing Felt..................	9DKA0-95	✓	✓	✓	✓		✓				2
	3256-E-5	1.35	Wrench, Rear Wheel Bearing Nut..................	16D0-41							✓			1
	3256-C-3	1.95	Wrench, Front and Rear Wheel Bearing Nut..................	4G0-696								✓	✓	1
	1805-Q-17	.20	Retainer, Rear Hub Bearing Felt..................	4DK0-183	✓	✓	✓	✓		✓				2
	1829-F-214	.23	Washer, Rear Hub Bearing Felt Retainer..................	4DK0-107	✓	✓	✓	✓		✓				2
	3897-H-190	12.10	Spacer, Rear Wheel..................	4UG0-559	✓	✓				✓				2
	532	5.04	Cup, Front Wheel Bearing—Inner..................	4Y2-137					✓					2
	539	6.99	Cone, Front Wheel Bearing—Inner..................	19T2-13					✓					2
	432	3.09	Cup, Front Wheel Bearing—Outer..................	4H2-137					✓					2
	444	5.20	Cone, Front Wheel Bearing—Outer..................	9N0-38					✓					2
	1244-E-83	1.10	Spacer, Front Wheel Bearing—Inner..................	9N0-525					✓					2
	A-1227-A-105	.85	Nut, Front Wheel Bearing Adjuster, & Dowel Assy. (1½" I.D.)..................	9TA0-130A					✓					2
	AT-1870	.85	Nut, Front Wheel Bearing Adjuster, & Dowel Assy. (1¼" I.D.)..................	9TA0-130					✓					2
	T-1860	.06	Dowel, Front Wheel Adjuster..................	4G0-639					✓					2
	1229-G-475	.21	Ring, Front Wheel Bearing Adjuster Nut Lock (1½"-12 Thds.)..................	9TA0-05A					✓					2

13-2 WHEEL ASSEMBLY — Master Parts List

THE AUTOCAR COMPANY, ARDMORE, PA.

CODE	MFR. PART No.	LIST PRICE	NAME OF PART	AUTOCAR PART No.	U-2044	U-4044	U-4144	U-4144-T	C-50	U-5044	U-7144-T	U-8144	U-8144-T	No. REQ.
1301	1229-L-246	.66	Ring, Front Wheel Bearing Adjuster Nut Lock (1¼″–7 Thds.)	9TA0-05					✓					2
	1229-B-106	.26	Nut, Front Wheel Bearing Adjuster Jam (1½″–12 Thds.)	9TA0-28A										2
	T-1872	.50	Nut, Front Wheel Bearing Adjuster Jam (1¼″–7 Thds.)	9TA0-28					✓					2
	1229-F-474	.07	Lock, Front Wheel Bearing Adjuster Jam Nut (1½″ I.D.)	9TA0-175A										2
	L-6088	.10	Lock, Front Wheel Bearing Adjuster Jam Nut (1¼″ I.D.)	9TA0-175					✓					2
	3256-J-10	.50	Wrench, Front Wheel Adjusting	16N0-117					✓					1
	1205-L-194	.21	Retainer, Front Wheel Bearing Oil Seal—Outer	9B0-17					✓					2
	1205-M-195	.17	Retainer, Front Wheel Bearing Oil Seal—Inner	9K0-17					✓					2
	5-X-181	.32	Felt, Front Wheel Bearing Oil Seal	9B0-95					✓					2
	749	15.60	Cone, Rear Wheel Bearing	4Y2-156					✓					4
	742	9.91	Cup, Rear Wheel Bearing	4Y2-157					✓					4
	......	.65	Washer, Rear Wheel Bearing Felt	4CHS2-526					✓					2
	......	.25	Washer, Rear Wheel Bearing	4CHS2-183					✓					2
	......	2.40	Nut, Rear Wheel Bearing Lock	4Y2-25					✓					2
	......	.50	Bolt, Rear Wheel Bearing Lock	4Y2-194					✓					2
	......	.04	Nut, Rear Wheel Bearing Lock Bolt	S-77					✓					2
	......	.01	Washer, Spring	S-1349					✓					4
	......	.30	Washer, Rear Wheel Bearing Lock	4Y2-166					✓					2
	......	.20	Key, Rear Wheel Bearing Lock Nut	4Y1-15					✓					2
1302	**Hubs and Drums**													
	A-322-A-105	28.80	Hub, Front Wheel, and Cup Assembly	9BP0-807	✓	✓	✓	✓		✓	✓			2
	A-322-C-107	46.75	Hub, Front Wheel, and Cup Assembly	9AK0-807								✓	✓	2
	20-X-20	.25	Stud, Front Wheel Hub—R. H.	4A3-489	✓	✓	✓	✓		✓	✓	✓	✓	10
	20-X-19	.25	Stud, Front Wheel Hub—L. H.	4A3-491	✓	✓	✓	✓		✓	✓	✓	✓	10
	1199-J-114	.32	Nut, Front and Rear Hub and Wheel Stud Cap—R. H.—Inner	4A0-505	✓	✓	✓	✓		✓				20
	1199-N-1054	.32	Nut, Front and Rear Wheel Hub Stud Cap—R. H.—Inner	4GKW0-505								✓	✓	20
	1199-N-118	.18	Nut, Front and Rear Hub and Wheel Stud Cap—R. H.—Outer	4A0-503	✓	✓	✓	✓		✓				20
	1199-L-1052	.18	Nut, Front and Rear Wheel Stud Cap—R. H.—Outer	4GLW0-503								✓	✓	20
	1199-K-115	.32	Nut, Front and Rear Hub and Wheel Stud Cap—L. H.—Inner	4A0-506	✓	✓	✓	✓		✓				20
	1199-M-1053	.32	Nut, Front and Rear Wheel Hub Stud Cap—L. H.—Inner	4GKW0-506								✓	✓	20
	1199-M-117	.18	Nut, Front and Rear Hub and Wheel Stud Cap—L. H.—Outer	4A0-504	✓	✓	✓	✓		✓				20
	1199-K-1051	.18	Nut, Front and Rear Wheel Hub Stud Cap—L. H.—Outer	4GLW0-504								✓	✓	20
	37888-E	.15	Nut, Front Wheel Budd Stud—R. H.	9A0-216	✓	✓	✓	✓		✓				10
	37889-E	.15	Nut, Front Wheel Budd Stud—L. H.	9A0-217	✓	✓	✓	✓		✓				10
	13332-E	.05	Nut, Front and Rear Hub Stud	4A0-499	✓	✓	✓	✓		✓				40
	13-X-28	.07	Nut, Front Wheel Hub and Drum Stud	4NFG0-1078							✓	✓	✓	20
	3819-K-11	12.40	Drum, Front Wheel Brake	9BP0-232	✓	✓	✓	✓		✓				2
	3819-M-169	19.40	Drum, Front Wheel Brake	9FKW0-232							✓			2
	3819-Q-17	16.80	Drum, Front Wheel Brake	9AK0-232								✓	✓	2
	1807-B-2	.06	Cover, Front Wheel Brake Drum Inspection	9SK0-514							✓	✓	✓	2
	S-244	.04	Screw, Hex. Cap (¼″–20 x ½″)	S-66							✓	✓	✓	2
	W-14	.01	Lockwasher, 9/32″ Spring	S-1340							✓	✓	✓	2
	......	41.20	Drum, Front Hub	9BP0-809	✓	✓	✓	✓		✓				2
	1898-B-210	.10	Drain, Front Hub Oil	4FKW0-1273							✓			2
	A-333-E-395	30.35	Hub, Rear, and Cups Assembly	4BP0-811	✓	✓	✓	✓		✓				2
	A-333-W-179	28.65	Hub, Rear Wheel, and Cup Assembly	4NFG0-811							✓			2
	A-133-A-157	38.65	Hub, Rear Wheel, and Cup Assembly	4G0-811A								✓	✓	2
	B31-333-W-179	41.50	Hub, Rear Wheel, and Stud Assembly—L. H.	4FKW0-801							✓			1
	A-79-333-A-157	50.75	Hub, Rear Wheel, and Stud Assembly—L. H.	4GKW0-801								✓	✓	1
	B30-333-W-179	41.50	Hub, Rear Wheel, and Stud Assembly—R. H.	4FKW0-801							✓			1
	A-78-333-A-157	50.75	Hub, Rear Wheel, and Stud Assembly—R. H.	4GKW0-801								✓	✓	1
	B33-333-W-179	66.05	Hub, Rear Wheel, and Drum Assembly—L. H.	4FKW0-809							✓			1
	A-81-333-A-157	73.40	Hub, Rear Wheel, and Drum Assembly—L. H.	4GKW0-809								✓	✓	1
	B32-333-W-179	66.05	Hub, Rear Wheel, and Drum Assembly	4FKW0-809							✓			1
	A-80-333-A-157	73.40	Hub, Rear Wheel, and Drum Assembly	4GKW0-809								✓	✓	1
	20-X-95	.50	Stud, Rear Wheel Hub—L. H.	4UK3-491	✓	✓	✓	✓		✓				10
	20-X-27	.49	Stud, Rear Wheel Hub—L. H.	4FKWA0-489							✓			10
	20-X-55	.40	Stud, Rear Wheel Hub—L. H.	4SD3-491A								✓	✓	10
	20-X-96	.50	Stud, Rear Wheel Hub—R. H.	4UK3-489	✓	✓	✓	✓		✓				10
	20-X-28	.49	Stud, Rear Wheel Hub—R. H.	4FKW0-489							✓			10
	20-X-54	.40	Stud, Rear Wheel Hub—R. H.	4SD3-489A								✓	✓	10

WHEEL ASSEMBLY

CODE	MFR. PART No.	LIST PRICE	NAME OF PART	AUTOCAR PART No.	MODEL									No. REQ.
					U-2044	U-4044	U-4144	U-4144-T	C-50	U-5044	U-7744-T	U-8144	U-8144-T	
1302	13-X-9	.06	**Nut,** Rear Wheel Hub and Drum Stud (¾"—16)............	S-847							✓	✓	✓	20
	1199-N-1054	.32	**Nut,** Rear Wheel Hub Stud Cap—R. H.—Inner............	4GKW0-505							✓			10
	1199-M-1053	.32	**Nut,** Rear Wheel Hub Stud Cap—L. H.—Inner............	4GKW0-506							✓			10
	1199-L-1052	.18	**Nut,** Rear Wheel Hub Stud Cap—R. H.—Outer............	4GKW0-503							✓			10
	1199-K-1051	.18	**Nut,** Rear Wheel Hub Stud Cap—L. H.—Outer............	4GKW0-504							✓			10
	......	14.85	**Drum,** Rear Wheel Brake...........................	4BP0-21	✓	✓	✓	✓		✓				2
	3219-D-1460	20.25	**Drum,** Rear Wheel Brake...........................	4FKW0-21							✓			2
	3219-K-583	18.75	**Drum,** Rear Wheel Brake...........................	4GK0-21								✓	✓	2
	3905-T-20	2.05	**Slinger,** Rear Wheel Brake Drum Oil.................	9UG0-513	✓	✓	✓	✓		✓				2
	3280-D-30	2.00	**Slinger,** Rear Wheel Brake Drum Oil.................	4NFG0-404							✓			2
	3280-K-505	1.60	**Slinger,** Rear Wheel Brake Drum Oil.................	4G0-404A								✓	✓	2
	1107-C-29	.06	**Cover,** Rear Wheel Brake Drum Inspection............	9N0-514							✓	✓	✓	2
	2X-19	.04	Screw, Hex. Cap (⅜"—16 x ½")..................	S-1461							✓	✓	✓	2
	X-529	.01	Lockwasher, ¹³⁄₃₂" Spring.........................	S-1347							✓	✓	✓	2
	......	1.15	**Wrench,** Front and Rear Wheel Nut...................	16SA3-744								✓	✓	1
	......	.85	**Handle,** Front and Rear Wheel Nut Wrench............	16UU0-743								✓	✓	1
	1199-J-114	.32	**Nut,** Front Hub and Wheel Stud Cap—R. H.—Inner.........	4A0-505							✓			10
	1199-N-118	.18	**Nut,** Front Hub and Wheel Stud Cap—R. H.—Outer.........	4A0-503							✓			10
	1199-K-115	.32	**Nut,** Front Hub and Wheel Stud Cap—L. H.—Inner.........	4A0-506							✓			10
	1199-M-117	.18	**Nut,** Front Hub and Wheel Stud Cap—L. H.—Outer.........	4A0-504							✓			10

THE AUTOCAR COMPANY, ARDMORE, PA.

Master Parts List

GROUP 14

ILLUSTRATIONS

GROUP 14

STEERING

GROUP 14

Master Parts List

THE AUTOCAR COMPANY, ARDMORE, PA.

As supplied on Autocar Models
U-8144   U-8144-T

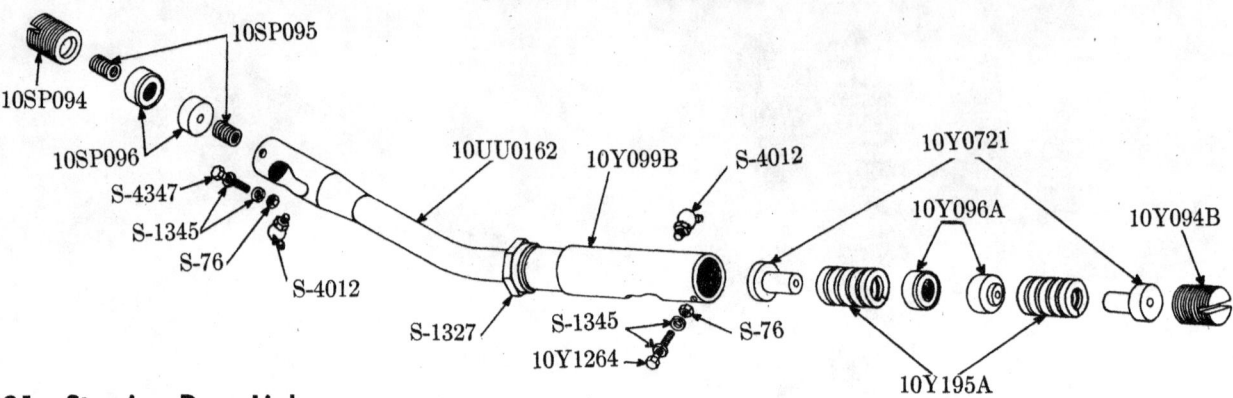

1401—Steering Drag Link

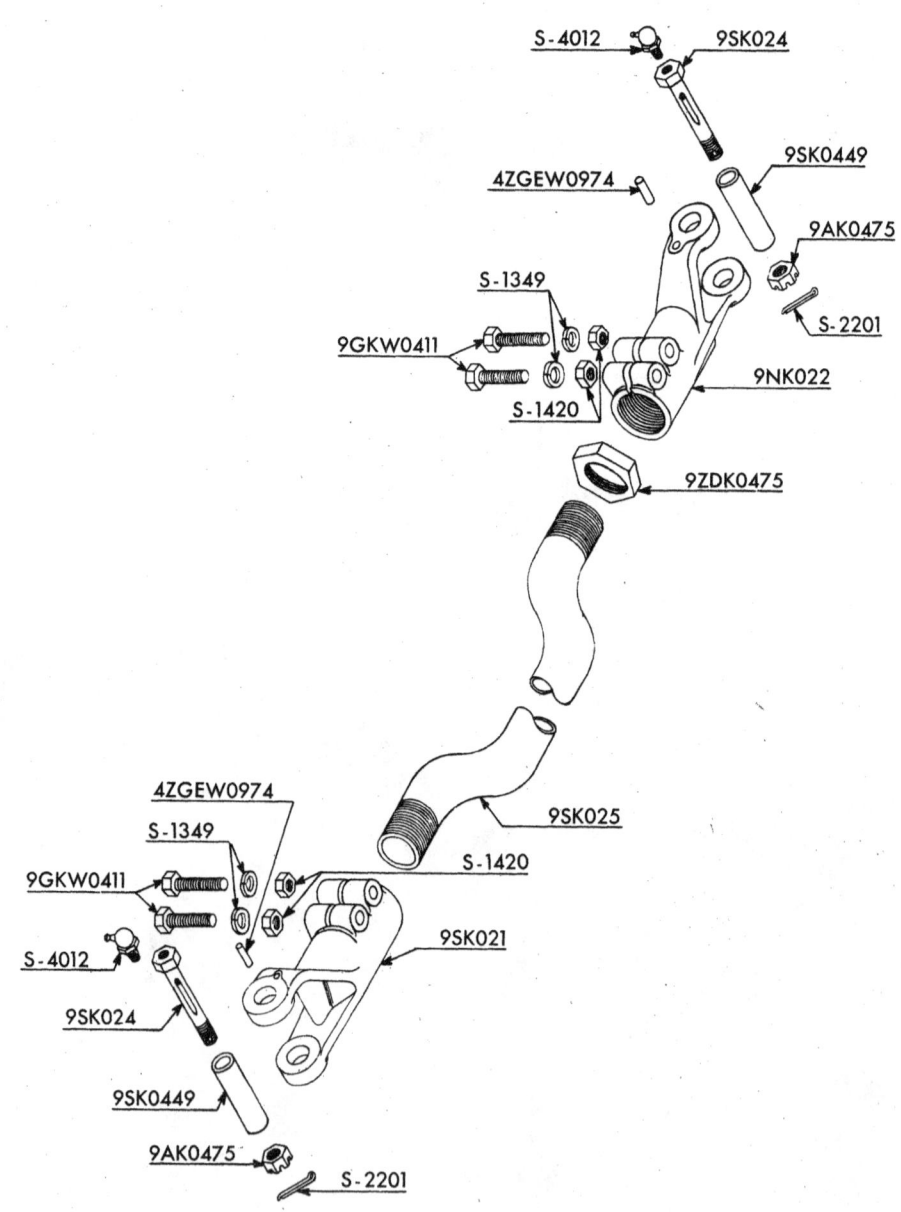

1402—Steering Tie Rod

THE AUTOCAR COMPANY,
ARDMORE, PA.

Master Parts List

GROUP 14

As supplied on Autocar Models
U-8144   U-8144-T

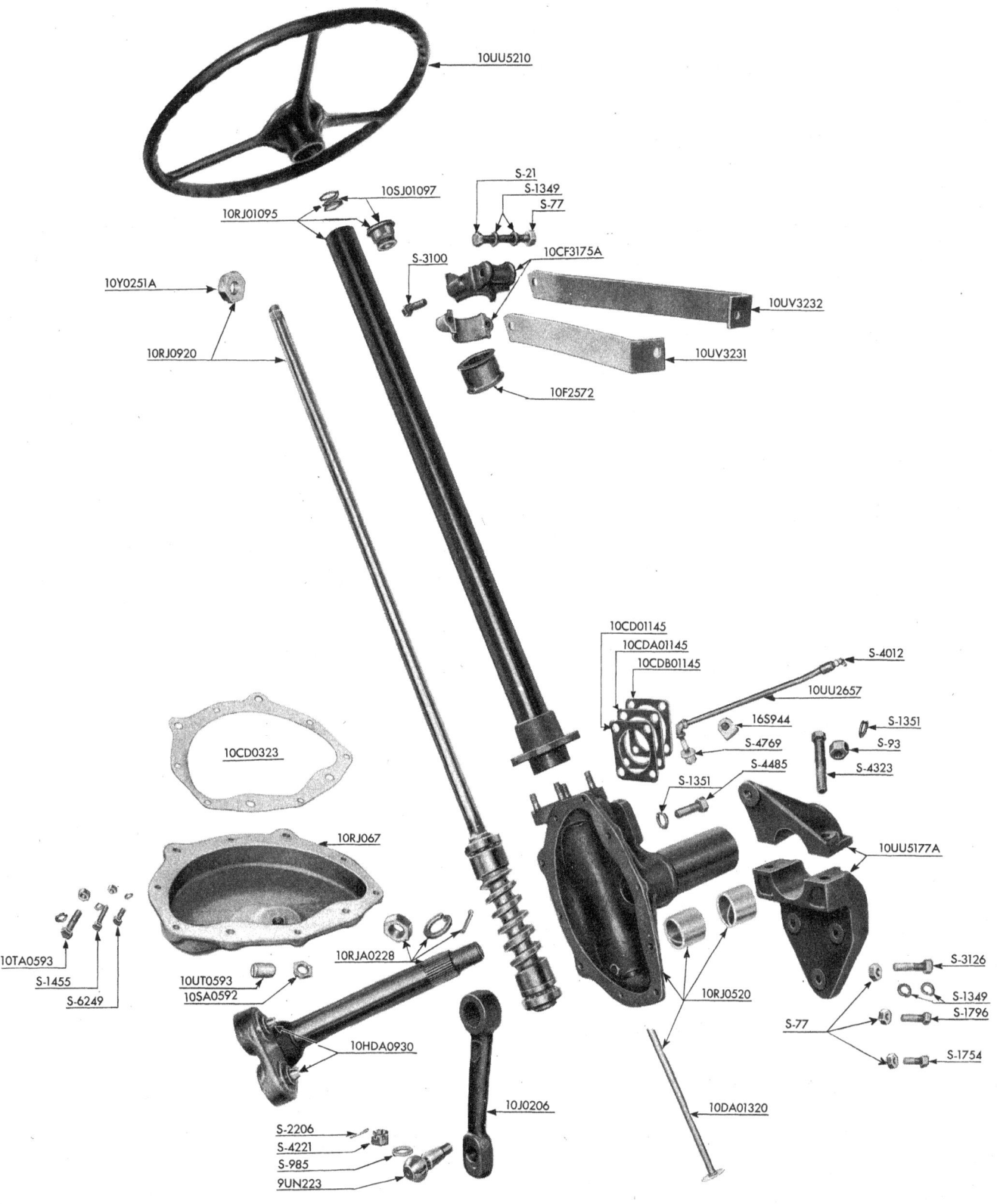

1403—Steering Gear Assembly

Master Parts List

CODE	MFR. PART No.	LIST PRICE	NAME OF PART	AUTOCAR PART No.	MODEL U-2044	U-4044	U-4144	U-4144-T	C-50	U-5044	U-7144-T	U-8144	U-8144-T	No. REQ.
1401	**Steering Drag Link**													
	......	14.00	**Link,** Drag, Assembly	**10UG3-10**	✓	✓	✓	✓		✓				1
	......	19.20	**Link,** Drag, Assembly	**10ZN4-10F**					✓					1
	......	20.00	**Link,** Drag, Assembly	**10UU3-10**							✓	✓	✓	1
	......	9.90	**Tube,** Drag Link	**10UG0-162**	✓	✓	✓	✓		✓				1
	......	12.00	**Tube,** Drag Link	**10ZN0-162F**					✓					1
	......	9.90	**Tube,** Drag Link	**10UU0-162**							✓	✓	✓	1
	......	.90	**Spring,** Drag Link Ball Seat	**10H1-95**	✓	✓	✓	✓	✓	✓				1
	......	.12	**Spring,** Drag Link Ball Seat	**10SP0-95**							✓	✓	✓	2
	......	.70	**Seat,** Drag Link Ball (1½" Dia. Ball)	**10H1-96**	✓	✓	✓	✓	✓	✓				2
	......	.55	**Seat,** Drag Link Ball (1¾" Dia. Ball)	**10SP0-96**							✓	✓	✓	2
	......	.90	**Plug,** Drag Link End	**10H1-94**	✓	✓	✓	✓	✓	✓				1
	......	.20	**Bolt,** Drag Link End	**10UF1-193A**	✓	✓	✓	✓	✓	✓				1
	......	.01	**Nut,** Drag Link End Bolt (¼"—20)	**S-76**	✓	✓	✓	✓	✓	✓				1
	......	.01	**Lockwasher,** ¼" Spring	**S-1345**	✓	✓	✓	✓	✓	✓				2
	......	.90	**Plug,** Drag Link End	**10SP0-94**							✓	✓	✓	1
	......	.06	**Bolt,** Drag Link End (¼"—20 x 2½")	**S-4347**							✓	✓	✓	1
	......	.01	**Lockwasher,** $\frac{9}{32}$" Spring	**S-1345**							✓	✓	✓	2
	......	.01	**Nut,** Drag Link End Bolt (¼"—20)	**S-76**							✓	✓	✓	1
	......	.06	**Nut,** Drag Link Adjustable End Lock (1¼"—18)	**S-2022**	✓	✓	✓	✓	✓	✓				1
	......	.25	**Nut,** Drag Link Adjustable End Lock (1½"—12)	**S-1327**							✓	✓	✓	1
	......	6.30	**End,** Drag Link Adjustable	**10ZN0-99**	✓	✓	✓	✓	✓	✓				1
	......	6.70	**Socket,** Drag Link	**10Y0-99B**							✓	✓	✓	1
	......	.85	**Bumper,** Spring	**10Y0-721**								✓	✓	2
	......	.90	**Spring,** Drag Link Ball Seat	**10H1-95**	✓	✓	✓	✓	✓	✓				2
	......	1.00	**Spring,** Drag Link Ball Seat	**10Y1-95A**							✓	✓	✓	2
	......	.70	**Seat,** Drag Link Ball (1½" Dia. Ball)	**10H1-96**	✓	✓	✓	✓	✓	✓				2
	......	.40	**Seat,** Drag Link Ball (1¾" Dia. Ball)	**10Y0-96A**							✓	✓	✓	2
	......	.90	**Plug,** Drag Link End	**10H1-94**	✓	✓	✓	✓	✓	✓				1
	......	.20	**Bolt,** Drag Link End	**10UF1-193A**	✓	✓	✓	✓	✓	✓				1
	......	.01	**Nut,** Drag Link End Bolt (¼"—20)	**S-76**	✓	✓	✓	✓	✓	✓				1
	......	.01	**Lockwasher,** ¼" Spring	**S-1345**	✓	✓	✓	✓	✓	✓				2
	......	.90	**Plug,** Drag Link End	**10Y0-94B**							✓	✓	✓	1
	......	.20	**Bolt,** Drag Link End (¼"—20 x 2¾")	**10Y1-264**							✓	✓	✓	1
	......	.01	**Lockwasher,** $\frac{9}{32}$" Spring	**S-1345**							✓	✓	✓	2
	......	.01	**Nut,** Drag Link End Bolt (¼"—20)	**S-76**							✓	✓	✓	1
	......	.20	**Fitting,** Alemite, 67½°	**S-4012**							✓	✓	✓	2
1402	**Steering Tie Rod**													
	3902-Z-52	16.20	**Rod,** Tie	**9UG0-25**	✓	✓	✓	✓		✓				1
	3902-R-18	20.45	**Tube,** Steering Cross	**9SK0-25**							✓	✓	✓	1
	3944-H-8	5.80	**End,** Steering Cross Tube (12 Threads)	**9SK0-21**	✓	✓	✓	✓		✓	✓	✓	✓	1
	3944-B-28	5.90	**End,** Steering Cross Tube (16 Threads)	**9NK0-22**	✓	✓	✓	✓		✓	✓	✓	✓	1
	1827-K-89	.55	**Nut,** Steering Cross Tube End Lock	**9ZDK0-475**	✓	✓	✓	✓		✓	✓	✓	✓	1
	1827-J-88	.25	**Nut,** Tie Rod Yoke	**9AK0-475**	✓	✓	✓	✓		✓				2
	S-1822	.07	**Bolt,** Steering Cross Tube End Clamp	**9GKW0-411**							✓	✓	✓	4
	N-18	.04	**Nut,** Steering Cross Tube End Clamp Bolt (½"—20)	**S-1420**							✓	✓	✓	4
	W-18	.01	**Lockwasher,** $\frac{17}{32}$" Spring	**S-1349**							✓	✓	✓	4
	10-X-216	1.35	**Pin,** Steering Cross Tube End	**9SK0-24**	✓	✓	✓	✓		✓	✓	✓	✓	2
	1246-L-272	.10	**Pin,** Steering Cross Tube End Pin Lock	**4AK0-974**							✓			2
	1246-L-220	.10	**Pin,** Steering Cross Tube End Pin Lock	**4ZGEW0-974**	✓	✓	✓	✓		✓		✓	✓	2
	1827-J-88	.25	**Nut,** Steering Cross Tube End Pin	**9AK0-475**							✓	✓	✓	2
	1825-S-71	.45	**Bushing,** Steering Cross Tube End	**9SK0-449**							✓	✓	✓	2
	3880-H-34	.35	**Slinger,** Front Dust Shield Oil	**9UG0-750**							✓			2
	A-3102-V-906	32.35	**Tube,** Steering Cross, Assembly (Incl. the following items)	**9AA0-20**					✓					1
	3102-V-906	7.05	**Tube,** Steering Cross—Only	**9AA0-25**					✓					1
	A-3144-J-114	10.25	**End,** Steering Cross Tube, Assembly—R. H. (1¼"—12 Thds.)	**9AA0-80**					✓					1
	A-3144-K-115	10.25	**End,** Steering Cross Tube, Assembly—L. H. (1¼"—14 Thds.)	**9AA0-90**					✓					1
	3144-J-114	10.25	**End,** Steering Cross Tube—R. H. Only (1¼"—12 Thds.)	**9A0-344**					✓					1
	3144-K-115	10.25	**End,** Steering Cross Tube—L. H. Only (1¼"—14 Thds.)	**9A0-261**					✓					1
	1246-A-261	.12	**Dowel,** Steering Cross Tube End	**9A0-545**					✓					2
	1108-P-42	.25	**Seat,** Steering Cross Tube End Ball	**9A0-84**					✓					2

THE AUTOCAR COMPANY
ARDMORE, PA.

Master Parts List

14-1 STEERING

CODE	MFR. PART No.	LIST PRICE	NAME OF PART	AUTOCAR PART No.	U-2044	U-4044	U-4144	U-4144-T	C-50	U-5044	U-7144-T	U-8144	U-8144-T	No. REQ.
1402	2258-E-317	.15	**Spring,** Steering Cross Tube End Ball	9B0-85					✓					2
	1250-F-110	.15	**Plug,** Steering Cross Tube End	9A0-76					✓					2
	1254-P-68	.18	**Lock,** Steering Cross Tube End Plug	9A0-515					✓					2
	1107-F-32	.06	**Cover,** Steering Cross Tube End Dust	9SA0-341					✓					2
	5-X-429	.12	**Rubber,** Steering Cross Tube End Dust	9A0-516					✓					2
	1107-H-60	.12	**Retainer,** Steering Cross Tube End Dust Rubber	9J0-17					✓					2
	1218-X-50	.07	**Spring,** Steering Cross Tube End Cover	9SA0-342					✓					2
	A-2110-D-56	2.35	**Stud,** Steering Cross Tube End, and Bearing Assembly	9A0-544					✓					2
	13453	.16	**Nut,** Steering Cross Tube End Ball Pin	9TA0-294					✓					2
	3-X-45	.09	**Bolt,** End Clamp	9TA0-237					✓					4
	13-X-7	.10	**Nut,** End Clamp Bolt	S-1311					✓					4
	X-533	.01	**Lockwasher,** Spring	S-1351					✓					4
1403	**Steering Gear Assembly**													
	T-71044	91.10	**Gear,** Steering, Complete (Ross)	10UK0-770	✓	✓	✓	✓		✓				1
	T-71049	82.15	**Gear,** Steering, Complete	10BCB0-770					✓					1
	T-72152	84.55	**Gear,** Steering, Complete	10SJ0-770							✓			1
	T-75104	140.77	**Gear,** Steering, Complete	10RJ0-770								✓	✓	1
	502955	16.06	**Housing,** Steering Gear, Assembly	10UK0-520	✓	✓	✓	✓		✓				1
	503084	14.76	**Housing,** Steering Gear, Assembly	10BCA0-520					✓					1
	069001	.30	**Bushing,** Housing—Inner	10TE0-215	✓	✓	✓	✓	✓	✓				1
	069002	.40	**Bushing,** Housing—Outer	10TE0-216	✓	✓	✓	✓	✓	✓				1
	7859-13	.40	**Tube,** End Cover and Oil Seal, Assembly	10UK0-1320	✓	✓	✓	✓	✓	✓				1
	503472	35.96	**Housing,** Steering Gear, Assembly	10SJ0-520							✓			1
	070001	.60	**Bushing,** Housing—Inner	10GE0-215							✓			1
	070002	.60	**Bushing,** Housing—Outer	10GE0-216							✓			1
	7859-17	.46	**Tube,** End Cover and Oil Seal, Assembly	10SJ0-1320							✓			1
	022023	.10	**Stud,** Housing (For Upper Cover)	10CD0-307							✓			4
	......	.04	**Nut,** Hex. (½"—20)	S-1420							✓			4
	......	.01	**Lockwasher,** $\frac{17}{32}$" Spring	S-1349							✓			4
	503743	30.76	**Housing,** Steering Gear, Assembly (Incl. next 4 items)	10RJ0-520								✓	✓	1
	022023	.10	**Stud,** Housing (For Upper Cover)	10CD0-307								✓	✓	4
	070001	.60	**Bushing,** Housing—Inner	10GE0-215								✓	✓	1
	070002	.60	**Bushing,** Housing—Outer	10GE0-216								✓	✓	1
	7859-18	.46	**Tube,** End Cover and Oil Seal, Assembly	10DA0-1320								✓	✓	1
	......	.14	**Pipe,** Oiler	10UU2-657								✓	✓	1
	......	.06	**Clip**	16S9-44								✓	✓	1
	......	.01	**Screw,** Round Head Machine (¼"—20 x ¾")	S-1614A								✓	✓	1
	......	.01	**Nut,** Hex. (¼"—20)	S-76								✓	✓	1
	......	.01	**Lockwasher,** $\frac{9}{32}$" Spring	S-1345								✓	✓	2
	......	.17	**Ell,** ⅛"—90° Pipe	S-4590								✓	✓	1
	......	.11	**Bushing,** ½" x ⅛" Pipe	S-4769								✓	✓	1
	......	.06	**Nipple,** ⅛" x 1½" Pipe	S-4144								✓	✓	1
	......	.04	**Coupling,** ⅛" Pipe	S-4664								✓	✓	1
	......	.20	**Fitting,** ⅛"—67½° Alemite	S-4012								✓	✓	1
	032074	.45	**Unit,** Housing Oil Seal	10UK0-1094	✓	✓	✓	✓	✓	✓				1
	032078	.45	**Seal,** Housing Leather Oil, Assembly	10CDKL0-1094							✓	✓	✓	1
	T-705000	2.40	**Cover,** Housing Side	10UK0-67	✓	✓	✓	✓	✓	✓				1
	021016	.18	**Screw,** Adjusting	10SA0-593	✓	✓	✓	✓	✓	✓				1
	025006	.06	**Nut,** Adjusting Screw Lock	10SA0-592	✓	✓	✓	✓	✓	✓				1
	T-709000	.04	**Gasket,** Side Cover	10UK0-323	✓	✓	✓	✓	✓	✓				1
	020040	.06	**Screw,** Hex. Cap (⅜"—24 x 1½")	S-1455	✓	✓	✓	✓	✓	✓				6
	T-73	.05	**Screw,** Hex. Cap (⅜"—24 x 1")	S-1445	✓	✓	✓	✓	✓	✓				4
	......	.02	**Nut,** Hex. (⅜"—24)	S-814	✓	✓	✓	✓	✓	✓				4
	......	.01	**Lockwasher,** $\frac{13}{32}$" Spring	S-1347	✓	✓	✓	✓	✓	✓				10
	......	.12	**Plug,** Pipe (⅜")	S-4004	✓	✓	✓	✓	✓	✓				2
	502480	2.76	**Cover,** Housing Side, Assembly	10SJ0-1790							✓			1
	T-715001	5.40	**Cover,** Housing Side	10HDA0-67A							✓			1
	021020	.30	**Screw,** Adjusting	10UT0-593							✓			1
	025006	.06	**Nut,** Adjusting Screw Lock	10SA0-592							✓			1
	T-719000	.07	**Gasket**	10HDA0-323							✓			1
	021102	.05	**Screw,** Hex. Cap (½"—20 x 1¾")	10TA0-593							✓			4

Master Parts List

CODE	MFR. PART No.	LIST PRICE	NAME OF PART	AUTOCAR PART No.	MODEL									No. REQ.
					U-2044	U-4044	U-4144	U-4144-T	C-50	U-5044	U-7144-T	U-8144	U-8144-T	
1403	......	.04	**Nut,** Hex. (½"—20)	**S-1420**							✓			4
	......	.01	**Lockwasher,** $\frac{17}{32}$" Spring	**S-1349**							✓			4
	020040	.06	**Screw,** Hex. Cap (⅜"—24 x 1½")	**S-1455**							✓			4
	021118	.03	**Nut,** Hex. (⅜"—24)	**S-6249**							✓			4
	......	.02	**Nut,** Hex. (⅜"—24)	**S-814**							✓			4
	......	.01	**Lockwasher,** $\frac{13}{32}$" Spring	**S-1347**							✓			6
	502520	6.36	**Cover,** Housing Side, Assembly (Next 3 items)	10RJ0-**1790**								✓	✓	1
	T-745012	6.00	**Cover,** Housing Side	10RJ0-**67**								✓	✓	1
	021020	.30	**Screw,** Adjusting	10UT0-**593**								✓	✓	1
	025006	.06	**Nut,** Adjusting Screw Lock	10SA0-**592**								✓	✓	1
	T-749000	.07	**Gasket**	10CD0-**323**								✓	✓	1
	021102	.05	**Screw,** Hex. Cap (½"—20 x 1¾")	10TA0-**593**								✓	✓	4
	......	.05	**Nut,** Hex. (½"—20)	**S-811**								✓	✓	4
	......	.01	**Lockwasher,** $\frac{17}{32}$" Spring	**S-1349**								✓	✓	4
	020040	.06	**Screw,** Hex. Cap (⅜"—24 x 1½")	**S-1455**								✓	✓	4
	......	.06	**Nut,** Hex. (⅜"—24)	**S-2020**								✓	✓	4
	021118	.03	**Screw,** Hex. Cap (⅜"—24 x 1⅛")	**S-6249**								✓	✓	2
	......	.01	**Lockwasher,** $\frac{13}{32}$" Spring	**S-1347**								✓	✓	6
	8030-50$\frac{1}{16}$	15.58	**Tube,** Cam and Wheel, Assembly with Bearings and Wheel Nut	10UK0-**920**	✓	✓	✓	✓		✓				1
	8029-50$\frac{1}{16}$	13.60	**Tube,** Cam and Wheel, Assembly with Wheel Nut	10UK0-**550**	✓	✓	✓	✓		✓				1
	8030-50$\frac{11}{16}$	....	**Tube,** Cam and Wheel, Assembly with Bearings and Wheel Nut	....					✓					1
	8029-50$\frac{11}{16}$	....	**Tube,** Cam and Wheel, Assembly with Wheel Nut	....					✓					1
	C-20	.10	**Nut,** Wheel	10SA0-**251**	✓	✓	✓	✓	✓	✓				1
	400021	.70	**Cup,** Ball	10UK0-**224**	✓	✓	✓	✓	✓	✓				2
	400005	.01	**Ring,** Ball Retaining	10SA0-**227**	✓	✓	✓	✓	✓	✓				2
	400014	.03	**Balls,** Steel (⅜")	**S-1670**	✓	✓	✓	✓	✓	✓				28
	8214-52$\frac{5}{16}$	27.34	**Tube,** Cam and Wheel, Assembly with Bearings and Wheel Nut	10SJ0-**920**							✓			1
	8213-52$\frac{5}{16}$	23.66	**Tube,** Cam and Wheel, Assembly with Wheel Nut	10SJ0-**550**							✓			1
	8395-52$\frac{11}{16}$	27.74	**Tube,** Cam and Wheel, Assembly with Bearings and Wheel Nut (Next 5 Items)	10RJ0-**920**								✓	✓	1
	8394-52$\frac{11}{16}$	24.86	**Tube,** Cam and Wheel, Assembly with Wheel Nut (Next Item)	10RJ0-**550**								✓	✓	1
	E-20	.16	**Nut,** Wheel	10Y0-**251A**							✓	✓	✓	1
	400002	1.00	**Cup,** Ball	10C0-**224**							✓	✓	✓	2
	400001	.02	**Ring,** Ball Retaining	10SCK0-**227**							✓	✓	✓	2
	400015	.04	**Ball** ($\frac{7}{16}$")	**S-1671**							✓	✓	✓	28
	7789-39	3.10	**Tube,** Jacket, and Bearing Assembly	10UK0-**1176**	✓	✓	✓	✓		✓				1
	7789-41⅝	3.00	**Tube,** Jacket, and Bearing Assembly	10BCA0-**1176**					✓					1
	065996	.70	**Bearing,** Jacket Tube	10UK0-**1097**	✓	✓		✓	✓	✓				1
	401105	.03	**Spring**	10UK0-**1178**	✓	✓		✓	✓	✓				1
	T-266000	2.60	**Cover,** Upper	10UK0-**1143**	✓	✓	✓	✓	✓	✓				1
	033042	.02	**Shim,** Upper Cover Brass (.002)	10UK0-**1145**	✓	✓	✓	✓	✓	✓				As
	033036	.03	**Shim,** Upper Cover Brass (.003)	10UKA0-**1145**	✓	✓	✓	✓	✓	✓				As
	033037	.03	**Shim,** Upper Cover Brass (.010)	10UKB0-**1145**	✓	✓	✓	✓	✓	✓				As
	7845-40	7.75	**Tube,** Jacket, Jacket Tube Ball Bear. Unit & Upper Cover Assy.	10SJ0-**1095**							✓			1
	7153-39½	5.35	**Tube,** Jacket, and Jacket Tube Ball Bearing Unit Assembly	10SJ0-**970**							✓			1
	7845-39⅛	7.75	**Tube,** Jacket, Jacket Tube Ball Bearing Unit and Upper Cover Assembly (Next 5 Items)	10RJ0-**1095**								✓	✓	1
	7153-38⅝	5.35	**Unit,** Jacket Tube and Jacket Tube Ball Bearing (Next 3 Items)	10RJ0-**970**								✓	✓	1
	066994	1.15	**Unit,** Jacket Tube Ball Bearing (Next 2 Items)	10SJ0-**1097**							✓	✓	✓	1
	401103	.04	**Spring**	10HDR0-**1178**							✓	✓	✓	1
	401076	.01	**Seat,** Spring	10BM0-**227**							✓	✓	✓	1
	T-716000	2.40	**Cover,** Upper	10CD0-**1143**							✓	✓	✓	1
	......	.04	**Nut,** Hex. (½"—20)	**S-1420**							✓	✓	✓	4
	......	.01	**Lockwasher,** $\frac{17}{32}$" Spring	**S-1349**							✓	✓	✓	4
	033059	.07	**Shim,** Brass (.002")	10CD0-**1145**							✓	✓	✓	As
	033060	.06	**Shim,** Steel (.003")	10CDA0-**1145**							✓	✓	✓	As
	033061	.06	**Shim,** Steel (.010")	10CDB0-**1145**							✓	✓	✓	As
	7920-9-¾	27.80	**Unit,** Lever Shaft & Roller Bearing—with Nut & Lockwasher	10UK0-**940**	✓	✓	✓	✓		✓				1
	7920-7-¾	27.80	**Unit,** Lever Shaft & Roller Bearing—with Nut & Lockwasher	10BCA0-**940**					✓					1
	7919-9¾	17.80	**Shaft,** Lever, with Nut and Lockwasher	10UK0-**1070**	✓	✓	✓	✓		✓				1
	7919-7-¾	17.80	**Shaft,** Lever, with Nut and Lockwasher	10BCA0-**1070**					✓					1
	025003	.28	**Nut,** Lever Shaft	10TE0-**314**	✓	✓	✓	✓	✓	✓				1

CODE	MFR. PART No.	LIST PRICE	NAME OF PART	AUTOCAR PART No.	MODEL									No. REQ.
					U-2044	U-4044	U-4144	U-4144-T	C-50	U-5044	U-7144-T	U-8144	U-8144-T	
1403	044987	10.00	**Unit,** Stud Roller Bearing	10UK0-930	√	√	√	√	√	√				2
	7847-11¾	32.62	**Units,** Levershaft, Stud—Roller Bearing, Nut & Lockwasher Assembly (Next 7 Items)	10SJ0-228							√			1
	7941-11¾	39.12	**Units,** Lever Shaft, Stud—Roller Bearing, Nut & Lockwasher Assembly (Next 7 Items)	10RJ0-228								√	√	1
	7846-11¾	22.62	**Levershaft,** Nut and Lockwasher Assembly	10SJA0-228							√			1
	7940-11¾	29.12	**Levershaft,** Nut and Lockwasher Assembly (Next 3 Items)	10RJA0-228								√	√	1
	I-61-L	.26	**Nut,** Levershaft	10Y3-14							√	√	√	1
	......	.16	**Lockwasher,** 1¼″ Spring	S-1358							√	√	√	1
	......	.01	**Cotter,** 3/16″ x 1¾″	S-2214							√	√	√	1
	044988	10.00	**Unit,** Stud—Roller Bearing	10HDA0-930							√	√	√	1 Pr.
			Note:—This bearing unit must be serviced with matched pairs. Each unit will have the number 1, 2, 3, 4, or 5, etched on it. The two units in a shaft must be of the same number. Two matched units will be packed together for service. Individual parts that are serviceable are only the 2 following:											
	025056	.35	**Nut**	10SJ0-314							√	√	√	2
	028017	.02	**Washer,** Pronged	10C0-598							√	√	√	2
	T-705000	14.75	**Ball,** Steering Arm and Ball Pin, Assembly	10UK0-1370	√	√	√	√		√				1
	......	4.65	**Arm,** Steering	10BCA0-206					√					1
	......	2.40	**Pin,** Ball	9ZA-223					√					1
	......	.12	**Nut,** Ball Pin (¾″—16)	S-4221					√					1
	......	6.50	**Arm,** Steering	10J0-206							√	√	√	1
	......	3.00	**Stud,** Ball	9UN2-23							√	√	√	1
	......	.12	**Nut,** Hex. (¾″—16)	S-4221							√	√	√	1
	......	.03	**Washer,** Flat St. (25/32″ x 1½″ x ⅛″)	S-985							√	√	√	1
	......	.01	**Cotter,** ⅛″ x 1½″	S-2206							√	√	√	1
1404	**Steering Wheel**													
	......	9.00	**Wheel,** Steering	10J4-210	√	√	√	√	√	√				1
	......	9.00	**Wheel,** Steering	10UU5-210							√	√	√	1
	C-20	.10	**Nut,** Steering Wheel	10SA0-251	√	√	√	√	√	√				1
	E-20	.10	**Nut,** Steering Wheel	10Y0-251A							√	√	√	1
	038601	.02	**Key,** Steering Wheel	10M2-254	√	√	√	√	√	√				1
	......	.02	**Key,** Steering Wheel	10M0-254							√	√	√	1
1405	**Steering Gear Brackets**													
	......	7.50	**Bracket,** Steering Gear Frame, and Cap	10UU5-177A							√	√	√	1
	......	.12	**Screw,** Hex. Cap (½″—13 x 1⅝″)	S-1754							√	√	√	1
	......	.12	**Screw,** Hex. Cap (½″—13 x 1⅞″)	S-1796							√	√	√	1
	......	.20	**Screw,** Hex. Cap (½″—13 x 2¼″)	S-3126							√	√	√	1
	......	.04	**Nut,** Hex. (½″—13)	S-77							√	√	√	3
	......	.01	**Lockwasher,** 17/32″ Spring	S-1349							√	√	√	6
	......	.25	**Screw,** Hex. Cap (⅝″—11 x 4¼″)	S-4323							√	√	√	2
	......	.20	**Screw,** Hex. Cap (⅝″—11 x 2″)	S-4485							√	√	√	1
	......	.05	**Nut,** Hex. (⅝″—11)	S-93							√	√	√	3
	......	.01	**Lockwasher,** 21/32″ Spring	S-1351							√	√	√	6
	......	3.60	**Bracket,** Steering Column Top	10CF3-175A							√	√	√	1
	......	.16	**Bushing,** Rubber	10F2-572							√	√	√	1
	......	.05	**Screw,** Hex. Cap (½″—13 x 1¾″)	S-3100							√	√	√	2
	......	.15	**Screw,** Hex. Cap (½″—13 x 3⅞″)	S-21							√	√	√	1
	......	.04	**Nut,** Hex. (½″—13)	S-77							√	√	√	1
	......	.01	**Lockwasher,** 17/32″ Spring	S-1349							√	√	√	4
	......	.45	**Brace,** Steering Column—R. H.	10UV3-231							√	√	√	1
	......	.45	**Brace,** Steering Column—L. H.	10UV3-232							√	√	√	1
	......	.09	**Screw,** Hex. Cap (½″—13 x 1⅜″)	S-1428							√	√	√	2
	......	.04	**Nut,** Hex. (½″—13)	S-77							√	√	√	2
	......	.01	**Lockwasher,** 17/32″ Spring	S-1349							√	√	√	4

THE AUTOCAR COMPANY, ARDMORE, PA.

Master Parts List

GROUP 15

ILLUSTRATIONS

GROUP 15

FRAME and BRACKETS

GROUP 15

Master Parts List

THE AUTOCAR COMPANY, ARDMORE, PA.

As supplied on Autocar Models
U-8144-T

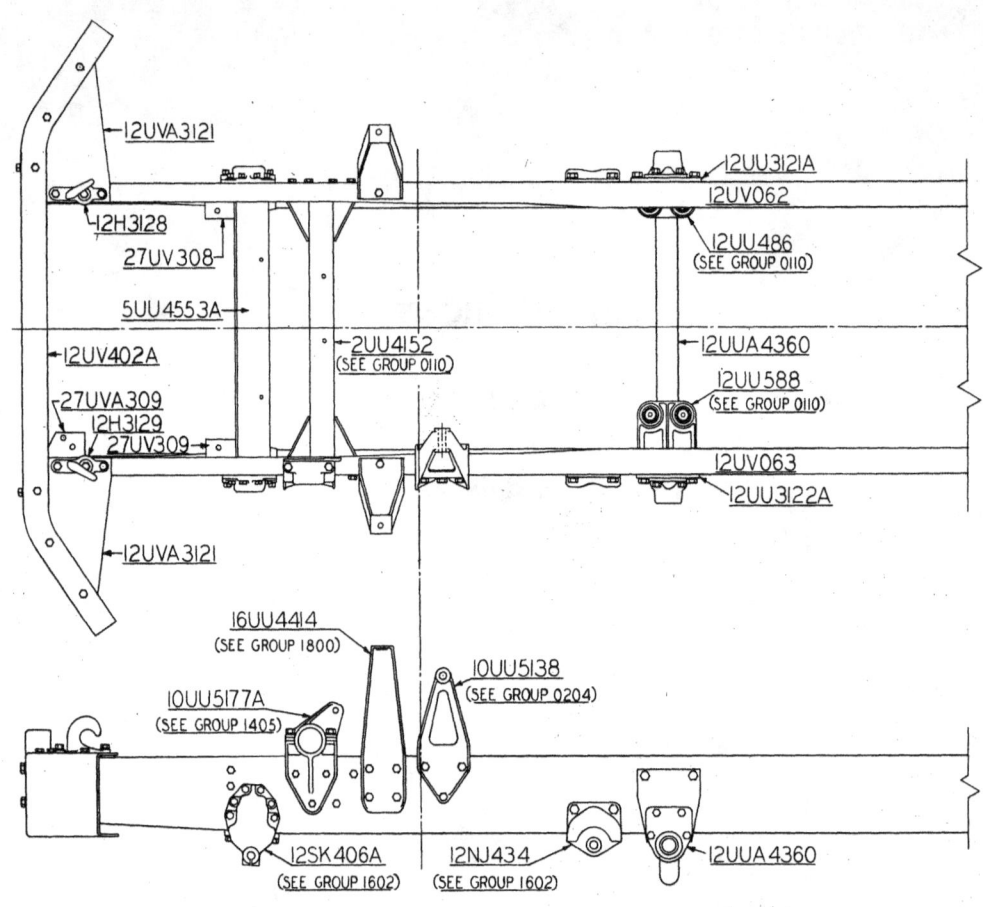

1500—Frame and Brackets—Front Half

THE AUTOCAR COMPANY, ARDMORE, PA.

Master Parts List

GROUP 15

As supplied on Autocar Models
U-8144-T

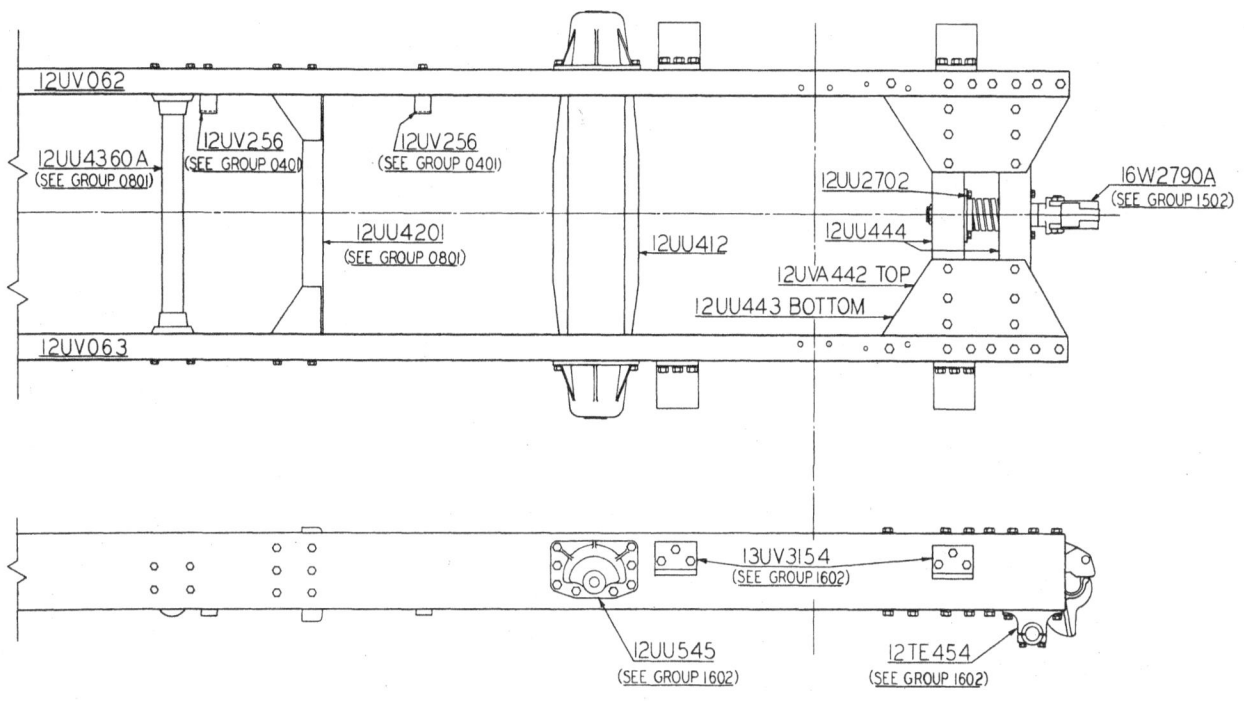

1500—Frame and Brackets – Rear Half

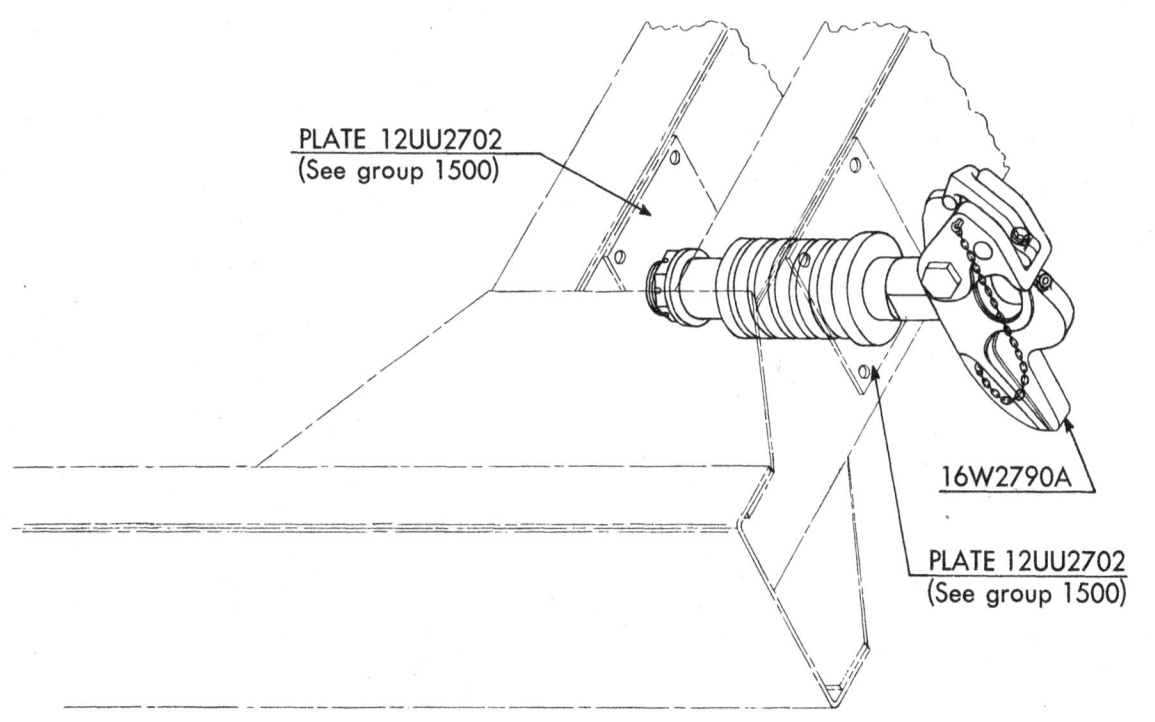

1502—Pintle Hook

15-0 FRAME

Master Parts List

THE AUTOCAR COMPANY, ARDMORE, PA.

CODE	MFR. PART No.	LIST PRICE	NAME OF PART	AUTOCAR PART No.	U-2044	U-4044	U-4144	U-4144-T	C-50	U-5044	U-7144-T	U-8144	U-8144-T	No. REQ.
1500			**Frame and Brackets**											
	.....	155.00	Frame with Front and Rear Cross Members only	12UK7-10B	✓									1
	.....	158.00	Frame with Front and Rear Cross Members only	12UG7-10B		✓								1
	.....	153.00	Frame with Front and Rear Cross Members only	12UG7-10C			✓							1
	.....	158.00	Frame with Front and Rear Cross Members only	12UGTA7-10				✓						1
	.....	163.50	Frame with Front and Rear Cross Members only	12UP7-10						✓				1
	.....	59.00	Rail, Frame—R. H. (Drilled)	12UK0-62B	✓									1
	.....	59.00	Rail, Frame—R. H. (Drilled)	12UG0-62		✓								1
	.....	59.00	Rail, Frame—R. H. (Drilled)	12UG0-62C			✓							1
	.....	59.00	Rail, Frame—R. H. (Drilled)	12UGTA0-62				✓						1
	.....	59.00	Rail, Frame—R. H. (Drilled)	12UP0-62						✓				1
	.....	66.00	Rail, Frame—R. H.	12UU0-62							✓			1
	.....	117.50	Rail, Frame—R. H.	12UV0-62								✓	✓	1
	.....	59.00	Rail, Frame—L. H. (Drilled)	12UK0-63B	✓									1
	.....	59.00	Rail, Frame—L. H. (Drilled)	12UG0-63		✓								1
	.....	59.00	Rail, Frame—L. H. (Drilled)	12UG0-63C			✓							1
	.....	59.00	Rail, Frame—L. H. (Drilled)	12UGTA0-63				✓						1
	.....	59.00	Rail, Frame—L. H. (Drilled)	12UP0-63						✓				1
	.....	66.00	Rail, Frame—L. H.	12UU0-63							✓			1
	.....	117.50	Rail, Frame—L. H.	12UV0-63								✓	✓	1
	.....	9.00	Bumper	12UK4-02	✓									1
	.....	9.00	Bumper	12UG4-02		✓	✓	✓		✓				1
	.....	25.60	Bumper, Front	12UU4-02							✓			1
	.....	13.50	Bumper, Front	12UV4-02A								✓	✓	1
	.....	.60	Bracket, Bumper—R. H.	12UK3-03	✓	✓	✓	✓		✓				1
	.....	.60	Bracket, Bumper—L. H.	12UK3-04	✓	✓	✓	✓		✓				1
	.....	.90	Bracket, Front Bumper	12UU3-03A							✓			2
	.....	.15	Screw, Hex. Cap (5/8"—11 x 3/4")	S-4073							✓			10
	.....	.05	Nut, Hex. (5/8"—11)	S-93							✓			10
	.....	.01	Lockwasher, 5/8" Spring	S-1351							✓			20
	.....	1.95	Bracket, Front Bumper	12UV3-03								✓	✓	2
	.....	.15	Screw, Hex. Cap (5/8"—11 x 2")	S-3387								✓	✓	8
	.....	.25	Screw, Hex. Cap (5/8"—11 x 2 1/4")	S-4350								✓	✓	2
	.....	.05	Nut, Hex. (5/8"—11)	S-93								✓	✓	10
	.....	.01	Lockwasher, 21/32" Spring	S-1351								✓	✓	20
	.....	3.45	Cross Member, Front	12UK3-82	✓									1
	.....	3.50	Cross Member, Front	12UG3-82		✓	✓	✓		✓				1
	.....	.40	Gusset, Front Cross Member—Top	12UG3-121	✓	✓	✓	✓		✓				2
	.....	.65	Gusset, Bumper—R. H.	12UUA3-121							✓			1
	.....	.35	Gusset, Front Cross Member—Bottom	12UGA3-122	✓	✓	✓	✓						2
	.....	.40	Gusset, Front Cross Member—Bottom	12UG3-122						✓				2
	.....	.65	Gusset, Bumper—R. H.	12UUA3-121							✓			1
	.....	.65	Gusset, Bumper—L. H.	12UUA3-122							✓			1
	.....	.09	Screw, Hex. Cap (1/2"—13 x 1 5/8")	S-1428							✓			6
	.....	.10	Screw, Hex. Cap (1/2"—13 x 1 5/8")	S-1839							✓			2
	.....	.04	Nut, Hex. (1/2"—13)	S-77							✓			8
	.....	.01	Lockwasher, 1/2" Spring	S-1349							✓			16
	.....	1.25	Gusset, Front Bumper	12UVA3-121								✓	✓	2
	.....	.09	Screw, Hex. Cap (1/2"—13 x 1 5/8")	S-1428								✓	✓	6
	.....	.04	Nut, Hex. (1/2"—13)	S-77								✓	✓	6
	.....	.01	Lockwasher, 17/32" Spring	S-1349								✓	✓	12
	.....	.80	Screw, Hex. Cap (5/8"—11 x 1 5/8")	S-1838								✓	✓	2
	.....	.05	Nut, Hex. (5/8"—11)	S-93								✓	✓	2
	.....	.01	Lockwasher, 21/32" Spring	S-1351								✓	✓	4
	.....	.85	Bracket, Winch Support—Front	27UVA3-09								✓	✓	1
	.....	.15	Screw, Hex. Cap (5/8"—11 x 1 3/4")	S-4073								✓	✓	1
	.....	.05	Nut, Hex. (5/8"—11)	S-93								✓	✓	1
	.....	.01	Lockwasher, 21/32" Spring	S-1351								✓	✓	2
	.....	1.05	Bracket, Winch Support—R. H.—Rear	27UV3-08								✓	✓	1
	.....	1.05	Bracket, Winch Support—L. H.—Rear	27UV3-09								✓	✓	1

THE AUTOCAR COMPANY
ARDMORE, PA.

Master Parts List

15-1 FRAME

CODE	MFR. PART No.	LIST PRICE	NAME OF PART	AUTOCAR PART No.	U-2044	U-4044	U-4144	U-4144-T	C-50	U-5044	U-7144-T	U-8144	U-8144-T	No. REQ.
1500		4.35	Cross Member, Radiator Support	5UU4-553A							✓	✓	✓	1
		.09	Screw, Hex. Cap (½″—13 x 1⅜″)	S-1428							✓	✓	✓	4
		.04	Nut, Hex. (½″—13)	S-77							✓	✓	✓	4
		.01	Lockwasher, $\frac{17}{32}$″ Spring	S-1349							✓	✓	✓	8
		2.40	Plate, Front Tubular Cross Member—R. H.	12UU3-121A							✓	✓	✓	1
		.10	Screw, Hex. Cap (½″—13 x 1⅝″)	S-1839							✓	✓	✓	2
		.04	Nut, Hex. (½″—13)	S-77							✓	✓	✓	2
		.01	Lockwasher, $\frac{17}{32}$″ Spring	S-1349							✓	✓	✓	4
		2.15	Plate, Front Tubular Cross Member—L. H.	12UU3-122A							✓	✓	✓	1
		27.00	Cross Member, Front Tubular	12UUA4-360							✓	✓	✓	1
		.20	Shim, Front Tubular Cross Member (.037″)	12UU9-206							✓	✓	✓	As
		.20	Shim, Front Tubular Cross Member (.062″)	12UUA9-206							✓	✓	✓	As
		.09	Screw, Hex. Cap (½″—13 x 1¼″)	S-25							✓	✓	✓	4
		.04	Nut, Hex. (½″—13)	S-77							✓	✓	✓	4
		.01	Lockwasher, $\frac{17}{32}$″ Spring	S-1349							✓	✓	✓	8
		10.00	Cross Member, Rear Spring Front Bracket	12UU4-12							✓	✓	✓	1
		.60	Shim, Rear Spring Front Bracket Cross Member (¼″)	12UU3-57							✓	✓	✓	2
		.50	Shim, Rear Spring Front Bracket Cross Member ($\frac{3}{16}$″)	12UU3-57A							✓	✓	✓	2
		5.55	Cross Member, Rear	12UL4-44	✓									1
		5.25	Cross Member, Rear	12UG4-44		✓	✓	✓	✓	✓				1
		5.55	Cross Member, Rear	12UU4-44							✓	✓	✓	2
		.40	Plate, Pintle Hook Anchor	12UG3-496		✓		✓	✓					1
		.30	Plate, Pintle Hook Bearing	12UU2-702							✓	✓	✓	2
		.09	Screw, Hex. Cap (½″—13 x 1½″)	S-1872							✓	✓	✓	8
		.04	Nut, Hex. (½″—13)	S-77							✓	✓	✓	8
		.01	Lockwasher, $\frac{17}{32}$″ Spring	S-1349							✓	✓	✓	16
		1.00	Gusset, Rear Cross Member	12UL3-42	✓									4
		.70	Gusset, Rear Cross Member—Top	12UG3-42		✓	✓	✓	✓					2
		.75	Gusset, Rear Cross Member—Bottom	12UG3-43		✓	✓	✓	✓					2
		1.75	Gusset, Rear Cross Member—Top	12UU4-42							✓			2
		.09	Screw, Hex. Cap (½″—13 x 1⅜″)	S-1428							✓			24
		.10	Screw, Hex. Cap (½″—13 x 1⅝″)	S-1839							✓			4
		.04	Nut, Hex. (½″—13)	S-77							✓			28
		.01	Lockwasher, ½″ Spring	S-1349							✓			56
		1.75	Gusset, Rear Cross Member—Top	12UVA4-42								✓	✓	2
		.05	Screw, Hex. Cap (½″—13 x 1¾″)	S-3100								✓	✓	22
		.10	Screw, Hex. Cap (½″—13 x 1⅝″)	S-1839								✓	✓	2
		.10	Screw, Hex. Cap (½″—13 x 2″)	S-922								✓	✓	2
		.04	Nut, Hex. (½″—13)	S-77								✓	✓	26
		.01	Lockwasher, $\frac{17}{32}$″ Spring	S-1349								✓	✓	52
		1.75	Gusset, Rear Cross Member—Bottom	12UU4-43							✓	✓	✓	2
		.09	Screw, Hex. Cap (½″—13 x 1⅜″)	S-1428							✓	✓	✓	20
		.01	Screw, Hex. Cap (½″—13 x 1⅝″)	S-1839							✓	✓	✓	2
		.04	Nut, Hex. (½″—13)	S-77							✓	✓	✓	22
		.01	Lockwasher, $\frac{17}{32}$″ Spring	S-1349							✓	✓	✓	44
		6.00	Cross Member, Pintle Hook	12UK4-44		✓		✓	✓					1
1501	Towing Attachments													
		2.30	Hook, Towing—R. H.	12CH3-128	✓	✓	✓	✓	✓					1
		2.30	Hook, Towing—L. H.	12CH3-129	✓	✓	✓	✓	✓					1
		2.00	Hook, Front Towing—R. H.	12W4-128							✓			1
		13.50	Hook, Front Towing—L. H.	12W4-129							✓			1
		1.75	Screw, Hex. Cap (¾″—16 x 3¼″)	S-6095							✓			2
		.07	Nut, Hex. (¾″—16)	S-847							✓			2
		.04	Lockwasher, ¾″ Spring	S-1353							✓			4
		.40	Screw, Hex. Cap (⅝″—18 x 2⅛″)	S-6235							✓			2
		.10	Nut, Hex. (⅝″—18)	S-1311							✓			2
		.01	Lockwasher, ⅝″ Spring	S-1351							✓			4
		2.30	Hook, Front Towing—R. H.	12H3-128								✓	✓	1

15-2
FRAME

Master Parts List

**THE AUTOCAR COMPANY
ARDMORE, PA.**

CODE	MFR. PART No.	LIST PRICE	NAME OF PART	AUTOCAR PART No.	MODEL									No. REQ.
					U-2044	U-4044	U-4144	U-4144-T	C-50	U-5044	U-7144-T	U-8144	U-8144-T	
1501	......	2.30	**Hook,** Front Towing—L. H.	12H3-129								✓	✓	1
	......	.15	**Screw,** Hex. Cap (⅝"—11 x 2")	S-900								✓	✓	2
	......	.20	**Screw,** Hex. Cap (⅝"—11 x 3")	S-901								✓	✓	2
	......	.05	**Nut,** Hex. (⅝"—11)	S-93								✓	✓	4
	......	.01	**Lockwasher,** 2½" Spring	S-1351								✓	✓	8
	......	.20	**Spacer,** Towing Hook—Rear	12UT3-108							✓			2
	......	.20	**Spacer,** Towing Hook—Front	12W2-108							✓			2
1502	**Pintle Hook**													
	110	29.40	**Hook,** Pintle Assembly	16W2-790A							✓	✓	✓	1

THE AUTOCAR COMPANY,
ARDMORE, PA.

Master Parts List

GROUP 16

ILLUSTRATIONS

GROUP 16

SPRINGS and SHOCK ABSORBERS

GROUP 16 — Master Parts List

THE AUTOCAR COMPANY, ARDMORE, PA.

As supplied on Autocar Models U-8144   U-8144-T

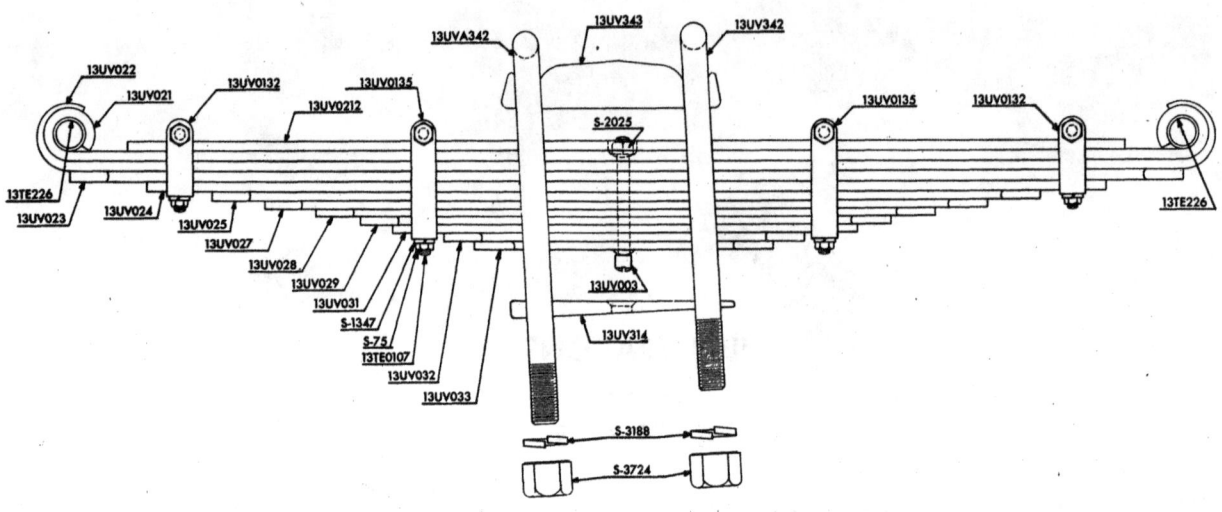

1601—Front Springs

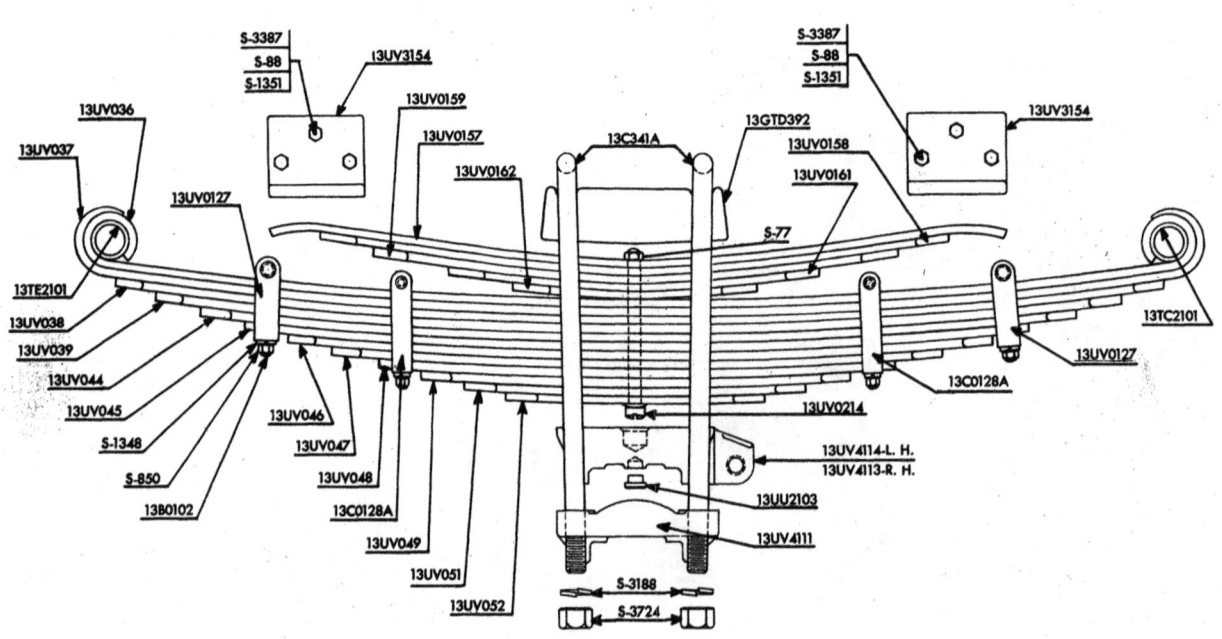

1601—Rear Springs

GROUP 16

THE AUTOCAR COMPANY, ARDMORE, PA.

Master Parts List

As supplied on Autocar Models
U-7144-T   U-8144   U-8144-T

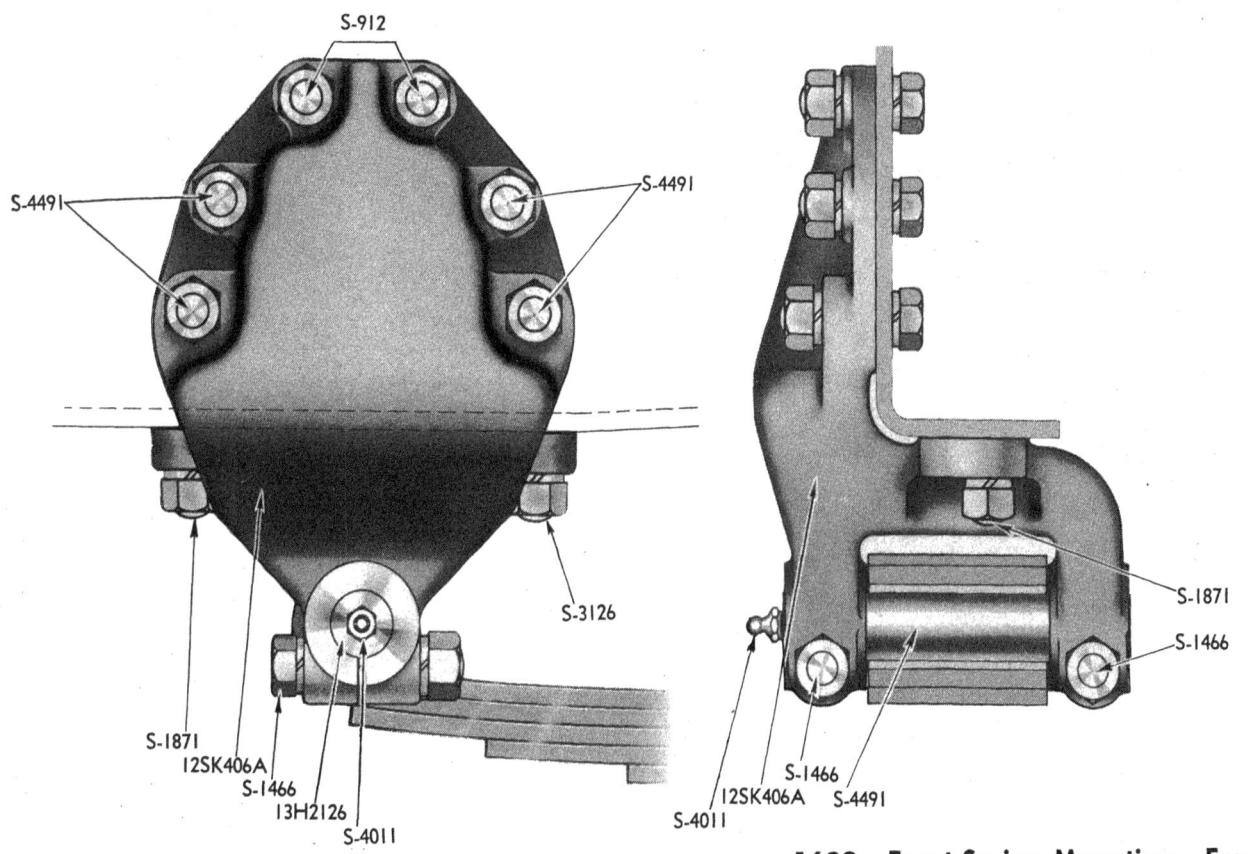

1602—Front Spring Mounting—Front

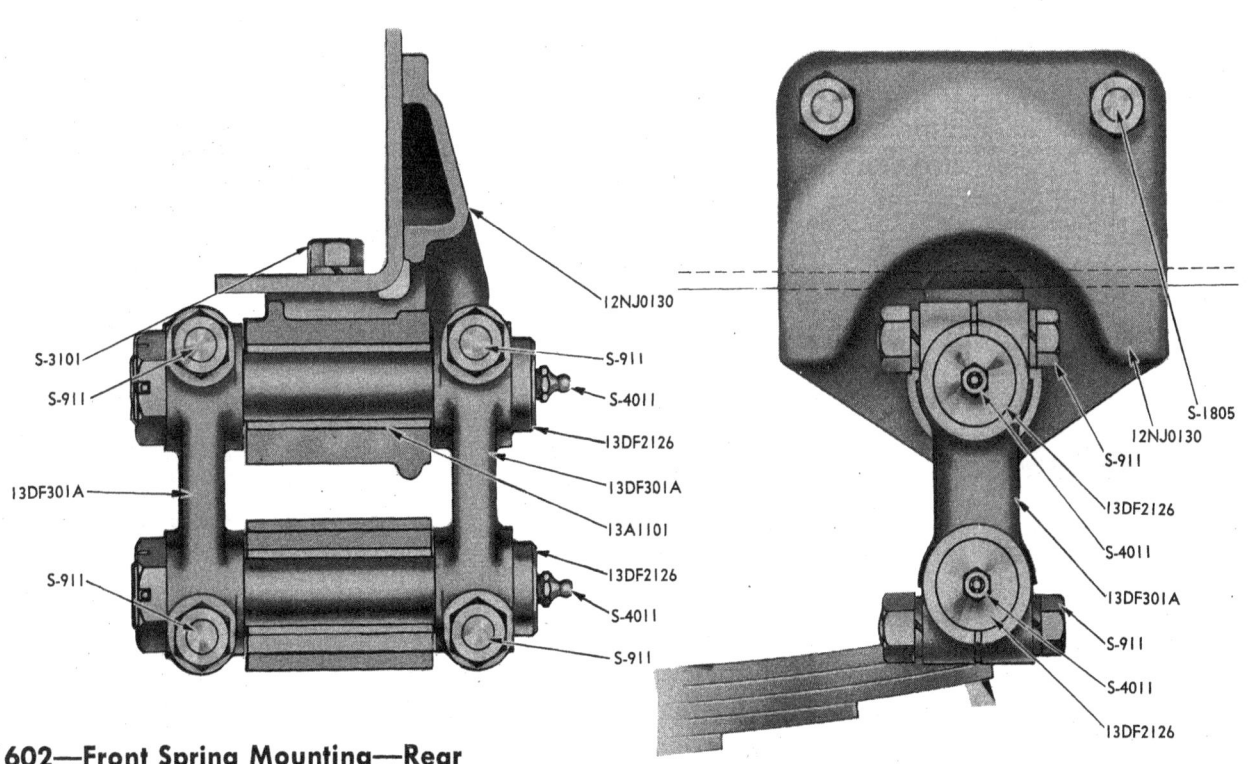

1602—Front Spring Mounting—Rear

GROUP 16

Master Parts List

THE AUTOCAR COMPANY, ARDMORE, PA.

As supplied on Autocar Models
U-7144-T U-8144 U-8144-T

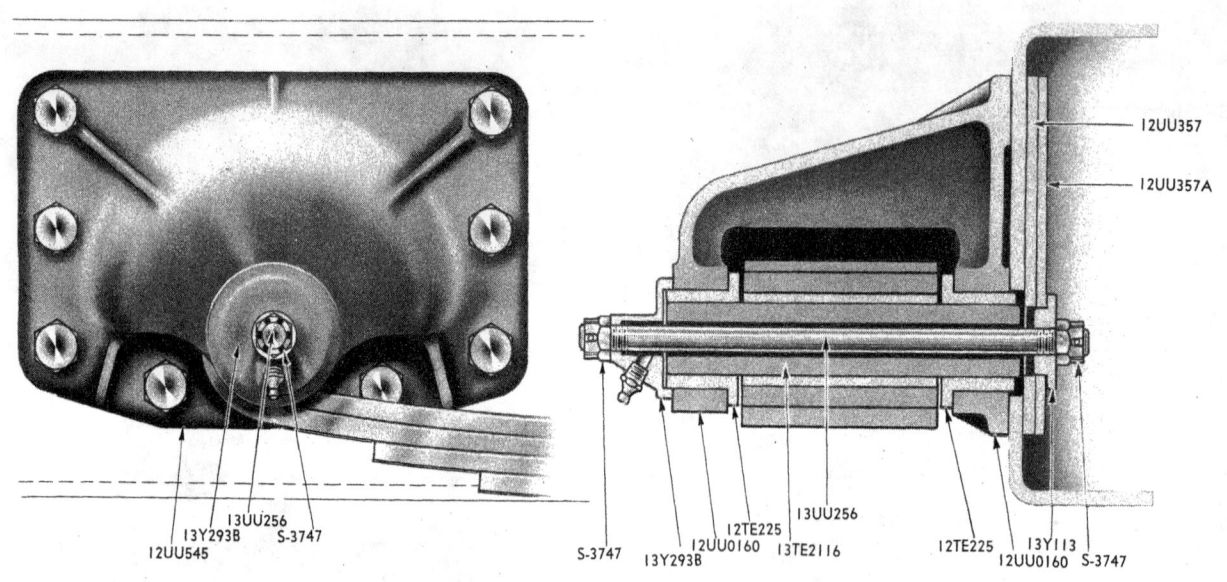

1602—Rear Spring Mounting—Front

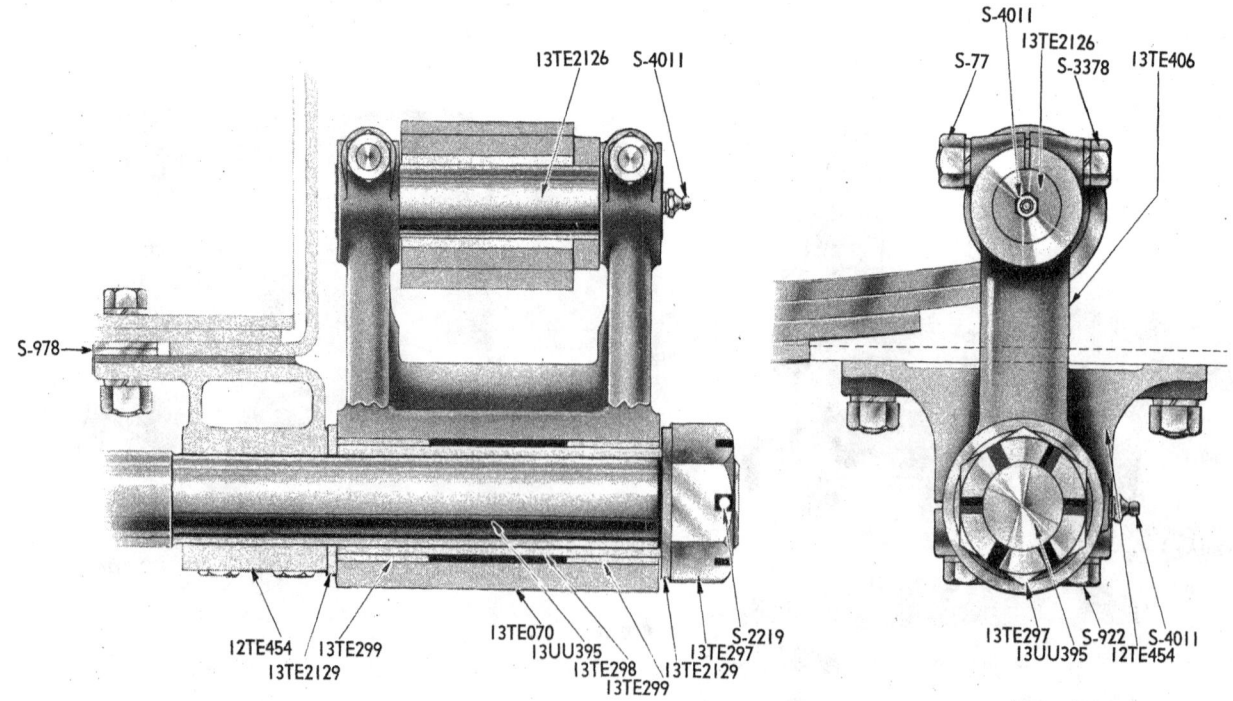

1602—Rear Spring Mounting—Rear

THE AUTOCAR COMPANY, ARDMORE, PA.

Master Parts List

GROUP 16

As supplied on Autocar Models
U-7144-T   U-8144   U-8144-T

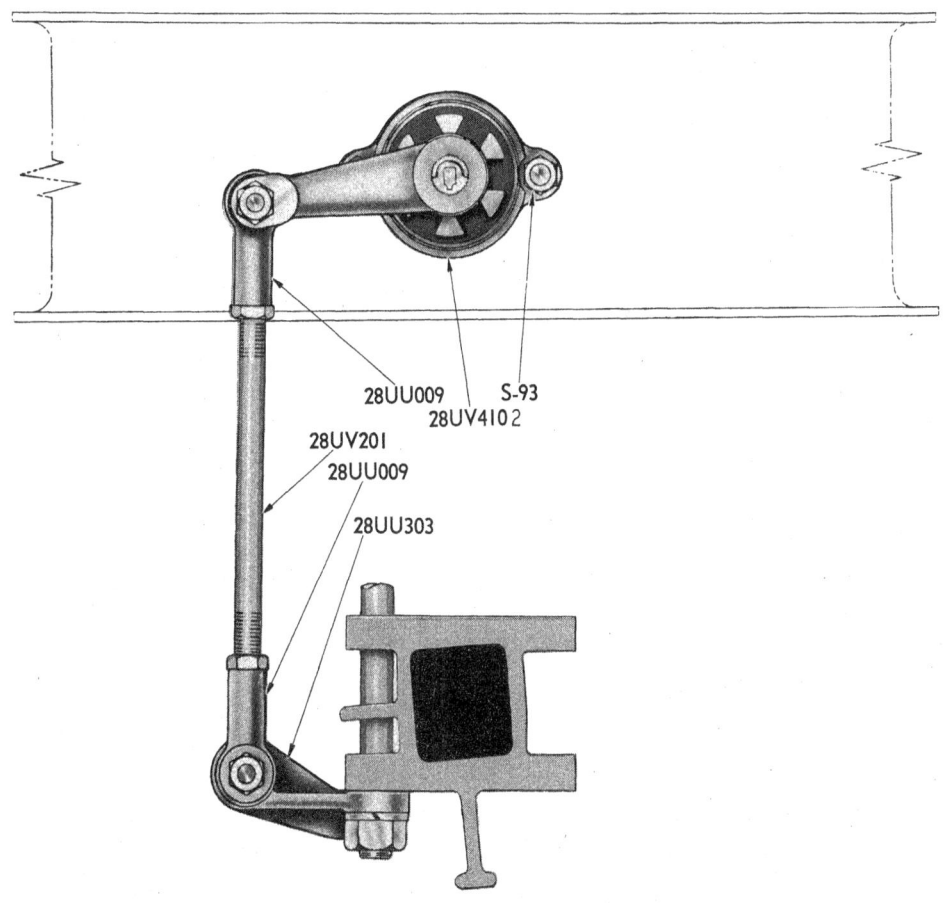

1603—Shock Absorber—Front

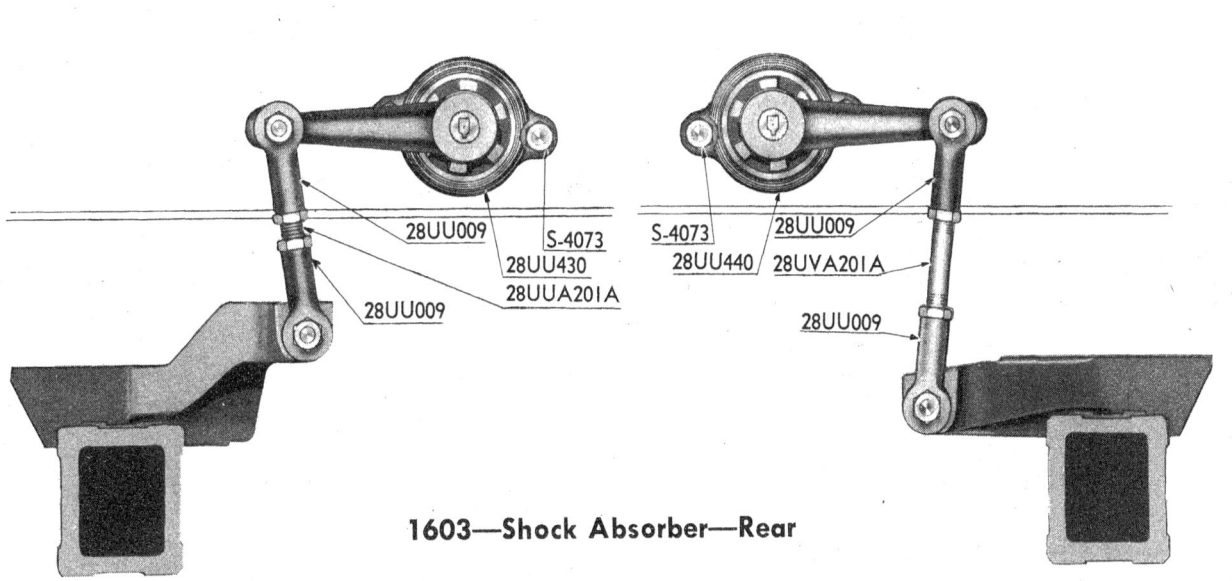

1603—Shock Absorber—Rear

16-0 SPRINGS

Master Parts List

THE AUTOCAR COMPANY, ARDMORE, PA.

CODE	MFR. PART No.	LIST PRICE	NAME OF PART	AUTOCAR PART No.	U-2044	U-4044	U-4144	U-4144-T	C-50	U-5044	U-7144-T	U-8144	U-8144-T	No. REQ.
1601			**Front and Rear Springs**											
	......	18.00	Spring, Front, Assembly	13UG4-09	✓	✓	✓	✓		✓				2
	......	12.80	Spring, Front, Assembly	13BF4-09					✓					2
	......	16.95	Spring, Front, Assembly	13UU4-09							✓			2
	......	20.25	Spring, Front, Assembly	13UV4-09								✓	✓	2
	......	1.25	Plate, Rebound	13BF0-65					✓					2
	......	1.70	Plate, Rebound	13UV0-212								✓	✓	2
	......	5.85	Spring, Front, No. 1 and No. 2 Assembled	13UG0-40	✓	✓	✓	✓		✓				2
	......	8.00	Spring, Front, No. 1 and No. 2 Assembled	13BF0-40					✓					2
	......	5.50	Spring, Front, No. 1 and No. 2 Assembled	13UU0-40							✓			2
	......	2.50	Spring, Front, No. 1 and Bushing Assembly	13UG0-21	✓	✓	✓	✓		✓				2
	......	4.20	Spring, Front, No. 1 and Bushing	13BF0-21					✓					2
	......	2.90	Spring, Front, No. 1 and Bushing Assembly	13UU0-21							✓			2
	......	3.05	Plate, No. 1 Spring	13UV0-21								✓	✓	2
	......	.35	Bushing, Front Spring	13TE2-26	✓	✓	✓	✓		✓	✓	✓	✓	4
	......	.60	Bushing, Front Spring	13A1-26					✓					2
	......	2.00	Plate, No. 2 Spring	13UG0-22	✓	✓	✓	✓		✓				2
	......	3.60	Plate, No. 2 Spring	13BF0-22					✓					2
	......	2.30	Plate, No. 2 Spring	13UU0-22							✓			2
	......	2.75	Plate, No. 2 Spring	13UV0-22								✓	✓	2
	......	1.65	Plate, No. 3 Spring	13UG0-23	✓	✓	✓	✓		✓				2
	......	1.25	Plate, No. 3 Spring	13BF0-23					✓					2
	......	1.95	Plate, No. 3 Spring	13UU0-23							✓			2
	......	2.45	Plate, No. 3 Spring	13UV0-23								✓	✓	2
	......	1.55	Plate, No. 4 Spring	13UG0-24	✓	✓	✓	✓		✓				2
	......	1.10	Plate, No. 4 Spring	13BF0-24					✓					2
	......	1.65	Plate, No. 4 Spring	13UU0-24							✓			2
	......	2.10	Plate, No. 4 Spring	13UV0-24								✓	✓	2
	......	1.45	Plate, No. 5 Spring	13UG0-25	✓	✓	✓	✓		✓				2
	......	.95	Plate, No. 5 Spring	13BF0-25					✓					2
	......	1.45	Plate, No. 5 Spring	13UU0-25							✓			2
	......	1.45	Plate, No. 5 Spring	13UV0-25								✓	✓	2
	......	1.25	Plate, No. 6 Spring	13UG0-27	✓	✓	✓	✓		✓				2
	......	.85	Plate, No. 6 Spring	13BF0-27					✓					2
	......	1.25	Plate, No. 6 Spring	13UU0-27							✓			2
	......	1.30	Plate, No. 6 Spring	13UV0-27								✓	✓	2
	......	1.20	Plate, No. 7 Spring	13UG0-28	✓	✓	✓	✓		✓				2
	......	.80	Plate, No. 7 Spring	13BF0-28					✓					2
	......	1.00	Plate, No. 7 Spring	13UU0-28							✓			2
	......	1.10	Plate, No. 7 Spring	13UV0-28								✓	✓	2
	......	1.10	Plate, No. 8 Spring	13UG0-29	✓	✓	✓	✓		✓				2
	......	.70	Plate, No. 8 Spring	13BF0-29					✓					2
	......	.70	Plate, No. 8 Spring	13UU0-29							✓			2
	......	.95	Plate, No. 8 Spring	13UV0-29								✓	✓	2
	......	1.00	Plate, No. 9 Spring	13UG0-31	✓	✓	✓	✓		✓				2
	......	.60	Plate, No. 9 Spring	13BF0-31					✓					2
	......	.85	Plate, No. 9 Spring	13UU0-31							✓			2
	......	.85	Plate, No. 9 Spring	13UV0-31								✓	✓	2
	......	.85	Plate, No. 10 Spring	13UG0-32	✓	✓	✓	✓		✓				2
	......	.55	Plate, No. 10 Spring	13BF0-32					✓					2
	......	.90	Plate, No. 10 Spring	13UV0-32								✓	✓	2
	......	.75	Plate, No. 11 Spring	13UG0-33	✓	✓	✓	✓		✓				2
	......	.40	Plate, No. 11 Spring	13BF0-33					✓					2
	......	.75	Plate, No. 11 Spring	13UV0-33								✓	✓	2
	......	.65	Plate, No. 12 Spring	13UG0-34	✓	✓	✓	✓		✓				2
	......	.35	Plate, No. 12 Spring	13BF0-34					✓					2
	......	.55	Plate, No. 13 Spring	13UG0-35	✓	✓	✓	✓		✓				2
	......	.25	Plate, No. 13 Spring	13BF0-35					✓					2
	......	.45	Plate, No. 14 Spring	13UG0-65	✓	✓	✓	✓		✓				2
	......	.30	Plate, No. 15 Spring	13UG0-67	✓	✓	✓	✓		✓				2
	......	.60	Clip, Front Spring Rebound, Assembly (Small)	13UG0-132	✓	✓	✓	✓		✓				4
	......	.60	Clip, Front Spring Rebound, Assembly (Small)	13RE0-132					✓					2

THE AUTOCAR COMPANY
ARDMORE, PA.

Master Parts List

16-1 SPRINGS

CODE	MFR. PART No.	LIST PRICE	NAME OF PART	AUTOCAR PART No.	U-2044	U-4044	U-4144	U-4144-T	C-50	U-5044	U-7144-T	U-8144	U-8144-T	No. REQ.
1601	......	.60	Clip, Front Spring Rebound, Assembly (Small)	13SK0-132							✓			4
	......	.60	Clip, Front Spring Rebound, Assembly (Small)	13UV0-132								✓	✓	4
	......	.60	Clip, Front Spring Rebound, Assembly (Large)	13UG0-135	✓	✓	✓	✓		✓				4
	......	.60	Clip, Front Spring Rebound, Assembly (Large)	13RL0-132					✓					2
	......	.60	Clip, Front Spring Rebound, Assembly (Large)	13SK0-135							✓			4
	......	.60	Clip, Front Spring Rebound, Assembly (Large)	13UV0-135								✓	✓	4
	......	.20	Bolt, Plow (3/8"—16)	13TE0-107	✓	✓	✓	✓		✓	✓	✓	✓	8
	......	.04	Nut, Plow Bolt (7/16"—14)	S-850	✓	✓	✓	✓	✓	✓	✓			8
	......	.03	Nut, Plow Bolt (3/8"—16)	S-75							✓	✓	✓	8
	......	.01	Lockwasher, 7/16" Spring	S-1348	✓	✓	✓	✓		✓	✓			8
	......	.01	Lockwasher, 13/32" Spring	S-1347							✓	✓	✓	8
	......	1.50	Bolt, Center	13BFA0-214					✓					2
	......	1.50	Bolt, Center (7/16" x 3 3/4")	13UU0-03							✓			2
	......	1.50	Bolt, Center	13UV0-03								✓	✓	2
	......	.02	Nut, Center Bolt	S-2025							✓	✓	✓	2
	......	2.15	Keeper, Front Spring Clip	13UU3-43							✓			2
	......	2.45	Keeper, Front Spring Clip	13UV3-43								✓	✓	2
	......	.14	Shim, Front Spring	13UU2-14							✓			2
	......	1.20	Shim, Front Spring	13UV3-14								✓	✓	2
	......	2.40	Clip, Front Spring—Front	13UU3-42							✓			2
	......	2.25	Clip, Front Spring—Front	13UVA3-42								✓	✓	2
	......	2.25	Clip, Front Spring—Rear	13UUA3-42							✓			2
	......	2.25	Clip, Front Spring—Rear	13UV3-42								✓	✓	2
	......	.15	Nut, Hex. (1"—14)	S-3724							✓	✓	✓	8
	......	.06	Lockwasher, 1 1/16" Spring	S-3188							✓	✓	✓	8
	......	34.25	Spring, Rear, Assembly	13UG4-11	✓	✓	✓	✓		✓				2
	......	32.50	Spring, Rear, Assembly	13BF3-11					✓					2
	......	35.85	Spring, Rear, Assembly	13UU4-11							✓			2
	......	53.50	Spring, Rear, Assembly	13UV4-11								✓	✓	2
	......	9.00	Spring, Rear, No. 1 and No. 2 Assembled	13UG0-50	✓	✓	✓	✓		✓				2
	......	11.00	Spring, Rear, No. 1 and No. 2 Assembled	13BF0-50					✓					2
	......	9.75	Spring, Rear, No. 1 and No. 2 Assembled	13UU0-50							✓			2
	......	11.25	Spring, Rear, No. 1 and No. 2 Assembled	13UV0-50								✓	✓	2
	......	3.60	Plate, No. 1 Spring—with Bushing	13UG0-36	✓	✓	✓	✓		✓				2
	......	7.45	Plate, No. 1 Spring—with Bushing	13BF0-36					✓					2
	......	5.00	Plate, No. 1 Spring	13UU0-36							✓			2
	......	5.75	Plate, No. 1 Spring	13UV0-36								✓	✓	2
	......	.35	Bushing, Rear Spring	13TE2-26	✓	✓	✓	✓		✓				4
	......	.60	Bushing, Rear Spring	13A1-101					✓					4
	......	.70	Bushing, Rear Spring	13TE2-101							✓			4
	......	1.20	Bushing, Rear Spring	13Y1-101A								✓	✓	4
	......	.25	Spacer, Rear Spring Pin	10AD2-135							✓			4
	......	3.60	Plate, No. 2 Spring	13UG0-37	✓	✓	✓	✓		✓				2
	......	4.50	Plate, No. 2 Spring	13BF0-37					✓					2
	......	4.50	Plate, No. 2 Spring	13UU0-37							✓			2
	......	5.25	Plate, No. 2 Spring	13UV0-37								✓	✓	2
	......	3.60	Plate, No. 3 Spring	13UG0-38	✓	✓	✓	✓		✓				2
	......	3.45	Plate, No. 3 Spring	13BF0-38					✓					2
	......	4.00	Plate, No. 3 Spring	13UU0-38							✓			2
	......	4.50	Plate, No. 3 Spring	13UV0-38								✓	✓	2
	......	3.25	Plate, No. 4 Spring	13UG0-39	✓	✓	✓	✓		✓				2
	......	3.10	Plate, No. 4 Spring	13BF0-39					✓					2
	......	3.45	Plate, No. 4 Spring	13UU0-39							✓			2
	......	4.15	Plate, No. 4 Spring	13UV0-39								✓	✓	2
	......	2.90	Plate, No. 5 Spring	13UG0-44	✓	✓	✓	✓		✓				2
	......	2.90	Plate, No. 5 Spring	13BF0-44					✓					2
	......	2.95	Plate, No. 5 Spring	13UU0-44							✓			2
	......	3.80	Plate, No. 5 Spring	13UV0-44								✓	✓	2
	......	2.50	Plate, No. 6 Spring	13UG0-45	✓	✓	✓	✓		✓				2
	......	2.60	Plate, No. 6 Spring	13BF0-45					✓					2
	......	2.50	Plate, No. 6 Spring	13UU0-45							✓			2
	......	3.40	Plate, No. 6 Spring	13UV0-45								✓	✓	2

16-2 SPRINGS

Master Parts List

THE AUTOCAR COMPANY, ARDMORE, PA.

CODE	MFR. PART No.	LIST PRICE	NAME OF PART	AUTOCAR PART No.	U-2044	U-4044	U-4144	U-4144-T	C-50	U-5044	U-7144-T	U-8144	U-8144-T	No. REQ.
1601	.....	1.90	**Plate,** No. 7 Spring	13UG0-46	✓	✓	✓	✓		✓				2
	.....	2.35	**Plate,** No. 7 Spring	13BF0-46					✓					2
	.....	2.00	**Plate,** No. 7 Spring	13UU0-46							✓			2
	.....	3.05	**Plate,** No. 7 Spring	13UV0-46								✓	✓	2
	.....	1.55	**Plate,** No. 8 Spring	13UG0-47	✓	✓	✓	✓		✓				2
	.....	2.05	**Plate,** No. 8 Spring	13BF0-47					✓					2
	.....	1.55	**Plate,** No. 8 Spring	13UU0-47							✓			2
	.....	2.70	**Plate,** No. 8 Spring	13UV0-47								✓	✓	2
	.....	1.25	**Plate,** No. 9 Spring	13UG0-48	✓	✓	✓	✓		✓				2
	.....	1.80	**Plate,** No. 9 Spring	13BF0-48					✓					2
	.....	1.00	**Plate,** No. 9 Spring	13UU0-48							✓			2
	.....	2.25	**Plate,** No. 9 Spring	13UV0-48								✓	✓	2
	.....	.90	**Plate,** No. 10 Spring	13UG0-49	✓	✓	✓	✓		✓				2
	.....	1.55	**Plate,** No. 10 Spring	13BF0-49					✓					2
	.....	1.85	**Plate,** No. 10 Spring	13UV0-49								✓	✓	2
	.....	.65	**Plate,** No. 11 Spring	13UG0-51	✓	✓	✓	✓		✓				2
	.....	1.35	**Plate,** No. 11 Spring	13BF0-51					✓					2
	.....	1.50	**Plate,** No. 11 Spring	13UV0-51								✓	✓	2
	.....	1.00	**Plate,** No. 12 Spring	13BF0-52					✓					2
	.....	1.10	**Plate,** No. 12 Spring	13UV0-52								✓	✓	2
	.....	.90	**Plate,** No. 13 Spring	13BF0-53					✓					2
	.....	.60	**Clip,** Rear Spring Rebound, Assembly (Small)	13UG0-127	✓	✓	✓	✓		✓				4
	.....	.60	**Clip,** Rear Spring Rebound, Assembly (Small)	13N0-127					✓					4
	.....	.60	**Clip,** Rear Spring Rebound, Assembly	13UU0-127							✓			4
	.....	.60	**Clip,** Rear Spring Rebound, Assembly (Small)	13UV0-127								✓	✓	4
	.....	.60	**Clip,** Rear Spring Rebound, Assembly (Large)	13UG0-128	✓	✓	✓	✓		✓				4
	.....	.60	**Clip,** Rear Spring Rebound, Assembly (Large)	13TAS0-128					✓					4
	.....	.60	**Clip,** Rear Spring Rebound, Assembly (Large)	13C0-128A								✓	✓	4
	.....	.20	**Bolt,** Plow ($\frac{7}{16}''-14$)	13B0-107	✓	✓	✓	✓	✓	✓	✓	✓		8
	.....	.04	**Nut,** Plow Bolt ($\frac{7}{16}''-14$)	S-850	✓	✓	✓	✓	✓	✓				8
	.....	.01	**Lockwasher,** $\frac{15}{32}''$ Spring	S-1348	✓	✓	✓	✓	✓	✓				8
	.....	1.50	**Bolt,** Center ($\frac{1}{2}''-13$)	13UV0-214								✓	✓	2
	.....	.04	**Nut,** Center Bolt ($\frac{1}{2}''-13$)	S-77								✓	✓	2
	.....	.20	**Bolt,** Plow	13UU0-107							✓			4
	.....	.04	**Nut,** Plow Bolt ($\frac{7}{16}''-14$)	S-850							✓			4
	.....	.01	**Lockwasher,** $\frac{7}{16}''$ Spring	S-1348							✓			4
	.....	1.50	**Bolt,** Center	13BF0-214					✓					2
	.....	.04	**Nut,** Hex. ($\frac{7}{16}''-14$)	S-850					✓					2
	.....	1.50	**Bolt,** Center	13UU0-214							✓			2
	.....	.04	**Nut,** Center Bolt ($\frac{1}{2}''-13$)	S-77							✓			2
	.....	7.90	**Spring,** Rear Auxiliary, Assembly	13UG4-156	✓	✓	✓	✓		✓				2
	.....	7.65	**Spring,** Rear Auxiliary, Assembly	13BF4-156					✓					2
	.....	7.50	**Spring,** Rear Auxiliary, Assembly	13UU4-156							✓			2
	.....	9.00	**Spring,** Rear Auxiliary, Assembly	13UV4-156								✓	✓	2
	.....	2.25	**Plate,** No. 1 Auxiliary Spring	13UG0-157	✓	✓	✓	✓		✓				2
	.....	1.70	**Plate,** No. 1 Auxiliary Spring	13BF0-157					✓					2
	.....	1.75	**Plate,** No. 1 Auxiliary Spring	13UU0-157							✓			2
	.....	2.80	**Plate,** No. 1 Auxiliary Spring	13UV0-157								✓	✓	2
	.....	2.10	**Plate,** No. 2 Auxiliary Spring	13UG0-158	✓	✓	✓	✓		✓				2
	.....	1.60	**Plate,** No. 2 Auxiliary Spring	13BF0-158					✓					2
	.....	1.75	**Plate,** No. 2 Auxiliary Spring	13UU0-158							✓			2
	.....	2.35	**Plate,** No. 2 Auxiliary Spring	13UV0-158								✓	✓	2
	.....	1.80	**Plate,** No. 3 Auxiliary Spring	13UG0-159	✓	✓	✓	✓		✓				2
	.....	1.45	**Plate,** No. 3 Auxiliary Spring	13BF0-159					✓					2
	.....	1.45	**Plate,** No. 3 Auxiliary Spring	13UU0-159							✓			2
	.....	1.85	**Plate,** No. 3 Auxiliary Spring	13UV0-159								✓	✓	2
	.....	1.45	**Plate,** No. 4 Auxiliary Spring	13UG0-161	✓	✓	✓	✓		✓				2
	.....	1.20	**Plate,** No. 4 Auxiliary Spring	13BF0-161					✓					2
	.....	1.15	**Plate,** No. 4 Auxiliary Spring	13UU0-161							✓			2
	.....	1.40	**Plate,** No. 4 Auxiliary Spring	13UV0-161								✓	✓	2
	.....	1.00	**Plate,** No. 5 Auxiliary Spring	13UG0-162	✓	✓	✓	✓		✓				2
	.....	.90	**Plate,** No. 5 Auxiliary Spring	13BF0-162					✓					2

Master Parts List — Springs 16-3

THE AUTOCAR COMPANY, ARDMORE, PA.

CODE	MFR. PART No.	LIST PRICE	NAME OF PART	AUTOCAR PART No.	U-2044	U-4044	U-4144	U-4144-T	C-50	U-5044	U-7144-T	U-8144	U-8144-T	No. REQ.
1601	......	.85	Plate, No. 5 Auxiliary Spring	13UU0-162							✓			2
	......	.90	Plate, No. 5 Auxiliary Spring	13UV0-162								✓	✓	2
	......	.65	Plate, No. 6 Auxiliary Spring	13UG0-167	✓	✓	✓	✓		✓				2
	......	.60	Plate, No. 6 Auxiliary Spring	13BF0-167					✓					2
	......	.60	Plate, No. 6 Auxiliary Spring	13UU0-167							✓			2
	......	2.90	Keeper, Rear Spring Clip	13UU3-92							✓			2
	......	4.50	Keeper, Rear Spring Clip	13GTD3-92								✓	✓	2
	......	8.50	Shim, Rear Spring—L. H.	13UU4-114							✓			1
	......	8.50	Shim, Rear Spring—L. H.	13UV4-114								✓	✓	1
	......	8.50	Shim, Rear Spring—R. H.	13UU4-113							✓			1
	......	8.50	Shim, Rear Spring—R. H.	13UV4-113								✓	✓	1
	......	.15	Dowel, Rear Spring Shim	13UG2-103							✓			2
	......	.15	Dowel, Rear Spring Shim	13UU2-103								✓	✓	2
	......	2.50	Clip, Rear Spring	13ZTE3-41C							✓			4
	......	2.70	Clip, Rear Spring	13C3-41C								✓	✓	4
	......	.20	Nut, Hex. (1"—14)	S-3724								✓	✓	8
	......	.06	Lockwasher, $1\frac{1}{16}$" Spring	S-3188								✓	✓	8
	......	.90	Plate, Rear Spring Clip	13UU4-111							✓			2
	......	5.50	Plate, Rear Spring Clip	13UV4-111								✓	✓	2
	......	1.80	Bracket, Auxiliary Spring	13DK3-154	✓	✓	✓	✓		✓				4
	......	1.70	Bracket, Auxiliary Spring	13UU3-154							✓			4
	......	1.90	Bracket, Auxiliary Spring	13UV3-154								✓	✓	4
	......	.15	Screw, Hex. Cap ($\frac{5}{8}$"—11 x 2")	S-3387							✓	✓	✓	12
	......	.06	Nut, Hex. ($\frac{5}{8}$"—11)	S-88							✓	✓	✓	12
	......	.01	Lockwasher, $\frac{21}{32}$" Spring	S-1351							✓	✓	✓	24
1602	Shackles and Spring Attaching Parts													
	......	5.50	Bracket, Front Spring—Front—R. H.	12UG4-34	✓	✓	✓	✓		✓				1
	......	5.50	Bracket, Front Spring—Front—L. H.	12UK4-34	✓	✓	✓	✓		✓				1
	......	.12	Screw, Hex. Cap ($\frac{1}{2}$"—13 x $1\frac{5}{8}$")	S-1754	✓	✓	✓	✓		✓				8
	......	.12	Screw, Hex. Cap ($\frac{1}{2}$"—13 x $2\frac{3}{4}$")	S-4077	✓	✓	✓	✓		✓				4
	......	.04	Nut, Hex. ($\frac{1}{2}$"—13)	S-77	✓	✓	✓	✓		✓				12
	......	.01	Lockwasher, $\frac{17}{32}$" Spring	S-1349	✓	✓	✓	✓		✓				24
	......	8.40	Bracket, Front Spring—R. H.	12RM5-03					✓					1
	......	8.40	Bracket, Front Spring—L. H.	12RM5-04					✓					1
	......	.03	Plug, Welch ($\frac{3}{4}$")	S-4208					✓					2
	......	.15	Screw, Hex. Cap ($\frac{5}{8}$"—11 x $1\frac{1}{2}$")	S-3098					✓					4
	......	.20	Screw, Hex. Cap ($\frac{5}{8}$"—11 x $1\frac{3}{4}$")	S-4328					✓					4
	......	.06	Nut, Hex. ($\frac{5}{8}$"—11)	S-88					✓					8
	......	.01	Lockwasher, $\frac{21}{32}$" Spring	S-1351					✓					16
	......	1.80	Cap, Front Spring Bracket	12RM4-478					✓					2
	......	.05	Screw, Hex. Cap ($\frac{1}{2}$"—13 x $1\frac{3}{4}$")	S-3100					✓					4
	......	.04	Nut, Hex. ($\frac{1}{2}$"—13)	S-77					✓					4
	......	.01	Lockwasher, $\frac{17}{32}$" Spring	S-1349					✓					8
	......	1.50	Plate, Front Spring Bracket Wearing	13RM3-209					✓					2
	......	.06	Dowel, Front Spring Bracket	12TL2-479					✓					2
	......	.60	Roller, Front Spring Bracket Rebound	13RM2-211					✓					2
	......	.30	Plate, Front Bumper Clamp	12RM2-262					✓					2
	......	.09	Screw, Hex. Cap ($\frac{1}{2}$"—13 x $1\frac{1}{4}$")	S-25					✓					4
	......	.01	Lockwasher, $\frac{17}{32}$" Spring	S-1349					✓					4
	......	1.20	Boot, Front Bumper Bracket	12RM3-491A					✓					2
	......	10.45	Bracket, Front Spring—Front	12SK4-06A							✓	✓	✓	2
	......	.10	Screw, Hex. Cap ($\frac{1}{2}$"—13 x 3")	S-1466							✓	✓	✓	4
	......	.12	Screw, Hex. Cap ($\frac{1}{2}$"—13 x $1\frac{3}{4}$")	S-4491							✓	✓	✓	8
	......	.10	Screw, Hex. Cap ($\frac{1}{2}$"—13 x 2")	S-912							✓	✓	✓	4
	......	.20	Screw, Hex. Cap ($\frac{1}{2}$"—13 x $2\frac{1}{4}$")	S-3126							✓	✓	✓	2
	......	.11	Screw, Hex. Cap ($\frac{1}{2}$"—13 x $2\frac{1}{2}$")	S-1871							✓	✓	✓	2
	......	.04	Nut, Hex. ($\frac{1}{2}$"—13)	S-77							✓	✓	✓	20
	......	.01	Lockwasher, $\frac{17}{32}$" Spring	S-1349							✓	✓	✓	40
	......	1.00	Pin, Front Spring Bracket	13H2-126	✓	✓	✓	✓		✓	✓	✓	✓	2
	......	.15	Fitting, Straight Alemite ($\frac{1}{8}$")	S-4011							✓	✓	✓	2

16-4 SPRINGS — Master Parts List — THE AUTOCAR COMPANY, ARDMORE, PA.

CODE	MFR. PART No.	LIST PRICE	NAME OF PART	AUTOCAR PART No.	U-2044	U-4044	U-4144	U-4144-T	C-50	U-5044	U-7144-T	U-8144	U-8144-T	No. REQ.
1602	......	8.50	Bracket, Front Spring, and Bushing Assembly—Rear	12NJ0-130	✓	✓	✓	✓		✓	✓	✓	✓	2
	......	7.80	Bracket, Front Spring—Rear	12NJ4-34	✓	✓	✓	✓		✓	✓	✓	✓	2
	......	.60	Bushing, Front Spring Bracket	13A1-101	✓	✓	✓	✓		✓	✓	✓	✓	2
	......	.12	Screw, Hex. Cap ($\frac{1}{2}$"—13 x $1\frac{3}{4}$")	S-3101	✓	✓	✓	✓		✓	✓	✓	✓	4
	......	.04	Nut, Hex. ($\frac{1}{2}$"—13)	S-77	✓	✓	✓	✓		✓	✓	✓	✓	4
	......	.01	Lockwasher, $\frac{17}{32}$" Spring	S-1349	✓	✓	✓	✓		✓	✓	✓	✓	8
	......	.30	Screw, Hex. Cap ($\frac{5}{8}$"—11 x $1\frac{3}{4}$")	S-1805	✓	✓	✓	✓		✓	✓	✓	✓	4
	......	.05	Nut, Hex. ($\frac{5}{8}$"—11)	S-93	✓	✓	✓	✓		✓	✓	✓	✓	4
	......	.01	Lockwasher, $\frac{21}{32}$" Spring	S-1351	✓	✓	✓	✓		✓	✓	✓	✓	8
	......	2.70	Shackle, Front Spring	13DF3-01A	✓	✓	✓	✓		✓	✓	✓	✓	4
	......	.12	Screw, Hex. Cap ($\frac{1}{2}$"—13 x $2\frac{5}{8}$")	S-911	✓	✓	✓	✓		✓	✓	✓	✓	8
	......	.04	Nut, Hex. ($\frac{1}{2}$"—13)	S-77	✓	✓	✓	✓		✓	✓	✓	✓	8
	......	.01	Lockwasher, $\frac{17}{32}$" Spring	S-1349	✓	✓	✓	✓		✓	✓	✓	✓	16
	......	1.70	Pin, Front Spring Shackle	13DF2-126	✓	✓	✓	✓		✓	✓	✓	✓	4
	......	.15	Fitting, Straight Alemite ($\frac{1}{8}$")	S-4011	✓	✓	✓	✓		✓	✓	✓	✓	4
	......	4.40	Bracket, Front Spring, and Bushing Assembly	12RL0-130A					✓					2
	......	3.90	Bracket, Front Spring—Rear	12RL4-34					✓					2
	......	.20	Bushing, Front Spring Bracket	13RL2-02					✓					4
	......	.12	Screw, Hex. Cap ($\frac{1}{2}$"—13 x $1\frac{5}{8}$")	S-1754					✓					12
	......	.04	Nut, Hex. ($\frac{1}{2}$"—13)	S-77					✓					12
	......	.01	Lockwasher, $\frac{17}{32}$" Spring	S-1349					✓					24
	......	1.20	Bolt, Front Spring	13RL2-03					✓					2
	......	.08	Nut, Hex. ($\frac{3}{4}$"—16)	S-2001					✓					2
	......	.06	Washer, Front Spring Stud Nut Lock	13A2-12					✓					2
	......	.60	Bushing, Front Spring	13A1-26					✓					2
	......	9.00	Bracket, Rear Spring—Front	12DK4-45A	✓	✓	✓	✓		✓				2
	......	2.25	Bolt, Rear Spring—Front	13A2-115	✓	✓	✓	✓		✓				2
	......	.12	Nut, Hex. ($\frac{3}{4}$"—16)	S-4221	✓	✓	✓	✓		✓				2
	......	4.60	Bracket, Rear Spring, and Bushing Assembly—Front—R. H.	12A0-160B						✓				1
	......	4.20	Bracket, Rear Spring—Front—R. H.	12A4-45B						✓				1
	......	4.60	Bracket, Rear Spring, and Bushing Assembly—Front—L. H.	12A0-170B						✓				1
	......	4.20	Bracket, Rear Spring—Front—L. H.	12A4-46B						✓				1
	......	.20	Bushing, Rear Spring Bracket	12A1-25						✓				4
	......	2.25	Bolt, Rear Spring—Front	13A2-115						✓				2
	......	.12	Nut, Hex. ($\frac{3}{4}$"—16)	S-4221						✓				2
	......	7.50	Bracket, Rear Spring, and Bushing Assembly—Front	12UU0-160							✓	✓	✓	2
	......	6.60	Bracket, Rear Spring—Front	12UU5-45							✓	✓	✓	2
	......	.40	Bushing, Rear Spring Bracket	12TE2-25							✓	✓	✓	4
	......	.20	Screw, Hex. Cap ($\frac{5}{8}$"—11 x $2\frac{1}{4}$")	S-4497							✓	✓	✓	4
	......	.20	Screw, Hex. Cap ($\frac{5}{8}$"—11 x $2\frac{1}{2}$")	S-902							✓	✓	✓	12
	......	.05	Nut, Hex. ($\frac{5}{8}$"—11)	S-93							✓	✓	✓	16
	......	.01	Lockwasher, $\frac{21}{32}$" Spring	S-1351							✓	✓	✓	32
	......	.30	Stud, Rear Spring	13UU2-56							✓	✓	✓	2
	......	.04	Nut, Rear Spring Stud Hex. ($\frac{1}{2}$"—20 Castle)	S-3747							✓	✓	✓	4
	......	.01	Pin, $\frac{3}{32}$" x $1\frac{1}{4}$", Cotter	S-2200							✓	✓	✓	4
	......	1.50	Tube, Rear Spring	13TE2-116							✓	✓	✓	2
	......	.60	Cap, Rear Spring Stud	13Y2-93B							✓	✓	✓	2
	......	.15	Fitting, Straight Alemite ($\frac{1}{8}$")	S-4011							✓	✓	✓	2
	......	.25	Washer, Rear Spring Stud	13Y1-13							✓	✓	✓	2
	......	.60	Plate, Rear Spring Bracket Reinforcement ($\frac{1}{4}$")	12UU3-57							✓	✓	✓	2
	......	.50	Plate, Rear Spring Bracket Reinforcement ($\frac{3}{16}$")	12UU3-57A							✓	✓	✓	2
	......	7.75	Bracket, Rear Spring Shackle	12UG4-54	✓	✓	✓	✓		✓				2
	......	3.60	Bracket, Rear Spring Shackle	12N4-54					✓					2
	......	6.60	Bracket, Rear Spring Shackle	12TE4-54							✓	✓	✓	2
	......	.05	Screw, Hex. Cap ($\frac{1}{2}$"—13 x $1\frac{3}{4}$")	S-3100							✓	✓	✓	6
	......	.10	Screw, Hex. Cap ($\frac{1}{2}$"—13 x 2")	S-922							✓	✓	✓	8
	......	.11	Screw, Hex. Cap ($\frac{1}{2}$"—13 x $2\frac{1}{4}$")	S-1880							✓	✓	✓	2
	......	.04	Nut, Hex. ($\frac{1}{2}$"—13)	S-77							✓	✓	✓	8
	......	.01	Lockwasher, $\frac{17}{32}$" Spring	S-1349							✓	✓	✓	24
	......	.25	Washer, Shackle Bracket	13TE2-129							✓	✓	✓	4
	......	.01	Washer, Flat Steel ($\frac{17}{32}$" x $1\frac{1}{4}$" x $\frac{1}{8}$")	S-978							✓	✓	✓	4
	......	.15	Fitting, Straight Alemite ($\frac{1}{8}$")	S-4011							✓	✓	✓	2

THE AUTOCAR COMPANY
ARDMORE, PA.

Master Parts List

16-5 SPRINGS

CODE	MFR. PART No.	LIST PRICE	NAME OF PART	AUTOCAR PART No.	U-2044	U-4044	U-4144	U-4144-T	C-50	U-5044	U-7144-T	U-8144	U-8144-T	No. REQ.
1602	......	7.15	Shaft, Rear Spring Cross	13DK3-95	✓		✓							2
	......	16.35	Shaft, Rear Spring Cross	13UU3-95							✓	✓	✓	1
	......	7.15	Pin, Rear Spring Shackle—Rear	13UG3-149		✓		✓		✓				2
	......	.40	Nut, Rear Spring Cross Shaft	13H2-97	✓	✓	✓	✓		✓				2
	......	.04	Washer, Flat Steel (1 5/32" x 2" x 1/8")	S-979	✓	✓	✓	✓		✓				2
	......	.25	Washer, Rear Spring Shackle—Inner	13RHP2-129	✓	✓	✓	✓		✓				2
	......	3.60	Pin, Rear Spring Shackle—Rear	13RH2-149					✓					2
	......	.20	Nut, Hex. (1" x 14 Castled)	S-1309					✓					2
	......	.05	Washer, Flat Steel (1 1/32" x 2 1/8" x 3/16")	S-719					✓					2
	......	.25	Washer, Rear Spring Shackle—Inner	13RH2-129					✓					2
	......	1.00	Nut, Rear Spring Cross Shaft	13TE2-97							✓	✓	✓	2
	......	.04	Pin, Cotter (1/4" x 2 3/4")	S-2219							✓	✓	✓	2
	......	9.40	Shackle, Rear Spring, and Bushing Assembly	13H0-70	✓	✓	✓	✓		✓				2
	......	8.40	Shackle, Rear Spring, and Bushing Assembly	13RH0-70					✓					2
	......	15.60	Shackle, Rear Spring, and Bushing Assembly	13TE0-70							✓	✓	✓	2
	......	8.40	Shackle, Rear Spring	13H3-06	✓	✓	✓	✓		✓				2
	......	7.20	Shackle, Rear Spring	13RH3-06					✓					2
	......	12.60	Shackle, Rear Spring	13TE4-06							✓	✓	✓	2
	......	.35	Bushing, Rear Spring Shackle	13H2-99A	✓	✓	✓	✓	✓	✓				4
	......	.50	Bushing, Rear Spring Shackle	13TE2-99							✓	✓	✓	4
	......	.90	Spacer, Rear Spring Shackle	13H2-98A	✓	✓	✓	✓		✓				2
	......	.12	Screw, Hex. Cap (1/2"—13 x 2 3/4")	S-4077	✓	✓	✓	✓		✓				4
	......	.04	Nut, Hex. (1/2—13)	S-77	✓	✓	✓	✓		✓				4
	......	.01	Lockwasher, 1/2" Spring	S-1349	✓	✓	✓	✓		✓				8
	......	.90	Spacer, Rear Spring Shackle	13RH2-98					✓					2
	......	.08	Screw, Hex. Cap (3/8"—16—2 1/4")	S-40					✓					4
	......	.03	Nut, Hex. (3/8"—16)	S-75					✓					4
	......	.01	Lockwasher, 3/8" Spring	S-1347					✓					8
	......	1.80	Spacer, Rear Spring Shackle	13TE2-98							✓	✓	✓	2
	......	.15	Screw, Hex. Cap (1/2"—13 x 3 1/4")	S-3378							✓	✓	✓	4
	......	.04	Nut, Hex. (1/2"—13)	S-77							✓	✓	✓	4
	......	.01	Lockwasher, 1/2" Spring	S-1349							✓	✓	✓	8
	......	1.00	Pin, Rear Spring	13H2-126	✓	✓	✓	✓		✓				2
	......	1.00	Pin, Rear Spring	13A2-126					✓					2
	......	1.50	Pin, Rear Spring	13TE2-126							✓	✓	✓	2
	......	.15	Fitting, Straight Alemite (1/8")	S-4011							✓	✓	✓	2
	......	.50	Spacer, Rear Spring Shackle Pin	13UG2-197		✓		✓		✓				2
	......	1.50	Bracket, Rear Spring Shackle Pin Anchor—R. H.	13UG3-248		✓		✓		✓				1
	......	1.50	Bracket, Rear Spring Shackle Pin Anchor—L. H.	13UGA3-248		✓		✓		✓				1
	......	.35	Nut, Hex. (1"—14)	S-844		✓		✓						2
1603	**Shock Absorbers and Mountings**													
	11406	5.00	Shock Absorber, Front Assembly—Monroe—(Complete)	28MA0-10	✓	✓	✓	✓	✓	✓				2
	76-A1	.75	Valve, Compression Assembly	76-A1	✓	✓	✓	✓	✓	✓				2
	71-65	.75	Valve, Relief Assembly	71-65	✓	✓	✓	✓	✓	✓				2
	10571	.10	Bushing	10571	✓	✓	✓	✓	✓	✓				4
	10639-B	.05	Intake, Spring Piston	10639-B	✓	✓	✓	✓	✓	✓				2
	10640-B	.05	Intake, Valve Piston	10640-B	✓	✓	✓	✓	✓	✓				2
	10641-2	.05	Rebound, Metering Spacer	10641-2	✓	✓	✓	✓	✓	✓				4
	10853	.10	Support, Washer Piston	10853	✓	✓	✓	✓	✓	✓				2
	10854	.01	Washer, Spacer Support	10854	✓	✓	✓	✓	✓	✓				2
	10873	.20	Piston	10873	✓	✓	✓	✓	✓	✓				2
	10874	.05	Support, Washer Gasket	10874	✓	✓	✓	✓	✓	✓				2
	10875	.05	Chamber, Gasket Reserve	10875	✓	✓	✓	✓	✓	✓				2
	12406	.80	Guide, Rod and Seal Assembly	12406	✓	✓	✓	✓	✓	✓				2
	12221	4.00	Head Assembly	12221	✓	✓	✓	✓	✓	✓				2
	12222	2.00	Base Assembly	12222	✓	✓	✓	✓	✓	✓				2
	T-317	1.35	Wrench	T-317	✓	✓	✓	✓	✓	✓				1
	T-347	.85	Thimble	T-347	✓	✓	✓	✓	✓	✓				1
	......	25.50	Shock Absorber, R. H. Front and Link Assembly—(Complete) (Houdaille "BBH")	28UU4-200							✓			1
	......	25.50	Shock Absorber, L. H. Front and Link Assembly—(Complete)	28UU4-100							✓			1

16-6 SPRINGS — Master Parts List — THE AUTOCAR COMPANY, ARDMORE, PA.

CODE	MFR. PART NO.	LIST PRICE	NAME OF PART	AUTOCAR PART No.	U-2044	U-4044	U-4144	U-4144-T	C-50	U-5044	U-7144-T	U-8144	U-8144-T	No. REQ.
1603	......	.10	**Screw,** Hex. Cap (½″–13 x 1⅝″)...............	S-1839							✓			4
	......	.04	**Nut,** Hex. (½″–13)............................	S-77							✓			4
	......	.01	**Lockwasher,** ½″ Spring.......................	S-1349							✓			8
	......	25.50	**Shock Absorber,** R. H. Front (Houdaille "BBH")	28UV4-10								✓	✓	1
	......	25.50	**Shock Absorber,** L. H. Front.................	28UV4-20								✓	✓	1
	......	.15	**Screw,** Hex. Cap (⅝″–11 x 2″)................	S-4073								✓	✓	1
	......	.05	**Nut,** Hex. (⅝″–11)..........................	S-93								✓	✓	1
	......	.01	**Lockwasher,** 21/32″ Spring...................	S-1351								✓	✓	2
	......	.25	**Stud,** Shock Absorber (Top).................	28M2-01	✓	✓	✓	✓	✓	✓				2
	......	.25	**Stud,** Shock Absorber (Bottom)..............	28MA2-01	✓	✓	✓	✓	✓	✓				2
	......	.04	**Nut,** Hex. (½″–13)..........................	S-77	✓	✓	✓	✓	✓	✓				8
	......	.01	**Lockwasher,** ½″ Spring......................	S-1349	✓	✓	✓	✓	✓	✓				4
	......	.03	**Washer** (17/32″ I.D.–1⅜″ O.D. x 1/16″ Thick)..	S-390	✓	✓	✓	✓	✓	✓				4
	......	.01	**Pin,** Cotter (3/32″ x ¾″)....................	S-2198	✓	✓	✓	✓	✓	✓				4
	......	1.00	**Plate,** R. H. Front Clip....................	28UG2-02	✓	✓	✓	✓	✓	✓				1
	......	1.00	**Plate,** L. H. Front Clip....................	28UG2-03	✓	✓	✓	✓	✓	✓				1
	......	2.25	**Plate,** R. H. Front Clip....................	28UU3-03							✓	✓	✓	1
	......	2.25	**Plate,** L. H. Front Clip....................	28UU3-02							✓	✓	✓	1
	......	.40	**Link,** Front................................	28UV2-01							✓	✓	✓	2
	......	.08	**Nut,** Hex. Jam (¾″–16)......................	S-857							✓	✓	✓	4
	......	3.55	**End,** Front Link Assembly...................	28UU0-09							✓	✓	✓	4
	......	.25	**Spacer,** Shock Absorber Stud................	28M2-05	✓	✓	✓	✓	✓	✓				4
	......	.75	**Bracket,** Shock Absorber Frame (R. H.)......	28UG2-11	✓	✓	✓	✓	✓	✓				1
	......	.75	**Bracket,** Shock Absorber Frame (L. H.)......	28UG2-12	✓	✓	✓	✓	✓	✓				1
	......	.09	**Screw,** Hex. Cap (½″–13 x 1½″)..............	S-1872	✓	✓	✓	✓	✓	✓				4
	......	.04	**Nut,** Hex. (½″–13)..........................	S-77	✓	✓	✓	✓	✓	✓				4
	......	.01	**Lockwasher,** ½″ Spring......................	S-1349	✓	✓	✓	✓	✓	✓				8
11406		5.00	**Shock Absorber,** Rear Assembly—Monroe—(Complete)	28MA0-10	✓	✓	✓	✓	✓	✓				2
76-A1		.75	**Valve,** Compression Assembly................	76-A1	✓	✓	✓	✓	✓	✓				2
71-65		.75	**Valve,** Relief Assembly.....................	71-65	✓	✓	✓	✓	✓	✓				2
10571		.10	**Bushing**....................................	10571	✓	✓	✓	✓	✓	✓				4
10639-B		.05	**Intake,** Spring Piston......................	10639-B	✓	✓	✓	✓	✓	✓				2
10640-B		.05	**Intake,** Valve Piston.......................	10640-B	✓	✓	✓	✓	✓	✓				2
10641-2		.05	**Rebound,** Metering Spacer...................	10641-2	✓	✓	✓	✓	✓	✓				4
10853		.10	**Support,** Washer Piston.....................	10853	✓	✓	✓	✓	✓	✓				2
10854		.01	**Washer,** Spacer Support.....................	10854	✓	✓	✓	✓	✓	✓				2
10873		.20	**Piston**.....................................	10873	✓	✓	✓	✓	✓	✓				2
10874		.05	**Support,** Washer Gasket.....................	10874	✓	✓	✓	✓	✓	✓				2
10875		.05	**Chamber,** Gasket Reserve....................	10875	✓	✓	✓	✓	✓	✓				2
12406		.80	**Guide,** Rod and Seal Assembly...............	12406	✓	✓	✓	✓	✓	✓				2
12221		4.00	**Head** Assembly..............................	12221	✓	✓	✓	✓	✓	✓				2
12222		2.00	**Base** Assembly..............................	12222	✓	✓	✓	✓	✓	✓				1
T-317		1.35	**Wrench**.....................................	T-317	✓	✓	✓	✓	✓	✓				1
T-347		.85	**Thimble**....................................	T-347	✓	✓	✓	✓	✓	✓				1
	......	25.50	**Shock Absorber,** R. H. Rear (Houdaille "BBH")	28UU4-40							✓	✓	✓	1
	......	25.50	**Shock Absorber,** L. H. Rear.................	28UU4-30							✓	✓	✓	1
	......	.15	**Screw,** Hex. Cap (⅝″–11 x 2″)...............	S-4073							✓	✓	✓	4
	......	.05	**Nut,** Hex. (⅝″–11)..........................	S-93							✓	✓	✓	4
	......	.01	**Lockwasher,** 21/32″ Spring..................	S-1351							✓	✓	✓	8
	......	.25	**Stud,** Shock Absorber (Top).................	28M2-01	✓	✓	✓	✓	✓	✓				4
	......	.25	**Stud,** Shock Absorber (Bottom)..............	28MA2-01	✓	✓	✓	✓	✓	✓				4
	......	.04	**Nut,** Hex. (½″–13)..........................	S-77	✓	✓	✓	✓	✓	✓				16
	......	.01	**Lockwasher,** ½″ Spring......................	S-1349	✓	✓	✓	✓	✓	✓				8
	......	.03	**Washer** (17/32″ I.D.–1⅜″ O.D. x 1/16″ Thick).	S-390	✓	✓	✓	✓	✓	✓				8
	......	.01	**Pin,** Cotter (3/32″ x ¾″)...................	S-2198	✓	✓	✓	✓	✓	✓				8
	......	.25	**Spacer,** Shock Absorber Stud................	28M2-05	✓	✓	✓	✓	✓	✓				8
	......	.75	**Bracket,** Shock Absorber Frame (R. H.)......	28UK2-11	✓	✓	✓	✓	✓	✓				2
	......	.75	**Bracket,** Shock Absorber Frame (L. H.)......	28UK2-12	✓	✓	✓	✓	✓	✓				2
	......	.09	**Screw,** Hex. Cap (½″–13 x 1½″)..............	S-1872	✓	✓	✓	✓	✓	✓				8
	......	.04	**Nut,** Hex. (½″–13″).........................	S-77	✓	✓	✓	✓	✓	✓				8
	......	.01	**Lockwasher,** ½″ Spring......................	S-1349	✓	✓	✓	✓	✓	✓				16
	......	4.80	**Shim,** Rear Spring..........................	13UG3-113	✓	✓	✓	✓	✓	✓				1

THE AUTOCAR COMPANY
ARDMORE, PA.

Master Parts List

16-7 SPRINGS

CODE	MFR. PART No.	LIST PRICE	NAME OF PART	AUTOCAR PART No.	MODEL U-2044	U-4044	U-4144	U-4144-T	C-50	U-5044	U-7144-T	U-8144	U-8144-T	No. REQ.
1603	......	4.80	**Shim,** Rear Spring	13UG3-114	✓	✓	✓	✓	✓	✓				1
	......	.25	**Link,** R. H. Rear	28UVA2-01A							✓	✓	✓	1
	......	.35	**Link,** L. H. Rear	28UUA2-01A							✓	✓	✓	1
	......	.08	**Nut,** Hex. Jam (¾"—16)	S-857							✓	✓	✓	4
	......	3.55	**End,** Link Assembly	28UU0-09							✓	✓	✓	4

GROUP 18

THE AUTOCAR COMPANY, ARDMORE, PA.

Master Parts List

ILLUSTRATIONS

GROUP 18

BODY

GROUP 18

Master Parts List

THE AUTOCAR COMPANY, ARDMORE, PA.

As supplied on Autocar Models
U-8144   U-8144-T

1800—Cab—Interior View

THE AUTOCAR COMPANY, ARDMORE, PA.

Master Parts List

GROUP 18

As supplied on Autocar Models U-8144-T

1800—Cab- Front View

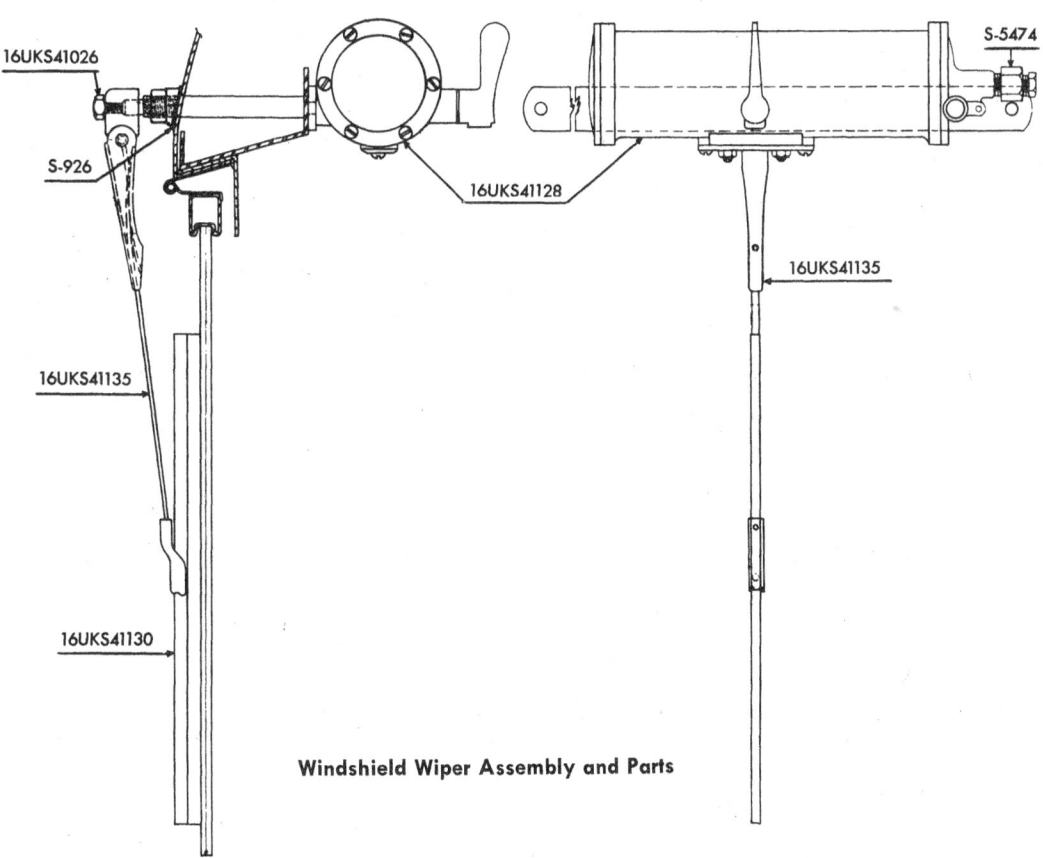

Windshield Wiper Assembly and Parts

GROUP 18

Master Parts List

THE AUTOCAR COMPANY, ARDMORE, PA.

As supplied on Autocar Models
U-8144-T

1800—Cab—Underside View

THE AUTOCAR COMPANY,
ARDMORE, PA.

Master Parts List

GROUP 18

As supplied on Autocar Models
U-8144-T

1800—Cab—Rear View

GROUP 18

Master Parts List
THE AUTOCAR COMPANY, ARDMORE, PA.

As supplied on Autocar Models
U-8144   U-8144-T

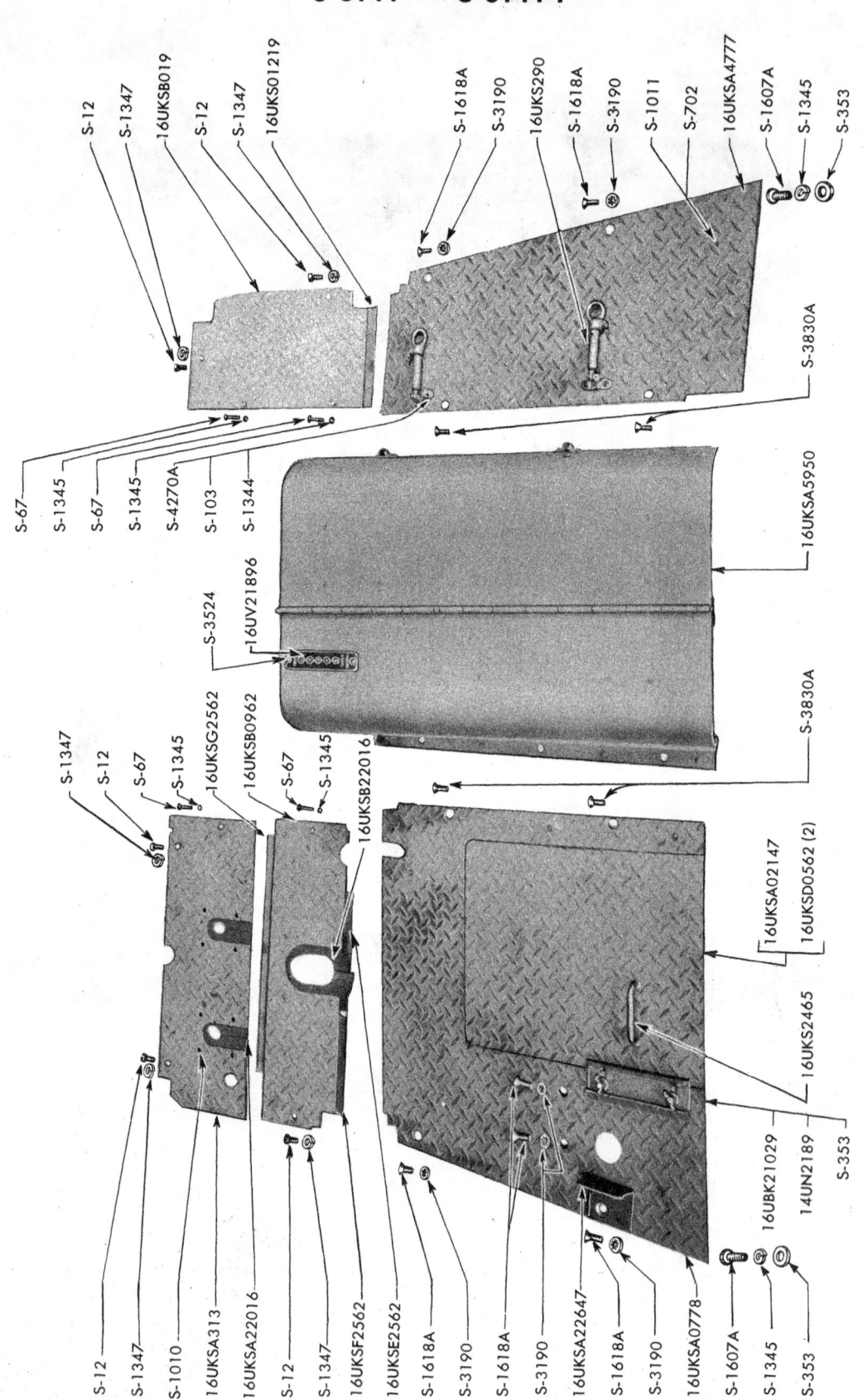

1800—Floorboards and Toeboards

THE AUTOCAR COMPANY, ARDMORE, PA.

Master Parts List

GROUP 18

As supplied on Autocar Models
U-8144-T

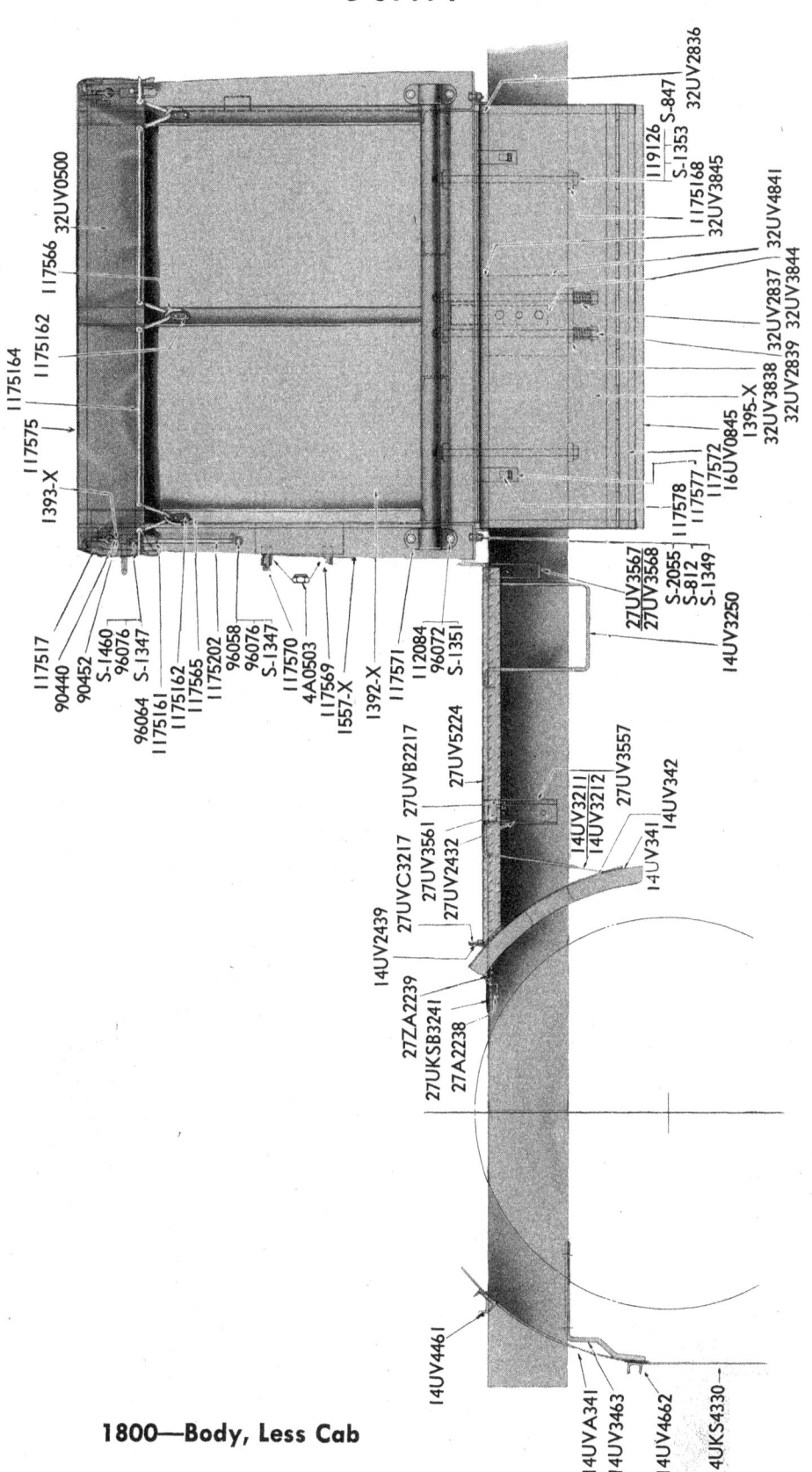

1800—Body, Less Cab

18-0 BODY

Master Parts List — THE AUTOCAR COMPANY, ARDMORE, PA.

CODE	MFR. PART No.	LIST PRICE	NAME OF PART	AUTOCAR PART No.	U-2044	U-4044	U-4144	U-4144-T	C-50	U-5044	U-7144-T	U-8144	U-8144-T	No. REQ.
1800	Cab and Body Assembly and		Miscellaneous Parts											
	.....	....	**Cab** Assembly, Includes.........	15UKSE0-700								✓	✓	1
	.....	....	Cowl.................	.....										..
	.....	....	Brackets, Headlamp and Blackout Lamp........	.....										..
	.....	....	Guard, Radiator and Lamp Brush........	.....										..
	.....	....	Fenders with Steps and Splash Shields.......	.....										..
	.....	....	Shield, Air—between Cowl and Radiator......	.....										..
	.....	....	Windshield and Brackets........	.....										..
	.....	....	Tubing, Windshield Wiper Only......	.....										..
	.....	....	Doors, Complete with Hardware........	.....										..
	.....	....	Handles, Grab........	.....										..
	.....	....	Seat Structure (Includes Battery Box)......	.....										..
	.....	....	Cushions, Back and Seat......	.....										..
	.....	....	Supports, Engine Hood......	.....										..
	.....	....	Wiring, Spotlight—Dash to Light Location.......	.....										..
	.....	....	Window, Rear, and Screen (Refer to various other groups if additional equipment within cab is required)......	.....										..
	.....	8.10	Bracket, Front Cab Support.......	16UU4-414								✓	✓	2
	.....	.12	Screw, Hex. Cap ($\frac{1}{2}"$—13 x $1\frac{5}{8}"$).....	S-1754								✓	✓	2
	.....	.04	Nut, Hex. ($\frac{1}{2}"$—13).....	S-77								✓	✓	2
	.....	.01	Lockwasher, $\frac{17}{32}"$ Spring......	S-1349								✓	✓	4
	.....	.15	Screw, Hex. Cap ($\frac{5}{8}"$—11 x $1\frac{3}{4}"$).....	S-4073								✓	✓	6
	.....	.05	Nut, Hex. ($\frac{5}{8}"$—11).....	S-93								✓	✓	6
	.....	.01	Lockwasher, $\frac{21}{32}"$ Spring.....	S-1351								✓	✓	12
	.....	.20	Bolts, Cab Mounting—Front ($\frac{5}{8}"$—11 x 4").....	S-6283								✓	✓	2
	.....	.05	Nut, Hex. ($\frac{5}{8}"$—11).....	S-4201								✓	✓	2
	.....	....	Spacers, Steel......	16UKS2-1127								✓	✓	2
	.....	....	Shims, Anti-Rattle......	16UD2-1127A								✓	✓	2
	.....	.06	Washer, Flat Steel ($\frac{1}{4}"$ x $1\frac{1}{16}"$ x $\frac{1}{8}"$).....	S-348								✓	✓	2
	.....	.25	Spring.....	15T2-274A								✓	✓	2
	.....	.01	Pin, Cotter ($\frac{1}{8}"$ x $1\frac{1}{4}"$).....	S-2205								✓	✓	2
	.....	29.25	Support, Rear Cab Assembly......	16UU5-415								✓	✓	1
	.....	.09	Screw, Hex. Cap ($\frac{1}{2}"$—13 x $1\frac{1}{2}"$).....	S-1872								✓	✓	6
	.....	.04	Nut, Hex. ($\frac{1}{2}"$—13).....	S-77								✓	✓	6
	.....	.01	Lockwasher, $\frac{17}{32}"$ Spring.....	S-1349								✓	✓	12
	.....	.15	Bolts, Cab Mounting—Rear ($\frac{5}{8}"$—11 x $8\frac{1}{2}"$).....	S-6287								✓	✓	2
	.....	.90	Shims, Anti-Rattle......	16UKS2-721								✓	✓	1
	.....	.05	Nut, Hex. ($\frac{5}{8}"$—11).....	S-4201								✓	✓	2
	.....	.25	Springs.....	15T2-274								✓	✓	4
	.....	.06	Washer, Flat Steel ($\frac{1}{4}"$ x $1\frac{1}{16}"$ x $\frac{1}{8}"$).....	S-348								✓	✓	2
	.....	.01	Pin, Cotter ($\frac{1}{8}"$ x $1\frac{1}{4}"$).....	S-2205								✓	✓	2
	.....	....	Door Assembly Complete with All Hardware—R. H.	15UKS0-890								✓	✓	1
	.....	....	Door Assembly Complete with All Hardware—L. H.	15UKS0-910								✓	✓	1
	.....	.45	Hinges, Door—R. H.......	15DLS2-24								✓	✓	3
	.....	....	Screw, Flat Head Machine ($\frac{1}{4}"$—20 x $\frac{1}{2}"$).....	S-3839A								✓	✓	24
	.....	....	Lockwasher, $\frac{9}{32}"$ Shakeproof External Csk...	S-3189								✓	✓	24
	.....	.45	Hinges, Door—L. H.......	15DLS2-147								✓	✓	3
	.....	....	Screw, Flat Head Machine ($\frac{1}{4}"$—20 x $\frac{1}{2}"$).....	S-3839A								✓	✓	24
	.....	....	Lockwasher, $\frac{9}{32}"$ Shakeproof External Csk...	S-3189								✓	✓	24
	.....	.50	Handle, Door—Outside.......	15UKS3-28								✓	✓	2
	.....	....	Screw, Flat Head Machine (#10—32 x $\frac{5}{8}"$).....	S-3811A								✓	✓	2
	.....	.35	Handle, Door—Inside.......	15NF2-29								✓	✓	2
	.....	.30	Escutcheon, Door Handle.......	15NF2-346								✓	✓	2
	.....	13.35	Glass, Door and Frame Assembly—R. H.......	15UKS3-1939								✓	✓	1
	.....	13.35	Glass, Door and Frame Assembly—L. H.......	15UKS3-2069								✓	✓	1
	.....	2.25	Lock, Door—R. H.......	15DLS3-26								✓	✓	1
	.....	.01	Screw, Round Head Machine (#10—24 x $\frac{1}{2}"$).....	S-4270A								✓	✓	6
	.....	....	Screw, Bind. Head Machine (#10—24 x $\frac{5}{8}"$).....	S-1606A								✓	✓	2
	.....	.01	Lockwasher $\frac{7}{32}"$ Spring......	S-1344								✓	✓	8
	.....	2.25	Lock, Door—L. H.......	15DLS3-27								✓	✓	1
	.....	.01	Screw, Round Head Machine (#10—24 x $\frac{1}{2}"$).....	S-4270A								✓	✓	6
	.....	....	Screw, Bind. Head Machine (#10—24 x $\frac{5}{8}"$).....	S-1606A								✓	✓	2

THE AUTOCAR COMPANY
ARDMORE, PA.

Master Parts List

18-1 BODY

CODE	MFR. PART No.	LIST PRICE	NAME OF PART	AUTOCAR PART No.	U-2044	U-4044	U-4144	U-4144-T	C-50	U-5044	U-7144-T	U-8144	U-8144-T	No. REQ.
1800	.....	.01	Lockwasher, $\frac{7}{32}''$ Spring........	S-1344								✓	✓	8
	.....	.12	Dovetail, Door (Male)........	15DLS2-146								✓	✓	4
	.....	....	Screw, Flat Head Machine (#12—24 x $\frac{5}{8}''$)........	S-1604A								✓	✓	8
	.....	2.15	Regulator, Door Window—R. H........	15U4-901A								✓	✓	1
	.....	....	Screw, Round Head Machine ($\frac{1}{4}''$—20 x $\frac{3}{4}''$)........	S-1614A								✓	✓	6
	.....	.01	Lockwasher, $\frac{7}{32}''$ Spring........	S-1345								✓	✓	6
	.....	2.15	Regulator, Door Window—L. H........	15U4-19A								✓	✓	1
	.....	....	Screw, Round Head Machine ($\frac{1}{4}''$—20 x $\frac{3}{4}''$)........	S-1614A								✓	✓	6
	.....	.01	Lockwasher, $\frac{9}{32}''$ Spring........	S-1345								✓	✓	6
	.....	.60	Handle, Door Window Regulator........	15NF2-182								✓	✓	2
	.....	.30	Escutcheon........	15NF2-346								✓	✓	2
	.....	....	Cover, Door Panel Hand Hole, Assy. (Includes Rubber Bumpers)	15UKSA3-3014								✓	✓	2
	.....	....	Screw, Round Head Machine ($\frac{1}{4}''$—20 x $\frac{5}{8}''$)........	S-3821A								✓	✓	4
	.....	18.75	Windshield Assembly—R. H........	15UKS4-97								✓	✓	1
	.....	....	Screw, Flat Head Machine (#10—32 x $\frac{3}{4}''$)........	S-191A								✓	✓	6
	.....	.01	Nut, Hex. (#10—32)........	S-3745								✓	✓	6
	.....	.01	Lockwasher, $\frac{7}{32}''$ Spring........	S-1344								✓	✓	6
	.....	18.75	Windhsield Assembly—L. H........	15UKS4-284								✓	✓	1
	.....	....	Screw, Flat Head Machine (#10—32 x $\frac{3}{4}''$)........	S-191A								✓	✓	6
	.....	.01	Nut, Hex. (#10—32 x $\frac{3}{4}''$)........	S-3745								✓	✓	6
	.....	.01	Lockwasher, $\frac{7}{32}''$ Spring........	S-1344								✓	✓	6
	.....	....	Bracket, Windshield—Center........	15UKS2-239								✓	✓	2
	.....	....	Screw, Flat Head Machine ($\frac{1}{4}''$—20 x $\frac{3}{4}''$)........	S-3843A								✓	✓	2
	.....	.01	Nut, Hex. ($\frac{1}{4}''$—20)........	S-76								✓	✓	2
	.....	....	Washer, Shakeproof ($\frac{1}{4}''$ Csk.)........	S-3189								✓	✓	2
	.....	.01	Lockwasher, $\frac{9}{32}''$ Spring........	S-1345								✓	✓	2
	.....	....	Bracket, Windshield—Side........	15UKS2-241								✓	✓	2
	.....	....	Screw, Flat Head Self-Tapping Machine (#10—32 x $\frac{3}{4}''$)...	S-3525								✓	✓	2
	.....	....	Washer, Shakeproof—External Teeth $\frac{3}{16}''$ Csk.	S-3287								✓	✓	2
	.....	.65	Arm, Windshield Quadrant........	15UKS2-244								✓	✓	4
	.....	.85	Lid, Cowl Side Vent, and Handle Assembly........	15UKS3-550								✓	✓	2
	.....	.06	Spring, Cowl Side Vent........	15T2-293								✓	✓	4
	.....	.03	Plunger, Cowl Side Vent Spring........	15T2-294								✓	✓	4
	.....	....	Washer, Flat ($\frac{5}{32}''$ x $\frac{1}{2}''$ x $\frac{1}{32}''$)........	S-742								✓	✓	2
	.....	.20	Rubber, Cowl Side Vent........	15DL3-295								✓	✓	2
	.....	....	Door, Radiator Filler........	15UKS2-2310								✓	✓	1
	.....	....	Bolt, Stove and Nut ($\frac{3}{16}''$ x $\frac{1}{2}''$)........	S-4270A								✓	✓	2
	.....	.01	Lockwasher, $\frac{7}{32}''$ Spring........	S-1344								✓	✓	4
	.....	....	Catch, Radiator Filler Door, and Bracket Assembly........	15UKS3-3033								✓	✓	1
	.....	.50	Windsplit, Radiator Filler Door........	16UKS3-181								✓	✓	1
	.....	....	Screw, Round Head Machine (#10—32 x $\frac{1}{2}''$)........	S-4277A								✓	✓	2
	.....	.01	Washer, Flat ($\frac{7}{32}''$ x $\frac{7}{16}''$ x $\frac{1}{32}''$)........	S-350								✓	✓	2
	.....	.01	Lockwasher, $\frac{7}{32}''$ Spring........	S-1344								✓	✓	2
	.....	6.35	Ventilator, Roof........	15URK2-395								✓	✓	1
	.....	....	Screw, Round Head Machine (#12—24 x $1\frac{1}{8}''$)........	S-6175								✓	✓	4
	.....	....	Screw, Oval Head Machine (#8—32 x $1\frac{1}{2}''$)........	S-6176								✓	✓	4
	.....	2.40	Screen, Rear Window, Assembly........	15UKS3-450								✓	✓	1
	.....	.02	Screw, Round Head Self-Tapping (#10—32 x 1'')........	S-3526								✓	✓	12
	.....	.85	Handle, Grab........	16URK3-165								✓	✓	2
	.....	....	Screw, Flat Head Machine ($\frac{1}{4}''$—20 x 1'')........	S-3843A								✓	✓	4
	.....	.01	Nut, Hex. ($\frac{1}{4}''$—20)........	S-76								✓	✓	4
	.....	.01	Lockwasher, $\frac{9}{32}''$ Spring........	S-1345								✓	✓	4
	.....	17.25	Hood, Engine, Assembly........	16UKS5-950								✓	✓	1
	.....	.01	Screw, Round Head Machine ($\frac{1}{4}''$—20 x 1'')........	S-1607A								✓	✓	3
	.....	.01	Rivet, Bifurcated ($\frac{3}{32}''$ x $\frac{5}{8}''$)........	S-1011								✓	✓	5
	.....	.01	Washer, Flat ($\frac{9}{32}''$ x $\frac{3}{4}''$ x $\frac{1}{16}''$)........	S-353								✓	✓	3
	.....	.03	Washer, Flat ($\frac{11}{64}''$ x $\frac{1}{8}''$ x $\frac{1}{16}''$)........	S-702								✓	✓	5
	.....	.01	Lockwasher, $\frac{9}{32}''$ Spring........	S-1345								✓	✓	3
	.....	.60	Catch, Hood........	16UKS2-90								✓	✓	2
	.....	....	Screw, Round Head Machine (#10—32 x $1\frac{1}{2}''$)........	S-1608A								✓	✓	2
	.....	.01	Nut, Hex. (#10—32)........	S-3745								✓	✓	2
	.....	.01	Washer, Flat ($\frac{7}{32}''$ x $\frac{7}{16}''$ x $\frac{1}{32}''$)........	S-350								✓	✓	2

18-2 BODY

Master Parts List — THE AUTOCAR COMPANY, ARDMORE, PA.

CODE	MFR. PART NO.	LIST PRICE	NAME OF PART	AUTOCAR PART NO.	U-2044	U-4044	U-4144	U-4144-T	C-50	U-5044	U-7144-T	U-8144	U-8144-T	NO. REQ.
1800	......	.01	Lockwasher, $\frac{7}{32}$" Spring............	S-1344								✓	✓	4
	......	.05	Hook, Hood Catch..........	16UBL2-243								✓	✓	2
	......	.01	Rivet, Button Head.........	S-1064								✓	✓	2
	......	.01	Washer, Flat ($\frac{7}{32}$" x $\frac{7}{16}$" x $\frac{1}{32}$").....	S-350								✓	✓	2
	......	11.75	Cover, Engine Fan, Assembly.....	16UKS5-828								✓	✓	1
	......	....	Screw, Round Head Machine ($\frac{1}{4}$"—20 x 1").....	S-1607A								✓	✓	6
	......	.01	Nut, Hex. ($\frac{1}{4}$"—20).........	S-76								✓	✓	2
	......	.01	Lockwasher, $\frac{9}{32}$" Spring.......	S-1345								✓	✓	6
	......	12.50	Cushion, Seat, Assembly—R. H........	16UKS3-2220								✓	✓	1
	......	12.50	Cushion, Seat, Assembly—L. H........	16UKS3-2230								✓	✓	1
	......	10.00	Cushion, Seat Back, Assembly—R. H.....	16UKS3-990								✓	✓	1
	......	10.00	Cushion, Seat Back, Assembly—L. H.....	16UKS3-2210								✓	✓	1
	......	.55	Bracket, Fire Extinguisher.........	16UKS2-2587								✓	✓	1
	......	....	Screw, Round Head Machine ($\frac{1}{4}$"—20 x $\frac{3}{4}$" Cad.)...	S-1614A								✓	✓	3
	......	.01	Nut, Hex. ($\frac{1}{4}$"—20).........	S-76								✓	✓	3
	......	.01	Lockwasher, $\frac{9}{32}$" Spring.......	S-1345								✓	✓	6
	......	18.35	Hood, Motor, Assembly........	16UKSA5-950								✓	✓	1
	......	.01	Screw, Round Head Machine ($\frac{1}{4}$"—20 x $\frac{7}{8}$").....	S-3147A								✓	✓	3
	......	.01	Lockwasher, $\frac{9}{32}$" Spring.......	S-1345								✓	✓	3
	......	.01	Washer, Flat ($\frac{9}{32}$" x $\frac{3}{4}$" x $\frac{1}{16}$").....	S-353								✓	✓	3
	......	.40	Plate, Power Take-Off Shift.......	16UV2-1896								✓	✓	1
	......	.01	Screw, Round Head Self-Tapping (#10 x $\frac{5}{8}$").....	S-3524								✓	✓	2
	......	....	Floorboard—R. H.............	16UKSA4-777								✓	✓	1
	......	.45	Asbestos............	16UKS0-1875								✓	✓	1
	......	.01	Screw, Round Head Machine ($\frac{3}{8}$"—16 x $\frac{3}{4}$").....	S-1618A								✓	✓	2
	......	.02	Screw, Flat Head Machine ($\frac{3}{8}$"—16 x 1").....	S-3830A								✓	✓	2
	......	.01	Screw, Round Head Machine ($\frac{1}{4}$"—20 x 1").....	S-1607A								✓	✓	1
	......	.01	Lockwasher, $\frac{13}{32}$" Shakeproof (Internal).....	S-3190								✓	✓	2
	......	.01	Lockwasher, $\frac{9}{32}$" Spring.......	S-1345								✓	✓	1
	......	.01	Washer, Flat ($\frac{1}{32}$" x $\frac{3}{4}$" x $\frac{1}{16}$").....	S-353								✓	✓	1
	......	.01	Rivet, Bifurcated ($\frac{5}{32}$" x $\frac{5}{8}$").....	S-1011								✓	✓	8
	......	.03	Washer, Flat ($\frac{13}{64}$" x $\frac{7}{8}$" x $\frac{1}{16}$").....	S-702								✓	✓	8
	......	.20	Catch, Hood...........	16UKS2-90								✓	✓	2
	......	.01	Screw, Round Head Machine (#10—24 x $\frac{1}{2}$").....	S-4270A								✓	✓	4
	......	.01	Nut, Hex. (#10—24).........	S-103								✓	✓	4
	......	.01	Lockwasher, $\frac{7}{32}$" Spring.......	S-1344								✓	✓	8
	......	....	Floorboard—L. H.............	16UKSA0-778								✓	✓	1
	......	.01	Screw, Round Head Machine ($\frac{3}{8}$"—16 x $\frac{3}{4}$").....	S-1618A								✓	✓	4
	......	.02	Screw, Flat Head Machine ($\frac{3}{8}$"—16 x 1").....	S-3830A								✓	✓	2
	......	.01	Screw, Round Head Machine ($\frac{1}{4}$"—20 x 1").....	S-1607A								✓	✓	1
	......	.01	Lockwasher, $\frac{13}{32}$" Shakeproof (Internal).....	S-3190								✓	✓	4
	......	.01	Lockwasher, $\frac{9}{32}$" Spring.......	S-1345								✓	✓	1
	......	.01	Washer, Flat ($\frac{9}{32}$" x $\frac{3}{4}$" x $\frac{1}{16}$").....	S-353								✓	✓	1
	......	.15	Plate, Floorboard Clamp.........	16UBK2-1029								✓	✓	1
	......	.25	Bolt, Wing.............	14UN2-189								✓	✓	2
	......	.01	Washer, Flat ($\frac{9}{32}$" x $\frac{3}{4}$" x $\frac{1}{16}$").....	S-353								✓	✓	2
	......	....	Guard, Starting Switch.........	16UKSA2-2647								✓	✓	1
	......	....	Insert, Floorboard..........	16UKSA0-2147								✓	✓	1
	......	....	Strip, Floorboard Insert Retainer.....	16UKSD0-562								✓	✓	2
	......	.10	Handle...........	16UKS2-465								✓	✓	1
	......	.03	Nut, Hex. ($\frac{5}{16}$"—18).........	S-82								✓	✓	2
	......	.01	Lockwasher, $\frac{11}{32}$" Spring.......	S-1346								✓	✓	2
	......	....	Toeboard—R. H...........	16UKSB0-19								✓	✓	1
	......	....	Support, Floorboard Front.......	16UKS0-1219								✓	✓	1
	......	.05	Screw, Hex. Cap ($\frac{3}{8}$"—16 x $\frac{3}{4}$").....	S-12								✓	✓	2
	......	.05	Screw, Hex. Cap ($\frac{1}{4}$"—20 x $1\frac{1}{4}$").....	S-67								✓	✓	2
	......	.01	Lockwasher, $\frac{9}{32}$" Spring.......	S-1345								✓	✓	2
	......	.01	Lockwasher, $\frac{11}{32}$" Spring.......	S-1347								✓	✓	2
	......	....	Toeboard—L. H...........	16UKSB0-962								✓	✓	1
	......	....	Strip, Toeboard Retaining (Top).....	16UKSG2-562								✓	✓	1
	......	....	Strip, Toeboard Retaining—L. H.....	16UKSF2-562								✓	✓	1
	......	....	Strip, Toeboard Retaining—R. H.....	16UKSE2-562								✓	✓	1

THE AUTOCAR COMPANY
ARDMORE, PA.

Master Parts List

18-3
BODY

CODE	MFR. PART No.	LIST PRICE	NAME OF PART	AUTOCAR PART No.	U-2044	U-4044	U-4144	U-4144-T	C-50	U-5044	U-7144-T	U-8144	U-8144-T	No. REQ.
1800	......	.40	Felt, Steering Column Finish Plate	16UKSB2-2016								✓	✓	1
	......	.05	Screw, Hex. Cap (3/8"—16 x 3/4")	S-12								✓	✓	1
	......	.05	Screw, Hex. Cap (1/4"—20 x 1 1/4")	S-67								✓	✓	1
	......	.01	Lockwasher, 13/32" Spring	S-1347								✓	✓	1
	......	.01	Lockwasher, 9/32" Spring	S-1345								✓	✓	1
	......	....	Insert, Toeboard	16UKSA3-13								✓	✓	1
	......	.05	Screw, Hex. Cap (3/8"—16 x 3/4")	S-12								✓	✓	2
	......	.05	Screw, Hex. Cap (1/4"—20 x 1 1/4")	S-67								✓	✓	1
	......	.01	Lockwasher, 13/32" Spring	S-1347								✓	✓	2
	......	.01	Lockwasher, 9/32" Spring	S-1345								✓	✓	1
	......	.15	Felt, Pedal Finish Plate	16UKSA2-2016								✓	✓	2
	......	....	Plate, Brake Pedal Finish	16UKS2-1927								✓	✓	2
	......	.01	Rivets, Bifurcated (5/32" x 1/2")	S-1010								✓	✓	8
	1192	200.00	Body, Complete, less tarpaulin and mounting equipment (Galion)	32UV0-500									✓	1
	1557-X	150.00	Body, less Endgates	1557-X								✓		1
	1392-X	28.75	Welded, Endgate, Assembly	1392-X								✓		2
	117571	2.75	Bearing, Endgate Hinge	117571								✓		4
	112084	.05	Screw, Hex. Cap (5/8"—18 x 1 1/2")	112084								✓		8
	96072	.05	Nut, Hex. (5/8"—18)	96072								✓		8
	95288	.01	Lockwasher, 21/32" Spring	S-1351								✓		16
	1393-X	1.25	Chain, Endgate, Assembly	1393-X								✓		4
	117517	.25	Hook, Endgate Chain	117517								✓		4
	90440	.05	Hook, Endgate Chain "S"	90440								✓		4
	90452	.25	Chain, Endgate	90452								✓		4
	1175161	.25	Cover, Endgate Chain	1175161								✓		4
	102203	.03	Screw, Hex. Cap (3/8"—24 x 1 1/4")	S-1460								✓		2
	102203	.03	Screw, Hex. Cap (3/8"—24 x 1")	102203								✓		2
	96076	.03	Nut, Hex. (3/8"—24)	96076								✓		4
	96068	.01	Lockwasher, 13/32" Spring	S-1347								✓		8
	96064	.03	Washer, Flat Steel (3/8")	90664								✓		4
	1175202	1.00	Handle, Grab	1175202								✓		2
	96058	.03	Screw, Hex. Cap (3/8"—24 x 1")	96058								✓		2
	96076	.03	Nut, Hex. (3/8"—24)	96076								✓		2
	96068	.01	Lockwasher, 13/32" Spring	S-1347								✓		4
	117575	15.00	Tarpaulin	117575								✓		1
	1175164	1.80	Drawrope, Tarpaulin	1175164								✓		1
	117566	.65	Hook, Lash—R. H.	117566								✓		8
	117565	.65	Hook, Lash—L. H.	117565								✓		6
	1175162	.05	Rivet, Lash Hook (5/16" x 1")	1175162								✓		28
	117569	.50	Stud, Spare Tire Mounting—Plain	117569								✓		2
	117570	.65	Stud, Spare Tire Mounting Padlock	117570								✓		1
	1175205	1.25	Sill, Wood Body—R. H.	32UV2-836								✓		1
	119126	.06	Bolt, Body Mounting—R. H. (3/4"—16 x 16")	119126								✓		4
	1175166	.03	Nut, Hex. (3/4"—16)	S-847								✓		4
	1175167	.01	Lockwasher, 25/32" Spring	S-1353								✓		8
	119157	1.25	Bolt, Body Mounting—L. H. (3/4"—16 x 18 1/16")	32UV2-839								✓		4
	......	.30	Nut, Hex. Self-Locking (3/4"—16)	S-2106								✓		4
	......	.06	Washer, Flat Steel (3/4")	S-391								✓		4
	1175168	.60	Plate, Body Mounting Bolt—R. H.	1175168								✓		2
	1175226	1.65	Plate, Body Mounting Bolt—L. H.	32UV3-838								✓		2
	1394-X	45.00	Box, Tool Assembly	16UV0-845								✓		1
	1395-X	7.50	Door, Tool Box Assembly	1395-X								✓		1
	117572	2.25	Hinge, Tool Box Door	117572								✓		1
	117577	.25	Hasp, Tool Box Door	117577								✓		2
	117578	.10	Staple, Tool Box Door Hasp	117578								✓		2
	1175169	.05	Screw, Hex. Cap (1/2"—20 x 1 1/4")	S-2055								✓		8
	96074	.03	Nut, Hex. (1/2"—20)	S-812								✓		8
	96069	.01	Lockwasher, 17/32" Spring	S-1349								✓		16
	1558-X	35.00	Welded, Deck Plate Assembly	27UV5-224								✓		1
	......	.12	Screw, Hex. Cap (1/2"—13 x 1 3/4")	S-3101								✓		4
	......	.04	Nut, Hex. (1/2"—13)	S-77								✓		4
	......	.01	Lockwasher, 17/32" Spring	S-1349								✓		8

18-4 BODY

Master Parts List — THE AUTOCAR COMPANY, ARDMORE, PA.

CODE	MFR. PART No.	LIST PRICE	NAME OF PART	AUTOCAR PART NO.	U-2044	U-4044	U-4144	U-4144-T	C-50	U-5044	U-7144-T	U-8144	U-8144-T	No. REQ.
1800	......	.06	Screw, Hex. Cap ($3/8''$—16 x $1¼''$)	S-64									✓	4
	......	.05	Screw, Hex. Cap ($3/8''$—16 x $1''$)	S-23									✓	4
	......	.03	Nut, Hex. ($3/8''$—16)	S-75									✓	8
	......	.01	Lockwasher, $13/32''$ Spring	S-1347									✓	16
	1175228	1.95	Outrigger, Deck Plate	27UV3-557									✓	1
	......	.12	Screw, Hex. Cap ($½''$—13 x $1¾''$)	S-3101									✓	2
	......	.04	Nut, Hex. ($½''$—13)	S-77									✓	2
	......	.01	Lockwasher, $17/32''$ Spring	S-1349									✓	4
	1559-X	1.75	Step, Stirrup Assembly	14UV3-250									✓	2
	......	.06	Screw, Hex. Cap ($3/8''$—16 x $1¼''$)	S-64									✓	8
	......	.03	Nut, Hex. ($3/8''$—16)	S-75									✓	8
	......	.01	Lockwasher, $13/32''$ Spring	S-1347									✓	16
	1175229	2.00	Reinforcement, Deck Plate—Rear	27UVC3-217									✓	1
	......	.06	Screw, Hex. Cap ($3/8''$—16 x $1¼''$)	S-64									✓	11
	......	.03	Nut, Hex. ($3/8''$—16)	S-75									✓	11
	......	.01	Lockwasher, $13/32''$ Spring	S-1347									✓	22
	1175204	4.50	Fender, Rear Quarter—Front—R. H.	14UV3-41									✓	1
	......	.05	Screw, Hex. Cap ($3/8''$—16 x $1''$)	S-23									✓	5
	......	.03	Nut, Hex. ($3/8''$—16)	S-75									✓	5
	......	.01	Lockwasher, $13/32''$ Spring	S-1347									✓	10
	1175203	4.75	Fender, Rear Quarter—Front—L. H.	14UV3-42									✓	1
	......	.05	Screw, Hex. Cap ($3/8''$—16 x $1''$)	S-23									✓	5
	......	.03	Nut, Hex. ($3/8''$—16)	S-75									✓	5
	......	.01	Lockwasher, $13/32''$ Spring	S-1347									✓	10
	1175230	1.10	Bracket, Quarter Deck Support—Upper	24UV2-439									✓	2
	......	.15	Nut, Budd Wheel Outer ($1 1/8''$—16)	4A0-503									✓	3
	......	5.00	Fender, Rear Quarter—Rear—R. H.	14UVA3-41									✓	1
	......	.05	Screw, Hex. Cap ($3/8''$—16 x $1''$)	S-23									✓	6
	......	.03	Nut, Hex. ($3/8''$—16)	S-75									✓	6
	......	.01	Lockwasher, $13/32''$ Spring	S-1347									✓	12
	......	5.00	Fender, Rear Quarter—Rear—L. H.	14UVA3-42									✓	1
	......	.05	Screw, Hex. Cap ($3/8''$—16 x $1''$)	S-23									✓	6
	......	.03	Nut, Hex. ($3/8''$—16)	S-75									✓	6
	......	.01	Lockwasher, $13/32''$ Spring	S-1347									✓	12
	......	.35	Iron, Rear Fender—R. H.	14UV3-211									✓	2
	......	.35	Iron, Rear Fender—L. H.	14UV3-212									✓	2
	......	1.05	Flap, Fender Assembly	14UKS4-330									✓	2
	......	5.25	Support, Rear Quarter Fender Rear—Upper	14UV4-461									✓	1
	......	.12	Screw, Hex. Cap ($½''$—13 x $1¾''$)	S-3101									✓	2
	......	.04	Nut, Hex. ($½''$—13)	S-77									✓	2
	......	.01	Lockwasher, $17/32''$ Spring	S-1349									✓	4
	......	3.65	Support, Rear Quarter Fender Rear Lower	14UV4-662									✓	1
	......	.09	Screw, Hex. Cap ($½''$—13 x $1½''$)	S-1872									✓	2
	......	.04	Nut, Hex. ($½''$—13)	S-77									✓	2
	......	.01	Lockwasher, $17/32''$ Spring	S-1349									✓	4
	......	1.50	Bracket, Rear Quarter Fender Lower Support	14UV3-463									✓	2
	......	.10	Screw, Cap Hex. ($½''$—13 x $2''$)	S-922									✓	2
	......	.04	Nut, Hex. ($½''$—13)	S-77									✓	2
	......	.01	Lockwasher, $13/32''$ Spring	S-1349									✓	4
	......	.20	Strap, Deck Plate Support—L. H.	27UVB2-217									✓	1
	......	1.50	Plate, Deck—Rear	27UKSB3-241									✓	1
	......	.30	Bracket, Deck Plate Clamp—Lower	27A2-238									✓	2
	......	.20	Spacer, Deck Plate Bracket	27ZA2-239									✓	2
	......	.35	Reinforcement, Deck Plate Outrigger	27UV2-432									✓	1
	......	1.50	Support, Deck Plate—L. H.	27UV3-561									✓	1
	......	.06	Screw, Hex. Cap ($3/8''$—16 x $1¼''$)	S-64									✓	3
	......	.03	Nut, Hex. ($3/8''$—16)	S-75									✓	3
	......	.01	Lockwasher, $13/32''$ Spring	S-1347									✓	6
	......	.55	Bracket, Deck Plate Mounting—Front—R. H.	27UV3-567									✓	1
	......	.55	Bracket, Deck Plate Mounting—Front—L. H.	27UV3-568									✓	1
	......	.30	Spring, Body Hold-Down—L. H.	32UV2-837									✓	4
	......	.75	Plate, Body Locating	32UV3-844									✓	1

THE AUTOCAR COMPANY
ARDMORE, PA.

Master Parts List

18-5 BODY

CODE	MFR. PART No.	LIST PRICE	NAME OF PART	AUTOCAR PART No.	U-2044	U-4044	U-4144	U-4144-T	C-50	U-5044	U-7144-T	U-8144	U-8144-T	No. REQ.
1800	......	.20	Screw, Hex. Cap (½"—13 x 4¾")..................	S-1889									✓	1
	......	.05	Screw, Hex. Cap (⅝"—11 x 1⅞")..................	S-6294									✓	2
	......	.04	Nut, Hex. (½"—13)..................	S-77									✓	1
	......	.05	Nut, Hex. (⅝"—11)..................	S-93									✓	2
	......	.01	Lockwasher, $\frac{17}{32}$" Spring..................	S-1349									✓	2
	......	.01	Lockwasher, $\frac{21}{32}$" Spring..................	S-1351									✓	4
	......	3.25	Sill, Body—L. H..................	32UV3-845									✓	1
1801	**Windshield Wiper Assembly and Parts**													
	FP-657	....	Wiper, Windshield Assembly Complete (Trico).............	16UKS4-823								✓	✓	2
	......	....	Bolt, Round Head Stove and Nut ($\frac{3}{16}$" x ¾")..........	S-2923A								✓	✓	2
	......	.01	Lockwasher, $\frac{7}{32}$" Spring..................	S-1344								✓	✓	2
	22141	....	Motor, Pressure Assembly..................	16UKS4-1128								✓	✓	2
	76570-51-ZE	....	Arm, Wiper Assembly..................	16UKS4-1135								✓	✓	2
	L-778-30-ZE	....	Blade, Wiper..................	16UKS4-1130								✓	✓	2
	75072-ZE	....	Nut, Arm..................	16UKS4-1026								✓	✓	2
	76115-ZE	....	Nut, Holding..................	16UKS4-1137								✓	✓	2
	77121-1	....	Washer, Leather..................	S-926								✓	✓	2
	75470-4-ZE	....	Valve, Pressure Regulating..................	16UKS0-1428								✓	✓	2
	375	....	Body, ¼" Plain Compresion..................	S-5474								✓	✓	2
	377	....	Nut, ¼" Plain Compression..................	S-5474								✓	✓	2

GROUP 19

THE AUTOCAR COMPANY, ARDMORE, PA.

Master Parts List

ILLUSTRATIONS

GROUP 19

WINCH and CONTROL LEVERS

GROUP 19

Master Parts List

THE AUTOCAR COMPANY, ARDMORE, PA.

As supplied on Autocar Models U-8144-T

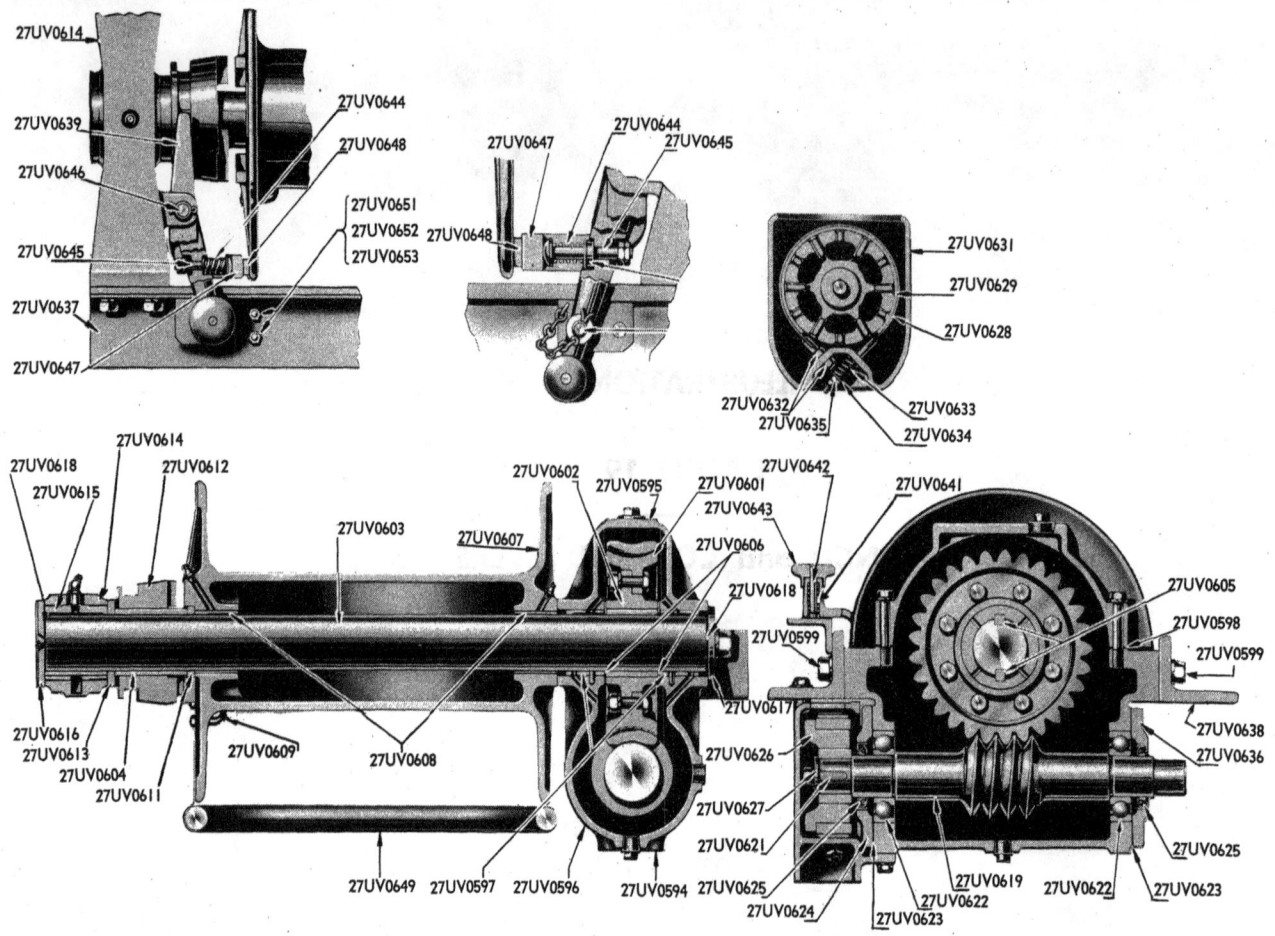

1900—Winch and Drive Shaft

THE AUTOCAR COMPANY, ARDMORE, PA.

Master Parts List

GROUP 19

As supplied on Autocar Models U-8144-T

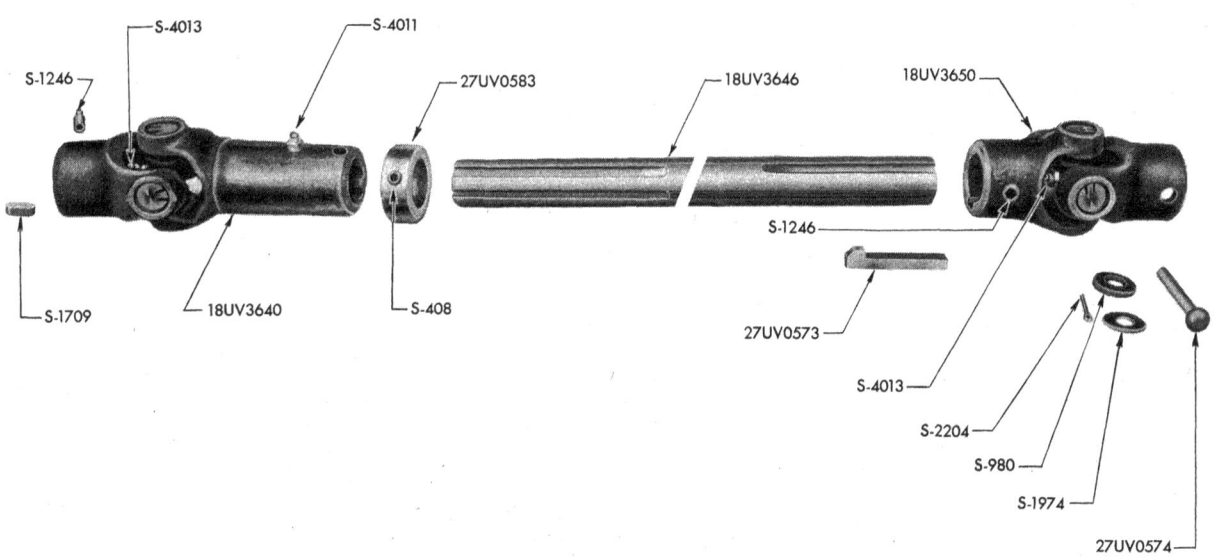

1900—Driveshaft and Universal Joints

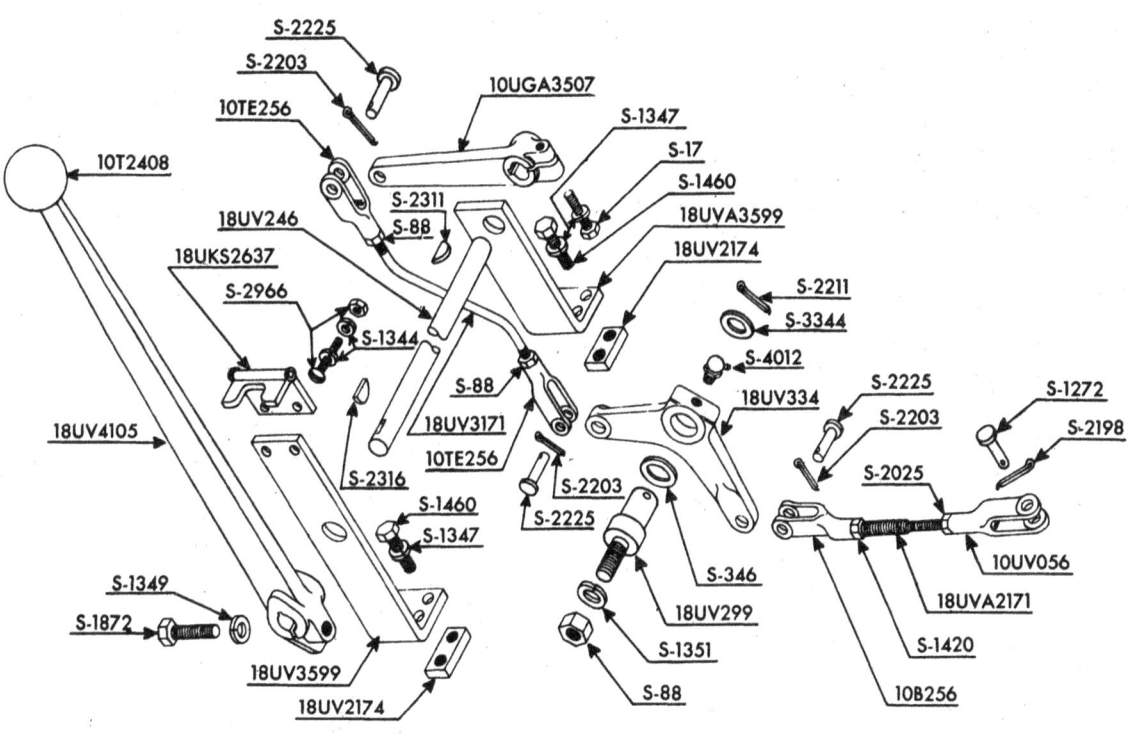

1901—Power Take-Off Control

GROUP 19

Master Parts List
THE AUTOCAR COMPANY, ARDMORE, PA.

As supplied on Autocar Models
U-8144-T

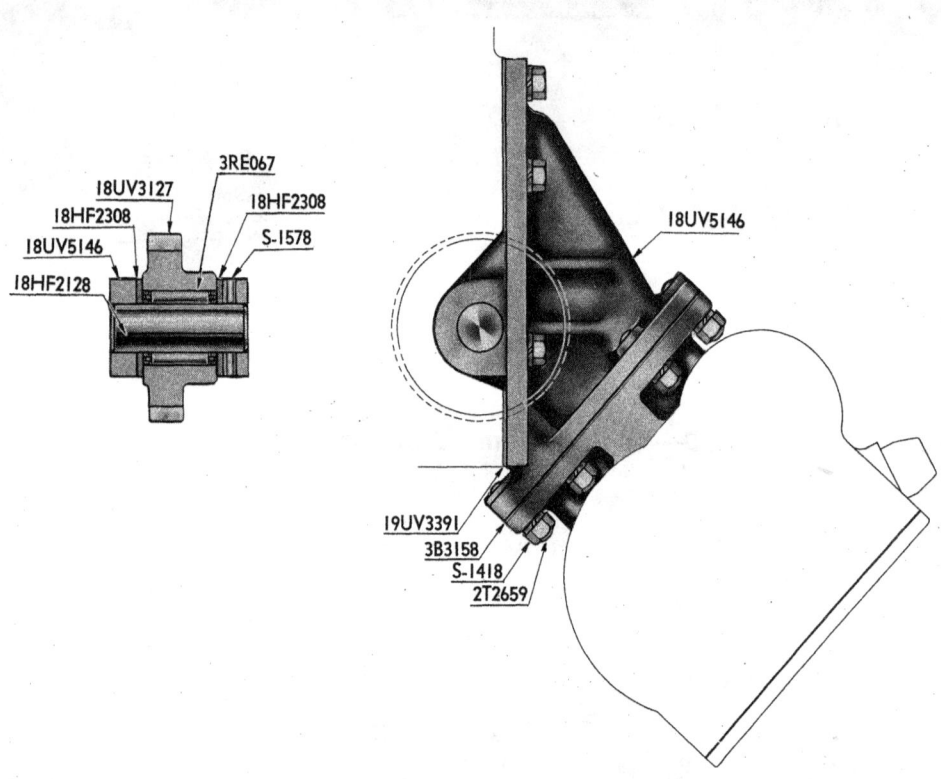

1911—Power Take-Off—End View

GROUP 19

As supplied on Autocar Models
U-8144-T

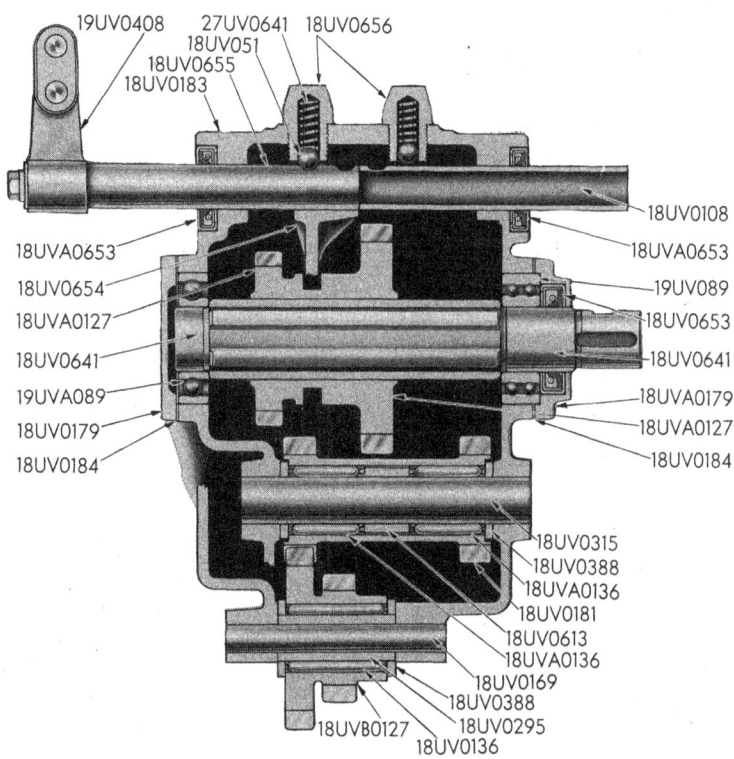

1911—Power Take-Off—Side View

19-0 WINCH

Master Parts List
THE AUTOCAR COMPANY, ARDMORE, PA.

CODE	MFR. PART No.	LIST PRICE	NAME OF PART	AUTOCAR PART No.	U-2044	U-4044	U-4144	U-4144-T	C-50	U-5044	U-7144-T	U-8144	U-8144-T	No. REQ.
1900	**Winch and Drive Shaft**													
	22Y5347	45.00	Case, Gear, Assembly Case and Cover (Gar Wood)	27UV0-594									✓	1
	22Y4201	10.00	Cover, Gear Case	27UV0-595									✓	1
	22Y5367	35.00	Case, Gear	27UV0-596									✓	1
	12Y4213	2.00	Bushing, Gear Case	27UV0-597									✓	2
	12Y4234	.05	Gasket, Cover	27UV0-598									✓	2
	12Y6229	.10	Screw, Special Cap	27UV0-599									✓	4
	12Y4312	21.50	Gear, Worm—L. H.	27UV0-601									✓	1
	22Y4310	12.50	Spider, Worm Gear, Assembly	27UV0-602									✓	1
	22Y5393A	15.00	Shaft, Drum, Assembly	27UV0-603									✓	1
	12Y6236	.30	Key, Clutch	27UV0-064									✓	2
	12Y5361	.10	Key, Worm Gear	27UV0-605									✓	2
	12Y4307	.65	Washer, Gear Thrust	27UV0-606									✓	2
	22Y5394	35.00	Drum Assembly	27UV0-607									✓	1
	12Y4249	1.00	Bushing, Drum	27UV0-608									✓	2
	22Y610	.50	Clamp, Rope	27UV0-609									✓	1
	12Y4206	1.25	Ring, Drum Thrust	27UV0-611									✓	1
	12Y4223	5.00	Clutch, Sliding	27UV0-612									✓	1
	12Y4216	1.00	Ring, Frame Thrust	27UV0-613									✓	1
	22Y5348	20.00	Frame, End, Assembly	27UV0-614									✓	1
	12Y3318	4.50	Sleeve, End Frame	27UV0-615									✓	1
	12Y6226	.25	Washer, Drum Shaft	27UV0-616									✓	1
	12Y6226	.25	Washer, Drum Shaft	27UV0-617									✓	1
	12Y3228	.03	Shim, Drum Shaft	27UV0-618									✓	6
	22Y5370	30.00	Worm Assembly	27UV0-619									✓	1
	10X212⅜	.10	Key, Brake Disc	27UV0-621									✓	1
	16X7309	6.50	Bearing, Ball (MRC-7309-Radio)	27UV0-622									✓	2
	12Y4242	.10	Gasket	27UV0-623									✓	2
	12Y5356	2.00	Cap, Worm—Brake End	27UV0-624									✓	1
	3Y393	1.00	Seal, Oil (Victor 60656)	27UV0-625									✓	2
	12Y4225	2.50	Disc, Brake	27UV0-626									✓	1
	7Y220	.10	Washer, Disc Retainer	27UV0-627									✓	1
	22Y5356	3.50	Band, Brake, Assembly	27UV0-628									✓	1
	22Y5344	1.00	Lining with Rivets	27UV0-629									✓	1
	22Y5351	5.00	Case, Brake, Assembly	27UV0-631									✓	1
	6X3⅜	.02	Nut, Brake Band—Dead End	27UV0-632									✓	2
	12Y2538	.05	Spring, Brake	27UV0-633									✓	1
	12Y2533	.15	Washer, Brake Spring	27UV0-634									✓	2
	6X3¼	.02	Nut, Brake Band—Spring End	27UV0-635									✓	2
	12Y4214	1.85	Cap, Worm—Joint End	27UV0-636									✓	1
	22Y7307	8.50	Angle, Base—Front	27UV0-637									✓	1
	......	.15	Screw, Hex. Cap (⅝"—11 x 2")	S-3387									✓	2
	......	.05	Nut, Hex. (⅝"—11)	S-93									✓	2
	......	.01	Lockwasher, 21/32" Spring	S-1351									✓	4
	12Y7307	5.50	Angle, Base—Rear	27UV0-638									✓	1
	......	.11	Screw, Hex. Cap (½"—13 x 2½")	S-1871									✓	2
	......	.04	Nut, Hex. (½"—13)	S-77									✓	2
	......	.01	Lockwasher, 17/32" Spring	S-1349									✓	4
	......	.15	Screw, Hex. Cap (⅝"—11 x 2")	S-3387									✓	2
	......	.05	Nut, Hex. (⅝"—11)	S-93									✓	2
	......	.01	Lockwasher, 21/32" Spring	S-1351									✓	4
	22Y6204R	2.75	Yoke, Clutch, Assembly	27UV0-639									✓	1
	7Y1037	.05	Spring, Poppet	27UV0-641									✓	1
	12Y6206	.35	Poppet	27UV0-642									✓	1
	12Y6208	.75	Knob, Poppet	27UV0-643									✓	1
	12Y2556	.10	Spring, Drag Brake	27UV0-644									✓	1
	12Y6224	.35	Bolt, Brake Spring	27UV0-645									✓	1
	22Y5357	.50	Pin, Yoke, Assembly	27UV0-646									✓	1
	22Y6207	1.50	Brake, Drag, Assembly	27UV0-647									✓	1
	12Y1531	.25	Lining with Rivets	27UV0-648									✓	1
	12Y5371	3.25	Guard, Rope	27UV0-649									✓	1
	3X316½	.07	Screw, Cap	27UV0-651									✓	4

CODE	MFR. PART No.	LIST PRICE	NAME OF PART	AUTOCAR PART No.	U-2044	U-4044	U-4144	U-4144-T	C-50	U-5044	U-7144-T	U-8144	U-8144-T	No. REQ.
1900	6X3½	.03	**Nut**	27UV0-652									✓	4
	8X4½	.02	**Lockwasher**	27UV0-653									✓	4
	6Y1911	7.50	**Joint,** Take-Off	18UV3-640									✓	1
	......	.10	**Key,** Feather (5/16″ x 5/16″ x 1⅜″)	S-1709									✓	1
	......	.10	**Screw,** Headless Dog Set (½″—13 x ½″)	S-1246									✓	1
	......	.25	**Fitting,** 90° Alemite (⅛″)	S-4013									✓	1
	......	.15	**Fitting,** Straight Alemite (⅛″)	S-4011									✓	1
	19X1¼	.55	**Collar,** Driveshaft	27UV0-583									✓	1
	......	.06	**Screw,** Headless Cup Set (⅜″—16 x ⅜″)	S-408									✓	1
	6Y1255A	6.00	**Driveshaft**	18UV3-646									✓	1
	10X320 5/16	.10	**Key,** Driveshaft	27UV0-573									✓	1
	6Y1931A	7.50	**Joint,** Worm	18UV3-650									✓	1
	......	.10	**Screw,** Headless Dog Set (½″—13 x ½″)	S-1246									✓	1
	......	.25	**Fitting,** 90° Alemite	S-4013									✓	1
	6Y1295	.09	**Pin,** Shear	27UV0-574									✓	1
	......	.03	**Washer,** Flat (13/32″ x 1″ x 3/16″)	S-980									✓	1
	......	.03	**Washer,** Flat (13/32″ x 1″ x 1/16″)	S-1974									✓	1
	......	.01	**Pin,** Cotter (⅛″ x 1″)	S-2204									✓	1
1910	**Control Levers and Rods**													
	......	6.00	**Lever,** P. T. O. Control Hand	18UV4-105									✓	1
	......	.90	**Ball,** P. T. O. Control Hand Lever	10T2-408									✓	1
	......	.09	**Screw,** Hex. Cap (½″—13 x 1½″)	S-1872									✓	1
	......	.01	**Lockwasher,** 17/32″ Spring	S-1349									✓	1
	......	.55	**Bracket,** P. T. O. Cross Shaft (Inner)	18UV3-599									✓	1
	......	.15	**Plate,** P. T. O. Steady Bearing Bracket	18UV2-174									✓	2
	......	.06	**Screw,** Hex. Cap (⅜″—24 x 1¼″)	S-1460									✓	4
	......	.01	**Lockwasher,** 13/32″ Spring	S-1347									✓	4
	......	.25	**Lock,** P. T. O. Lever Hinge Assembly	18UKS2-637									✓	1
	......	.03	**Bolt,** Round Head Stove (3/16″—1″)	S-2966									✓	2
	......	.01	**Lockwasher,** 7/32″ Spring	S-1344									✓	4
	......	.65	**Shaft,** P. T. O. Control Cross	18UV2-46									✓	1
	......	.06	**Key,** #15 Whitney (¼″—1″)	S-2316									✓	1
	......	.06	**Key,** #11 Whitney (3/16″—⅞″)	S-2311									✓	1
	......	.40	**Bracket,** P. T. O. Cross Shaft (Outer)	18UVA3-599									✓	1
	......	1.80	**Lever,** P. T. O. Transfer	10UGA3-507									✓	1
	......	.06	**Screw,** Hex. Cap (⅜″—16 x 1½″)	S-17									✓	1
	......	.01	**Lockwasher,** 13/32″ Spring	S-1347									✓	1
	......	.45	**Rod,** P. T. O. Shift (Front)	18UV3-171									✓	1
	......	.06	**Nut,** Hex. (⅝″—11)	S-88									✓	2
	......	.90	**Clevis**	10TE2-56									✓	2
	......	.12	**Pin,** Clevis (½″ x 1 13/32″)	S-2225									✓	2
	......	.01	**Pin,** Cotter (⅛″ x ¾″)	S-2203									✓	2
	......	2.90	**Crank,** Bell	18UV3-34									✓	1
	......	.20	**Fitting,** 67½° Alemite (⅛″)	S-4012									✓	1
	......	.45	**Stud,** P. T. O. Control Lever	18UV2-99									✓	1
	......	.01	**Lockwasher,** 21/32″ Spring	S-1351									✓	1
	......	.06	**Nut,** Hex. (⅝″—11)	S-88									✓	1
	......	.04	**Washer,** Flat (⅞″ x 1¾″ x 1/16″)	S-346									✓	1
	......	.03	**Washer,** Flat (⅞″ x 1¾″ x ⅛″)	S-3344									✓	1
	......	.01	**Pin,** Cotter (5/32″ x 1½″)	S-2211									✓	1
	......	.10	**Rod,** P. T. O. Shift (Rear)	18UVA2-171									✓	1
	......	.02	**Nut,** Hex. (7/16″—20)	S-2025									✓	1
	......	.04	**Nut,** Hex. (½″—20)	S-1420									✓	1
	......	1.15	**Clevis**	10B2-56									✓	1
	......	1.00	**Clevis**	10UV0-56									✓	1
	......	.12	**Pin,** Clevis (½″ x 1 13/32″)	S-2225									✓	1
	......	.06	**Pin,** Clevis (7/16″ x 1⅜″)	S-1272									✓	1
	......	.01	**Pin,** Cotter (⅛″ x ¾″)	S-2203									✓	1
	......	.01	**Pin,** Cotter (3/32″ x ¾″)	S-2198									✓	1

19-2 WINCH

Master Parts List — The Autocar Company, Ardmore, PA.

CODE	MFR. PART No.	LIST PRICE	NAME OF PART	AUTOCAR PART No.	U-2044	U-4044	U-4144	U-4144-T	C-50	U-5044	U-7144-T	U-8144	U-8144-T	No. REQ.
1911	**Power Take-Off**													
	.....	17.85	**Adapter**, Power Take-Off Assembly................	18UV0-530									✓	1
	.....	6.25	**Bracket**, Power Take-Off Adapter................	18UV5-146									✓	1
	.....	.06	**Screw**, Hex. Cap (3/8″—16 x 1 1/8″).............	S-46									✓	5
	.....	.01	**Lockwasher**, 13/32″ Spring......................	S-1347									✓	5
	.....	.15	**Gasket**, Adapter...............................	19UV3-391									✓	1
	.....	.04	**Stud**, Adapter (Trans. Case)....................	19UV2-352									✓	1
	.....	.05	**Nut**, Hex. (3/8″—24)............................	S-809									✓	1
	.....	2.10	**Shaft**, Power Take-Off Adapter Gear..............	18HF2-128									✓	1
	.....	.03	**Pin**, Steel (3/16″ x 2 1/2″).....................	S-1578									✓	1
	.....	5.55	**Gear**, Power Take-Off Adapter (22T).............	18UV3-127									✓	1
	.....	.80	**Washer**, Power Take-Off Adapter Gear Thrust.....	18HF2-308									✓	2
	.....	1.20	**Bearing**, Power Take-Off Adapter Gear—(Rollway No. 160426)	3RE0-67									✓	1
	.....	.12	**Stud**, Power Take-Off Adapter...................	2T2-659									✓	6
	.....	.03	**Nut**, Hex. (3/8″—24)............................	S-1418									✓	6
	.....	.01	**Lockwasher**, 13/32″ Spring......................	S-1347									✓	6
	.....	.10	**Gasket**..	3B3-158									✓	1
	77Y6000	181.50	**Take-Off**, Power Assembly.......................	18UV0-550									✓	1
	7Y1600	22.00	**Take-Off** Case................................	18UV0-183									✓	1
	7Y1660	10.00	**Gear**, Intermediate 22/14T......................	18UVB0-127									✓	1
	7Y1661	12.50	**Gear**, Idler 14/14T.............................	18UV0-181									✓	1
	21X94536	.85	**Bearing**, Roller...............................	18UV0-136									✓	1
	7Y1566	2.00	**Sleeve**, Bearing...............................	18UV0-295									✓	1
	7Y1032	.50	**Washer**, Thrust................................	18UV0-388									✓	4
	7Y1004	1.50	**Pin**, Intermediate Gear Assembly................	18UV0-169									✓	1
	21X94526	.75	**Bearing**, Roller...............................	18UVA0-136									✓	1
	7Y1003	3.30	**Pin**, Idler Gear Assembly.......................	18UV0-315									✓	1
	7Y1602	10.00	**Shaft**, Take-Off Assembly.......................	18UV0-641									✓	1
	7Y1662	12.50	**Gear**, Sliding 28/20T...........................	18UVA0-127									✓	1
	16X5207	6.60	**Bearing**, Ball—Double Row......................	19UV0-89									✓	1
	16X207M	3.45	**Bearing**, Ball—Single Row......................	19UVA0-89									✓	1
	7Y1643	.20	**Spacer**, Bearing...............................	18UV0-613									✓	1
	7Y1022	1.30	**Cap**, Driveshaft...............................	18UV0-179									✓	1
	7Y1039	.50	**Seal**, Oil.....................................	18UV0-653									✓	1
	7Y1029	2.00	**Cover**, Oil Seal...............................	18UVA0-179									✓	1
	7Y1043	.05	**Gasket**, Cover.................................	18UV0-184									✓	2
	7Y1601	6.00	**Cover**...	18UV0-102									✓	1
	7Y1513	4.20	**Yoke**, Shift...................................	18UV0-654									✓	1
	7Y1603	4.75	**Rod**, Shift....................................	18UV0-655									✓	1
	7Y1642	2.50	**Tube**, Shift Rod...............................	18UV0-108									✓	1
	7Y1040	.45	**Seal**, Oil.....................................	18UVA0-653									✓	2
	7Y1027	1.50	**Arm**, Shift Rod................................	19UV0-408									✓	1
	7Y521	.05	**Pawl**, Gear Shift..............................	18UV0-51									✓	2
	7Y1037	.05	**Spring**, Pawl..................................	27UV0-641									✓	2
	7Y1034	.75	**Poppet**..	18UV0-656									✓	2
	7Y1646	.15	**Gasket**, Cover.................................	18UV0-249									✓	1
	.....	16.60	**Adapter**, Power Take-Off Assembly...............	18RL0-530					✓					1
	.....	6.25	**Gear**, Power Take-Off Adapter...................	18RL2-127					✓					1
	.....	.45	**Washer**, Power Take-Off Adapter Gear Thrust.....	18RL2-308					✓					2
	.....	.60	**Shaft**, Power Take-Off Adapter Gear.............	18RL0-128					✓					1
	.....	.03	**Lockwasher** (1 3/32″ x 1 3/8″ x 1/16″)..........	S-3359					✓					1
	.....	.04	**Screw**, Lock...................................	S-2065					✓					1
	.....	1.55	**Bearing**, Power Take-Off Adapter Gear Shaft.....	18RL0-136					✓					1
	.....	6.30	**Bracket**, Power Take-Off Adapter................	18RL4-146					✓					1
	.....	.30	**Gasket**, Power Take-Off Adapter Bracket.........	19A3-391					✓					1
	.....	.06	**Screw**, Hex. Cap (3/8″—16 x 1 1/8″).............	S-46					✓					6
	.....	.01	**Lockwasher**, 13/32″ Spring......................	S-1347					✓					6

THE AUTOCAR COMPANY, ARDMORE, PA.

Master Parts List

GROUP 21

ILLUSTRATIONS

GROUP 21

BUMPERS and GUARDS

GROUP 21

Master Parts List

THE AUTOCAR COMPANY, ARDMORE, PA.

As supplied on Autocar Models
U-8144   U-8144-T

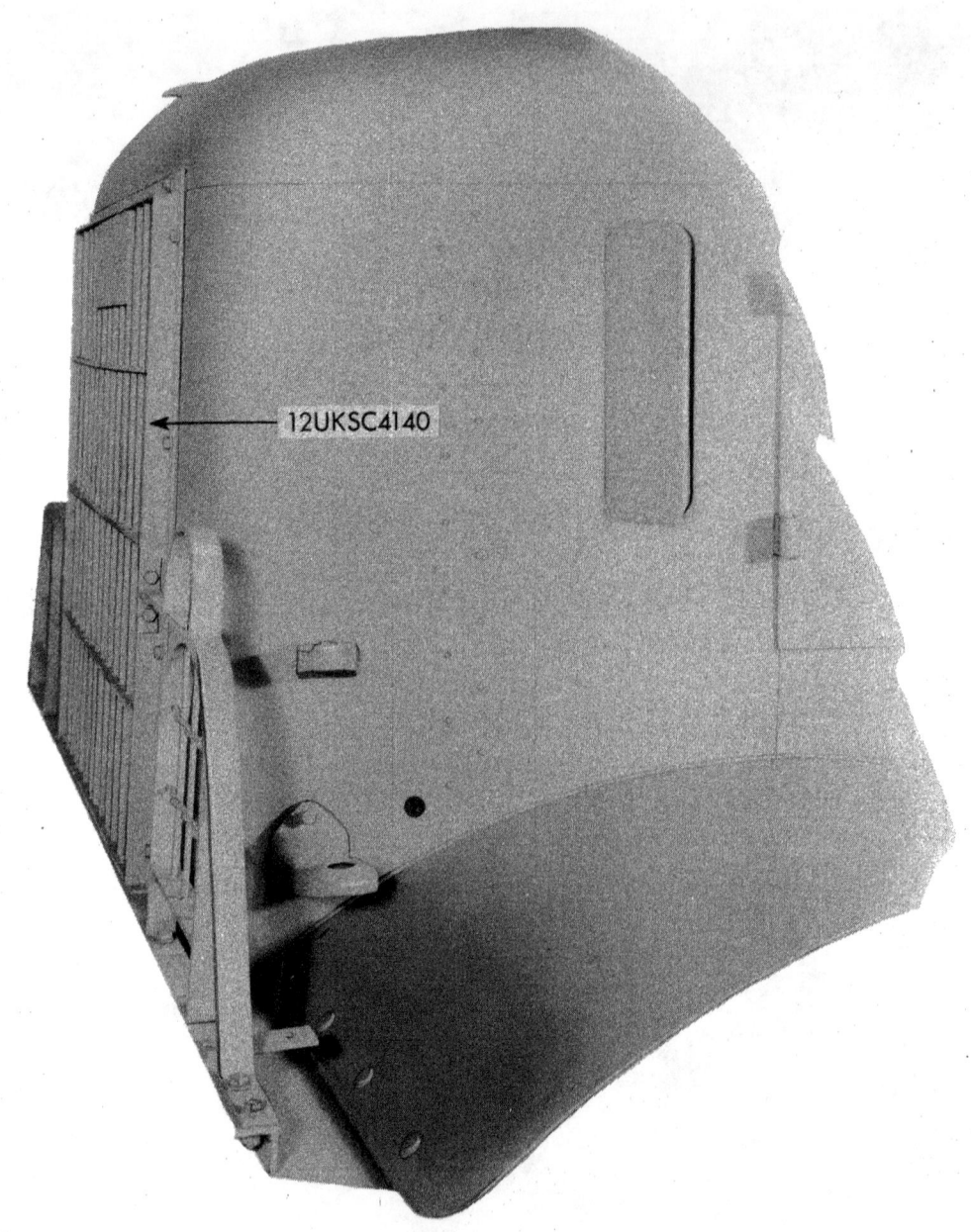

2103—Radiator Guard

Master Parts List

THE AUTOCAR COMPANY
ARDMORE, PA.

21-1 BUMPER

CODE	MFR. PART No.	LIST PRICE	NAME OF PART	AUTOCAR PART No.	U-2044	U-4044	U-4144	U-4144-T	C-50	U-5044	U-7144-T	U-8144	U-8144-T	No. REQ.
2101	Bumpers (see Group 1500)													
2103	Radiator Guard													
	......	....	**Guard,** Radiator and Headlamp Brush Assembly	12UKSC4-140								✓	✓	1
	......	.06	**Screw,** Hex. Cap (⅜"—16 x 1¼")	S-64								✓	✓	11
	......	.03	**Nut,** Hex. (⅜"—16)	S-75								✓	✓	11
	......	.01	**Lockwasher,** 1½" Spring	S-1347								✓	✓	22

THE AUTOCAR COMPANY
ARDMORE, PA.

Master Parts List

22-1
MISC.

CODE	MFR. PART NO.	LIST PRICE	NAME OF PART	AUTOCAR PART No.	U-2044	U-4044	U-4144	U-4144-T	C-50	U-5044	U-7144-T	U-8144	U-8144-T	No. REQ.
2200	Identification and Caution Plates													
	......	.80	Plate, Name	16UKA2-2285	✓									1
	......	.60	Plate, Name	16UGA2-2285		✓	✓	✓						1
	......	1.10	Plate, Name	16UPA2-2285						✓				1
	......	.60	Plate, Name	16UU2-2285							✓			1
	......	.70	Plate, Name	16UVA2-2285A								✓	✓	1
	......	.75	Plate, Caution	16UK2-2285	✓									1
	......	.50	Plate, Caution	16UG2-2285		✓	✓	✓						1
	......	.85	Plate, Caution	16UP2-2285						✓				1
	......	.30	Plate, Caution (Air Pressure)	16UUA2-2285							✓	✓	✓	1
	......	.60	Plate, Shift (Main Trans.)	16UG2-1896	✓									1
	......	.95	Plate, Shift (Main Trans.)	16UK2-1896		✓	✓	✓						1
	......	.90	Plate, Shift (Main Trans.)	16UP2-1896						✓				1
	......	.85	Plate, Shift (Transfer Case)	16UL2-1896A	✓	✓	✓	✓		✓				1
	......	.30	Plate, Caution (Engine Speed)	16UUB2-2285							✓	✓	✓	1
	......	.20	Plate, Caution (Drain)	16UUC2-2285							✓	✓	✓	1
	......	.15	Plate, Publications Reference	16UU2-2842							✓	✓	✓	1
	......	.90	Plate, Gearshift Instruction	16UU3-1896							✓	✓	✓	1
	......	.40	Plate, P. T. O. Instruction	16UV2-1896							✓	✓	✓	1
2201	Rear View Mirrors													
	......	3.00	Mirror, Rear View Assembly—Outside	16URK3-768A	✓	✓								2
	......	3.00	Mirror, Rear View Assembly—Outside	16URK3-768C			✓	✓						2
	......	3.00	Mirror, Rear View Assembly—Outside	16URK3-768						✓				2
	......	3.00	Mirror, Rear View Assembly—Outside	16UKS4-768							✓	✓	✓	2
	......	.65	Mirror, Rear View Assembly—Inside	16URKA2-768		✓				✓				1
	......	1.00	Mirror, Rear View Assembly—Inside	16URKB2-768				✓						1
	......	.90	Mirror, Rear View Assembly—Inside	16UKSA3-768							✓	✓	✓	1
2202	Tarpaulins, Bows and End Curtains (See Group 1800)													
2203	Speedometer and Parts													
	590-X	12.00	Head, Speedometer Assembly (Stewart-Warner)	16UKS3-760								✓	✓	1
	......	5.85	Cable, Speedometer	16UV0-800								✓	✓	1
	......	.06	Clip, Speedometer Cable	16S2-44								✓	✓	2
	......	.04	Screw, Hex. Cap (¼"—20 x ¾")	S-30								✓	✓	2
	......	.01	Nut, Hex. (¼"—20)	S-76								✓	✓	2
	......	.01	Lockwasher, $\frac{9}{32}$" Spring	S-1345								✓	✓	4
	......	.01	Washer, Flat Steel ($\frac{9}{32}$" x ¾" x $\frac{1}{16}$")	S-353								✓	✓	2
	......	.10	Bracket, Speedometer Cable Clip	16UUA2-2192								✓	✓	1
	......	.06	Clip, Speedometer Cable	16H2-44								✓	✓	3
	......	.04	Screw, Hex. Cap (¼"—20 x ¾")	S-30								✓	✓	3
	......	.01	Nut, Hex. (¼"—20)	S-76								✓	✓	3
	......	.01	Lockwasher, $\frac{9}{32}$" Spring	S-1345								✓	✓	6
	......	.01	Clip, Speedometer Cable	6W2-754								✓	✓	2
	......	.04	Screw, Hex. Cap (¼"—20 x ¾")	S-30								✓	✓	2
	......	.01	Nut, Hex. (¼"—20)	S-76								✓	✓	2
	......	.01	Lockwasher, $\frac{9}{32}$" Spring	S-1345								✓	✓	4
	......	.10	Bracket, Speedometer Cable Clip	16UU2-2192								✓	✓	1
	......	.10	Bracket, Speedometer Cable Clip	16UUB3-2192								✓	✓	1
	......	.25	Extension, Speedometer Cable	16SK0-1706								✓	✓	1
	......	.85	Sleeve, Speedometer Cable Extension	16SK0-1705								✓	✓	1
	......	4.25	Adapter, Speedometer	16SCM0-603								✓	✓	1
	NPN	40.00	Head, Tachometer Assembly (Stewart-Warner)	16UKS3-1210								✓	✓	1
	......	4.00	Lock, Tachometer	16UKS3-1844								✓	✓	1
	......	6.65	Cable, Tachometer	16UU0-1220								✓	✓	1
	......	.01	Clip, Tachometer Cable	6W2-754								✓	✓	2
	......	.04	Screw, Hex. Cap (¼"—20 x ¾")	S-30								✓	✓	2
	......	.01	Nut, Hex. (¼"—20)	S-76								✓	✓	2
	......	.01	Lockwasher, $\frac{9}{32}$" Spring	S-1345								✓	✓	4
	......	.07	Bracket, Tachometer Cable Clip	16UU2-3026								✓	✓	1
	......	.10	Bracket, Tachometer Cable Clip	16UUA2-3026								✓	✓	1

THE AUTOCAR COMPANY, ARDMORE, PA.
Master Parts List

23-1 MISC.

CODE	MFR. PART No.	LIST PRICE	NAME OF PART	AUTOCAR PART No.	U-2044	U-4044	U-4144	U-4144-T	C-50	U-5044	U-7144-T	U-8144	U-8144-T	No. REQ.
2300			**Miscellaneous Tools, Tire Chains and Extinguishers**											
	.....	....	Tire, 9.00-20 Spare and Tube	....	✓	✓	✓	✓		✓	✓			1
	.....	....	Tire, 12.00-20 Spare and Tube	....								✓	✓	1
	.....	26.00	Wheel, Budd (Type "L" #33287)	9BJ0-403	✓	✓	✓	✓		✓				1
	.....	26.00	Wheel, Budd (#44470—L—Rim)	9KA4-403							✓			2
	.....	37.00	Wheel, Budd (B-45520)	9CJ4-403								✓	✓	1
	.....	2.20	Padlock	16UKS2-06								✓	✓	1
	.....	2.15	Wrench, Front Wheel Bearing	16UG3-41	✓	✓	✓	✓		✓				1
	.....	1.35	Wrench, Front Wheel Bearing Adjusting Nut	16B0-117							✓			1
	.....	1.35	Wrench, Rear Wheel Bearing Adjusting Nut	16D0-41							✓			1
	.....	1.95	Wrench, Wheel Bearing Adjusting Nut	4G0-696								✓	✓	1
	.....	.75	Wrench, Water Pump Spanner	X-23-072	✓									1
	.....	.70	Wrench, Water Pump Nut	16TE2-134							✓	✓	✓	1
	.....	7.55	Wrench, Auto	16A0-28	✓	✓	✓	✓		✓	✓	✓	✓	1
	.....	1.00	Wrench, Spark Plug	16A0-88	✓	✓	✓	✓		✓	✓	✓	✓	1
	.....	1.80	Wrench—Crescent Type	16A0-2555	✓	✓	✓	✓		✓	✓	✓	✓	1
	.....	.82	Wrench—$\frac{1}{2}''$—$\frac{19}{32}''$	16A0-2556	✓	✓	✓	✓		✓	✓	✓	✓	1
	.....	.58	Wrench—$\frac{3}{8}''$—$\frac{7}{16}''$	16B9-2556	✓	✓	✓	✓		✓	✓	✓	✓	1
	.....	1.00	Wrench—$\frac{9}{16}''$—$\frac{11}{16}''$	16A0-2557	✓	✓	✓	✓		✓	✓	✓	✓	1
	.....	1.24	Wrench—$\frac{5}{8}''$—$\frac{25}{32}''$	16A0-2558	✓	✓	✓	✓		✓	✓	✓	✓	1
	.....	1.65	Wrench—$\frac{3}{4}''$—$\frac{7}{8}''$	16A0-2559	✓	✓	✓	✓		✓	✓	✓	✓	1
	.....	1.70	Crank, Starting Assembly	1ZN2-10	✓						✓			1
	.....	1.35	Crank, Starting Assembly	1TE2-10		✓	✓	✓		✓				1
	.....	2.15	Crank, Starting	1UV2-10								✓	✓	1
	.....	.80	Package, Padlock	16URK2-06		✓								1
	.....	2.00	Package, Padlock	16URKB2-06				✓						1
	.....	.60	Package, Padlock	16URKA2-06						✓				2
	.....	2.20	Padlock	16UKS2-06							✓			2
	.....	16.20	Jack, Lifting and Handle	16RE0-33	✓	✓	✓	✓		✓				1
	.....	21.30	Jack, Hydraulic	16SCH0-33							✓	✓	✓	1
	.....	6.75	Gun, Alemite Grease	16A0-30	✓	✓	✓	✓		✓	✓	✓	✓	1
	.....	.30	Adapter, Grease Gun	16UU0-2844							✓	✓	✓	1
	.....	1.35	Can, Oil	16A0-34	✓	✓	✓	✓		✓	✓	✓	✓	1
	.....	1.60	Pliers—Slip Joint	16A0-198	✓	✓	✓	✓		✓	✓	✓	✓	1
	.....	2.10	Hammer—16-oz. Ball Pein	16A0-218	✓	✓	✓	✓		✓	✓	✓	✓	1
	.....	1.20	Driver, Screw—6"	16A0-226	✓	✓	✓	✓		✓	✓	✓	✓	1
	.....	.50	Driver, Screw—Phillips #2	16C0-226	✓	✓	✓	✓		✓	✓	✓	✓	1
	.....	1.10	Driver, Screw—Phillips #3	16D0-226	✓	✓	✓	✓		✓	✓	✓	✓	1
	.....	22.75	Chains, Tire (Front Type D)	16A0-645	✓	✓	✓	✓		✓				1
	.....	41.60	Chains, Tire (Rear Type N)	16B0-645	✓	✓	✓	✓		✓				1
	..... Per Pr.	31.80	Chains, Tire	16G0-645							✓	✓	✓	4
	.....	14.30	Extinguisher, Fire	16URK2-646	✓	✓	✓	✓		✓				1
	.....	14.00	Extinguisher, Fire	16UKS2-646							✓	✓	✓	1
	.....	49.75	Chain, Towing Assembly	16UU0-2736							✓	✓	✓	1
	.....	16.00	Hose, Trailer Assembly	25B0-350							✓	✓	✓	1
	.....	9.75	Hose, Tire Inflating	25C9-730	✓	✓	✓	✓		✓				1
	.....	17.15	Hose, Tire Inflation Assembly	25DB9-730							✓	✓	✓	1
	.....	.55	Hose, Trailer	25C0-37		✓		✓						2
	.....	13.30	Hose, Trailer	25A0-350				✓						1
	.....	134.50	Cable, Winch Assembly	27UU0-556								✓	✓	1
	.....	61.00	Block, Snatch Assembly	27UV0-680								✓	✓	1
	.....	.09	Pin, Winch Drive Shaft Shear	27UV0-674								✓	✓	2
	.....	3.75	Bag, Tool	16URK2-2793	✓	✓				✓				1
	.....	5.70	Bag, Tool	16URKA2-2793			✓	✓						1
	.....	6.00	Bag, Tool	16URKA3-2793							✓	✓	✓	1
	.....	.08	Container, Key	16HF2-2794	✓	✓	✓	✓		✓	✓	✓	✓	1
	.....	3.40	Wrench, Socket (Rear Wheel)	16HD3-41	✓	✓	✓	✓		✓				1
	.....	1.50	Wrench, Oil Pressure Crow Foot (Hercules #2268-A)	2BL0-1215	✓									1
	.....	2.15	Wrench, Oil Pressure "T" (Hercules #X5800)	2BL0-1214	✓									1
	.....	15.00	Box, Chain	16URK4-2775		✓		✓						1
	.....	18.25	Cable, Trailer Light	16URK3-1570		✓		✓						1
	.....	....	Book, Parts and Instruction	....	✓	✓	✓	✓		✓				1

Master Parts List

THE AUTOCAR COMPANY, ARDMORE, PA.

CODE	MFR. PART No.	LIST PRICE	NAME OF PART	AUTOCAR PART No.	MODEL									No. REQ.
					U-2044	U-4044	U-4144	U-4144-T	C-50	U-5044	U-7144-T	U-8144	U-8144-T	
2300	......	....	Manual, Parts	TM-10-1116							✓			1
	......	....	Manual, Parts	TM-10-1118								✓	✓	1
	......	....	Manual, Maintenance	TM-10-1117							✓			1
	......	....	Manual, Maintenance	TM-10-1119								✓	✓	1

THE AUTOCAR COMPANY
ARDMORE, PA.

Master Parts List

23-3 MISC.

A Consolidation of All Ball and Roller Bearings Serviced by Autocar for U. S. Government

Government Code	Major Unit	Description	Type	Autocar Part Number	Manufacturer's Identification	U-2044	U-4044	U-4144	U-4144-T	C-50	U-5044	U-7144-T	U-8144	U-8144-T	Quantity Used
0203	Clutch	Release Bearing	Ball	3BM2-169A	Aetna A-959-1	✓	✓	✓	✓	✓	✓	✓	✓	✓	1
0205		Pilot Bearing	Ball	3A0-243	SKF 6205-Z	✓	✓	✓	✓						1
		Pilot Bearing	Ball	3D0-243	SKF 6304-Z					✓	✓				1
		Pilot Bearing	Ball	3B0-243A	SKF FLB-30							✓	✓	✓	1
0504	Fan	Spindle Bearing—Cup	Roller	5SCM0-256	Timken 09194	✓	✓	✓	✓	✓	✓				2
		Spindle Bearing—Cup	Roller	5DW0-132	Timken 07204							✓	✓	✓	2
		Spindle Bearing—Cone	Roller	5SCM0-48	Timken 09074	✓	✓	✓	✓	✓	✓				2
		Spindle Bearing—Cone	Roller	5DW0-133	Timken 07098							✓	✓	✓	2
0601	Generator	Commutator End Bearing	Ball	16BLU0-2883	Auto-Lite X-295	✓	✓	✓	✓			✓	✓	✓	1
		Commutator End Bearing	Ball		Delco 908503						✓				1
		Drive End Bearing	Ball	16BLU0-2889	Auto-Lite X-298	✓	✓	✓	✓			✓	✓	✓	1
		Drive End Bearing	Ball		Delco 901305						✓				1
0703	Transmission	Mainshaft Pilot Bearing	Roller	3BLA0-147	Hyatt 94622	✓	✓	✓	✓						1
		Mainshaft Pilot Bearing	Roller	3RL0-147	Hyatt 1306-TM					✓	✓				1
		Mainshaft Pilot Bearing	Roller	3TF0-147	SKF WJM-30							✓	✓	✓	1
		Mainshaft Rear Bearing	Ball	3BLA0-251	New Dept. 43308	✓	✓	✓	✓						1
		Mainshaft Rear Bearing—Cone	Roller	3T2-61	Timken 336					✓	✓				2
		Mainshaft Rear Bearing—Cup	Roller	3T2-62	Timken 332					✓	✓				2
		Mainshaft Rear Bearing—Cone	Roller	3TF2-61	Timken 377							✓	✓	✓	2
		Mainshaft Rear Bearing—Cup	Roller	3TF2-62	Timken 372-A							✓	✓	✓	2
		Countershaft Front Bearing	Roller	3BL0-142	Hyatt 1206-TS	✓									1
		Countershaft Front Bearing	Roller	3BLA0-142	Hyatt 1207-TS		✓	✓	✓						1
		Countershaft Front and Rear Bearing—Cone	Roller	3D2-36	Timken 449					✓	✓				2
		Countershaft Front and Rear Bearing—Cup	Roller	4H2-137	Timken 432					✓	✓				2
		Countershaft Front Bearing	Roller	3TF0-142B	Hyatt R-1309-TS							✓	✓	✓	1
		Countershaft Rear Bearing—Cone	Roller	3T2-61	Timken 336							✓	✓	✓	2
		Countershaft Rear Bearing—Cup	Roller	3T2-62	Timken 332							✓	✓	✓	2
		Driveshaft Bearing	Ball	3BLA0-79	New Dept. 47511	✓	✓	✓	✓						1
		Driveshaft Bearing	Ball	3D2-248A	SKF 6309Z-C-003					✓	✓				1
		Driveshaft Bearing	Ball	3T0-79	New Dept. 7409-XL							✓			1
		Driveshaft Bearing	Ball	3TF0-79A	SKF T-76409								✓	✓	1
0702		Reverse Gear Bearing	Roller	3BLA0-147	Hyatt 94622		✓	✓	✓						2
		Reverse Gear Needle Bearing	Roller	3TFB2-1073	SKF							✓	✓	✓	106
0802	Transfer Case	Mainshaft Rear Bearing	Ball	19DKB0-362	SKF 1407	✓	✓	✓	✓		✓				1
		Mainshaft Rear Bearing	Ball	19FK0-367	SKF 7609							✓	✓	✓	1
		Mainshaft Front Bearing	Ball	3CKA2-285	SKF 1309	✓	✓	✓	✓						1
		Mainshaft Front Bearing	Ball	19FK0-361	SKF 7610							✓	✓	✓	1
		Mainshaft Drive Idler Bearing	Ball	3CKA2-285	SKF 1309	✓	✓	✓	✓		✓				2
		Mainshaft Drive Idler Bearing	Ball	19FK0-421	SKF 7209							✓	✓	✓	2
0803		Driven Shaft Bearing—Cone	Roller	4ZGE0-653	Timken 438	✓	✓	✓	✓		✓				2
		Driven Shaft Bearing—Cup	Roller	4H2-137	Timken 432	✓	✓	✓	✓		✓				2
		Driven Shaft Front Bearing—Cone	Roller	19FK0-442	Timken 3979							✓	✓	✓	1
		Driven Shaft Front Bearing—Cup	Roller	4UF1-826	Timken 3920							✓	✓	✓	1
		Driven Shaft Rear Bearing—Cone	Roller	19FK0-636	Timken 49585							✓	✓	✓	1
		Driven Shaft Rear Bearing—Cup	Roller	19FK0-637	Timken 49520							✓	✓	✓	1
		Declutch Shaft Bearing	Ball	3SA2-285	SKF 1308	✓	✓	✓	✓		✓				1
		Declutch Shaft Bearing	Ball	3FK0-285	SKF 7309							✓	✓	✓	1
0804		Idler Shaft Bearing—Cone	Roller	19FK0-636	Timken 49585							✓	✓	✓	2
		Idler Shaft Bearing—Cup	Roller	19FK0-637	Timken 49520							✓	✓	✓	2

23-4 MISC.
Master Parts List
THE AUTOCAR COMPANY, ARDMORE, PA.

A Consolidation of All Ball and Roller Bearings Serviced by Autocar for U. S. Government

Government Code	Major Unit	Description	Type	Autocar Part Number	Manufacturer's Identification		U-2044	U-4044	U-4144	U-4144-T	C-50	U-5044	U-7144-T	U-8144	U-8144-T	Quantity Used
0902	Driveshaft	Needle Bearing Assembly	Roller	3GE0-635A	Spicer	K-6-6-48							✓			8
		Needle Bearing Assembly	Roller	3RG0-635A	Spicer	K-4-6-68X							✓	✓	✓	8
		Needle Bearing Assembly	Roller	3UN0-635	Spicer	K-5-6-48							✓	✓	✓	16
		Needle Bearing Assembly	Roller	3RG0-635	Spicer	K-4-6-28					✓					4
		Steady Bearing	Ball	3SA2-285	SKF	1308-F					✓					1
1002	Front Axle	Differential Bearing—Cup	Roller	4D2-157	Timken	472-A	✓	✓	✓	✓		✓				2
		Differential Bearing—Cone	Roller	19SKB0-443	Timken	482	✓	✓	✓	✓		✓				2
		Differential Bearing	Ball	4NFG0-567	New Dept.	1214							✓			2
		Differential Bearing—Cup	Roller	4CGA2-211	Timken	572								✓	✓	2
		Differential Bearing—Cone	Roller	4CGA2-209	Timken	581								✓	✓	2
1003		Pinion Bearing Cone—Front	Roller	4UG0-1232	Timken	49175	✓	✓	✓	✓		✓				1
		Pinion Bearing Cup—Front	Roller	4RL0-137	Timken	49368	✓	✓	✓	✓		✓				1
		Pinion Bearing Cone—Rear	Roller	4DF0-138	Timken	59200	✓	✓	✓	✓		✓				1
		Pinion Bearing Cup—Rear	Roller	4RL0-142	Timken	59412	✓	✓	✓	✓		✓				1
		Spur Pinion Bearing—Cone	Roller	4NK0-136	Timken	615							✓			2
		Spur Pinion Bearing—Cup	Roller	4FD0-219	Timken	612							✓			2
		Spur Pinion Bearing—Cone—R. H.	Roller	4FD0-358	Timken	623								✓	✓	1
		Spur Pinion Bearing—Cup—R. H.	Roller	4FD0-219	Timken	612								✓	✓	1
		Spur Pinion Bearing—Cone—L. H.	Roller	4GKW0-138	Timken	65225								✓	✓	1
		Spur Pinion Bearing—Cup—L. H.	Roller	4DFL0-137	Timken	65500								✓	✓	1
		Bevel Gear Bearing Cone	Roller	4RL0-141	Timken	59175	✓	✓	✓	✓		✓				2
		Bevel Gear Bearing Cup	Roller	4RL0-142	Timken	59412	✓	✓	✓	✓		✓				2
		Bevel Pinion Rear Bearing—Cone	Roller	4ZGEW0-156	Timken	621							✓			1
		Bevel Pinion Rear Bearing—Cup	Roller	4FD0-219	Timken	612							✓			1
		Bevel Pinion Rear Bearing—Cone	Roller	4FD0-358	Timken	623								✓	✓	1
		Bevel Pinion Rear Bearing—Cup	Roller	4FD0-219	Timken	612								✓	✓	..
		Bevel Pinion Front Bearing—Cone	Roller	4NFG0-138	Timken	527							✓			1
		Bevel Pinion Front Bearing—Cup	Roller	3T2-37	Timken	522							✓			1
		Bevel Pinion Front Bearing—Cone	Roller	4DFL0-138	Timken	59200								✓	✓	1
		Bevel Pinion Front Bearing—Cup	Roller	4RL0-142	Timken	59412								✓	✓	1
1006		Steering Gear Jacket Tube Bearing	Ball	10UK0-1097	Ross	065996	✓	✓	✓	✓		✓				1
		Steering Knuckle Thrust Bearing	Roller	9N0-51	Timken	T-138					✓					..
		Steering Knuckle Bearing—Cone	Roller	4ZDK0-138	Timken	41125							✓			4
		Steering Knuckle Bearing—Cup	Roller	4ZDK0-139	Timken	41286							✓			4
		Steering Knuckle Bearing—Cone	Roller	4A0-138	Timken	53176								✓	✓	4
		Steering Knuckle Bearing—Cup	Roller	4SA2-139	Timken	53387								✓	✓	4
1301		Front Wheel Bearing Cone—Inner	Roller	4FD0-138	Timken	594	✓	✓	✓	✓		✓				2
		Front Wheel Bearing Cone—Inner	Roller	19T2-13	Timken	539					✓					2
		Front Wheel Bearing Cone—Inner	Roller	4FD0-138	Timken	594							✓			2
		Front Wheel Bearing Cone—Inner	Roller	9SK0-92	Timken	71450								✓	✓	2
		Front Wheel Bearing Cup—Inner	Roller	4FD0-139	Timken	592-A	✓	✓	✓	✓		✓				2
		Front Wheel Bearing Cup—Inner	Roller	4Y2-137	Timken	532					✓					2
		Front Wheel Bearing Cup—Inner	Roller	9SKA0-825	Timken	71750								✓	✓	2
		Front Wheel Bearing Cone—Outer	Roller	19SKB0-445	Timken	498	✓	✓	✓	✓		✓				2
		Front Wheel Bearing Cone—Outer	Roller	9N0-38	Timken	444					✓					2
		Front Wheel Bearing Cone—Outer	Roller	9SK0-39	Timken	52400								✓	✓	2
		Front Wheel Bearing Cup—Outer	Roller	19SKB0-444	Timken	493	✓	✓	✓	✓		✓				2
		Front Wheel Bearing Cup—Outer	Roller	4H2-137	Timken	432					✓					2
		Front Wheel Bearing Cup—Outer	Roller	9SK0-825	Timken	52637								✓	✓	2
1105	Rear Axle	Differential Bearing—Cup	Roller	4D2-157	Timken	472-A	✓	✓	✓	✓		✓				2
		Differential Bearing—Cup	Roller	4Y2-211	Timken	563						✓				2
		Differential Bearing—Cup	Roller	4CGA2-211	Timken	572								✓	✓	2
		Differential Bearing—Cone	Roller	19SKB0-443	Timken	482	✓	✓	✓	✓		✓				2

THE AUTOCAR COMPANY, ARDMORE, PA.
Master Parts List
23-5 MISC.

A Consolidation of All Ball and Roller Bearings Serviced by Autocar for U. S. Government

Government Code	Major Unit	Description	Type	Autocar Part Number	Manufacturer's Identification	U-2044	U-4044	U-4144	U-4144-T	C-50	U-5044	U-7144-T	U-8144	U-8144-T	Quantity Used
1105	Cont.	Differential Bearing—Cone	Roller	4Y2-209	Timken 567					✓					2
		Differential Bearing—Cone	Roller	4CGA2-209	Timken 581								✓	✓	2
		Differential Bearing	Ball	4NFG0-567	New Dept. 1214							✓			2
		Pinion Bearing Cone—Front	Roller	4UG0-1232	Timken 49175	✓	✓	✓	✓		✓				1
		Pinion Bearing Cup—Front	Roller	4RL0-137	Timken 49368	✓	✓	✓	✓		✓				1
		Pinion Bearing Cone—Rear	Roller	4DFL0-138	Timken 59200	✓	✓	✓	✓		✓				1
		Pinion Bearing Cone—Rear	Roller	4RL0-142	Timken 59412	✓	✓	✓	✓		✓				1
		Spur Pinion Bearing—Cup	Roller	9UF1-825	Timken 3720							✓			2
		Spur Pinion Bearing—Cup	Roller	4H2-137	Timken 432								✓	✓	2
		Spur Pinion Bearing—Cone	Roller	4A0-568	Timken 3780							✓			2
		Spur Pinion Bearing—Cone	Roller	4ZGE0-653	Timken 438								✓	✓	2
		Bevel Gear Bearing Cone	Roller	4RL0-141	Timken 59175	✓	✓	✓	✓		✓				2
		Bevel Gear Bearing Cup	Roller	4RL0-142	Timken 59412	✓	✓	✓	✓		✓				2
		Spur Pinion Bevel Gear Bearing	Roller	4B0-1091	Hyatt U-1217-TAM							✓			1
		Spur Pinion Bevel Gear Bearing	Roller	4GTC0-1066	Hyatt U-1218-TAM								✓	✓	1
		Thru Shaft Front Bearing—Cone	Roller	4Y2-141	Timken 560							✓			1
		Thru Shaft Front Bearing—Cone	Roller	19D0-13	Timken 570								✓	✓	1
		Thru Shaft Front Bearing—Cup	Roller	4Y2-142	Timken 552-A							✓			1
		Thru Shaft Front Bearing—Cup	Roller	4Y2-211	Timken 563								✓	✓	1
		Thru Shaft Rear Bearing—Cone	Roller	4NFG0-138	Timken 527							✓			1
		Thru Shaft Rear Bearing—Cone	Roller	19C2-305	Timken 536								✓	✓	1
		Thru Shaft Rear Bearing—Cup	Roller	3T2-37	Timken 522							✓			1
		Thru Shaft Rear Bearing—Cup	Roller	4Y2-137	Timken 532-A								✓	✓	1
1301		Rear Wheel Bearing Cone—Inner	Roller	4UG0-141	Timken 566	✓	✓	✓	✓		✓				2
		Rear Wheel Bearing Cup—Inner	Roller	4Y2-211	Timken 563	✓	✓	✓	✓		✓				2
		Rear Wheel Bearing Cone—Outer	Roller	4Y2-141	Timken 560	✓	✓	✓	✓		✓				2
		Rear Wheel Bearing Cup—Outer	Roller	4NFG0-142	Timken 552-A	✓	✓	✓	✓		✓				2
		Rear Wheel Bearing Cone	Roller	4Y2-156	Timken 749					✓					4
		Rear Wheel Bearing Cup	Roller	4Y2-157	Timken 742					✓					4
		Rear Wheel Bearing—Cone	Roller	4NFG0-156	Timken 5557							✓			4
		Rear Wheel Bearing—Cone	Roller	4G0-156	Timken 596								✓	✓	4
		Rear Wheel Bearing—Cup	Roller	4NFG0-157	Timken 5520							✓			4
		Rear Wheel Bearing—Cup	Roller	4FD0-139	Timken 592-A								✓	✓	4
1403	Steering Gear	Jacket Tube and Bearing Assembly	Ball	10UK0-1176	Ross 7789-39	✓	✓	✓	✓		✓				1
		Steering Cross Tube End Stud & Bearing Assy.	Ball	9A0-544	Timken A-2110-D-56					✓					2
		Jacket Tube Bearing Unit	Ball	10SJ0-1097	Ross 066994							✓	✓	✓	1
		Stud Bearing Unit	Roller	10HDA0-930	Ross 044988							✓	✓	✓	2
		Cam and Wheel Tube Ball Cup	Ball	10UK0-224	Ross 400021	✓	✓	✓	✓		✓				2
		Cam and Wheel Tube Ball Cup	Ball	10C0-224	Ross 400002							✓	✓	✓	2
		Cam and Wheel Tube Ball Cup Retaining Ring	Ball	10SA0-227	Ross 400005	✓	✓	✓	✓		✓				2
		Cam and Wheel Tube Ball Cup Retaining Ring	Ball	10SCK0-227	Ross 400001							✓	✓	✓	2
		Cam and Wheel Tube Balls	Ball	S-1670	Ross 400014	✓	✓	✓	✓	✓	✓				28
		Cam and Wheel Tube Balls	Ball	S-1671	Ross 400015							✓	✓	✓	28
1911	Power Take-Off	Adapter Gear Bearing	Roller	18RL0-136	Hyatt 94632					✓					1
		Adapter Gear Bearing	Roller	3RE0-67	Rollway 160426							✓		✓	1
		Intermediate Gear Bearing	Roller	18UV0-136	Hyatt 21-X-94536									✓	1
		Idler Gear Bearing	Roller	18UVA0-136	Hyatt 21-X-94526									✓	2
		Driving Shaft Bearing	Ball	19UVA0-89	MRC 16-X-207M									✓	1
		Driving Shaft Bearing	Ball	19UV0-89	SKF 16-X-5207									✓	1

Master Parts List

23-6 MISC.
Code 2304

THE AUTOCAR COMPANY ARDMORE, PA.

A Consolidation of All Parts Common Serviced by Autocar

Description	Part No.	Quantity Required
STEEL BALLS		
3/8"	S-1670	33
7/16"	S-1671	36
1/2"	S-1672	2
3/4"	S-1676	2
25/32"	S-1678	1
SPRING BOLTS		
3/8"—16 x 1 1/4"	S-2042	8
PHILLIPS FLAT HEAD STOVE BOLTS WITH NUTS		
1/4" x 1 1/8"	S-3053	2
1/4" x 1 1/4"	S-3086A	2
PHILLIPS ROUND HEAD STOVE BOLTS WITH NUTS		
3/16" x 3/4"	S-2923A	2
3/16" x 1"	S-2966A	7
1/4" x 3/4"	S-2942A	16
GALVANIZED PIPE REDUCING BUSHINGS		
1/4" x 1/8"	S-4759	4
3/8" x 1/4"	S-4762	3
1/2" x 1/8"	S-4769	1
1/2" x 1/4"	S-4772	1
3/4" x 1/2"	S-4774	1

Description	Part No.	Quantity Required
GALVANIZED HOSE CLAMPS		
1 1/2" I.D. 5 Ply	S-2392	4
1 3/4" I.D. 5 Ply	S-2398	8
2 3/4" I.D. 5 Ply	S-2400	1
CYLINDER DRAIN COCK		
1/8"	S-2500	1
CONNECTOR BODY		
1/4" x 3/8"	S-5919	2
LONG NUT COMPRESSION CONNECTORS		
1/4" x 3/8"	S-5865	1

Master Parts List

23-7 MISC.

THE AUTOCAR COMPANY
ARDMORE, PA.

Description	Part No.	Quantity Required
IMPERIAL "HI-DUTY" CONNECTORS		
5/16" x 1/4"	S-5474	1
PIPE REDUCING COUPLING		
1/2" x 1/4"	S-4791	1
PIPE COUPLING		
1/8" Galv.	S-4664	1
COMPRESSION ELBOWS		
1/4" x 3/8"	S-5878	3
FLARED TYPE ELBOWS		
5/16" x 1/8"	S-5065	1
3/8" x 1/4"	S-5066	1
ELLS		
1/8"—90° Brass	S-4501	1
1/8"—90° Galv.	S-4590	1
1/4"—45° Brass	S-4821	4
1/4"—45° Galv.	S-4823	2
1/4"—90° Brass	S-4617	1
1/4"—90° Galv.	S-4171	1
1/2"—45° Brass	S-4822	1
1/2"—45° Galv.	S-4610	1
1/2"—90° Galv.	S-4620	1
STREET ELLS		
1/4"	S-3443	1
WEATHERHEAD BRASS STREET ELLS		
45° #3350X4	S-8026	1
45° Male—Female	S-4021	1
STRAIGHT ALEMITE FITTINGS		
#1610	S-4011	16
67 1/2° ALEMITE FITTINGS		
#1612	S-4012	16
90° ALEMITE FITTINGS		
#1613	S-4013	4
ALEMITE MEASURING FITTINGS		
#1719	S-4019	6

23-8 MISC.
Master Parts List
THE AUTOCAR COMPANY, ARDMORE, PA.

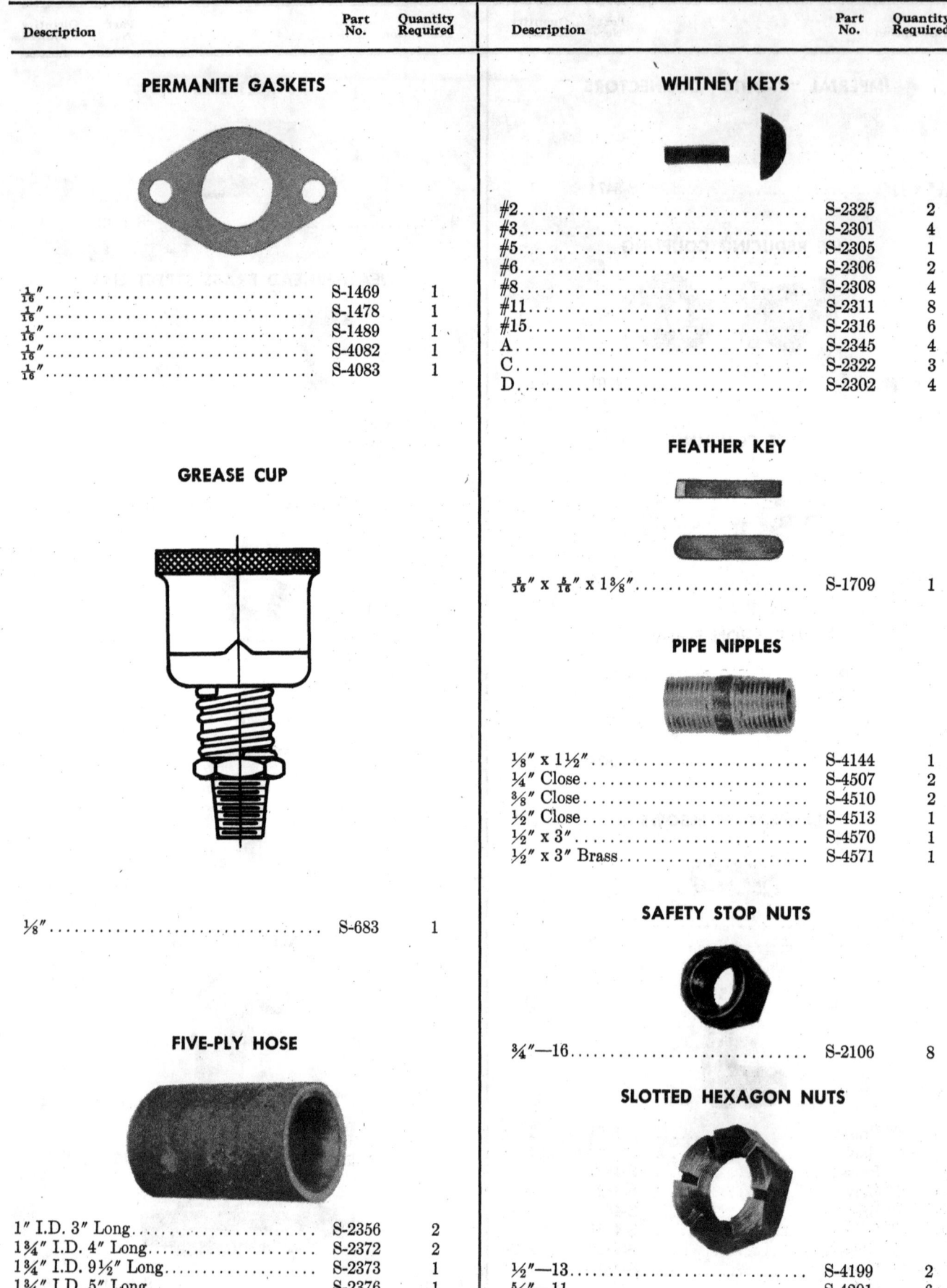

Description	Part No.	Quantity Required
PERMANITE GASKETS		
1/16"	S-1469	1
1/16"	S-1478	1
1/16"	S-1489	1
1/16"	S-4082	1
1/16"	S-4083	1
GREASE CUP		
1/8"	S-683	1
FIVE-PLY HOSE		
1" I.D. 3" Long	S-2356	2
1¾" I.D. 4" Long	S-2372	2
1¾" I.D. 9½" Long	S-2373	1
1¾" I.D. 5" Long	S-2376	1

Description	Part No.	Quantity Required
WHITNEY KEYS		
#2	S-2325	2
#3	S-2301	4
#5	S-2305	1
#6	S-2306	2
#8	S-2308	4
#11	S-2311	8
#15	S-2316	6
A	S-2345	4
C	S-2322	3
D	S-2302	4
FEATHER KEY		
5/16" x 5/16" x 1 3/8"	S-1709	1
PIPE NIPPLES		
1/8" x 1½"	S-4144	1
¼" Close	S-4507	2
3/8" Close	S-4510	2
½" Close	S-4513	1
½" x 3"	S-4570	1
½" x 3" Brass	S-4571	1
SAFETY STOP NUTS		
¾"—16	S-2106	8
SLOTTED HEXAGON NUTS		
½"—13	S-4199	2
5/8"—11	S-4201	6

Master Parts List

HEXAGON NUTS

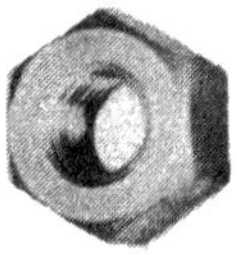

Description	Part No.	Quantity Required
#10—24	S-103	4
#10—32	S-3745	16
¼"—20	S-76	69
¼"—20	S-834	2
¼"—28	S-1416	7
¼"—28	S-1417	2
¼"—28	S-2023	20
5/16"—18	S-82	36
5/16"—18	S-1301	4
5/16"—24	S-2021	3
5/16"—24	S-2024	2
5/16"—24	S-3715	8
⅜"—16	S-75	110
⅜"—16	S-78	34
⅜"—16	S-441	2
⅜"—16	S-829	4
⅜"—16	S-4197	2
⅜"—24	S-805	18
⅜"—24	S-809	2
⅜"—24	S-814	32
⅜"—24	S-1418	95
⅜"—24	S-2020	4
⅜"—24	S-6249	6
7/16"—14	S-431	2
7/16"—14	S-850	26
7/16"—14	S-3720	2
7/16"—20	S-1419	28
7/16"—20	S-2025	21
7/16"—20	S-3756	24
½"—13	S-77	369
½"—13	S-1300	5
½"—13	S-1310	2
½"—13	S-4199	12
½"—20	S-811	30
½"—20	S-812	8
½"—20	S-830	29
½"—20	S-1420	136
½"—20	S-3723	1
½"—20	S-3742	23
½"—20	S-3747	29
9/16"—18	S-84	88
9/16"—18	S-836	24
⅝"—11	S-88	23
⅝"—11	S-93	89
⅝"—11	S-808	2
⅝"—11	S-2004	5
⅝"—18	S-810	11
⅝"—18	S-1311	44
¾"—10	S-4203	5
¾"—16	S-97	20
¾"—16	S-801	8
¾"—16	S-847	26
¾"—16	S-857	8
¾"—16	S-1408	5
¾"—16	S-2001	4
¾"—16	S-3753	4

HEXAGON NUTS—Continued

Description	Part No.	Quantity Required
¾"—16	S-4221	6
⅞"—14	S-1409	1
⅞"—14	S-4201	4
1"—14	S-844	2
1"—14	S-1309	2
1"—14	S-3724	16
1"—14	S-3736	8
1"—16	S-3713	2
1"—16 L.H.	S-3714	4
1"—20	S-87	1
1¼"—18	S-2022	1
1½"—12	S-1327	1
1½"—16	S-862	1

FLARED COPPER TUBE FITTINGS (Brass)
SHORT UNION NUT

Description	Part No.	Quantity Required
¼" x ⅛"	S-4302	1
5/16" x ⅛"	S-5013	2
⅜" x ¼" Standard	S-5005	2

GITS OILERS

Description	Part No.	Quantity Required
Style "B" #110	S-3423	1

OIL SEALS

Description	Part No.	Quantity Required
#60861 Victor	S-6106	2
#60506 Victor	S-6108	1
#60970 Victor	S-7214	1

Master Parts List

23-10 MISC. — THE AUTOCAR COMPANY, ARDMORE, PA.

Description	Part No.	Quantity Required
CLEVIS PINS		
$\frac{1}{4}''\times\frac{31}{32}''$	S-2228	4
$\frac{7}{16}''\times 1\frac{3}{8}''$	S-1272	1
$\frac{3}{8}''\times 1\frac{3}{16}''$	S-2257	4
$\frac{3}{8}''\times 1\frac{9}{32}''$	S-2258	2
$\frac{3}{8}''\times 1\frac{11}{32}''$	S-2259	2
$.496\times 1\frac{37}{64}''$	S-1285	4
$\frac{1}{2}''\times 1\frac{9}{16}''$	S-2225	14
$\frac{1}{2}''\times 2\frac{1}{4}''$	S-1274	2
$.605\times 1\frac{29}{32}''$	S-1292	4
COTTER PINS		
$\frac{3}{32}''\times\frac{3}{4}''$	S-2198	22
$\frac{3}{32}''\times 1''$	S-2199	4
$\frac{3}{32}''\times 1\frac{1}{4}''$	S-2200	10
$\frac{1}{8}''\times\frac{3}{4}''$	S-2203	9
$\frac{1}{8}''\times 1''$	S-2204	29
$\frac{1}{8}''\times 1\frac{1}{4}''$	S-2205	16
$\frac{1}{8}''\times 1\frac{1}{2}''$	S-2206	11
$\frac{1}{8}''\times 2''$	S-2208	5
$\frac{1}{8}''\times 2\frac{1}{2}''$	S-2216	1
$\frac{5}{32}''\times 1\frac{1}{4}''$	S-2221	4
$\frac{5}{32}''\times 1\frac{1}{2}''$	S-2211	1
$\frac{3}{16}''\times 2''$	S-2214	1
$\frac{1}{4}''\times 2\frac{3}{4}''$	S-2219	2
STEEL PINS		
$\frac{1}{8}''\times\frac{3}{4}''$	S-1499	2
$\frac{1}{8}''\times\frac{7}{8}''$	S-1546	3
$\frac{1}{8}''\times 1''$	S-3253	4
$\frac{5}{32}''\times 1\frac{1}{16}''$	S-3192	1
$\frac{3}{16}''\times 1\frac{1}{2}''$	S-1506	4
$\frac{3}{16}''\times 1\frac{3}{16}''$	S-3235	1
$\frac{3}{16}''\times 1\frac{1}{4}''$	S-1528	1
$\frac{3}{16}''\times 1\frac{5}{16}''$	S-3237	1
$\frac{3}{16}''\times 1\frac{9}{16}''$	S-1558	1
$\frac{3}{16}''\times 2\frac{1}{2}''$	S-1578	1
$\frac{1}{4}''\times 1\frac{5}{8}''$	S-1584	1
$\frac{5}{16}''\times 2''$	S-3217	1
TAPER PINS		
$.2513\times 1\frac{1}{4}''$	S-1593	2
$\frac{9}{32}''\times 1\frac{1}{4}''$	S-1599	4
HOLLOW PIPE PLUGS		
$\frac{3}{8}''$	S-4005	4
$\frac{1}{2}''$	S-4172	1
$\frac{3}{4}''$	S-617	4
$1''$	S-4189	1
$1\frac{1}{4}''$	S-4830	1
SLOTTED PIPE PLUGS		
$\frac{1}{8}''$	S-4745	2
$\frac{1}{8}''$—Brass	S-4747	2
$\frac{1}{4}''$	S-4748	5
SQUARE HEAD PIPE PLUGS		
$\frac{1}{8}''$	S-4291	1
$\frac{1}{4}''$	S-629	3
$\frac{1}{4}''$	S-4737	2
$\frac{3}{8}''$	S-4004	3
$\frac{1}{2}''$	S-2490	5
$\frac{3}{4}''$	S-4131	3
$1''$	S-4156	7
$1\frac{1}{4}''$	S-4157	6
WELSH PLUGS		
$\frac{5}{8}''$	S-4207	6
$\frac{3}{4}''$	S-4208	4
$\frac{13}{16}''$	S-4235	4
$\frac{7}{8}''$	S-4209	5
$1\frac{1}{8}''$	S-4236	1
$1\frac{1}{4}''$	S-4212	13

THE AUTOCAR COMPANY
ARDMORE, PA.

Master Parts List

23-11 MISC.

Description	Part No.	Quantity Required
BIFURCATED RIVETS		
5/32" x 1/2"	S-1010	8
5/32" x 5/8"	S-1011	13
5/32" x 7/8"	S-1015	20
SOFT STEEL BUTTON HEAD RIVETS		
3/16" x 5/16"	S-1064	2
5/16" x 1"	S-563	12
1/2" x 2"	S-1083	28
1/2" x 3"	S-1084	16
FLAT HEAD BRASS TUBULAR RIVETS		
1/8" x 5/16"	S-1018	18
3/16" x 7/16"	S-1020	112
HEXAGON HEAD CAP SCREWS		
1/4"—20 x 3/8"	S-4342	1
1/4"—20 x 1/2"	S-66	4
1/4"—20 x 5/8"	S-36	2
1/4"—20 x 3/4"	S-30	23
1/4"—20 x 1"	S-3118	10
1/4"—20 x 1 1/4"	S-67	8
1/4"—20 x 2"	S-71	2
1/4"—20 x 2 1/2"	S-4347	1
1/4"—28 x 3/4"	S-1427	2
1/4"—28 x 1 1/4"	S-7437	19
5/16"—18 x 9/16"	S-3377	2
5/16"—18 x 5/8"	S-16	12
5/16"—18 x 21/32"	S-6277	6
5/16"—18 x 3/4"	S-14	26
5/16"—18 x 7/8"	S-15	5
5/16"—18 x 7/8"	S-1782	2
5/16"—18 x 1"	S-3120	4
5/16"—18 x 1 1/8"	S-27	2
HEXAGON HEAD CAP SCREWS—Continued		
5/16"—18 x 1 1/4"	S-32	6
5/16"—18 x 1 1/4"	S-7440	6
5/16"—18 x 1 1/2"	S-31	4
5/16"—18 x 2"	S-3103	3
5/16"—18 x 2 1/2"	S-3112	2
5/16"—24 x 3/4"	S-1783	4
5/16"—24 x 7/8"	S-4428	8
3/8"—16 x 1/2"	S-1461	24
3/8"—16 x 5/8"	S-53	12
3/8"—16 x 5/8"	S-2065	11
3/8"—16 x 3/4"	S-12	66
3/8"—16 x 3/4"	S-6259	1
3/8"—16 x 7/8"	S-50	54
3/8"—16 x 1"	S-23	128
3/8"—16 x 1"	S-1762	4
3/8"—16 x 1 1/8"	S-46	37
3/8"—16 x 1 1/8"	S-6257	20
3/8"—16 x 1 1/4"	S-64	73
3/8"—16 x 1 1/4"	S-7438	4
3/8"—16 x 1 1/2"	S-17	13
3/8"—16 x 1 5/8"	S-4053	2
3/8"—16 x 1 5/8"	S-6285	2
3/8"—16 x 1 3/4"	S-1879	4
3/8"—16 x 1 3/4"	S-2057	2
3/8"—16 x 1 3/4"	S-6289	2
3/8"—16 x 2"	S-3383	2
3/8"—16 x 2"	S-4051	1
3/8"—16 x 2 1/4"	S-40	8
3/8"—16 x 2 1/2"	S-411	2
3/8"—16 x 2 3/4"	S-6097	2
3/8"—16 x 3 1/4"	S-4439	1
3/8"—16 x 3 1/2"	S-1436	4
3/8"—24 x 1"	S-1445	4
3/8"—24 x 1 1/8"	S-6249	2
3/8"—24 x 1 1/4"	S-1460	6
3/8"—24 x 1 1/2"	S-1455	14
3/8"—24 x 1 3/4"	S-1457	3
3/8"—24 x 2"	S-3094	2
7/16"—14 x 1"	S-919	2
7/16"—14 x 1 1/8"	S-6085	2
7/16"—14 x 1 1/4"	S-1881	6
7/16"—14 x 1 3/8"	S-2087	29
7/16"—14 x 1 1/2"	S-923	8
7/16"—14 x 2"	S-3373	8
7/16"—14 x 2 3/4"	S-1901	2
7/16"—14 x 3 1/2"	S-3104	1
7/16"—20 x 1 1/2"	S-6226	20
7/16"—20 x 1 5/8"	S-7405	18
7/16"—20 x 2 1/8"	S-4332	4
1/2"—13 x 3/4"	S-4064	5
1/2"—13 x 7/8"	S-2086	9
1/2"—13 x 1"	S-20	3
1/2"—13 x 1 1/8"	S-19	16
1/2"—13 x 1 1/4"	S-25	18
1/2"—13 x 1 1/4"	S-3127	10
1/2"—13 x 1 1/4"	S-7441	1
1/2"—13 x 1 3/8"	S-1428	83
1/2"—13 x 1 1/2"	S-1872	50
1/2"—13 x 1 1/2"	S-7442	1
1/2"—13 x 1 5/8"	S-1754	23
1/2"—13 x 1 5/8"	S-1839	19

MISC. Master Parts List — THE AUTOCAR COMPANY, ARDMORE, PA.

HEXAGON HEAD CAP SCREWS—Continued

Description	Part No.	Quantity Required
½″—13 x 1¾″	S-1463	2
½″—13 x 1¾″	S-3100	39
½″—13 x 1¾″	S-3101	28
½″—13 x 1¾″	S-4491	8
½″—13 x 1⅞″	S-1796	1
½″—13 x 1⅞″	S-4076	16
½″—13 x 1⅞″	S-4329	8
½″—13 x 2″	S-912	4
½″—13 x 2″	S-922	27
½″—13 x 2″	S-6293	2
½″—13 x 2″	S-6294	2
½″—13 x 2¼″	S-44	1
½″—13 x 2¼″	S-1880	6
½″—13 x 2¼″	S-3126	3
½″—13 x 2½″	S-1871	11
½″—13 x 2½″	S-1877	2
½″—13 x 2⅝″	S-911	8
½″—13 x 2¾″	S-4077	11
½″—13 x 3″	S-1466	4
½″—13 x 3″	S-3108	12
½″—13 x 3″	S-6099	4
½″—13 x 3⅛″	S-3106	26
½″—13 x 3¼″	S-3378	5
½″—13 x 3¼″	S-6280	2
½″—13 x 3½″	S-1459	4
½″—13 x 3½″	S-4063	1
½″—13 x 3⅝″	S-1440	32
½″—13 x 3¾″	S-1888	8
½″—13 x 3⅞″	S-21	1
½″—13 x 4¾″	S-1889	1
½″—20 x 1¼″	S-2055	15
½″—20 x 1½″	S-1429	8
½″—20 x 1¾″	S-1430	2
½″—20 x 2″	S-1438	8
½″—20 x 2¼″	S-1868	1
⅝″—11 x 1″	S-4321	6
⅝″—11 x 1⅜″	S-900	2
⅝″—11 x 1½″	S-1870	12
⅝″—11 x 1½″	S-3098	4
⅝″—11 x 1⅝″	S-1838	2
⅝″—11 x 1¾″	S-1805	4
⅝″—11 x 1¾″	S-4073	28
⅝″—11 x 1¾″	S-4328	4
⅝″—11 x 2″	S-3387	28
⅝″—11 x 2″	S-4485	1
⅝″—11 x 2¼″	S-4350	2
⅝″—11 x 2¼″	S-4497	4
⅝″—11 x 2½″	S-902	12
⅝″—11 x 2¾″	S-1874	4
⅝″—11 x 3″	S-901	2
⅝″—11 x 3½″	S-4346	8
⅝″—11 x 4″	S-6283	2
⅝″—11 x 4″	S-6284	4
⅝″—11 x 4¼″	S-4323	2
⅝″—11 x 4¼″	S-7403	4
⅝″—11 x 8½″	S-6287	2
⅝″—18 x 1½″	S-2085	4
⅝″—18 x 2⅛″	S-6235	2
¾″—16 x 3¼″	S-6095	2

HEADLESS CUP SET SCREW

Description	Part No.	Quantity Required
⅜″—16 x ⅜″	S-408	1

HEADLESS DOG SET SCREW

Description	Part No.	Quantity Required
½″—13 x ½″	S-1246	2

HEADLESS SET SCREWS

Description	Part No.	Quantity Required
5⁄16″—18 x ⅜″	S-422	1
5⁄16″—18 x 1″	S-1235	11
5⁄16″—18 x 1½″	S-1232	5
5⁄16″—18 x 1⅞″	S-1236	1
5⁄16″—18 x 2¼″	S-1233	1
⅜″—16 x ⅜″	S-408	1

SQUARE HEAD SET SCREWS

Description	Part No.	Quantity Required
¼″—20 x 1⅜″	S-425	1
⅜″—16 x 1″	S-411	2

PHILLIPS BINDING HEAD MACHINE SCREW

Description	Part No.	Quantity Required
#10—24 x ½″	S-1606A	4

FLAT HEAD MACHINE SCREWS

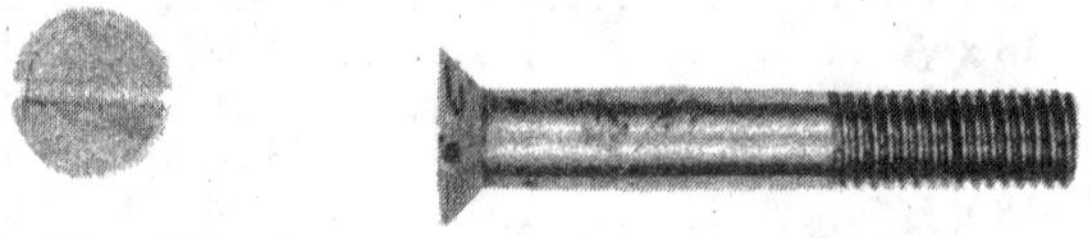

Description	Part No.	Quantity Required
¼″—20 x ⅝″	S-4265	4
5⁄16″—18 x 1⅛″	S-3132	4
5⁄16″—18 x ¾″	S-4280	8
½″—13 x 1¼″	S-3826	2

THE AUTOCAR COMPANY
ARDMORE, PA.

Master Parts List

23-13
MISC.

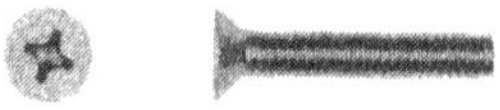

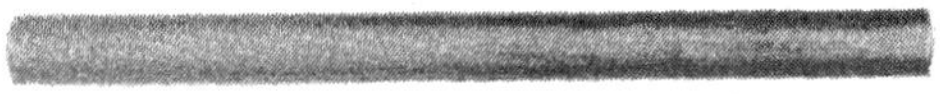

Description	Part No.	Quantity Required
ROUND HEAD MACHINE SCREWS		
#6—32 x ⅜"	S-6132	2
#8—32 x ⅜"	S-3139	3
#8—32 x ¼"	S-208	1
¼"—20 x ½"	S-6140	2
¼"—20 x 1"	S-6156	2
PHILLIPS FLAT HEAD MACHINE SCREWS		
#10—24 x ⅝"	S-1632A	2
#10—32 ⅝"	S-3811A	2
#10—32 x ¾"	S-191A	12
#12—24 x ⅝"	S-1604A	8
¼"—20 x ½"	S-3839A	48
¼"—20 x ¾"	S-3843A	8
⅜"—16 x 1"	S-3830A	4
⅜"—16 x 1¼"	S-4261A	2
⁷⁄₁₆"—20 x 1¾"	S-4042A	2
PHILLIPS OVAL HEAD MACHINE SCREWS		
¼"—20 x 1¾"	S-195A	2
PHILLIPS ROUND HEAD MACHINE SCREWS		
#10—24 x ½"	S-4270A	16
#10—32 x ½"	S-4277A	2
#10—32 x 1½"	S-1608A	2
¼"—20 x ⅝"	S-3821A	4
¼"—20 x ¾"	S-1614A	28
¼"—20 x ⅞"	S-3147A	3
¼"—20 x 1"	S-1607A	15
¼"—20 x 1¼"	S-1629A	2
⅜"—16 x ¾"	S-1618A	6
PHILLIPS FLAT HEAD SELF-TAPPING SCREWS		
#10—32 x ¾"	S-3525	2
#10—32 x 1"	S-3526	12
PHILLIPS ROUND HEAD SELF-TAPPING SCREWS		
#10—32 x ⅝"	S-3524	2
OVAL HEAD WOOD SCREWS		
#12 x 1"	S-329	3
TEES		
⅜" Galv.	S-4699	1
COPPER TUBING		
⁵⁄₁₆" O.D. #20 Ga.	S-5150	102 Ft.
⅜" O.D. #19 Ga.	S-5146	428 Ft.
½" O.D. #18 Ga.	S-5149	80 Ft.
¾" O.D. #20 Ga.	S-5147	48 Ft.
FLARED TYPE SINGLE UNIONS		
¼"—⅛"	S-5023	1
⁵⁄₁₆"—⅛"	S-5024	1
⁵⁄₁₆"—¼"	S-5024A	1
½"—⅜"	S-5027	3

Master Parts List

THE AUTOCAR COMPANY, ARDMORE, PA.

STEEL SPRING WASHERS

Description	Part No.	Quantity Required
#6	S-1343	3
#8	S-3164	4
#8 Cad.	S-3293	1
$\frac{7}{32}''$	S-1344	70
$\frac{9}{32}''$	S-1345	180
$\frac{11}{32}''$	S-1346	80
$\frac{13}{32}''$	S-1347	706
$\frac{15}{32}''$	S-1348	95
$\frac{17}{32}''$	S-1349	974
$\frac{19}{32}''$	S-1350	72
$\frac{21}{32}''$	S-1351	315
$\frac{25}{32}''$	S-1353	26
$1\frac{1}{16}''$	S-3188	16
$1\frac{1}{32}''$	S-1357	8

FLAT STEEL WASHERS

I.D.	O.D.	THICK		Part No.	Quantity Required
$\frac{5}{32}''$	$\frac{5}{16}''$	$\frac{1}{16}''$ Brass		S-726	1
$\frac{5}{32}''$	$\frac{1}{2}''$	$\frac{1}{32}''$		S-742	2
$\frac{3}{16}''$	$\frac{3}{8}''$	$\frac{1}{32}''$		S-987	1
$\frac{13}{64}''$	$\frac{7}{8}''$	$\frac{1}{16}''$		S-702	11
$\frac{7}{32}''$	$\frac{7}{16}''$	$\frac{1}{32}''$		S-350	6
$\frac{1}{4}''$	$\frac{11}{16}''$	$\frac{1}{8}''$		S-348	4
$\frac{9}{32}''$	$\frac{3}{4}''$	$\frac{1}{32}''$		S-720	2
$\frac{9}{32}''$	$\frac{3}{4}''$	$\frac{1}{16}''$		S-353	12
$\frac{5}{16}''$	$\frac{3}{4}''$	$\frac{1}{16}''$		S-395	2
$\frac{3}{8}''$	$\frac{3}{4}''$	$\frac{1}{16}''$		S-334	1
$\frac{25}{64}''$	$\frac{5}{8}''$	$\frac{3}{64}''$ Leather		S-926	2
$\frac{13}{32}''$	$\frac{3}{4}''$	$\frac{1}{16}''$		S-387	2
$\frac{13}{32}''$	$\frac{3}{4}''$	$\frac{1}{16}''$ Copper		S-735	4
$\frac{13}{32}''$	$1''$	$\frac{1}{16}''$		S-1974	1
$\frac{13}{32}''$	$1''$	$\frac{3}{16}''$		S-980	1
$\frac{13}{32}''$	$1\frac{5}{32}''$	$\frac{1}{16}''$		S-722	4
$\frac{13}{32}''$	$1\frac{5}{32}''$	$\frac{1}{16}''$		S-722	4
$\frac{13}{32}''$	$1\frac{3}{8}''$	$\frac{1}{8}''$		S-3359	1
$\frac{15}{32}''$	$\frac{23}{32}''$	$\frac{1}{8}''$		S-3329	2
$\frac{15}{32}''$	$\frac{7}{8}''$	$\frac{1}{16}''$		S-3326	10
$\frac{17}{32}''$	$\frac{7}{8}''$	$\frac{3}{16}''$		S-704	12
$\frac{17}{32}''$	$1''$	$\frac{3}{16}''$		S-374	1
$\frac{17}{32}''$	$1\frac{1}{8}''$	$\frac{1}{16}''$		S-709	4
$\frac{17}{32}''$	$1\frac{1}{4}''$	$\frac{1}{8}''$		S-978	8
$\frac{17}{32}''$	$1\frac{3}{8}''$	$\frac{1}{16}''$		S-390	12
$\frac{17}{32}''$	$1\frac{3}{8}''$	$\frac{1}{4}''$		S-347	2
$\frac{21}{32}''$	$1\frac{5}{8}''$	$\frac{1}{8}''$		S-927	8
$\frac{11}{16}''$	$1\frac{1}{4}''$	$\frac{1}{8}''$		S-3345	4
$\frac{25}{32}''$	$1\frac{1}{2}''$	$\frac{1}{8}''$		S-985	1
$\frac{25}{32}''$	$1\frac{3}{4}''$	$\frac{1}{4}''$		S-391	4
$\frac{7}{8}''$	$1\frac{1}{2}''$	$\frac{1}{8}''$		S-951	4
$\frac{7}{8}''$	$1\frac{3}{4}''$	$\frac{1}{16}''$		S-346	3
$\frac{7}{8}''$	$1\frac{3}{4}''$	$\frac{1}{8}''$		S-3344	1

FLAT STEEL WASHERS—Continued

I.D.	O.D.	THICK	Part No.	Quantity Required
$1\frac{1}{32}''$	$1\frac{3}{4}''$	$\frac{3}{32}''$	S-3316	2
$1\frac{1}{32}''$	$2\frac{1}{8}''$	$\frac{3}{16}''$	S-719	2
$1\frac{1}{16}''$	$1\frac{13}{16}''$	$\frac{1}{4}''$	S-3188	16
$1\frac{5}{32}''$	$2''$	$\frac{1}{8}''$	S-979	2
$1\frac{11}{32}''$	$2\frac{1}{4}''$	$\frac{1}{16}''$	S-2536	8
$1\frac{5}{8}''$	$2''$	$\frac{1}{16}''$	S-380	2
$2\frac{13}{16}''$	$3\frac{5}{8}''$	$\frac{5}{16}''$ Felt	S-2544	1

SHAKEPROOF EXTERNAL WASHERS

Description	Part No.	Quantity Required
$\frac{3}{16}''$	S-3287	2
$\frac{9}{32}''$	S-3172	2
$\frac{9}{32}''$	S-3189	50
$\frac{3}{8}''$	S-3175	8
$\frac{17}{32}''$	S-3179	19
$\frac{21}{32}''$	S-3173	4
$\frac{29}{32}''$	S-3174	10

SHAKEPROOF INTERNAL WASHERS

Description	Part No.	Quantity Required
$\frac{9}{32}''$ Cad.	S-3296	45
$\frac{11}{32}''$	S-7500	12
$\frac{3}{8}''$	S-3190	40
$\frac{13}{32}''$ Cad.	S-3297	6
$\frac{17}{32}''$	S-3294	2
$\frac{17}{32}''$ Cad.	S-3295	3
$\frac{1}{4}''$	S-3184	1
$\frac{5}{16}''$	S-3283	2
$\frac{3}{8}''$	S-3185	6

SHAKEPROOF INTERNAL & EXTERNAL WASHERS

Description	Part No.	Quantity Required
$\frac{9}{32}''$	S-3298	21
$\frac{13}{32}''$	S-3299	6
$\frac{17}{32}''$	S-3300	3

SOFT IRON WIRE

Description	Part No.	Quantity Required
#16 Ga. 13" Long	S-4284	4
#16 Ga. 15" Long	S-4106	1

ALPHABETICAL INDEX—MASTER PARTS LIST

	Group
Adapter, Carburetor	0301
Adapter, Distributor	0603
Adapter, Flywheel Bell Housing Starter	0109
Adapter, Grease Gun	2300
Adapter, Power Take-Off Assembly	1911
Adapter, Speedometer	2203
Adapter, Steering Arm Nut Pin (1¼" Ball)	1007
Adapter, Tachometer (Stewart-Warner)	0603
Adapter, Viscometer	0107
Adjuster, Differential Bearing	1105
Adjuster, Front Brake Slack—L. H.	1203
Adjuster, Front Brake Slack—R. H.	1203
Adjuster, Front Brake Slack (Bendix-Westinghouse)	1203
Adjuster, Front Brake Slack—L. H. (Bendix-Westinghouse)	1203
Adjuster, Rear Brake Slack—R. H. (Bendix-Westinghouse)	1203
Adjuster, Rear Brake Slack—L. H. (Bendix-Westinghouse)	1203
Adjuster, Rear Brake Shaft Slack	1203
Adjuster, Rear Slack—L. H. (Bendix-Westinghouse)	1203
Air Cleaner Assembly—Complete (United)	0301
Armature	0602
Armature Assembly	0601
Ammeter (Stewart-Warner)	0605
Ammeter	0605
Ammeter, Auxiliary (Stewart-Warner)	0605
Angle, Base—Front	1900
Anti-Rattle, Gas Tank Bracket	0300
Arm, Advance Control	0603
Arm, Advance Control, Assembly	0603
Arm, Breaker Point and, Assembly	0603
Arm, Brush	0601
Arm, Disc Brake Lever—Rear—L. H.	1201
Arm, Disc Brake Lever—Rear—R. H.	1201
Arm, Disc Brake Link Lever—Front—L. H.	1201
Arm, Disc Brake Link Lever—Front—R. H.	1201
Arm, Distributor Clamp, and Dial Assembly	0603
Arm, Fuel Pump Rocker	0302
Arm, Lever—Front L. H.	1201
Arm, Lever—Rear L. H.	1201
Arm, Overdrive Shift Rod	0706
Arm, Reverse Gearshift Rod	0706
Arm, Shift Rod	1911
Arm, Steering	1006
Arm, Steering	1403
Arm, Steering—L. H. (Double)	1007
Arm, Steering—R. H. (Single)	1007
Arm, Windshield Quadrant	1800
Arm, Wiper Assembly	1801
Arrestor, Flame	0402
Asbestos	1800
Axle, Carburetor Float	0301
Axle, Driving	1102
Axle, Driving, and Pilot Assembly—L. H. (Inner)—Short	1007
Axle, Driving, and Pilot Seat Assembly—R. H. (Inner)—Long	1007
Axle, Front—Assembly	1000
Axle, Front—Assembly (Ratio 8.148-1) (Timken)	1000
Axle, Front—Assembly—Complete	1000
Axle, Front Driving—Long	1007
Axle, Front Driving—Short	1007
Axle, Rear, Assembly (Timken)	1100
Axle, Rear, Assembly, Ratio 8.148, (Timken)	1100
Axle, Rear, Housing—Bare	1101
Bag, Tool	2300
Ball, Clutch Release Sleeve Fulcrum Ring	0203
Ball, Declutch Shift Hand Lever	0805
Ball, Gear Shift Hand Lever	0805
Ball, Gearshift Lever	0706
Ball, Gearshift—Front	0706

	Group
Ball, Pressure Lever Locking	0203
Ball, P. T. O. Control Hand Lever	1910
Ball, Steering Arm and Ball Pin, Assembly	1403
Ball, Universal Drive	1007
Ball, Universal Joint Cage	1007
Band, Brake, Assembly	1900
Band, Commutator Cover	0602
Band, Cover	0602
Band, Head, Assembly	0601
Bar, Differential Bearing Adjuster Lock	1105
Bar, Valve Cover Plate Clamp	0101
Base Assembly	1603
Base, Distributor, Assembly	0603
Base, Oil Filter	0107
Base, Oil Filter, and Shell Assembly	0107
Battery (Exide)	0610
Battery 6-V (Exide)	0610
Battery to Instrument Panel Cable Assembly	0610
Bearing, Absorbent Bronze	0603
Bearing, Ball—Double Row	1911
Bearing, Ball (MRC-7309-Radio)	1900
Bearing, Ball—S.A.E. #203	0601
Bearings, Ball—S.A.E. #204	0601
Bearing, Ball—Single Row	1911
Bearing, Bevel Gear	1105
Bearing, Camshaft Center	0106
Bearing, Camshaft Front & Rear	0106
Bearing, Camshaft Front	0106
Bearing, Camshaft Idler Gear Shaft	0106
Bearing, Camshaft Intermediate	0106
Bearing, Camshaft Rear	0106
Bearing (C.E.)	0602
Bearing, Clutch Pilot (S.K.F.)	0205
Bearing, Clutch Release (Aetna)	0203
Bearing, Connecting Rod (Standard)	0104
Bearing, Connecting Rod (.010 U/S)	0104
Bearing, Connecting Rod (.020 U/S)	0104
Bearing, Connecting Rod (.030 U/S)	0104
Bearing, Connecting Rod (.040 U/S)	0104
Bearing, Countershaft Front	0703
Bearing, Countershaft Front (Hyatt #R-1309-TS)	0703
Bearing, Countershaft Rear	0703
Bearing, Crankshaft Center Main (Standard)	0102
Bearing, Crankshaft Center Main (.010 U/S)	0102
Bearing, Crankshaft Center Main (.020 U/S)	0102
Bearing, Crankshaft Center Main (.030 U/S)	0102
Bearing, Crankshaft Center Main (.040 U/S)	0102
Bearing, Crankshaft Front Main (Standard)	0102
Bearing, Crankshaft Front Main (.010 U/S)	0102
Bearing, Crankshaft Front Main (.020 U/S)	0102
Bearing, Crankshaft Front Main (.030 U/S)	0102
Bearing, Crankshaft Front Main (.040 U/S)	0102
Bearing, Crankshaft Intermediate Main (Standard)	0102
Bearing, Crankshaft Intermediate Main (.010 U/S)	0102
Bearing, Crankshaft Intermediate Main (.020 U/S)	0102
Bearing, Crankshaft Intermediate Main (.030 U/S)	0102
Bearing, Crankshaft Intermediate Main (.040 U/S)	0102
Bearing, Crankshaft Rear Main (Standard)	0102
Bearing, Crankshaft Rear Main (.010 U/S)	0102
Bearing, Crankshaft Rear Main (.020 U/S)	0102
Bearing, Crankshaft Rear Main (.030 U/S)	0102
Bearing, Crankshaft Rear Main (.040 U/S)	0102
Bearing, Countershaft Rear (Marlin-Rockwell)	0703
Bearing (D.E.)	0602
Bearing, Declutch Shaft	0803
Bearing, Declutching Shaft (S.K.F.)	0803
Bearing, Differential	1002
Bearing, Differential	1105
Bearing, Endgate Hinge	1800
Bearing, Gearshift Connecting Tube—Front	0706
Bearing, Gearshift Connecting Tube—Rear	0706
Bearing, Governor Drive Shaft Ball	0305
Bearing, Governor Thrust Ball	0305
Bearing, Idler Shaft	0106

1

ALPHABETICAL INDEX—MASTER PARTS LIST

Group

Bearing, Idler Shaft (S.K.F.) 0804
Bearing, Jacket Tube 1403
Bearing, Main Drive Gear Idler 0802
Bearing, Main Drive Gear (S.K.F. #1749) 0703
Bearing, Main Drive Shaft Front (S.K.F.) 0802
Bearing, Main Drive Shaft Rear (S.K.F.) 0802
Bearing, Main Pilot (S.K.F. #WJM-30) 0703
Bearing, Mainshaft Pilot 0703
Bearing, Main Shaft Front 0802
Bearing, Mainshaft Rear 0703
Bearing, Main Shaft Rear 0802
Bearing, Needle, Assembly 0902
Bearing, Power Take-Off Adapter Gear—
(Rollway No. 160426) 1911
Bearing, Power Take-Off Adapter Gear Shaft 1911
Bearings, Reverse Gear Needle 0703
Bearing, Roller 1911
Bearing, Steady 0902
Bearing, Steering Knuckle Thrust 1007
Bearing, Transmission Driveshaft 0703
Belt, Fan 0503
Belt, Fan (Hercules) 0503
Biscuit, Rear Engine Support Bracket Rubber 0110
Biscuit, Rear Engine Support Rubber 0110
Blade, Fan, Assembly—20" with 4 Blades 0503
Blade, Fan, Assembly—21" with 6 Blades 0503
Blade, Fan, Assembly—22" with 6 Blades 0503
Blade, Wiper 1801
Block, Clutch Release Trunnion 0203
Block, Junction—to Inst. Panel Cable Assembly 0606
Block, Junction—7 Pole (Auto-Lite) 0606
Block, Snatch Assembly 2300
Body Assembly 0607
Body Assembly 0608
Body Assembly—Prime 0607
Body, Carburetor Throttle, Assembly 0301
Body, Carburetor Union 0301
Body, Complete, less tarpaulin and mounting
equipment (Galion) 1800
Body, Fuel Pump 0302
Body, Governor 0305
Body, Headlamp—Assembly 0607
Body, less Endgates 1800
Body, Oil Filler 0107
Body, Oil Filler 0110
Body, Oil Pump 0107
Body, Oil Pump (Lower) 0107
Body, Oil Pump (Upper) 0107
Body, Water Pump 0503
Body, Water Pump, and Bushing Assembly 0503
Bolt, Body Mounting—L. H. (¾"—16 x 18 1/16") 1800
Bolt, Body Mounting—R. H. (¾"—16 x 16") 1800
Bolt, Brake Spring 1900
Bolt, Center 1601
Bolt, Center (½"—13) 1601
Bolt, Center (7/16" x 3¾") 1601
Bolt, Connecting Rod Bearing 0104
Bolt, Crankshaft Flywheel 0102
Bolt, Differential 1002
Bolt, Differential and Spur Gear 1103
Bolt, Differential Cap 1103
Bolt, Differential Case 1002
Bolt, Differential Case 1103
Bolt, Distributor Clamp Arm and, Assembly 0603
Bolt, Drag Link End 1401
Bolt, Drag Link End (¼"—20 x 2¾") 1401
Bolt, End Clamp 1402
Bolt, First and Second Speed Slide Gear Sleeve 0702
Bolt, Flywheel 0102
Bolt, Front Brake Shoe Link 1203
Bolt, Front Brake Slack Adjuster Clamp 1203
Bolt, Front Spring 1602
Bolt, Housing and Carrier Through 1001
Bolt, Oil Filter Case Clamp (Fram #11735) 0107

Group

Bolt, Plow 1601
Bolt, Plow (7/16"—14) 1601
Bolt, Plow (⅜"—16) 1601
Bolt, Rear Brake Camshaft Support Bracket 1203
Bolt, Rear Brake Shoe Lining 1202
Bolt, Rear Brake Spider to Housing 1203
Bolt, Rear Engine Support 0110
Bolt, Rear Spring—Front 1602
Bolt, Rear Wheel Bearing Lock 1301
Bolt, Spoon (¼"—20 x 1 33/64") 1201
Bolt, Steady Bearing Housing 0902
Bolt, Steering Cross Tube End Clamp 1402
Bolt, Through 0602
Bolt, Through (½"—13 x 7 3/16") 1201
Bolt, Through (½"—13 x 7 3/16") 1201
Bolt, Wing 1800
Bond (Cab to Frame) 0611
Bond (2—Cab to Cab Support Rear (2—
Body to Cab Support) 0611
Bond (Crankcase to Frame) 0611
Bond (Exhaust Pipe to Frame) 0611
Bond (Radiator to Cross Member) 0611
Bond (R. H. Engine Support to Frame) 0611
Bond (Splash Guard to Frame) 0611
Bond (Steering Column Bracket) 0611
Book, Parts and Instruction 2300
Boot, Front Bumper Bracket 1602
Boot, Gearshift Universal Lever 0706
Box, Chain 2300
Box, Governor Valve (Casting Only) 0305
Box, Intake Manifold Heat 0108
Box, Tool Assembly 1800
Bowl, Carburetor Fuel, Assembly 0301
Bowl, Fuel Filter 0302
Bowl, Fuel Pump Glass 0302
Bowl, Fuel Pump Metal 0302
Brace, Oil Filler Breather Pipe 0107
Brace, Radiator Top 0501
Brace, Steering Column—L. H. 1405
Brace, Steering Column—R. H. 1405
Bracket, Accelerator Cross Shaft 0303
Bracket, Accelerator Treadle 0303
Bracket, Air Line Rubber Mounting 1206
Bracket, Air Reservoir Tank 1205
Bracket, Air Tank 1205
Bracket, Anchor 1201
Bracket, Auxiliary Spring 1601
Bracket, Brake Application Valve 1205
Bracket, Bumper—L. H. 1500
Bracket, Bumper—R. H. 1500
Bracket, Brake Application Valve 1205
Bracket, Carburetor Air Shutter, Assembly 0301
Bracket, Carburetor Float 0301
Bracket, Choke and Throttle Control 0303
Bracket, Clutch and Brake Pedal 0204
Bracket, Compressor Mounting 1205
Bracket, Control 0805
Bracket, Deck Plate Clamp—Lower 1800
Bracket, Deck Plate Mounting—Front—R. H. 1800
Bracket, Deck Plate Mounting—Front—L. H. 1800
Bracket, Disc Brake Anchor 1201
Bracket, Fan 0503
Bracket, Filter to Pump Titeflex Gas Line Clip 0304
Bracket, Fire Extinguisher 1800
Bracket, Front Brake Shaft, and Bushing Assembly 1203
Bracket, Front Bumper 1500
Bracket, Front Cab Support 1800
Bracket, Front Engine Support 0110
Bracket, Front Spring, and Bushing Assembly 1602
Bracket, Front Spring, and Bushing Assembly—Rear 1602
Bracket, Front Spring—Front 1602
Bracket, Front Spring—L. H. 1602
Bracket, Front Spring—Front—L. H. 1602
Bracket, Front Spring—Front—R. H. 1602

ALPHABETICAL INDEX—MASTER PARTS LIST

	Group
Bracket, Front Spring—Rear	1602
Bracket, Front Spring—R. H.	1602
Bracket, Gas Tank	0300
Bracket, Gas Tank—Front	0300
Bracket, Gas Tank—Rear	0300
Bracket, Gearshift Connecting Tube—Retainer, Felt	0706
Bracket, Gearshift Lever	0706
Bracket, Gearshift Lever Mounting	0706
Bracket, Gearshift Universal Lever	0706
Bracket, Hand Brake Lever Fulcrum	1201
Bracket, Hose Coupling Frame—L. H.—Front	1206
Bracket, Hose Coupling Frame—L. H.—Rear	1206
Bracket, Hose Coupling Frame—R. H.—Front	1206
Bracket, Hose Coupling Frame—R. H.—Rear	1206
Bracket, Hose Spring	1206
Bracket, Muffler	0401
Bracket, Muffler—Front	0401
Bracket, Muffler—Rear	0401
Bracket, Oil Filter	0107
Bracket, Oil Filter Steady	0107
Bracket, Pedal	0204
Bracket, Pedal, and Bushing Assembly	0204
Bracket, Pivot Anchor	0706
Bracket, Power Take-Off Adapter	1911
Bracket, P. T. O. Cross Shaft (Inner)	1910
Bracket, P. T. O. Cross Shaft (Outer)	1910
Bracket, Pump to Carb. Gas Line Clip	0304
Bracket, Quarter Deck Support—Upper	1800
Bracket, Radiator Support	0501
Bracket, Radiator Top Brace	0501
Bracket, Rear Brake Shaft, and Bushing Assembly—R. H.	1203
Bracket, Rear Brake Camshaft and Diaphragm—L. H.	1203
Bracket, Rear Brake Camshaft and Diaphragm—R. H.	1203
Bracket, Rear Brake Camshaft Support, and Bushing Assembly	1203
Bracket, Rear Brake Shaft, and Bushing Assembly—L. H.	1203
Bracket, Rear Engine Support	0110
Bracket, Rear Quarter Fender Lower Support	1800
Bracket, Rear Spring, and Bushing Assembly—Front	1602
Bracket, Rear Spring, and Bushing Assembly—Front—L. H.	1602
Bracket, Rear Spring, and Bushing Assembly—Front—R. H.	1602
Bracket, Rear Spring—Front	1602
Bracket, Rear Spring—Front—L. H.	1602
Bracket, Rear Spring—Front—R. H.	1602
Bracket, Rear Spring Shackle	1602
Bracket, Rear Spring Shackle Pin Anchor—L. H.	1602
Bracket, Rear Spring Shackle Pin Anchor—R. H.	1602
Bracket, Relay Valve	1205
Bracket, Shaft Brake Anchor	1201
Brackets, Headlamp and Blackout Lamp	1800
Bracket, Shock Absorber Frame (L. H.)	1603
Bracket, Shock Absorber Frame (R. H.)	1603
Bracket, Speedometer Cable Clip	2203
Bracket, Starting Switch, and Nut Assembly	0602
Bracket, Starting Switch, Assembly	0602
Bracket, Starting Switch (Part of 16UBL2443A)	0602
Bracket, Steering Column Top	1405
Bracket, Steering Gear Frame, and Cap	1405
Bracket, Tachometer Cable Clip	2203
Bracket, Tank to Filter Gas Line Clip	0304
Bracket, Tubing Clip	1209
Bracket (Weld to Exhaust Pipe)	0611
Bracket, Winch Support—Front	1500
Bracket, Winch Support—L. H.—Rear	1500
Bracket, Winch Support—R. H.—Rear	1500
Bracket, Windshield—Center	1800
Bracket, Windshield—Side	1800
Brake, Disc, Assembly (American Cable)	1201
Brake, Drag, Assembly	1900
Brake, Rear, Spider and Bushing Assembly	1101
Brake, Rear, Spider and Sleeve Assembly—Long	1101
Brake, Rear, Spider and Sleeve Assembly—Short	1101
Breaker, Horn Circuit (Klix-on 30-Amp.)	0609
Breaker, Klix-on Automatic Reset Circuit	0609
Breaker, Starting Switch Circuit (Klix-on 60-Amp.)	0602
Breather, Housing Oil	1001
Breather, Housing Oil	1101
Bridge, Top Cover	0107
Brush—Main	0601
Brush, Motor	0602
Bulb, Blackout Parking Lamp—#64	0607
Bumper	1500
Bumper, Front	1500
Bumper, Spring	1401
Bushing	1201
Bushing	1204
Bushing	1603
Bushing, Accelerator Cross Shaft Tube	0303
Bushing, Accelerator Treadle Bracket	0303
Bushing, Anchor Bracket	1201
Bushing, Bendix Housing	0602
Bushing, Blackout Switch	0606
Bushing, Brake Shoe	1201
Bushing, Brush Holder Rivet Insulating	0602
Bushing, Camshaft Idler Gear Shaft	0106
Bushing, Camshaft, #1	0106
Bushing, Camshaft, #2	0106
Bushing, Camshaft, #4	0106
Bushing, Camshaft, #5	0106
Bushing, Carburetor Throttle Lever	0301
Bushing, Carbureor Throttle Shaft	0301
Bushing, Center Plate	0602
Bushing, Clutch Pedal Lever	0204
Bushing, Clutch Pedal Lever, and Assembly	0204
Bushing, Clutch Throwout Shaft	0203
Bracket, Coil Mounting	0603
Bushing, Declutching Shaft	0803
Bushing, Drive Housing	0602
Bushing, Driven Shaft	0803
Bushing, Drum	1900
Bracket, Flame Arrestor	0402
Bushing, Front Brake Camshaft	1203
Bushing, Front Brake Shaft Bracket—Short	1203
Bushing, Front Brake Shoe	1202
Bushing, Front Engine Support Bracket Rubber	0110
Bushing, Front Spring	1601
Bushing, Front Spring	1602
Bushing, Front Spring Bracket	1602
Bushing, Gear Case	1900
Bushing, Gear Shift Hand Lever	0805
Bushing, Gearshift Lever Ball	0706
Bushing, Gearshift Lever Ball Split	0706
Bushing, Gearshift Lever Socket	0706
Bushing, Gearshift Lever Socket Split	0706
Bushing, Governor Lower Drive Shaft	0305
Bushing, Governor Upper Drive Shaft	0305
Bushing, Hand Brake Lever	1201
Bushing, Housing	0602
Bushing, Housing Breather Nipple	1101
Bushing, Housing—Inner	1403
Bushing, Housing—Outer	1403
Bushing, Instrument Panel Light Switch	0606
Bushing, Insulating	0601
Bushing, Lever Arm	1201
Bushing, Main Drive Shaft Direct Drive Gear	0802
Bushing, Mainshaft Fourth Gear Bearing	0702
Bushing, Mainshaft Overdrive Gear	0702
Bushing, Mainshaft Third Speed Gear	0702
Bushing, Mainshaft Third Speed Gear Roller	0702
Bushing, Motor Terminal Stud Insulation ($\frac{7}{16}$")	0602
Bushing, Oil Pump Body (Lower)	0107
Bushing, Oil Pump Body (Upper)	0107

ALPHABETICAL INDEX—MASTER PARTS LIST

	Group
Bushing, Oil Pump Cover	0107
Bushing, Oil Pump Idler Gear	0107
Bushing, Overdrive Countershaft Gear	0702
Bushing, Pedal Bracket	0204
Bushing, Piston Pin	0104
Bushing, Piston Pin	0110
Bushing, Front Brake Shaft Bracket—Long	1203
Bushing, Rear Brake Camshaft Support Bracket	1203
Bushing, Rear Brake Shoe	1202
Bushing, Rear Brake Shaft Bracket—Thin	1203
Bushing, Rear Brake Spider	1101
Bushing, Rear Brake Spider	1203
Bushing, Rear Engine Support Bracket Rubber	0110
Bushing, Rear Engine Support Rubber	0110
Bushing, Rear Spring	1601
Bushing, Rear Spring Bracket	1602
Bushing, Rear Spring Shackle	1602
Bushing, Reverse Gear	0702
Bushing, Rubber	1405
Bushing, Speedometer Driven Gear	0705
Bushing, Speedometer Driven Gear (S-W)	0804
Bushing, Steering Cross Tube End	1402
Bushing, Steering Knuckle	1006
Bushing, Steering Knuckle (With Oil Groove)	1006
Bushing, Terminal Screw	0603
Bushing, Tie Rod Bolt	1007
Bushing, Universal Drive	1007
Bushing, Water Pump Body	0503
Bushing, Water Pump Cover	0503
Bushing, Water Pump Head	0503
Bushing, Water Pump Packing	0503
Button, Oil Pressure Regulator Spring	0107
Button, Pressure Spring Insulator	0205
Buzzer, Low Pressure Indicator	1205
Cab Assembly	1800
Cable, Battery Ground, Assembly	0610
Cable, Battery Ground, Assembly (Positive)	0610
Cable, Battery to Starting Switch, Assembly	0602
Cable, Generator to Regulator, Assembly	0601
Cable, Magnetic Switch to Battery, Assembly	0602
Cable, Magnetic Switch to Motor, Assembly	0602
Cables, Ignition—Assembled	0604
Cable, Speedometer	2203
Cable, Starting Switch to Battery, Assembly (Negative)	0602
Cable, Starting Switch to Battery, Assembly (Positive)	0602
Cable, Tachometer	2203
Cable, Tail Lamp, Assembly	0608
Cable, Tail Lamp, Assembly (Chassis)	0608
Cable, Trailer Light	2300
Cable, Tail Light, Assembly	0608
Cable, Trailer Lighting—Assembly (Warner Elec.)	0606
Cable, Winch Assembly	2300
Cage, Bevel Pinion Bearing, and Cup Assembly	1003
Cage, Bevel Pinion Forward Bearing, and Cup Assembly	1003
Cage, Drive Pinion, and Cup Assembly	1105
Cage, Drive Pinion, Assembly	1105
Cage, Main Shaft Front Bearing	0802
Cage, Spur Pinion Bearing	1105
Cage, Pinion Bearing, and Cap Assembly	1003
Cape, Pinion Bearing, and Cup Assembly	1105
Cage, Spur Pinion Bearing	1105
Cage, Spur Pinion Bearing, and Cap Assembly	1003
Cage, Spur Pinion Bearing—Assembly	1003
Cage, Universal Drive Ball	1007
Cage, Universal Drive—Outer	1007
Cage, Universal Joint	1007
Cam	0603
Camshaft	0106
Camshaft Assembly with Gear	0106
Camshaft, Front Brake—L. H.	1203
Camshaft, Front Brake—R. H.	1203
Camshaft, Rear Brake—L. H.	1203
Camshaft, Rear Brake—R. H.	1203
Can, Oil	2300
Cap, Countershaft Front Bearing	0703
Cap, Countershaft Rear Bearing	0703
Cap, Declutch Shaft Bearing	0803
Cap, Declutching Shaft Bearing, Assembly	0803
Cap, Distributor	0603
Cap, Distributor, Assembly	0603
Cap, Drive Pinion Cage	1105
Cap, Driven Shaft Rear Bearing	0803
Cap, Driveshaft	1911
Cap, Dust (⅞ O.D. & C.E.)	0602
Cap, Fan Hub	0503
Cap, Front Spring Bracket	1602
Cap, Fuel Pump Spring	0302
Cap, Gas Tank, and Chain	0300
Cap, Gas Tank, Assembly	0300
Cap, Gearshift Finger Housing	0706
Cap, Governor Body	0305
Cap, Governor Throttle Lever Body	0305
Cap, Idler Shaft Front Bearing	0804
Cap, Idler Shaft Rear Bearing	0804
Cap, Main Drive Gear Bearing	0703
Cap, Mounting Side, and Cage Stud	1103
Cap, Mounting Side, and Cup Assembly	1003
Cap, Mounting Side, and Cup Assembly	1103
Cap, Needle Bearing	0902
Cap, Oil Filler, Assembly	0107
Cap, Oil Filler, Assembly	0110
Cap, Oil Filler Breather, Assembly	0107
Cap, Oil Filter Pressure Relief Valve Adjusting Screw	0107
Cap, Radiator Filler	0501
Cap, Rear Spring Stud	1602
Cap, Worm—Brake End	1900
Cap, Worm—Joint End	1900
Cap, Sleeve Dust	0902
Cap, Steering Knuckle Bearing—Lower	1006
Cap, Steering Knuckle Bearing—Upper	1006
Cap, Transmission Driveshaft Bearing	0703
Carburetor Assembly (Zenith)	0301
Carburetor Complete (Stromberg)	0301
Carburetor (Stromberg SF3)	0301
Carburetor Complete (Stromberg)	0301
Carrier, Declutching Shaft	0803
Carrier, Declutch Shaft Bearing	0803
Carrier, Differential—and Cap Assembly	1002
Carrier, Differential, and Cap Assembly	1103
Carrier, Differential—and Cap Stud	1002
Carrier, Differential—Assembly—Complete	1002
Carrier, Differential, Assembly—Complete (Ratio 8.148)	1103
Carrier, Differential, Assembly—Complete (Ratio 8.43)	1103
Carrier, Differential—to Housing Gasket	1002
Carrier, Gear, Assembly	1103
Carrier, Gear, with Caps	1103
Cartridge, Oil Filter	0107
Case, Brake, Assembly	1900
Case, Differential—Assembly	1002
Case, Differential Assembly	1103
Case, Differential—L. H.	1103
Case, Differential—R. H.	1103
Case, Gear	1900
Case, Gear, Assembly Case and Cover (Gar Wood)	1900
Case, Oil Filter—Assembly (Fram #5267)	0107
Case, Speedometer Drive Gear, and Bushing Assembly	0705
Case, Speedometer Drive Gear, Assembly	0705
Case, Transfer—and Cover Assembly	0801
Case, Transfer—Assembly, Complete (Wisconsin)	0800
Case, Transfer—Assembly (with Declutching Unit)	0800

4

ALPHABETICAL INDEX—MASTER PARTS LIST

	Group
Catch, Hood	1800
Catch, Radiator Filler Door, and Bracket Assembly	1800
Center, Front Axle—Only	1807
Chain, Brass—4½" Long	0110
Chain, Endgate	1800
Chain, Endgate, Assembly	1800
Chains, Tire	2300
Chains, Tire (Front Type D)	2300
Chains, Tire (Rear Type N)	2300
Chain, Towing Assembly	2300
Chamber, Front Brake—L. H.	1207
Chamber, Front Brake—R. H.	1207
Chamber, Gasket Reserve	1603
Chamber, Rear Brake—L. H.	1207
Chamber, Rear Brake—R. H.	1207
Channel, Front Engine Support	0110
Clamp, Accelerator Cross Shaft	0303
Clamp, Carburetor Air Shutter Bracket Wire	0301
Clamp, Flame Arrestor	0402
Clamp, Jackshaft Bearing	1105
Clamp, Oil Pump Suction Line Flange	0107
Clamp, Rope	1900
Clamp, Steady Bearing Housing Flinger	0902
Clamp, Terminal	0603
Cleaner, Air (United Specialties)	0301
Clevis	1201
Clevis	1910
Clevis, Brake Control Rod	1204
Clevis, Accelerator Control Rod	0303
Clevis, Carburetor Control Rod	0303
Clevis, Clutch Control Rod	0203
Clevis, Clutch Control Rod Adjusting	0204
Clevis, Control	0108
Clevis, Declutch Shift Control Rod	0805
Clevis, Gear Shift Control Rod	0805
Clevis, Hand Brake Shaft Lever	1201
Clevis, Transfer Lever	1201
Clip	1403
Clip, Accelerator Spring	0303
Clip, Bendix Spring Support	0602
Clip, Brake Control Rod Return Spring	1204
Clip, Breaker Arm Spring	0603
Clip, Breaker Lever Retainer	0603
Clip, Brush Lead	0602
Clip, Brush Lead Terminal	0602
Clip, Cable	0110
Clip, Choke and Throttle Control	0303
Clip, Drain Tube—Front	0110
Clip, Drain Tube—Rear	0110
Clip, Field Lead	0602
Clip, Front Spring—Front	1601
Clip, Front Spring Rebound, Assembly (Small)	1601
Clip, Front Spring Rebound, Assembly (Large)	1601
Clip, Fuel Pump Link Pin	0302
Clip, Gas Line	0304
Clip, Hose	1206
Clip, Hose Spring	1206
Clip—Loom	0607
Clip Loom (Tinned)	0611
Clip, Muffler	0401
Clip, Rear Spring	1601
Clip, Rear Spring Rebound, Assembly	1601
Clip, Rear Spring Rebound, Assembly (Large)	1601
Clip, Rear Spring Rebound, Assembly (Small)	1601
Clip, Speedometer Cable	2203
Clip, Tachometer Cable	2203
Clip, Tubing	1209
Cluster, Valve Tappet	0105
Clutch Assembly (Long) (Does not include Housing and Control)	0200
Clutch Assembly (W. C. Lipe)	0200
Clutch Assembly (W. C. Lipe) (Does not include Housing and Control)	0200
Clutch, Declutch Driving	0803
Clutch, Declutch Sliding	0803
Clutch, Main Shaft Power Takeoff	0802
Clutch, Sliding	0803
Clutch, Sliding	1900
Cocks, Cut-Off	1209
Cock, Cylinder Block Drain (⅜") Pipe Thread	0101
Cock, Drain	1205
Cock, Radiator Drain	0501
Cock, Radiator Drain, ⅜" (Weatherhead #270)	0501
Cocks, Shut-Off	1209
Cock, Shut Off (¼" x ¼") (45-C-1966)	0304
Coil (Auto-Lite)	0604
Coil, Field, Assembly—Complete	0601
Coil, Field—Assembly—L. H.	0602
Coil, Field—Assembly (Lower)	0602
Coil, Field—Assembly—R. H.	0602
Coil, Field—Assembly (Upper)	0602
Coil, Field—Left	0601
Coil, Field—Right	0601
Coil, Ignition	0603
Collar, Control Hand Lever Shaft	0805
Collar, Declutch Control Cross Shaft	0805
Collar, Driveshaft	1900
Collar, Governor Spring	0305
Collar, Oil Pump	0107
Collar, Pedal Lever Shaft	0204
Collar, Rear Brake Camshaft	1203
Compressor	1205
Compressor, Air	1205
Condenser	0603
Condenser Assembly	0603
Condenser, Generator	0611
Condenser, Ignition Coil	0611
Cone, Bevel Gear Bearing	1003
Cone, Bevel Gear Bearing	1105
Cone, Bevel Pinion Forward Bearing	1003
Cone, Bevel Pinion Rear Bearing	1003
Cone, Countershaft Front Bearing	0703
Cone, Countershaft Rear Bearing	0703
Cone, Countershaft Rear Bearing (Timken #336)	0703
Cone, Differential Bearing	1002
Cone, Driven Shaft Bearing (Timken)	0803
Cone, Driven Shaft Front Bearing	0803
Cone, Driven Shaft Rear Bearing	0803
Cone, Drive Pinion Bearing—Front	1105
Cone, Drive Pinion Bearing—Rear	1105
Cone, Fan Spindle Bearing (Timken)	0503
Cone, Front Wheel Bearing—Inner	1301
Cone, Front Wheel Bearing—Outer	1301
Cone, Idler Shaft Front Bearing	0804
Cone, Idler Shaft Rear Bearing	0804
Cone, Jackshaft Bearing—L. H.	1105
Cone, Jackshaft Bearing—R. H.	1105
Cone, Mainshaft Bearing (Timken #377)	0703
Cone, Mainshaft Rear Bearing	0703
Cone, Pinion Bearing—Front	1003
Cone, Pinion Bearing—Front	1105
Cone, Pinion Bearing—Rear	1003
Cone, Pinion Bearing—Rear	1105
Cone, Rear Wheel Bearing	1301
Cone, Rear Wheel Bearing—Inner	1301
Cone, Rear Wheel Bearing—Outer	1301
Cone, Spur Pinion Bearing	1003
Cone, Spur Pinion Bearing	1105
Cone, Spur Pinion—L. H.	1003
Cone, Spur Pinion—R. H.	1003
Cone, Steering Knuckle Bearing	1006
Cone, Thru Shaft Forward Bearing	1105
Connector and Wiring Assembly	0607
Connector, Flexible Hose	1209
Connector, Frame Tube and Hose	1209
Connector, Hose	1209

ALPHABETICAL INDEX—MASTER PARTS LIST

Part	Group
Connectors, Chamber Hose	1209
Connector, ⅜" Tube	1209
Connector, ½" Tube	1209
Connector, ¾" Tube	1209
Container, Key	2300
Core, Radiator, and Tanks	0501
Cotter	1002
Cotter, Crankshaft Flywheel Bolt	0102
Cotter, Oil Pump Suction Line Flange Clamp Screw	0107
Coupling, Distributor	0603
Coupling, Distributor Drive	0603
Coupling, Distributor Drive (Grooved)	0603
Coupling, Dummy	1206
Coupling, Dummy Hose	1209
Coupling, Dummy Hose (No Vent)	1209
Coupling, Dummy Hose (Rear of Body)	1206
Coupling, Dummy Hose (With Bleeder Vent)	1209
Coupling, Hose	1209
Countershaft	0704
Countershaft Assembly—Gears and Bearings	0702
Countershaft (16 and 10 Teeth)	0704
Cover	1911
Cover, Bevel Pinion Forward Bearing, Assembly	1003
Cover, Commutator End Cap	0601
Cover, Cylinder Block Fuel Pump Opening	0101
Cover, Cylinder Block Gear Case	0101
Cover, Cylinder Valve	0101
Cover, Differential Carrier Top	1103
Cover, Distributor Cap	0603
Cover, Door Panel Hand Hole, Assembly (Includes Rubber Bumpers)	1800
Cover, Driven Shaft Rear, Assembly	0803
Cover, Endgate Chain	1800
Cover, Engine Fan, Assembly	1800
Cover, Front Wheel Brake Drum Inspection	1302
Cover, Fuel Pump Bottom	0302
Cover, Fuel Pump Opening	0101
Cover, Fuel Pump Top Screw	0302
Cover, Gear Case	1900
Cover, Gearshift Lever Dust	0706
Cover, Gearshift Lever Socket	0706
Cover, Gearshift Top Control	0706
Cover, Housing Side	1403
Cover, Hand Hole, and Baffle Assembly	1103
Cover, Housing Side, Assembly	1403
Cover, Idler Shaft Front Bearing	0804
Cover, Idler Shaft Rear Bearing	0804
Cover, Main Drive Front	0802
Cover, Main Drive Shaft Rear Bearing	0802
Cover, Main Shaft Rear Bearing	0802
Cover, Oil Filter Case (Fram #11559)	0107
Cover, Oil Filter Shell	0107
Cover, Oil Pump	0107
Cover, Oil Pump, and Bushing Assembly	0107
Cover, Oil Seal	1911
Cover, Power Takeoff Opening	0701
Cover and Pressure Plate Assembly (Including Cover Plate, Springs and Pressure Plate)	0202
Cover, Rear Axle Housing Rear	1101
Cover, Rear Axle Housing Rear, and Carrier Gasket	1101
Cover, Reverse Gearshift Rod Plunger	0706
Cover, Rear Gearshift Rod—Side	0706
Cover, Rear Gearshift Rod—Top	0706
Cover, Rear Wheel Brake Drum Inspection	1302
Cover, Shifter	0805
Cover, Spur Pinion Bearing	1105
Cover, Spur Pinion Bearing Cage	1105
Cover, Spur Pinion Bearing—L. H.	1003
Cover, Spur Pinion Bearing—R. H.	1003
Cover, Steady Bearing Housing End	0902
Cover, Steering Cross Tube End Dust	1402
Cover, Thru Shaft Forward Bearing, Assembly	1105
Cover, Thru Shaft Rear Bearing	1105
Cover, Top	0107
Cover, Transmission—and Dowel Assembly	0701
Cover, Transmission—Assembly (Including Top Cover Control)	0701
Cover, Transmission Case	0701
Cover, Upper	1403
Cover, Upper	1403
Cover, Valve	0105
Cover, Water Pump	0503
Cowl	1800
Crank, Bell	1910
Crank, Governor Valve Box Bell	0305
Crank, Starting	2300
Crank, Starting Assembly	2300
Crankcase Assembly	0101
Crankshaft	0102
Crankshaft and Gear Assembly	0110
Cross, Differential	1103
Cross Member, Front	1500
Cross Member, Front Engine Support	0110
Cross Member, Front Tubular	1500
Cross Member, Pintle Hook	1500
Cross Member, Radiator Support	1500
Cross Member, Rear	1500
Cross Member, Rear Spring Front Bracket	1500
Cup, Ball	1403
Cup, Bevel Gear Bearing	1003
Cup, Bevel Gear Bearing	1105
Cup, Bevel Pinion Forward Bearing	1003
Cup, Bevel Pinion Rear Bearing	1003
Cup, Countershaft Front Bearing	0703
Cup, Countershaft Rear Bearing	0703
Cup, Countershaft Rear Bearing (Timken #332)	0703
Cup, Differential Bearing	1105
Cup, Differential Bearing	1002
Cup, Differential Bearing	1105
Cup, Driven Shaft Bearing (Timken)	0803
Cup, Driven Shaft Front Bearing	0803
Cup, Driven Shaft Rear Bearing	0803
Cup, Drive Pinion Bearing—Front	1105
Cup, Fan Spindle Bearing (Timken)	0503
Cup, Front Wheel Bearing—Inner	1301
Cup, Front Wheel Bearing—Outer	1301
Cup, Grease	0603
Cup, Idler Shaft Front Bearing	0804
Cup, Idler Shaft Rear Bearing	0804
Cup, Jackshaft Bearing—L. H.	1105
Cup, Jackshaft Bearing—R. H.	1105
Cup, Mainshaft Bearing (Timken #372-A)	0703
Cup, Mainshaft Rear Bearing	0703
Cup, Pinion Bearing—Front	1003
Cup, Pinion Bearing—Front	1105
Cup, Pinion Bearing—Rear	1003
Cup, Pinion Bearing—Rear	1105
Cup, Rear Wheel Bearing	1301
Cup, Rear Wheel Bearing—Inner	1301
Cup, Rear Wheel Bearing—Outer	1301
Cup, Spring	0205
Cup, Spur Pinion Bearing	1105
Cup, Spur Pinion—L. H.	1003
Cup, Spur Pinion—R. H.	1003
Cup, Steering Knuckle Bearing	1006
Cup, Thru Shaft Forward Bearing	1105
Cup, Thru Shaft Rear Bearing	1105
Cup, Valve Spring	0110
Cup, Valve Spring—Latest Design (Use with 2D210A—2D2710)	0110
Cup, Valve Spring—Original Design (Use with 2SA210C—2SA2710A)	0110
Cup, Water Pump Grease	0503
Cushions, Back and Seat	1800
Cushion, Seat, Assembly—L. H.	1800
Cushion, Seat, Assembly—R. H.	1800

ALPHABETICAL INDEX—MASTER PARTS LIST

	Group
Cushion, Seat Back, Assembly—L. H.	1800
Cushion, Seat Back, Assembly—R. H.	1800
Cylinder and Crankcase Assembly (Including Valve Seats, Main Bearings and Valve Guides)	0101
Cylinder Block Assembly (Including Valve Seats and all Studs)	0101
Cylinder Block (Includes Bearing Caps, Valve Guides, Camshaft Bearings and Idler Bearing)	0101
Cylinder, Filter	0107
Cylinder, Filter (4" O.D. at Base)	0107
Cylinder, Filter—4" O.D. at Base (Part of 2E0807)	0107
Cylinder, Filter—4⅛" O.D. at Base (Part of 2A0807)	0107
Cylinder Head	0101
Cylinder Head Assembly	0101
Cylinder Head—Front	0101
Cylinder Head—Rear	0101
Deflector, Air	0601
Dial, Distributor Clamp Arm	0603
Diaphragm, Front Brake, Assembly—L. H.	1207
Diaphragm, Front Brake, Assembly—R. H.	1207
Diaphragm, Front Brake—L. H.	1207
Diaphragm, Front Brake—R. H.	1207
Diaphragm, Fuel Pump, Set (5 Pieces)	0302
Diaphragm, Rear Brake	1207
Diaphragm, Rear Brake—L. H.	1207
Diaphragm, Rear Brake—R. H.	1207
Differential and Ring Gear Assembly	1103
Differential and Gear Assembly (8.43 Ratio)	1103
Differential Assembly	1002
Differential Assembly	1103
Disc Assembly	0201
Disc, Brake	1900
Disc, Brake (14")	1201
Disc, Clutch with Facing	0201
Disc, Driveshaft Brake	1201
Disc, Water Pump Lead	0503
Distributor and Gear Assembly	0603
Distributor and Gear Assembly (Delco)	0603
Distributor Assembly (Auto-Lite)	0603
Dome, Fuel Pump Air	0302
Door Assembly	0607
Door Assembly	0608
Door Assembly Complete with All Hardware—L. H.	1800
Door Assembly Complete with All Hardware—R. H.	1800
Door Assembly—Less Lens	0607
Doors, Complete with Hardware	1800
Door, Radiator Filler	1800
Door, Tool Box Assembly	1800
Dovetail, Door (Male)	1800
Dowel	1102
Dowel	1103
Dowel, Connecting Rod	0104
Dowel, Crankshaft	0102
Dowel, Crankshaft Flywheel	0102
Dowel, Crankshaft Rear Main Bearing Flange	0102
Dowel, Cylinder Block Pressure Regulator	0101
Dowel, Flywheel	0102
Dowel, Front Driving Axle Flange	1007
Dowel, Front Spring Bracket	1602
Dowel, Front Wheel Adjuster	1301
Dowel, Gear Carrier Cap	1103
Dowel, Gearshift Lever Bracket	0706
Dowel, Rear Axle Housing Rear Cover	1101
Dowel, Rear Spring Shim	1601
Dowel, Steering Cross Tube End	1402
Dowel, Transmission Case Cover	0701
Dowel, Valve Tappet Cluster	0105
Dowel, Water Pump	0503
Drain, Front Hub Oil	1302
Drawrope, Tarpaulin	1800
Drive, Bendix—Assembly (Eclipse Machine Co.)	0602
Drive, Bendix—Head	0602

	Group
Drive, Universal, Assembly	1007
Drive, Universal only	1007
Driven Member Assembly (Including Hub and Plate Assembly)	0201
Driver, Screw—6"	2300
Driver, Screw—Phillips #2	2300
Driver, Screw—Phillips #3	2300
Driver, Third and Fourth Speed Clutch	0702
Driveshaft	1900
Drive Shaft and Governor Assembly	0603
Driveshaft, Front	0901
Driveshaft, Front—Series 1500	0901
Driveshaft, Intermediate	0901
Driveshaft, Intermediate—Series 1600	0901
Driveshaft, Rear	0901
Driveshaft, Rear—Series 1600	0901
Driveshaft, Transmission, and Gear	0703
Driveshaft, Transmission, and Gear (29 Teeth)	0703
Drum Assembly	1900
Drum, Front Hub	1302
Drum, Front Wheel Brake	1302
Drum, Rear Wheel Brake	1302
Elbow, Carburetor	0301
Elbow, Cylinder Head Water Outlet	0101
Elbow, Gas Tank Inlet, Assembly	0300
Elbow, Housing Filler	1101
Elbow, Oil Line (¼")	0107
Elbow, Radiator Inlet	0501
Elbow, Radiator Outlet	0501
Elbow, Water Pump Inlet	0505
Elbow, ⅜"	1209
Elbow, ½"	1209
Elbow, ¾"	1209
Element, Fuel Filter	0302
Extension, Speedometer Cable	2203
End, Drag Link Adjustable	1401
End, Front Link Assembly	1603
End, Gearshift Lever	0706
End, Link Assembly	1603
End, Pawl Rod	1201
End, Steering Cross Tube, Assembly—R. H. (1¼"—12 Threads)	1402
End, Steering Cross Tube—L. H. Only (1¼"—14 Threads)	1402
End, Steering Cross Tube—R. H. Only (1¼"—12 Threads)	1402
End, Steering Cross Tube (12 Threads)	1402
End, Steering Cross Tube (16 Threads)	1402
Engine Assembly (Autocar)	0100
Engine Assembly (Hercules)	0100
Escutcheon	1800
Escutcheon, Door Handle	1800
Extinguisher, Fire	2300
Facing, Clutch Disc	0201
Facing, Disc	0201
Facing, Friction	0201
Fan Assembly (Schwitzer-Cummins)—Inc. Bracket	0503
Fan Assembly (Schwitzer-Cummins)—Less Bracket	0503
Fan Assembly—20" with 4 Blades	0503
Fan Assembly (Schwitzer-Cummins)— 21" with 6 Blades	0503
Fan and Hub Assembly (Schwitzer-Cummins)— 21" with 6 Blades	0503
Fan, Ventilating	0601
Felt, Front Wheel Bearing	1301
Felt, Front Wheel Bearing Oil Seal	1301
Felt, Gearshift Connecting Tube Bracket	0706
Felt, Pedal Finish Plate	1800
Felt, Rear Wheel Bearing	1301
Felt, Steering Column Finish Plate	1800
Felt, Steering Knuckle	1006

ALPHABETICAL INDEX—MASTER PARTS LIST

	Group
Felt, Steering Knuckle Pin	1007
Felt, Trunnion Socket	1006
Fender, Rear Quarter—Front—R. H.	1800
Fender, Rear Quarter—Rear—L. H.	1800
Fenders with Steps and Splash Shields	1800
Ferrule, Oil Line (¼")	0107
Filler, Housing Oil	1001
Filler, Transfer Case Oil	0801
Filter, Field, Assembly	0611
Filter, Fuel, Assembly (Zenith Model F328)	0302
Filter, Oil, Assembly—Complete	0107
Filter, Oil, Assembly—RPM Type with 4" O.D. at Base of Filter Cylinder	0107
Filter, Oil Assembly—TRM Type with 4" O.D. at Base of Filter Cylinder	0107
Filter, Oil—Element Assembly (Fram Model C-31)	0107
Filter, Oil—Unit Assembly (Fram Model F-36)	0107
Filter to Ammeter Jumper Assembly	0611
Filter to Coil Jumper Assembly	0611
Filter, Two Unit, Assembly	0611
Finger, Gearshift	0706
Finger, Gearshift Universal Lever	0706
Finger, Shift Lever	0706
Fitting, Excess Grease Relief	1203
Fitting, Oil Filter Shell Cover	0107
Fitting, Rear Brake Spider Grease Relief, Assembly	1101
Fitting, Universal Drive Relief	????
Flange, Axle	1102
Flange, Crankshaft Rear Main Bearing	0102
Flange, Declutch Shaft	0803
Flange, Drive	1105
Flange, Driven Shaft	0803
Flange, Driving	1007
Flange, Exhaust Pipe	0402
Flange, Exhaust Pipe (Hercules #40314A)	0402
Flange, Fixed Joint	0902
Flange, Fixed Joint (with 6⅝" Dia. Outside Pilot)	0902
Flange, Front Driving Axle	1007
Flange, Intake & Exhaust Manifold	0108
Flange, Joint (5¾" O.D.)	0902
Flange, Joint (With 2¾" Dia. Outside Pilot)	0902
Flange, Joint (With 3¾" Dia. Outside Pilot)	0902
Flange, Joint (5¾" O.D.)	0902
Flange, Main Shaft	0802
Flange, Mainshaft Driveshaft	0704
Flange, Slip Joint (With 6⅝" Dia. Outside Pilot)	0902
Flange, Steering Knuckle, Assembly—L. H.	1006
Flange, Steering Knuckle, Assembly—R. H.	1006
Flange, Steering Knuckle Companion, Assembly	1006
Flange, Water Outlet Hose	0505
Flap, Fender Assembly	1800
Flinger, Steady Bearing Housing	0902
Float, Carburetor, Assembly	0301
Floorboard—L. H.	1800
Floorboard—R. H.	1800
Flywheel	0109
Flywheel and Ring Gear Assembly	0110
Fork, Declutching Shifter	0803
Fork, Declutch Shift	0803
Fork, First and Reverse, Second and Third Speed Gearshift	0706
Fork, First and Second Speed Gearshift	0706
Fork, Fourth and Overdrive Gearshift Rod	0706
Fork, Gear Shift	0805
Fork, Overdrive Shift	0706
Fork, Reverse Gearshift	0706
Fork, Shifter	0805
Fork, Third and Fourth Speed Shift	0706
Frame and Field Assembly	0602
Frame, Battery Hold Down, Assembly	0610
Frame, End, Assembly	1900
Frame, End (C.E.)	0602
Frame with Front and Rear Cross Members only	1500

	Group
Gasket	0601
Gasket	1102
Gasket	1201
Gasket	1403
Gasket	1900
Gasket	1911
Gasket, Adapter	1911
Gasket, Bevel Pinion Bearing Cage	1003
Gasket, Bevel Pinion Forward Bearing Cage	1003
Gasket, Bevel Pinion Forward Bearing Cover	1003
Gasket, Bottom Cylinder	0107
Gasket, Bottom Cylinder (4" O.D.)	0107
Gasket, Bottom Cylinder (4⅛" O.D.)	0107
Gasket, Carburetor	0301
Gasket,* Carburetor Bowl to Body	0301
Gasket, Carburetor, Set	0301
Gasket, Crankshaft Rear Bearing Seal	0102
Gasket, Crankshaft Rear Bearing Seal	0110
Gasket, Crankshaft Rear Bearing Seal—Center	0102
Gasket, Crankshaft Rear Bearing Seal—Center	0110
Gasket, Cover	1900
Gasket, Cover	1911
Gasket, Compressor Mounting Bracket	1205
Gasket, Countershaft Front Bearing Cap	0703
Gasket, Countershaft Rear Bearing Cap	0703
Gasket, Cylinder Block	0101
Gasket, Cylinder Block Gear Case Cover	0101
Gasket, Cylinder Block Fuel Pump Opening Cover	0101
Gasket, Cylinder Block Oil Orifice	0101
Gasket, Cylinder Head	0101
Gasket, Cylinder Water Jacket Plate	0101
Gasket, Declutching Carrier	0803
Gasket, Declutching Shaft Bearing Cap	0803
Gasket, Declutching Shaft Bearing Carrier	0803
Gasket, Differential Carrier Inspection Plug	1103
Gasket, Differential Carrier to Housing	1002
Gasket, Differential Carrier to Housing	1103
Gasket, Differential Carrier Top Cover	1103
Gasket, Door	0607
Gasket, Drive Gear Case Inspection Cover	1103
Gasket, Drive Pinion Cage Cap	1105
Gasket, Driven Shaft Rear Bearing Cover	0803
Gasket, Drive Shaft Retainer	1007
Gasket, Driving Axle Flange	1102
Gasket, Driving Flange	1007
Gasket, End Cap Cover	0601
Gasket, Exhaust Manifold Attaching—Inner	0108
Gasket, Exhaust Manifold Attaching—Outer	0108
Gasket, Exhaust Manifold—End	0108
Gasket, Exhaust Manifold—Large	0108
Gasket, Exhaust Manifold—Small	0108
Gasket, Exhaust Pipe Flange	0402
Gasket, Exhaust Pipe Flange (Hercules #40028A)	0402
Gasket, Fan Bearing Oil	0503
Gasket, Fan Hub	0503
Gasket, Fan Spacer	0503
Gasket, Flywheel Bell Housing	0109
Gasket, Front Driving Axle Oil Seal Retainer	1007
Gasket, Fuel Pump Opening Cover	0101
Gasket, Front Driving Axle Flange	1007
Gasket, Fuel Filter Bowl	0302
Gasket, Fuel Filter Bowl Nut	0302
Gasket, Fuel Pump Valve Plug	0302
Gasket, Fuel Pump Bottom Cover	0302
Gasket, Fuel Pump Bowl	0302
Gasket, Gear Carrier and Housing	1103
Gasket, Fuel Pump Pull Rod	0302
Gasket, Gear Carrier Inspection Plug	1003
Gasket, Gearshift Finger Housing	0706
Gasket, Gearshift Finger Housing Cap	0706
Gasket, Gearshift Finger Housing Cover Plate	0706
Gasket, Gearshift Gate	0706
Gasket, Gearshift Top Control Cover	0706

ALPHABETICAL INDEX—MASTER PARTS LIST

	Group
Gasket, Governor Body	0305
Gasket, Governor Valve Box Cover Plate	0305
Gasket, Housing	1101
Gasket, Housing and Carrier	1002
Gasket, Housing Oil Filter Plug	1001
Gasket, Idler Shaft Front and Rear Bearing Cover	0804
Gasket, Idler Shaft Front Bearing Cap	0804
Gasket, Idler Shaft Rear Bearing Cap	0804
Gasket, Intake Manifold	0108
Gasket, Intake Manifold Attaching	0108
Gasket, Intake Manifold Heat Box (To Exhaust Manifold)	0108
Gasket, Intake Manifold Heat Box (To Intake Manifold)	0108
Gasket, Intake Manifold Stove—Bottom	0108
Gasket, Intake Manifold to Governor	0108
Gasket, Intake Pipe	0108
Gasket, Journal	0902
Gasket, Main Drive Front and Rear Bearing Cover	0802
Gasket, Main Drive Gear Bearing Cap	0703
Gasket, Main Shaft Front Bearing Cage	0802
Gasket, Main Shaft Rear Bearing Cover	0802
Gasket, Manifold Hot Spot	0108
Gasket, Mounting Side Cap—R. H.	1003
Gasket, Mounting Side Cap—R. H.	1103
Gasket, Oil Filter Attaching	0107
Gasket, Oil Filter Base	0107
Gasket, Oil Filter Case (Fram #11639)	0107
Gasket, Oil Filter Case Cover Screw (Fram #11581)	0107
Gasket, Oil Filter Case Cover (Fram #11582)	0107
Gasket, Oil Filter Mounting	0107
Gasket, Oil Filter Pressure Relief Valve Adjusting Screw	0107
Gasket, Oil Filter Shell Cover	0107
Gasket, Oil Filter Shell Cover Fitting	0107
Gasket, Oil Pan	0107
Gasket, Oil Pan Drain Plug	0107
Gasket, Oil Pan (Half)	0107
Gasket, Oil Pan—Large	0110
Gasket, Oil Pan Oil Strainer Cap	0107
Gasket, Oil Pan (Small)	0107
Gasket, Oil Pressure Relief Valve	0107
Gasket, Oil Pump Main Discharge Pipe	0107
Gasket, Oil Pump Suction Line Flange	0107
Gasket, Oil Relief Valve Cap	0110
Gasket, Power Take-Off Adapter Bracket	1911
Gasket, Power Takeoff Opening Cover	0701
Gasket, Radiator Filler Cap	0501
Gasket, Rear Brake Spider to Housing	1208
Gasket, Rear Gearshift Rod Cover	0706
Gasket, Reverse Gearshift Rod Plunger Cover	0706
Gasket, Shifting Cover	0805
Gasket, Side Cover	1403
Gasket, Speedometer Drive Gear Case	0705
Gasket, Spur Pinion Bearing Cage	1003
Gasket, Spur Pinion Bearing Cage and Cover	1105
Gasket, Spur Pinion Bearing Cover	1105
Gasket, Spur Pinion Bearing Cover—L. H.	1003
Gasket, Spur Pinion Bearing Cover—R. H.	1003
Gasket, Thermostat Housing	0505
Gasket, Thermostat Rubber Seal	0502
Gasket, Thru Shaft Forward Bearing Cover	1105
Gasket, Thru Shaft Rear Bearing	1105
Gasket, Top	0107
Gasket, Transfer Case and Cover	0801
Gasket, Transmission Case Cover	0701
Gasket, Transmission Driveshaft Bearing Cap	0703
Gasket, Valve Cover	0105
Gasket, Valve Cover Plate	0101
Gasket, Water Outlet Manifold	0505
Gasket, Water Pump	0503
Gasket, Water Pump Cover	0503
Gasket, Water Pump Discharge Manifold	0505
Gate, Gearshift	0706
Gauge, Air (Stewart-Warner)	0605
Gauge, Air Pressure, Assembly	1209
Gauge, Crankcase Oil Level, Unit	0107
Gauge, Gas	0605
Gauge, Gas Tank	0300
Gauge, Gas Tank, Unit (In Tank)	0300
Gauge, Oil	0605
Gauge, Oil Level, Assembly	0107
Gauge, Oil Pan Bayonet, Assembly	0107
Gauge, Oil (Stewart-Warner)	0605
Gear	0602
Gear, Bevel	1003
Gear, Bevel	1105
Gear, Bevel (21 Teeth)	1105
Gear, Bevel (21-T)	1003
Gear, Idler Shaft	0106
Gear, Camshaft	0106
Gear, Camshaft Idler	0106
Gear, Camshaft Idler Assembly with Shaft	0106
Gear, Camshaft Idler Shaft, Plunger and Plug Assembly	0106
Gear, Countershaft Drive	0702
Gear, Countershaft Drive (Use with 3TF2916 Washer)	0702
Gear, Crankshaft	0102
Gear, Differential and—Assembly (8.148 Ratio)	1002
Gear, Differential and—Assembly (8.435 Ratio)	1002
Gear, Differential Bevel Side	1002
Gear, Differential Pinion	1002
Gear, Differential Pinion	1103
Gear, Differential Pinion	1104
Gear, Differential & Ring—Assembly (Specify Ratio)	1002
Gear, Differential Side	1002
Gear, Differential Side	1003
Gear, Differential Side	1104
Gear, Distributor	0603
Gear, Distributor Drive	0603
Gear, Driven	0803
Gear, Driven Shaft (37 Teeth)	0803
Gear, First and Reverse Slide	0702
Gear, First and Second Speed Slide—Assembly	0702
Gear, First and Second Speed Slide (40 Teeth)	0702
Gear, First Speed Slide	0702
Gear, Flywheel	0110
Gear, Flywheel Ring	0109
Gear, Flywheel (Teeth) 126	0109
Gear, Governor	0305
Gear, Idler	0804
Gear, Idler 14/14T	1911
Gear, Idler Shaft Driven (37 Teeth)	0804
Gear, Idler Shaft Low Speed (41 Teeth)	0804
Gear, Intermediate 22/14T	1911
Gear, Jackshaft Bevel (26 Teeth)	1105
Gear, Jackshaft Spur and Bevel, Assembly	1105
Gear, Jackshaft Spur (6.4 Ratio—17 Teeth)	1105
Gear, Jackshaft Spur (6.94 Ratio—16 Teeth)	1105
Gear, Jackshaft Spur (7.56 Ratio—15 Teeth)	1105
Gear, Jackshaft Spur (8.27 Ratio—14 Teeth)	1105
Gear, Jackshaft Spur (9.00 Ratio—13 Teeth)	1105
Gear, Low Speed	0804
Gear, Main Drive	0703
Gear, Main Drive	0802
Gear, Main Drive Shaft Direct Drive (37 Teeth)	0802
Gear, Main Drive Shaft Slide (21 Teeth)	0802
Gear, Mainshaft First and Reverse Slide	0702
Gear, Main Shaft Low Speed Sliding	0802
Gear, Mainshaft Overdrive, and Bushing Assembly	0702
Gear, Mainshaft Overspeed	0702
Gear, Mainshaft Overspeed Slide (19 Teeth)	0702
Gear, Mainshaft Second and Third Speed Slide	0702
Gear, Mainshaft Second Speed	0702
Gear, Mainshaft Third Speed	0702
Gear, Mainshaft Third Speed—and Bushing Assembly	0702

ALPHABETICAL INDEX—MASTER PARTS LIST

	Group
Gear, Mainshaft Third Speed Idler	0702
Gear, Oil Pump	0107
Gear, Oil Pump Drive	0107
Gear, Oil Pump Idler	0107
Gear, Oil Pump Idler, and Bushing Assembly	0107
Gear, Oil Pump Idler Shaft	0107
Gear, Overdrive Countershaft	0702
Gear, Overdrive Countershaft, and Bushing Assembly	0702
Gear, Overdrive Countershaft (37 Teeth)	0702
Gear, Power Take-Off Adapter	1911
Gear, Power Take-Off Adapter (22T)	1911
Gear, Power Takeoff Driving	0702
Gear, Reverse	0702
Gear, Reverse, Assembly with Bushing	0702
Gear, Reverse, Assembly with Bushing (14 and 18 Teeth)	0702
Gear, Reverse, Assembly (With Needle Bearings)	0702
Gear, Second and Reverse Speed Countershaft	0702
Gear, Second Speed Slide	0702
Gear, Sliding 28/20T	1911
Gear, Speedometer Drive	0705
Gear, Speedometer Driven (13 Teeth)	0705
Gear, Speedometer Driven (14 Teeth)	0705
Gear, Speedometer Driven (15 Teeth)	0705
Gear, Speedometer Driven (16 Teeth)	0705
Gear, Speedometer Driven (17 Teeth)	0705
Gear, Speedometer Driven (S-W)	0804
Gear, Spur Driving (6.4 Ratio—46 Teeth)	1105
Gear, Spur Driving (9.00 Ratio—50 Teeth)	1105
Gear, Spur Driving (6.94 Ratio—47 Teeth)	1105
Gear, Spur (6.62 Ratio)	1103
Gear, Spur Driving (7.56 Ratio—48 Teeth)	1105
Gear, Spur (8.148 Ratio)	1003
Gear, Spur (8.148 Ratio)	1103
Gear, Spur (8.148 Ratio) (See also Group 1103)	1105
Gear, Spur (8.21 Ratio)	1103
Gear, Spur Driving (8.27 Ratio—49 Teeth)	1105
Gear, Spur (8.43 Ratio)	1103
Gear, Spur (8.43 Ratio) (See also Group 1103)	1105
Gear, Spur (8.435 Ratio)	1003
Gear, Spur (Ring Gear) (8.21 Ratio)	1003
Gear, Steering, Complete	1403
Gear, Steering, Complete (Ross)	1403
Gear, Third and Fourth Speed Slide (28 Teeth)	0702
Gear, Third Speed Countershaft	0702
Gear, Worm—L. H.	1900
Gears, Oil Pump Shaft	0107
Generator, Assembly	0601
Gland, Water Pump	0503
Gland, Water Pump Packing	0503
Glass, Door and Frame Assembly—L. H.	1800
Glass, Door and Frame Assembly—R. H.	1800
Governor	1205
Governor (Handy 703—722—138)	0305
Governor (Handy V—6—83)	0305
Governor (Pierce)	0305
Grommet	0604
Grommet	0607
Grommet (For Regulator Cable)	0611
Grommet, Ignition Wires Support	0603
Guard, Flat Oil	0601
Guard, Radiator and Headlamp Brush Assembly	2103
Guard, Radiator and Lamp Brush	1800
Guard, Rope	1900
Guard, Starting Switch	1800
Guide, Exhaust Valve	0105
Guide, Hand Brake Lever Pawl Rod	1201
Guide, Intake Valve	0105
Guide, Oil Relief Valve Spring	0107
Guide, Rod and Seal Assembly	1603
Guide, Water Pump Paddle Seal Spring	0503
Guide, Valve Plunger	0110
Gun, Alemite Grease	2300
Gusset, Bumper—L. H.	1500
Gusset, Bumper—R. H.	1500
Gusset, Front Bumper	1500
Gusset, Front Cross Member—Bottom	1500
Gusset, Front Cross Member—Top	1500
Gusset, Rear Cross Member	1500
Gusset, Rear Cross Member—Bottom	1500
Gusset, Rear Cross Member—Top	1500
Hammer—16-oz. Ball Pein	2300
Handle	1800
Handle, Door—Inside	1800
Handle, Door—Outside	1800
Handle, Door Window Regulator	1800
Handle, Front and Rear Wheel Nut Wrench	1302
Handle, Grab	1800
Hasp, Tool Box Door	1800
Head Assembly	1603
Head, Drive End	0601
Head, Drive End, Assembly	0601
Head, Fuel Filter	0302
Head, Speedometer Assembly (Stewart-Warner)	2203
Head, Tachometer Assembly (Stewart-Warner)	2203
Head, Water Pump	0503
Head, Water Pump, and Bushing Assembly	0503
Headlamp Assembly	0607
Headlamp Assembly—Complete 6-V	0607
Headlamp Assembly (Guide)	0607
Headlamp to Dimmer Switch Cable Assembly	0607
Hinges, Door—L. H.	1800
Hinges, Door—R. H.	1800
Hinge, Tool Box Door	1800
Hood, Engine, Assembly	1800
Hood, Motor, Assembly	1800
Hook, Endgate Chain	1800
Hook, Endgate Chain "S"	1800
Hook, Front Towing—L. H.	1501
Hook, Front Towing—R. H.	1501
Hook, Hood Catch	1800
Hook, Ignition Wire	0603
Hook, Lash—L. H.	1800
Hook, Lash—R. H.	1800
Hook, Pintle Assembly	1502
Hook, Towing—L. H.	1501
Hook, Towing—R. H.	1501
Horn, Dual Electric, Assembly (Delco)	0609
Horn, Electric, Assembly (Dual) (E. A. Lab.)	0609
Hose, Air, Assembly (162½")	1209
Hose, Air Gauge, Assembly	1209
Hose Assembly	1209
Hose Assembly (100" Each)	1209
Hose Assembly (112½") for Semi-Trailer	1209
Hose Assembly (216" Each)	1209
Hose Assembly (216") for Emerg. Tow	1209
Hose, Chamber (⅜" I.D.)	1209
Hose, Dummy	1209
Hose, Tire Inflating	2300
Hose, Tire Inflation Assembly	2300
Hose, Tire Inflation, Assembly (30 Ft.)	1209
Hose, Trailer	2300
Hose, Trailer Assembly	2300
Hose (⅜" I.D.) (105")	1209
Hose (⅜" I.D.) (208")	1209
Housing and Studs Assembly (Without Tubes)	1101
Housing and Tubes Assembly	1101
Housing Assembly	1001
Housing Assembly	1101
Housing, Bendix	0602
Housing, Countershaft Rear Bearing	0703
Housing, Distributor	0603
Housing, Flywheel Bell	0109
Housing, Gear	0602
Housing, Gearshift Finger	0706

ALPHABETICAL INDEX—MASTER PARTS LIST

	Group
Housing, Gearshift Finger—Assembly	0706
Housing, Mainshaft Rear Bearing	0703
Housing, Steady Bearing	0902
Housing, Steady Bearing, Assembly	0902
Housing, Steady Bearing, Assembly (Use with 30CD-5000—1500 Series Driveshaft Assembly)	0902
Housing, Steering Gear, Assembly	1403
Housing, Thermostat	0505
Hub, Fan, and Pulley	0503
Hub, Fourth and Overdrive Shift	0702
Hub, Front Wheel, and Cup Assembly	1302
Hub, Rear, and Cups Assembly	1302
Hub, Rear Wheel, and Cup Assembly	1302
Hub, Rear Wheel, and Drum Assembly	1302
Hub, Rear Wheel, and Drum Assembly—L. H.	1302
Hub, Rear Wheel, and Stud Assembly—L. H.	1302
Hub, Rear Wheel, and Stud Assembly—R. H.	1302
Hub, Third and Fourth Speed Shift	0702
Impeller, Water Pump	0503
Indicator, Heat	0605
Indicator, Heat (Stewart-Warner)	0605
Indicator, Low Pressure	1205
Insert, Floorboard	1800
Insert, Toeboard	1800
Inserts, Exhaust Valve Seat	0101
Inserts, Valve Seat	0101
Insulation, Transfer Case Bracket	0801
Insulator, Field Connection	0601
Insulator, Radiator Support—Bottom	0501
Insulator, Radiator Support—Top	0501
Insulator, Rubber—Bottom	0110
Insulator, Rubber—Top	0110
Insulator, Terminal Post Bottom	0601
Insulator, Terminal Post Top	0601
Insulator, Transfer Case	0801
Intake, Spring Piston	1603
Intake, Valve Piston	1603
Iron, Rear Fender—L. H.	1800
Iron, Rear Fender—R. H.	1800
Jack, Hydraulic	2300
Jack, Lifting and Handle	2300
Jaw, Crankshaft Starting Crank	0102
Jaw, First and Second Speed Gearshift Rod	0706
Jet, Carburetor Accelerating (#25)	0301
Jet, Carburetor Cap (38-1)	0301
Jet, Carburetor Compensator (#32)	0301
Jet, Carburetor Idling (#16)	0301
Jet, Carburetor Main, Adjustment Assembly	0301
Jet, Carburetor Main (#32)	0301
Joint, Ball	0108
Joint, Fixed, Assembly (Flange Type with Outside Pilot)	0902
Joint, Slip, Assembly	0902
Joint, Slip, Assembly (Flange Type with Outside Pilot)	0902
Joint, Take-Off	1900
Joint, Worm	1900
Journal Assembly	0902
Keeper, Front Spring Clip	1601
Keeper, Gearshift Lever Socket Spring	0706
Keeper, Rear Spring Clip	1601
Key, Bevel Gear	1105
Key, Bevel Pinion	1003
Key, Brake Disc	1900
Key, Camshaft Gear	0106
Key, Clutch	1900
Key, Countershaft Drive Gear	0702
Key, Crankshaft Gear	0102
Key, Drive Pinion Shaft Flange	1105
Key, Driven Gear	0803

	Group
Key, Driven Shaft	0803
Key, Driven Shaft End	0803
Key, Driveshaft	1900
Key, Fan Drive Pulley (Hercules)	0503
Key, Governor Castle Nut Cotter (Drive Shaft Gear)	0305
Key, Governor Gear (Woodruff)	0305
Key, Governor Throttle Rod End (Valve Box) Cotter	0305
Key, Idler Gear	0804
Key, Idler Shaft Gear	0106
Key, Ignition	0604
Key, Low Speed Gear	0804
Key, Mainshaft Driving Flange	0704
Key, Main Shaft Flange	0802
Key, Oil Pump Drive Gear	0107
Key, Oil Pump Idler Shaft Gear	0107
Key, Overdrive Countershaft Gear	0702
Key, Pinion Shaft	1003
Key, Pinion Shaft	1105
Key, Power Takeoff Driving Gear	0702
Key, Rear Wheel Bearing Lock Nut	1301
Key, Reverse Gear Sleeve	0702
Key, Steering Arm	1007
Key, Steering Knuckle Pin Draw	1007
Key, Steering Wheel	1404
Key, Stub Shaft (1½" Taper Shaft)	0902
Key, Third Speed Countershaft Gear	0702
Key, Thru Shaft	1105
Key, Valve Stem	0110
Key, Valve Stem (Use with 2SA210C—2SA2710A)	0110
Key, Valve Stem (Use with 2D210A—2D2710)	0110
Key, Water Pump Impeller Shaft	0503
Key, #3 Woodruff	0601
Key, Woodruff—No. 6	0602
Key, #8 Woodruff	0601
Key, Worm Gear	1900
Knob, Blackout Switch	0606
Knob, Instrument Panel Light Switch	0606
Knob, Poppet	1900
Knuckle, Steering, and Bushing Assembly	1006
Knuckle, Steering—Assembly	1006
Knuckle, Steering, and Bushing Assembly—L. H. (1½"—12 Thds.)	1006
Knuckle, Steering, and Bushing Assembly—R. H. (1½"—12 Thds.)	1006
Lamp, Blackout Parking—Assembly	0607
Lamp, Blackout Tail and Stop, Assembly—R. H.	0608
Lamp, Blackout Tail and Stop—R. H. (Painted Yellow Green)	0608
Lamp, Blackout Tail—L. H.	0608
Lamp, Blackout Tail—R. H.	0608
Lamp, Head—Sealed Beam Unit—6-Volt	0607
Lamp, Stop, Unit Assembly	0608
Lamp, Tail, Unit Assembly	0608
Lead Assembly	0601
Lead, Brush Connector	0602
Lead, Brush and Field Connector	0602
Lead, Field Brush	0602
Lead, Ground Brush	0602
Leadwasher, Valve Cover Screw	0105
Lever, Accelerator Cross Shaft	0303
Lever, Brake	1201
Lever, Brake Operating (L. H. offset)	1201
Lever, Brake Pedal	1204
Lever, Breaker, and Point	0603
Lever, Carburetor Air Shutter, Assembly	0301
Lever, Carburetor Control Rod	0303
Lever, Carburetor Throttle and Swivel Assembly	0301
Lever, Carburetor Throttle Shaft and Assembly	0301
Lever, Carburetor Throttle Stop	0301
Lever, Choke Control	0108
Lever, Clutch Control	0203
Lever, Clutch Control Shaft and Control, Assembly	0203

ALPHABETICAL INDEX—MASTER PARTS LIST

	Group
Lever, Clutch Pedal	0204
Lever, Clutch Pressure	0203
Lever, Clutch Throwout Shaft	0203
Lever, Clutch Throwout Shaft Stop	0203
Lever, Clutch Trunnion	0203
Lever, Declutch Shift	0805
Lever, Declutch Shift Control	0805
Lever, Declutch Shift Hand	0805
Lever, Declutch Shift Transfer	0805
Lever, Disc Brake Link Brake Operating	1201
Lever, Disc Brake Operating (L. H. Off Set)	1201
Lever, Disc Brake Operating (R. H. Off Set)	1201
Lever, Fuel Pump Priming	0302
Lever, Gearshift	0706
Lever, Gearshift, and Bracket Assembly	0706
Lever, Gearshift, Assembly	0706
Lever, Gearshift—Bare	0706
Lever, Gear Shift Hand	0805
Lever, Gearshift Hand	0805
Lever, Gearshift—Only	0706
Lever, Gearshift Universal	0706
Lever, Governor Throttle	0305
Lever, Hand Brake and Bushing Assembly—Bare	1201
Lever, Hand Brake, Assembly	1201
Lever, Hand Brake, Assembly—with Pawl Rod, Etc.	1201
Lever, Hand Brake—Only	1201
Lever, Hand Brake Shaft	1201
Lever, Hand Brake Transfer	1201
Lever, Heat Control	0108
Lever, Pressure	0203
Lever, Pressure Plate and, Unit	0205
Lever, P. T. O. Control Hand	1910
Lever, P. T. O. Transfer	1910
Lever, Release	0205
Levershaft, Nut and Lockwasher Assembly	1403
Lid, Battery	0610
Lid, Cowl Side Vent, and Handle Assembly	1800
Light, Blackout Tail and Stop, Assembly—L. H.	0608
Light, Blackout Tail and Stop—L. H. (Painted Olive Drab)	0608
Light, Blackout Tail and Stop—L. H. (Painted Yellow Green)	0608
Light, Blackout Tail and Stop—R. H. (Painted Olive Drab)	0608
Line, Clutch Trunnion Oil, Assembly (1¼" I.D.)	0204
Line, Compressor Water, Assembly to Manifold	0505
Line, Compressor Water, Assembly to Pump	0505
Line, Filter to Pump Titeflex Gas	0304
Line, Gearshift Front Bearing Oil	0706
Line, Gearshift Rear Bearing Oil	0706
Line, Oil, Assembly—Crankcase to Governor	0107
Line, Oil, Assembly—Visco. Unit to Oil Pressure Line	0107
Line, Oil Pump Front Suction, Assembly	0107
Line, Oil Pump Rear Suction	0107
Line, Tank to Filter Gas (Bundy Tubing)	0304
Line, Titeflex (To Pump)—12" Long	0300
Lining, Brake Shoe	1201
Lining, Disc Brake Shoe	1201
Lining, Front Brake Shoe	1202
Lining, Front Brake Shoe (Undrilled)	1202
Lining, Rear Brake Shoe	1202
Lining, Rear Brake Shoe (Undrilled)	1202
Lining with Rivets	1900
Link, Accelerator Spring	0303
Link, Brake Control Rod Return	1204
Link, Clutch Pedal Return Spring—Lower	0203
Link, Clutch Pedal Return Spring—Upper	0203
Link, Clutch Release Trunnion Return Spring	0203
Link, Clutch Throwout Shaft Lever Spring	0203
Link, Declutch Shift Control	0805
Link, Drag, Assembly	1401
Link, Front	1603
Link, Front Brake Shoe	1203

	Group
Link, Fuel Pump	0302
Link, L. H. Rear	1603
Link, R. H. Rear	1603
Lock, Bevel Pinion Forward Bearing Jam Nut	1003
Lock, Differential Bearing Adjusting	1105
Lock, Differential Bearing Adjusting Nut	1002
Lock, Differential Bearing Adjusting Nut	1105
Lock, Differential Bearing Adjusting Ring	1002
Lock, Door—L. H.	1800
Lock, Door—R. H.	1800
Lock, Front Wheel Bearing Adjuster Jam Nut (1¼" I.D.)	1301
Lock, Front Wheel Bearing Adjuster Jam Nut 1½" I.D.	1301
Lock, Front Wheel Bearing Adjusting Jam Nut	1301
Lock, Gear Shaft Bearing Stud	1003
Lock, Gear Shaft Bearing Stud	1105
Lock, Ignition Switch and	0604
Lock, Pinion Bearing	1003
Lock, Pinion Bearing	1105
Lock, P. T. O. Lever Hinge Assembly	1910
Lock, Rear Brake Shoe Anchor Pin	1203
Lock, Reverse Gear Shaft	0704
Lock, Spur Pinion Bearing	1003
Lock, Steering Cross Tube End Plug	1402
Lock, Tachometer	2203
Lock, Water Pump Packing Nut	0503
Lockpin, Mainshaft Third Speed Gear Bushing	0702
Lockscrew, Declutch Driving Clutch	0803
Lockwasher	1900
Lockwasher (⁵⁄₁₆")	1201
Lockwasher, Bendix Housing Screw	0602
Lockwasher, Bendix Spring Screw	0602
Lockwasher, Brush Field Screw	0602
Lockwasher, Brush Ground Lead Screw	0602
Lockwasher, Brush Holder Screw	0602
Lockwasher, Carburetor Venturi Screw	0301
Lockwasher, Countershaft Front Bearing	0703
Lockwasher, Cylinder Block Gear Case Cover	0101
Lockwasher, Declutch Shift Lock Spring Screw	0803
Lockwasher, Drive Pinion Shaft Flange Nut	1105
Lockwasher, Field Terminal Stud	0602
Lockwasher, Fuel Pump Pull Rod Nut	0302
Lockwasher, Gear Case Screw and Through Bolt	0602
Lockwasher, Gear Shaft Bearing	1105
Lockwasher, Lug	1001
Lockwasher, Mainshaft Drive Thrust Nut	0704
Lockwasher, Mainshaft Driving Flange Nut	0704
Lockwasher, Oil Pump Attaching Screw	0107
Lockwasher, Pinion Bearing	1105
Lockwasher—Sheet Metal	0402
Lockwasher, ⅜" Spring	0704
Lockwasher, Square	0402
Lockwasher, Terminal Screw	0603
Lockwire, Connecting Rod Piston Pin Clamp Screw	0104
Lockwire, Crankshaft Main Bearing Screw	0102
Lockwire, Oil Pump Attaching Screw	0107
Lockwire, Oil Pump Body Attaching Screw	0107
Lockwire, Oil Pump Cover Screw	0107
Loom	0300
Loom (⁷⁄₁₆" I.D.)	1209
Loom (¹³⁄₁₆" I.D.)	1209
Loom (1⅛" I.D.)	1209
Loom, Filter to Pump Titeflex Gas Line	0304
Loom, Pump to Carb. Gas Line and, Assembly	0304
Loom—Wire Cover	0607
Lug, First and Reverse Speed Gearshift Rod	0706
Lug, First and Second Speed Gearshift	0706
Lug, Fourth and Overdrive Gearshift	0706
Lug, Reverse Gearshift	0706
Lug, Second and Third Speed Gearshift	0706
Lug, Third and Fourth Speed Gearshift	0706

Group

Mainshaft—Bare 0704
Manifold, Exhaust 0108
Manifold, Exhaust and Intake Assembly 0108
Manifold, Intake 0108
Manifold, Intake and Exhaust 0108
Manifold, Water Outlet 0505
Manifold, Water Pump Discharge 0505
Manual, Maintenance 2300
Manual, Parts 2300
Member, Transfer Case Cross—Assembly 0801
Member, Transfer Case Support—Rear 0801
Mirror, Rear View Assembly—Inside 2201
Mirror, Rear View Assembly—Outside 2201
Molding Assembly—Prime 0607
Molding, Headlamp Assembly 0607
Motor, Pressure Assembly 1801
Motor, Starting—Assembly (Delco) 0602
Mounting Assembly (Gear Carrier Assembly) 1003
Mounting Assembly (Gear Carrier Assembly) 1103
Mounting with Caps (Gear Carrier) 1003
Mounting with Caps (Gear Carrier) 1103
Muffler Assembly 0401

Neck, Housing Filler 1101
Nipple, Gas Tank Outlet, Assembly 0300
Nipple, Ignition Wire 0604
Nipple, Housing Breather 1001
Nipple, Housing Breather 1101
Nut 1403
Nut (½″—20) 1201
Nut, Adjusting Screw Lock 1403
Nut, Advance Control Arm Clamp Screw 0603
Nut, Advance Control Screw 0603
Nut, Arm 1801
Nut, Armature Shaft 0601
Nut, Bevel Pinion 1003
Nut, Bevel Pinion Forward Adjusting 1003
Nut, Bevel Pinion Forward Bearing Jam 1003
Nut, Bevel Pinion Shaft 1003
Nut, Bevel Pinion Shaft 1105
Nut, Brake Band—Dead End 1900
Nut, Brake Band—Spring End 1900
Nut, Budd Wheel Outer (1⅛″—16) 1800
Nut, Camshaft Gear 0106
Nut, Camshaft Gear Clamp 0106
Nut, Camshaft Thrust Plunger Adjusting 0106
Nut, Carburetor Air Shutter Shaft 0301
Nut, Carburetor Tube Clamp Screw 0301
Nut, Clutch Trunnion Oil Line Clamp 0204
Nut, Connecting Rod Bearing Bolt 0104
Nut, Connecting Rod Cap 0104
Nut, Countershaft Front Bearing Lock 0703
Nut, Countershaft Rear Bearing Lock 0703
Nut, Crankshaft Flywheel Bolt 0102
Nut, Declutch Shaft 0803
Nut, Differential Bearing Adj 1003
Nut, Differential Bearing Adjusting 1003
Nut, Differential Bearing Adjusting 1103
Nut, Differential Carrier and Cap Stud 1103
Nut, Driven Shaft 0803
Nut, Drive Pinion Bearing Lock 1105
Nut, Drive Pinion Shaft Flange 1105
Nut, Exhaust Manifold Stud 0108
Nut, Fan Adjusting Screw 0503
Nut, Fan Adjusting Screw Lock 0503
Nut, Fan Bracket Clamp 0503
Nut, Field Terminal 0602
Nut, Front and Rear Hub and Wheel Stud Cap—L. H.—Inner 1302
Nut, Front and Rear Hub and Wheel Stud Cap—L. H.—Outer 1302
Nut, Front and Rear Hub and Wheel Stud Cap—R. H.—Inner 1302
Nut, Front and Rear Hub and Wheel Stud Cap—R. H.—Outer 1302
Nut, Front and Rear Hub Stud 1302
Nut, Front and Rear Wheel Hub Stud Cap—L. H.—Inner 1302
Nut, Front and Rear Wheel Hub Stud Cap—L. H.—Outer 1302
Nut, Front and Rear Wheel Hub Stud Cap—R. H.—Inner 1302
Nut, Front and Rear Wheel Stud Cap—R. H.—Outer 1302
Nut, Front Hub and Wheel Stud Cap—L. H.—Inner 1302
Nut, Front Hub and Wheel Stud Cap—L. H.—Outer 1302
Nut, Front Hub and Wheel Stud Cap—R. H.—Inner 1302
Nut, Front Hub and Wheel Stud Cap—R. H.—Outer 1302
Nut, Front Wheel Bearing Adjusting 1301
Nut, Front Wheel Bearing Adjuster, and Dowel Assembly (1¼″ I.D.) 1301
Nut, Front Wheel Bearing Adjuster, and Dowel Assembly (1½″ I.D.) 1301
Nut, Front Wheel Bearing Adjusting Jam 1301
Nut, Front Wheel Bearing Lock 1301
Nut, Front Wheel Bearing Adjuster Jam (1¼″—7 Thds.) 1301
Nut, Front Wheel Bearing Adjuster Jam (1½″—12 Thds.) 1301
Nut, Front Wheel Budd Stud—L. H. 1302
Nut, Front Wheel Budd Stud—R. H. 1302
Nut, Front Wheel Hub and Drum Stud 1302
Nut, Fuel Filter Bowl 0302
Nut, Fuel Filter Element 0302
Nut, Fuel Pump Bail Thumb 0302
Nut, Fuel Pump Pull Rod 0302
Nut, Gearshift Lockout Plunger Spring 0706
Nut, Governor Adjusting Screw Lock 0305
Nut, Governor Bumper Screw Lock (⅜″—24) 0305
Nut, Governor Drive Shaft Castle (⅜″—24) 0305
Nut, Governor Throttle Rod Tube Lock 0305
Nut, Hex. 1006
Nut, Hex. (⅜″—24) 1800
Nut, Hex. (⅝″—18) 1800
Nut, #6—32 Hex. 0601
Nut, #10—32 Hex. 0601
Nut, #14—24 Hex. 0601
Nut, Hex. (7/16″—13 Brass) 0402
Nut, Holding 1801
Nut, Idler Shaft Gear Thrust Plunger Adjusting 0106
Nut, Intake and Exhaust Manifold Attaching Stud 0108
Nut, Intake and Exhaust Manifold Stud 0108
Nut, Intake Manifold Heat Box Attaching Stud 0108
Nut, Levershaft 1403
Nut, Lock (⅜″—16) 1201
Nut, Main Bearing Stud 0102
Nut, Main Drive and Driven Shaft 0802
Nut, Main Shaft 0802
Nut, Mainshaft Drive Thrust 0704
Nut, Mainshaft Driving Flange 0704
Nut, Mainshaft Rear Bearing Lock 0703
Nut, Motor Terminal (⅜″—16) 0602
Nut, Motor Terminal Stud 0602
Nut, Oil Filter Pressure Relief Valve Adjusting Screw Lock 0107
Nut, Oil Line (¼″) 0107
Nut, Oil Pressure Regulator 0107
Nut, Oil Pressure Relief Valve Cap 0107
Nut, Oil Pressure Relief Valve Lock 0107
Nut, Oil Pump Suction Line Flange Clamp Screw 0107
Nut, Oil Relief Valve Cap 0110
Nut, Oil Relief Valve Lock 0110
Nut, Pinion Bearing Adjusting 1003
Nut, Pinion Bearing Adjusting 1105
Nut, Rear Brake Shoe Anchor Pin 1203
Nut, Rear Spring Cross Shaft 1602
Nut, Rear Wheel Bearing Adjusting 1301

ALPHABETICAL INDEX—MASTER PARTS LIST

	Group
Nut, Rear Wheel Bearing Jam	1301
Nut, Rear Wheel Bearing Lock	1301
Nut, Rear Wheel Hub Stud Cap—L. H.—Inner	1302
Nut, Rear Wheel Hub Stud Cap—L. H.—Outer	1302
Nut, Rear Wheel Hub Stud Cap—R. H.—Inner	1302
Nut, Rear Wheel Hub Stud Cap—R. H.—Outer	1302
Nut, Release Lever Adjusting	0205
Nut, ½"—20 Slotted	0601
Nut, Spherical	1201
Nut, Steady Bearing Housing Lock	0902
Nut, Steering Arm	1007
Nut, Steering Arm Ball Stud	1006
Nut, Steering Arm Stop Screw	1007
Nut, Steering Cross Tube End Ball Pin	1402
Nut, Steering Cross Tube End Lock	1402
Nut, Steering Cross Tube End Pin	1402
Nut, Steering Knuckle Stop Screw	1006
Nut, Steering Wheel	1404
Nut, #10—32 Square	0601
Nut, Terminal Screw	0603
Nut, Thru Shaft	1105
Nut, Thru Shaft—Forward	1105
Nut, Tie Rod Yoke	1402
Nut, Transmission Driveshaft Bearing Retaining	0703
Nut, Transmission Driveshaft Lock	0703
Nut, Valve Adjusting Screw Cap	0107
Nut, Valve Plunger Adjusting Screw Lock	0110
Nut, Valve Tappet Adjusting Screw	0105
Nut, Water Pump Packing	0503
Nut, Water Pump Packing—L. H.	0503
Nut, Water Pump Packing—R. H.	0503
Nut, Wheel	1403
Oiler	0602
Oiler (in Gear Housing)	0602
Oiler, Magazine	0204
Oiler, Press-in Sleeve	0603
Oiler, ¼"—Press-in Type	0601
Oil Line, Clutch, Assembly	0204
Oil Seal, Crankshaft	0102
Oil Seal, Cylinder Block Gear Case Cover (Accessory Shaft)	0101
Oil Seal, Cylinder Block Generator	0101
Oil Thrower, Crankshaft	0102
Orifice, Cylinder Block Oil	0101
Outrigger, Deck Plate	1800
Package, Padlock	2300
Packing, Carburetor Throttle Shaft	0301
Packing, Declutch Shaft	0803
Packing, Declutching Shifter Shaft	0803
Packing, Gear Shift Shaft	0805
Packing, Universal Joint	1007
Packing, Water Pump	0503
Packing, Water Pump (4 Pieces per Box)	0503
Pad, Brake Pedal	1204
Pad, Clutch Pedal	0204
Paddle, Water Pump	0503
Padlock	2300
Pan, Oil	0107
Pan, Oil—Large and Small Assembly	0110
Pan, Oil—Large and Small Assembly (Conventional Models)	0110
Pan, Oil—Large Assembly	0110
Pan, Oil—Large Assembly (Conventional Models)	0110
Pan, Oil (Small)	0107
Pawl, Gear Shift	1911
Pawl, Hand Brake Lever	1201
Pedal, Clutch	0204
Pilot, Universal Drive	1007
Pilot, Universal Joint	1007
Pin, Accelerator Treadle	0303
Pin, Anchor	1201

	Group
Pin, Ball	1403
Pin, Brake Shoe	1201
Pin, Brake Shoe Link	1201
Pin, Brush Holder	0602
Pin, Brush Holder—and Insulation	0602
Pin, Brush Holder Stop	0602
Pin, Carburetor Bracket	0301
Pin, Carburetor Lever Bushing Taper	0301
Pin, Carburetor Stop Lever Taper	0301
Pin, Carburetor Throttle Stop	0301
Pin, Carburetor Thrust Washer Taper	0301
Pin, Clevis (½")	1201
Pin, Clutch Pressure Plate Retracting Spring—Ret.	0202
Pin, Clutch Release Sleeve Spring Stop	0203
Pin, $\frac{3}{32}$" x 1" Cotter	0601
Pin, Crankshaft Knurled	0102
Pin, Crankshaft Rear Bearing Dowel	0110
Pin, Crankshaft Starting Crank	0102
Pin, Cylinder Block Fuel Pump Drive	0101
Pin, Disc Brake Shoe	1201
Pin, Distributor Coupling	0603
Pin, Distributor Gear	0603
Pin, Dowel	0601
Pin, Dowel (D.E.)	0602
Pin, Dowel (⅛—D.E. and C.E.)	0602
Pin, Fan Support Bracket Cotter	0503
Pin, Front Brake Shoe Anchor	1203
Pin, Front Brake Shoe Wear Plate	1202
Pin, Front Spring Bracket	1602
Pin, Front Spring Shackle	1602
Pin, Fuel Pump Link	0302
Pin, Fuel Pump Rocker Arm	0302
Pin, Gearshift Connecting Rod Yoke Clevis	0706
Pin, Gearshift Lever Socket	0706
Pin, Gearshift Lever Socket Center	0706
Pin, Gearshift Rod Interlock	0706
Pin, Governor Bell Crank to Throttle Rod	0305
Pin, Governor Spider (⅛" x ¾")	0305
Pin, Governor Throttle Rod End (Valve Box) ($\frac{3}{32}$" x $\frac{7}{16}$")	0305
Pin, Governor Valve Box Bell Crank ($\frac{3}{32}$" x ¾")	0305
Pin, Governor Weight	0305
Pin, Governor Yoke (#1 x ¾" Taper)	0305
Pin, Hand Brake Lever Pawl	1201
Pin, Hinge	0603
Pin, Housing Front Cover Dowel	1103
Pin, Idler Gear Assembly	1911
Pin, Intermediate Gear Assembly	1911
Pin, Main Bearing Dowel	0102
Pin, Main Drive Shaft Direct Drive Gear Bushing	0802
Pin, Off-Set Clevis	1204
Pin, Oil Pump Drive Gear	0107
Pin, Overdrive Shift Rod Arm and Fork Connecting	0706
Pin, Piston	0103
Pin, Piston (.003 O/S)	0103
Pin, Piston (.005 O/S)	0103
Pin, Piston (.010 O/S)	0103
Pin, Piston (Standard)	0103
Pin, Pull Back Spring	0202
Pin, Rear Brake Shoe Anchor	1203
Pin, Rear Brake Shoe Spring	1203
Pin, Rear Spring	1602
Pin, Rear Spring Shackle—Rear	1602
Pin, Reverse Gearshift Rod Arm and Fork Connecting	0706
Pin, Release Lever	0205
Pin, Release Lever Adjusting Yoke	0205
Pin, Shear	1900
Pin, Steering Arm Ball	1007
Pin, Steering Arm Ball (1½")	1007
Pin, Steering Cross Tube End	1402
Pin, Steering Cross Tube End Pin Lock	1402
Pin, Steering Knuckle (Without Oil Groove)	1006
Pin, Transfer Case Dowel	0801

14

ALPHABETICAL INDEX—MASTER PARTS LIST

	Group
Pin, Trunnion Socket Bearing (Spindle Pin)	1006
Pin, Universal Drive Pilot	1007
Pin, Universal Joint Pilot	1007
Pin, Valve Spring Seat	0105
Pin, Valve Stem	0110
Pin, Winch Drive Shaft Shear	2300
Pin, Yoke, Assembly	1900
Pinion, Bevel	1003
Pinion, Bevel	1105
Pinion, Bevel, and Thru Shaft Assembly	1105
Pinion, Differential	1103
Pinion, Differential	1104
Pinion, Differential Bevel Side	1002
Pinion, Differential Bevel Side	1103
Pinion, Differential Bevel Side	1104
Pinion, Drive, and Shaft (11 Teeth)	1105
Pinion, Spur (8.43 Ratio)	1105
Pinion, Spur (8.148 Ratio)	1003
Pinion, Spur (8.148 Ratio)	1105
Pinion, Spur (8.435 Ratio)	1003
Pipe, Air Cleaner and Assembly	0301
Pipe, Exhaust	0402
Pipe, Intake	0108
Pipe, Oiler	1403
Pipe, Oil Filler Breather, Assembly	0107
Pipe, Oil Overflow, Assembly	0107
Pipe, Oil Pan Sump Vent	0107
Pipe, Oil Pump Intake	0110
Pipe, Oil Pump Main Discharge, Assembly	0107
Pipe, Oil Pump Outlet, Assembly	0107
Pipe, Oil Pump, Union Assembly	0107
Pipe, Radiator Outlet (1¾" O.D. x 14⅞")	0505
Pipe, Tail	0402
Pipe, Water Pump By-Pass	0505
Piston	1603
Piston, Carburetor Pump and Vacuum, Assembly	0301
Piston (.010 O/S)	0103
Piston (.010 O/S) (4" Dia. Aluminum)	0103
Piston (.020 O/S)	0103
Piston (.020 O/S) (4" Dia. Aluminum)	0103
Piston (.030 O/S)	0103
Piston (.030 O/S) (4" Dia. Aluminum)	0103
Piston (.040 O/S)	0103
Piston (Semi-Finished)	0103
Piston (Standard)	0103
Piston (Standard) (4" Dia. Aluminum)	0103
Plate, Adjuster	0202
Plate, Air Line Rubber Mounting Bracket	1206
Plate, No. 1 Auxiliary Spring	1601
Plate, No. 2 Auxiliary Spring	1601
Plate, No. 3 Auxiliary Spring	1601
Plate, No. 4 Auxiliary Spring	1601
Plate, No. 5 Auxiliary Spring	1601
Plate, No. 6 Auxiliary Spring	1601
Plate, Body Locating	1800
Plate, Body Mounting Bolt—L. H.	1800
Plate, Body Mounting Bolt—R. H.	1800
Plate, Brake Pedal Finish	1800
Plate, Brake Shoe Wear	1202
Plate, Breaker	0603
Plate, Breaker, Assembly	0603
Plate, Breaker, Part Assembly	0603
Plate, Cam and Stop	0603
Plate, Carburetor Air Shutter	0301
Plate, Carburetor Throttle	0301
Plate, Caution	2200
Plate, Caution (Air Pressure)	2200
Plate, Caution (Drain)	2200
Plate, Caution (Engine Speed)	2200
Plate, Center Bearing (D.E.)	0602
Plate, Clutch Adjusting	0202
Plate, Clutch Pressure	0202
Plate, Commutator End, Assembly	0601
Plate, Convex Bottom	0107
Plate, Cover, Bare	0205
Plate, Cover, Unit (Including Riveted Spring Cups)	0205
Plate, Cylinder Water Jacket (Front)	0101
Plate, Cylinder Water Jacket (Rear)	0101
Plate, Cylinder Water Jacket (Pressed Steel)	0101
Plate, Deck—Rear	1800
Plate, Engine Serial Number	0101
Plate, Drive Housing Cover	0602
Plate, Floorboard Clamp	1800
Plate, Flywheel Bell Housing Timing Hole Cover	0109
Plate, Flywheel Bell Housing Timing Hole Cover—L.W.	0109
Plate, Front Brake Dust Shield Mounting	1208
Plate, Front Brake Shoe Wear	1202
Plate, Front Bumper Clamp	1602
Plate, L. H. Front Clip	1603
Plate, R. H. Front Clip	1603
Plate, Front Spring Bracket Wearing	1602
Plate, Front Tubular Cross Member—L. H.	1500
Plate, Front Tubular Cross Member—R. H.	1500
Plate, Gearshift Finger Housing Cover—and Stop Assembly	0706
Plate, Gearshift Instruction	2200
Plate, Governor Valve Box Cover	0305
Plate, Ignition Wire Support	0604
Plate, Intake and Exhaust Filler	0108
Plate, Mainshaft and Weight	0603
Plate, Name	2200
Plate, Needle Bearing Cap Screw Lock	0902
Plate, Oil Filter Case Retaining (Fram #11734)	0107
Plate, Pintle Hook Anchor	1500
Plate, Pintle Hook Bearing	1500
Plate, P. T. O. Instruction	2200
Plate, Power Take-Off Shift	1800
Plate, P. T. O. Steady Bearing Bracket	1910
Plate, Pressure	0202
Plate, Pressure, Bare	0205
Plate, Publications Reference	2200
Plate, Rear Brake Show Wear	1202
Plate, Rear Engine Support Cover—Front	0110
Plate, Rear Engine Support Cover—Rear—L. H.	0110
Plate, Rear Engine Support Cover—Rear—R. H.	0110
Plate, Rear Spring Bracket Reinforcement (³⁄₁₆")	1602
Plate, Rear Spring Bracket Reinforcement (¼")	1602
Plate, Rear Spring Clip	1601
Plate, Rebound	1601
Plate, Shift (Main Trans.)	2200
Plate, Shift (Transfer Case)	2200
Plate, No. 1 Spring	1601
Plate, No. 2 Spring	1601
Plate, No. 3 Spring	1601
Plate, No. 4 Spring	1601
Plate, No. 5 Spring	1601
Plate, No. 6 Spring	1601
Plate, No. 7 Spring	1601
Plate, No. 8 Spring	1601
Plate, No. 9 Spring	1601
Plate, No. 10 Spring	1601
Plate, No. 11 Spring	1601
Plate, No. 12 Spring	1601
Plate, No. 13 Spring	1601
Plate, No. 14 Spring	1601
Plate, No. 15 Spring	1601
Plate, No. 1 Spring—with Bushing	1601
Plate, Steady Bearing Housing Felt Seal	0902
Plate, Valve Cover	0101
Plates, Cushion and Facing Unit Assembly	0201
Pliers—Slip Joint	2300
Plug, Bevel Pinion Bearing Cage	1003
Plug, Carrier	1003
Plug, Carburetor Filter	0301
Plug, Carrier	1105

ALPHABETICAL INDEX—MASTER PARTS LIST

Group

Plug, Countershaft Front Bearing Welch ... 0703
Plug, Cylinder Block Expansion (⅝") ... 0101
Plug, Cylinder Block Expansion (1¼") ... 0101
Plug, Cylinder Block Oil Orifice ... 0101
Plug, Cylinder Block Pipe (¾") ... 0101
Plug, Cylinder Block Pipe (1") ... 0101
Plug, Cylinder Head Pipe ... 0101
Plug, Cylinder Oil Header Pipe ... 0101
Plug, Declutching Carrier Oil Filler ... 0803
Plug, Declutch Shaft Bearing Carrier Oil ... 0803
Plug, Differential Carrier Inspection ... 1103
Plug, Drag Link End ... 1401
Plug, End (C.E.) ... 0602
Plug, Fuel Filter Drain ... 0302
Plug, Fuel Filter ¼" Pipe ... 0302
Plug, Fuel Pump Valve ... 0302
Plug, Gear Carrier Inspection ... 1003
Plug, Gearshift Rod Interlock ... 0706
Plug, Governor Welch (9/16") at Rocker Shaft ... 0305
Plug, Housing Drain ... 1001
Plug, Housing Drain ... 1101
Plug, Housing Filler Neck ... 1101
Plug, Housing Oil Filler ... 1001
Plug, Intake Manifold Pipe (⅛") ... 0108
Plug, Intake Manifold Pipe (½") ... 0108
Plug, Oil Filter Bracket Pipe (⅛") ... 0107
Plug, Oil Filter Case Drain (Fram #11584) ... 0107
Plug, Oil Filter Drain (¼") ... 0107
Plug, Oil Hole ... 0602
Plug, Oil Pan Drain ... 0107
Plug, Oil Pan Drain (⅜") ... 0107
Plug, Oil Pan Drain (⅞") ... 0107
Plug, Oil Pump Drain (⅜") ... 0110
Plug, Oil Well (C.E.) ... 0602
Plug, Reverse Gearshift Rod—Lower ... 0706
Plug, Spark (Champion #8) ... 0604
Plug, Spark (Champion H-10) ... 0604
Plug, Spark (Champion J-10) ... 0604
Plug, Speedometer Hole ... 0705
Plug, Steering Knuckle Bottom ... 1007
Plug, Transfer Case Drain Pipe (1") ... 0801
Plunger, Camshaft ... 0106
Plunger, Camshaft Idler Gear ... 0106
Plunger, Camshaft Thrust ... 0106
Plunger, Contact Spring ... 0603
Plunger, Cowl Side Vent Spring ... 1800
Plunger, First and Reverse Gearshift Latch ... 0706
Plunger, Gearshift Lockout (Reverse) ... 0706
Plunger, Gear Shift Lock Spring ... 0805
Plunger, Gearshift Rod ... 0706
Plunger, Gearshift Rod Interlock ... 0706
Plunger, Idler Shaft Gear Thrust ... 0106
Plunger, Reverse Gearshift Latch ... 0706
Plunger, Reverse Gearshift Rod—Lower ... 0706
Plunger, Universal Drive Pilot Pin ... 1007
Plunger, Valve ... 0110
Plunger, Valve, Assembly ... 0110
Pole Piece ... 0601
Poppet ... 1900
Post, Return Spring ... 0205
Post, Terminal ... 0601
Pulley, Compressor ... 1205
Pulley, Fan Drive ... 0503
Pulley, Fan Drive (Hercules) ... 0503
Pump, Fuel, Assembly (A C Type "D") ... 0302
Pump, Oil, Assembly ... 0107
Pump, Water, Assembly ... 0503

Race, Universal Drive, Inner ... 1007
Race, Universal Joint Cage Ball, Inner ... 1007
Radiator Assembly ... 0501
Rail, Frame—L. H. ... 1500
Rail, Frame—L. H. (Drilled) ... 1500

Group

Rail, Frame—R. H. ... 1500
Rail, Frame—R. H. (Drilled) ... 1500
Ratchet, Crankshaft Starting ... 0102
Rebound, Metering Spacer ... 1603
Reducer, Housing Breather Nipple ... 1001
Reflector, Amber (K-D Lamp) ... 0608
Reflector, Red (K-D Lamp) ... 0608
Regulator, Door Window—L. H. ... 1800
Regulator, Door Window—R. H. ... 1800
Regulator to Filter Jumper Assembly ... 0611
Regulator to Generator Cable Assembly ... 0611
Reinforcement, Deck Plate Outrigger ... 1800
Reinforcement, Deck Plate—Rear ... 1800
Reinforcement, Gas Tank Bracket Frame ... 0300
Relay, Electric Horn (Delco) ... 0609
Relay, Electric Horn (E. A. Lab.) ... 0609
Relay, Electric Horn (In Horn) ... 0609
Release, Spring Lever Arm ... 1201
Retainer, Axle Shaft Oil Seal ... 1101
Retainer, Bearing ... 0601
Retainer, Brake Shoe Pin ... 1201
Retainer, Declutching Shifter Shaft Packing ... 0803
Retainer, Declutch Shaft Packing ... 0803
Retainer, Disc Brake Shoe Pin ... 1201
Retainer, Drive End Packing ... 0601
Retainer, Fan Bearing Cork ... 0503
Retainer, Front Driving Axle Oil Seal ... 1007
Retainer, Front Wheel Bearing Felt ... 1301
Retainer, Front Wheel Bearing Felt Washer ... 1301
Retainer, Front Wheel Bearing Oil Seal—Inner ... 1301
Retainer, Front Wheel Bearing Oil Seal—Outer ... 1301
Retainer, Gear Shift Shaft Packing ... 0805
Retainer, Jackshaft Bearing—L. H. ... 1105
Retainer, Jackshaft Bearing—R. H. ... 1105
Retainer, Jackshaft Bearing, and Cup Assembly—L. H. 1105
Retainer, Jackshaft Bearing, and Cup Assembly—R. H. 1105
Retainer, Journal Gasket ... 0902
Retainer, Main Drive Gear Bearing ... 0703
Retainer, Rear Hub Bearing Felt ... 1301
Retainer, Rear Wheel Bearing Felt—Inner ... 1301
Retainer, Rear Wheel Bearing Felt—Outer ... 1301
Retainer, Steering Cross Tube End Dust Rubber ... 1402
Retainer, Steering Knuckle Felt ... 1006
Retainer, Steering Knuckle Pin Felt ... 1007
Retainer, Steering Knuckle Thrust Bearing ... 1007
Retainer, Transmission Driveshaft Bearing ... 0703
Retainer, Universal Drive and Shaft ... 1007
Retainer, Universal Joint and Shaft ... 1007
Ring, Ball Retaining ... 1403
Ring, Bendix Take-Up ... 0602
Ring, Bevel Pinion Forward Bearing Spinner ... 1003
Ring, Clutch Flywheel ... 0202
Ring, Clutch Release Sleeve Fulcrum ... 0203
Ring, Clutch Release Sleeve Snap ... 0203
Ring, Countershaft Front Bearing Welch Plug Retaining ... 0703
Ring, Countershaft Gear Retaining ... 0702
Ring, Countershaft Gear Snap ... 0703
Ring, Countershaft Rear Bearing Spacer ... 0703
Ring, Differential Bearing Adjusting ... 1002
Ring, Differential Bearing Adjusting ... 1105
Ring, Differential Bearing Retainer ... 1002
Ring, Differential Bearing Retaining ... 1105
Ring, Drum Thrust ... 1900
Ring, Flywheel ... 0202
Ring, Frame Thrust ... 1900
Ring, Front Brake Shaft Snap ... 1203
Ring, Front Wheel Bearing Adjuster Nut Lock (1¼"—7 Thds.) ... 1301
Ring, Front Wheel Bearing Adjuster Nut Lock (1½"—12 Thds.) ... 1301
Ring, Fulcrum ... 0203
Ring, Gearshift Universal Lever Snap ... 0706

ALPHABETICAL INDEX—MASTER PARTS LIST

	Group
Ring, Generator Oil Seal (Duprene)	0101
Ring, Headlamp Retaining	0607
Ring, Idler Shaft Front Bearing Snap	0804
Ring, Lock	0204
Ring, Lock	1204
Ring, Lock Spring	0603
Ring, Main Drive Shaft Front Bearing	0802
Ring, Needle Bearing Lock	0902
Ring, Oil Pump Idler Shaft Snap	0107
Ring, Oil Pump Snap	0107
Ring, Overdrive Countershaft Gear Clutch	0702
Ring, Pinion Bearing Spinner	1003
Ring, Pinion Bearing Spinner	1105
Ring, Piston Compression (Standard)	0103
Ring, Piston Compression (.010 O/S)	0103
Ring, Piston Compression (.020 O/S)	0103
Ring, Piston Compression (.030 O/S)	0103
Ring, Piston Compression (.040 O/S)	0103
Ring, Piston Compression (.050 O/S)	0103
Ring, Piston Compression (.060 O/S)	0103
Ring, Piston Compression (Standard) Complete Set	0103
Ring, Piston Compression (.010 O/S) Complete Set	0103
Ring, Piston Compression (.020 O/S) Complete Set	0103
Ring, Piston Compression (.030 O/S) Complete Set	0103
Ring, Piston Compression (.040 O/S) Complete Set	0103
Ring, Piston Compression (.050 O/S) Complete Set	0103
Ring, Piston Compression (.060 O/S) Complete Set	0103
Ring, Piston Compression (Standard) (4") 2nd & 3rd Groove	0103
Ring, Piston Compression (.010 O/S) (4") 2nd & 3rd Groove	0103
Ring, Piston Compression (.020 O/S) (4") 2nd & 3rd Groove	0103
Ring, Piston Compression (.030 O/S) (4") 2nd & 3rd Groove	0103
Ring, Piston Compression (Standard) (4") Top Groove	0103
Ring, Piston Compression (.010 O/S) (4") Top Groove	0103
Ring, Piston Compression (.020 O/S) (4") Top Groove	0103
Ring, Piston Compression (.030 O/S) (4") Top Groove	0103
Ring, Piston Oil Regulator (Standard)	0103
Ring, Piston Oil Regulator (.010 O/S)	0103
Ring, Piston Oil Regulator (.020 O/S)	0103
Ring, Piston Oil Regulator (.030 O/S)	0103
Ring, Piston Oil Regulator (.040 O/S)	0103
Ring, Piston Oil Regulator (.050 O/S)	0103
Ring, Piston Oil Regulator (.060 O/S)	0103
Ring, Piston Oil Regulator (Standard) (4")	0103
Ring, Piston Oil Regulator (.010 O/S) (4")	0103
Ring, Piston Oil Regulator (.020 O/S) (4")	0103
Ring, Piston Oil Regulator (.030 O/S) (4")	0103
Ring, Piston Oil Regulator (Standard) Complete Set	0103
Ring, Piston Oil Regulator (.010 O/S) Complete Set	0103
Ring, Piston Oil Regulator (.020 O/S) Complete Set	0103
Ring, Piston Oil Regulator (.030 O/S) Complete Set	0103
Ring, Piston Oil Regulator (.040 O/S) Complete Set	0103
Ring, Piston Oil Regulator (.050 O/S) Complete Set	0103
Ring, Piston Oil Regulator (.060 O/S) Complete Set	0103
Ring, Piston Pin Lock	0110
Ring, Pivot Anchor Bracket Snap	0706
Ring, Rear Brake Slack Adjuster Retainer Snap	1203
Ring, Retaining	0607
Ring, Retaining—Mainshaft 3rd Speed Gear Roller Bearing	0702
Ring, Reverse Gear Needle Bearing Separator	0703
Ring, Reverse Gear Needle Bearing Snap	0703
Ring, Snap	0203
Ring, Steady Bearing Housing Felt	0902
Ring, Third and Fourth Speed Clutch	0702
Ring, Universal Joint Retainer Snap	1007
Ring, Water Pump Packing Bearing Lock	0503
Ring, Water Pump Paddle Seal Clamp	0503
Ring, Water Pump Snap	0503
Rivet, Brake Shoe	1201
Rivets, Brake Shoe Lining	1201
Rivet, Brush Holder	0602
Rivet, Clutch Disc Facing	0201
Rivet, Cushion Plates and Facing Unit	0201
Rivet, Disc Brake Shoe Lining	1201
Rivet, Distributor Drive	0603
Rivet, Friction Facing	0201
Rivet, Front Brake Shoe Anchor Pin Bracket	1203
Rivet, Front Brake Shoe Lining	1202
Rivet, Lash Hook (5/16" x 1")	1800
Rivet, Spring Cup	0205
Rod, Accelerator Control	0303
Rod, Brake	1204
Rod, Brake Control	1204
Rod, Carburetor Control	0303
Rod, Clutch Control	0203
Rod, Clutch Control, Assembly	0203
Rod, Clutch Control (6½" Long)	0203
Rod, Connecting—Assembly	0104
Rod, Connecting—Assembly (Standard)	0104
Rod, Connecting—Assembly (.010 U/S Bearings)	0104
Rod, Connecting—Assembly (.020 U/S Bearings)	0104
Rod, Connecting—Assembly (.030 U/S Bearings)	0104
Rod, Connecting—Assembly (.040 U/S Bearings)	0104
Rod, Connecting, with Bearings	0104
Rod, Declutch Shift Control—Assembly	0805
Rod, Disc Brake Lever Arm Tie	1201
Rod, Disc Brake Link Tie (Lever Arm)	1201
Rod, Disc Brake Push	1201
Rod, First and Reverse Speed Gearshift	0706
Rod, First and Second Speed Gearshift	0706
Rod, Fourth and Overdrive Gearshift	0706
Rod, Fuel Pump Pull	0302
Rod, Gearshift Connecting, and Yoke Assembly	0706
Rod, Gear Shift Control—Assembly	0805
Rod, Gearshift Finger Shift	0706
Rod, Governor Throttle	0305
Rod, Governor Throttle, End	0305
Rod, Governor Throttle, End (Valve Box)	0305
Rod, Hand Brake—Forward	1201
Rod, Hand Brake Lever Pawl	1201
Rod, Hand Brake—Rear	1201
Rod, Oil Filter Center	0107
Rod, Overdrive Shift	0706
Rod, Overdrive Shift—Lower	0706
Rod, Overdrive Shift—Upper	0706
Rod, P. T. O. Shift (Front)	1910
Rod, P. T. O. Shift (Rear)	1910
Rod, Reverse Gearshift	0706
Rod, Reverse Gearshift—Lower	0706
Rod, Reverse Gearshift—Upper	0706
Rod, Second and Third Speed Gearshift	0706
Rod, Shift	1911
Rod, Third and Fourth Speed Gearshift	0706
Rod, Tie	1201
Rod, Tie	1402
Roller, Front Spring Bracket Rebound	1602
Rollers, Mainshaft Second Speed Gear	0702
Rollers, Mainshaft Third Speed Gear	0702
Roller, Needle	0205
Rotor	0603
Rubber, Air Line Mounting	1206
Rubber, Cowl Side Vent	1800
Rubber, Front Engine Support	0110
Rubber, Steering Cross Tube End Dust	1402
Screen, Carburetor Filter	0301
Screen, Fuel Pump	0302
Screen, Rear Window, Assembly	1800
Screw	0603
Screw, Adjusting	1201
Screw, Adjusting	1403
Screw, Advance Arm Dial	0603

ALPHABETICAL INDEX—MASTER PARTS LIST

	Group
Screw, Advance Control	0603
Screw, Advance Control Arm	0603
Screw, Advance Control Arm Clamp	0603
Screw, Air Compressor Coupling Cap	0110
Screw, Anchor Pin Set	1201
Screw, Axle Flange Jack	1102
Screw, Axle Shaft Oil Seal Retainer Set	1101
Screw, Bendix Head Spring	0602
Screw, Bendix Housing	0602
Screw, Bendix Housing & E. Frame Fasten'g (#10—32 x 1⅛")	0602
Screw, Bendix Shaft Spring	0602
Screw, Bevel Gear	1105
Screw, Bottom Cylinder	0107
Screw, Breaker Plate and Cap Spring Support	0603
Screw, Bridge	0107
Screw, Brush Connector Lead	0602
Screw, Brush Ground	0602
Screw, Brush Lead to Field	0602
Screw, Brush to Holder	0602
Screw, Camshaft Bushing	0106
Screw, Camshaft Idler Gear Cap	0106
Screw, Camshaft Thrust Plunger Adjusting	0106
Screw, Cap	1803
Screw, Carburetor Air Shutter Lever Swivel	0301
Screw, Carburetor Air Shutter—L. W.	0301
Screw, Carburetor Air Shutter Plate	0301
Screw, Carburetor Bowl to Body	0301
Screw, Carburetor Fuel Inlet Channel	0301
Screw, Carburetor Idle Adjusting	0301
Screw, Carburetor Throttle Lever Clamp	0301
Screw, Carburetor Throttle Lever Swivel	0301
Screw, Carburetor Throttle Plate	0301
Screw, Carburetor Throttle Stop	0301
Screw, Carburetor Venturi	0301
Screw, Carburetor Wire Clamp	0301
Screw, Center Bearing Plate	0602
Screw, Compressor Mounting Bracket	1205
Screw, Condenser Attaching	0603
Screw, Connecting Rod Piston Pin Clamp	0104
Screw, Contact Support Fastening	0603
Screw, Countershaft Bearing Cap	0703
Screw, Countershaft Front Bearing Lock	0703
Screw, Cover Band Bt. Head (#10—32 x 1¼")	0602
Screw, Crankshaft Center & Rear Main Bearing Cap	0102
Screw, Crankshaft Front & Intermediate Main Bearing Cap	0102
Screw, Crankshaft Starting Crank Pin Set	0102
Screw, Cylinder Block Fuel Pump Opening Cover (⁵⁄₁₆"—18 x ½")	0101
Screw, Cylinder Block Fuel Pump Opening Cover L. W. (⁵⁄₁₆")	0101
Screw, Cylinder Block Oil Orifice (⁵⁄₁₆"—18 x ¾")	0101
Screw, Cylinder Block Oil Orifice L. W. (⁵⁄₁₆")	0101
Screw, Cylinder Head Cap	0101
Screw, Cylinder Water Jacket Pet Cock Cap	0101
Screw, Declutching Shifter Lever	0803
Screw, Declutching Shifter Lock Spring	0803
Screw, Declutch Shift Fork	0803
Screw, Declutch Shift Lock Spring	0803
Screw, Differential Bearing Adjusting Lock	1105
Screw, Differential Bearing Adjusting Ring Lock	1002
Screw, Disc Brake Shoe Adjusting	1201
Screw, Distributor Cap Cover	0603
Screw, Distributor Cap Spring Support	0603
Screw, Door	0607
Screw, Door	0608
Screw, Fan	0503
Screw, Fan Adjusting	0503
Screw, Fan Adjusting, Assembly	0503
Screw, Fan Support Bracket (Hercules)	0503
Screw, First and Second Speed Gearshift Fork	0706
Screw, Flywheel Bell Housing Oil Seal	0109
Screw, Flywheel Bell Housing Timing Hole Cover Plate	0109
Screw, Flywheel Bell Housing to Crankcase—L. W.	0109
Screw, Flywheel Bell Housing to Crankcase (½" x 1")	0109
Screw, Flywheel Bell Housing to Crankcase (½"x1¼")	0109
Screw, Frame	0601
Screw, Front Driving Axle Oil Seal Retainer	1007
Screw, Fuel Pump Bail and	0302
Screw, Fuel Pump Bottom Cover	0302
Screw, Gearshift Arm and Fork	0706
Screw, Gearshift Fork and Lug	0706
Screw, Gearshift Universal Lever Finger Lock	0706
Screw, Generator Lock	0101
Screw, Governor Adjusting	0305
Screw, Governor Body Cap (#10—24 x ⅝")	0305
Screw, Governor Body Cap (Seal Wire) (#10—24x½")	0305
Screw, Governor Bumper Spring Adjusting	0305
Screw, Governor Throttle Lever Clevis	0305
Screw, Governor Throttle Lever Set	0305
Screw, Governor Valve Box Cover and Valve Shaft (#6—32)	0305
Screw, Headless Set (⁵⁄₁₆"—24 x 1¼")	0706
Screw, Hex. Cap (⅝"—18 x 1½")	1800
Screw, Hex. Cap (⁵⁄₁₆"—18 x)	1006
Screw, Hex. Cap (½"—20 x 1¾")	1403
Screw, Hex. Cap (⅜"—16 x 2⅜")	0505
Screw, Hex. Cap (⅜"—16 x 3⅛")	0505
Screw, Hex. Cap (⅝"—24 x 1")	1800
Screw, Hex. Head Cap (⁵⁄₁₆"—18 x ⅝")	1201
Screw, Housing Sleeve Retaining	1101
Screw, Idler Shaft Gear Thrust Plunger Adjusting, Assembly	0106
Screw, Intake & Exhaust Manifold Set	0108
Screw, Intake Manifold Heat Box Attaching (To Intake Manifold)	0108
Screw, Molding	0607
Screw, Needle Bearing Cap	0902
Screw, Oil Filter Attaching (½" x 1¼")	0107
Screw, Oil Filter Attaching (½" x 2¾")	0107
Screw, Oil Filter Case Cover (Fram #11580)	0107
Screw, Oil Filter Pressure Relief Valve Adjusting	0107
Screw, Oil Jet Set	0110
Screw, Oil Overflow Pipe Attaching	0107
Screw, Oil Overflow Pipe Attaching, L. W.	0107
Screw, Oil Pan Attaching (½"—13 x 1")	0107
Screw, Oil Pan Attaching, L. W.	0107
Screw, Oil Pan (Front End)	0107
Screw, Oil Pan Oil Strainer Cap	0107
Screw, Oil Pan Oil Strainer Cap, L. W.	0107
Screw, Oil Pressure Relief Valve Adjusting	0107
Screw, Oil Pump Attaching	0107
Screw, Oil Pump Body Attaching	0107
Screw, Oil Pump Cover	0107
Screw, Oil Pump Main Discharge Pipe	0107
Screw, Oil Pump Suction Line Flange	0107
Screw, Oil Pump Suction Line Flange Clamp	0107
Screw, Oil Relief Valve	0107
Screw, Oil Relief Valve	0110
Screw, Piston Pin Clamp	0104
Screw, Pole Piece	0601
Screw, Pole Piece	0602
Screw, Pole Piece Attaching (⁵⁄₁₆"—18 x ⅝")	0602
Screw, Pole Piece Attaching (⅜"—24 x ⅝")	0602
Screw, Puller	1102
Screw, Rear Axle Tube Set	1101
Screw, Rear Brake Camshaft Collar	1203
Screw, Rear Brake Spider and Housing	1101
Screw, Rear Brake Spider and Housing	1203
Screw, Rear Cover Cap	1101
Screw, Retaining Ring	0607
Screw, Reverse Gearshift Fork	0706
Screw, Reverse Gearshift Fork Guide	0706
Screw, Set	1201
Screw, Set (⁵⁄₁₆"—24 x 1¼")	1201

ALPHABETICAL INDEX—MASTER PARTS LIST

	Group
Screw, Set (Shoe Adjusting)	1201
Screw, Shifter Fork	0805
Screw, Shift Lever Finger Lock	0706
Screw, Special Cap	1900
Screw, Spur Pinion Bearing Washer	1105
Screw, Steering Arm Stop	1007
Screw, Steering Knuckle Stop	1006
Screw, Terminal	0603
Screw, Third and Fourth Speed Gearshift Fork	0706
Screw, Transfer Case Cover	0801
Screw, Universal Drive and Shaft Retainer	1007
Screw, Universal Joint Retainer	1007
Screw, Valve Adjusting	0107
Screw, Valve Cover	0105
Screw, Valve Plunger Adjusting	0110
Screw, Valve Tappet Adjusting	0105
Screw, Valve Tappet Cluster	0105
Screw, Water Pump Cover	0503
Screw ($\frac{7}{16}$"—14 x 1$\frac{3}{8}$")	0706
Screw ($\frac{7}{16}$"—14 x 1$\frac{5}{8}$")	0706
Screw ($\frac{7}{16}$"—20 x 1$\frac{3}{8}$")	0706
Screw, #6—32 x $\frac{7}{16}$" Hex. Head	0603
Screw, #6—32 x $\frac{3}{32}$" Fill. Head	0603
Screw, #6—32 x $\frac{5}{16}$" Flat Head	0601
Screw, #8—32 x $\frac{7}{16}$" Fill. Head	0603
Screw, #8—32 x $\frac{1}{4}$" Round Head	0601
Screw, #10—32 x $\frac{1}{2}$" Flat Head	0601
Screw, #10—32 x $\frac{5}{8}$" Flat Head	0601
Screw, #10—32 x $\frac{1}{4}$" Fill. Head	0603
Screw, #10—32 x $\frac{7}{16}$" Round Head	0601
Screw, #10—32 x $\frac{5}{8}$" Round Head	0601
Screw, #10—32 x 1$\frac{1}{4}$" Round Head	0601
Seal, Axle Shaft Oil	1101
Seal, Axle Shaft Oil, Assembly	1101
Seal, Bevel Pinion Bearing Cage Oil	1003
Seal, Bevel Pinion Forward Bearing Oil	1003
Seal, Crankshaft Rear Bearing	0110
Seal, Crankshaft Rear Bearing and Gasket Assembly	0110
Seal, Crankshaft Rear Bearing & Gasket Assembly	0102
Seal, Declutching Shaft Bearing and Cap Oil	0803
Seal, Declutch Shaft Oil	0803
Seal, Driven Shaft Rear Bearing Cap Oil	0803
Seal, Driven Shaft Oil	0803
Seal, Driving Axle Oil	1102
Seal, Driving Axle Oil, Assembly	1007
Seal, Flywheel Bell Housing Oil	0109
Seal, Front Driving Axle Oil	1007
Seal, Gearshift Lever Socket Oil	0706
Seal, Housing Leather Oil, Assembly	1403
Seal, Main Drive Front Oil	0802
Seal, Main Shaft Front Bearing Cage Oil	0802
Seal, Oil	1911
Seal, Oil Pump to Oil Pan Felt	0107
Seal, Oil (Victor 60656)	1900
Seal, Pinion Bearing Cage Oil	1003
Seal, Pinion Bearing Cage Oil	1005
Seal, Rear Engine Support Oil	0110
Seal, Speedometer Drive Gear Case Oil	0705
Seal, Thru Shaft Forward Bearing Cover Oil	1105
Seal, Water Pump Paddle and, Assembly	0503
Seal, Water Pump Paddle Carbon	0503
Seal, Water Pump Paddle Flexible	0503
Seat, Drag Link Ball (1$\frac{1}{2}$" Dia. Ball)	1401
Seat, Drag Link Ball (1$\frac{3}{4}$" Dia. Ball)	1401
Seat, Fuel Pump Bowl	0302
Seat, Fuel Pump Top Cover and Valve, Assembly	0302
Seat, Spring	1403
Seat, Steering Cross Tube End Ball	1402
Seat Structure (Includes Battery Box)	1800
Seat, Universal Drive Pilot Pin	1007
Seat, Universal Joint Pilot Pin	1007
Seat, Valve Spring	0105
Sector, Hand Brake Lever	1201

	Group
Set, Engine Gasket	0101
Shackle, Front Spring	1602
Shackle, Rear Spring	1602
Shackle, Rear Spring, and Bushing Assembly	1602
Shaft, Accelerator Cross	0303
Shaft, Axle	1102
Shaft, Axle, and Flange Assembly	1102
Shaft, Axle, Assembly	1102
Shaft, Axle, Assembly—Long	1102
Shaft, Axle, Assembly—Short	1102
Shaft, Bevel Pinion (10-T)	1003
Shaft, Bevel Pinion (10 Teeth)	1105
Shaft, Camshaft Idler Gear	0106
Shaft, Carburetor Air Shutter	0301
Shaft, Carburetor Throttle	0301
Shaft, Clutch and Brake Pedal	0204
Shaft, Clutch Control Cross	0203
Shaft, Clutch Throwout	0203
Shaft, Control Hand Lever	0805
Shaft, Declutch	0803
Shaft, Declutch Control Cross	0805
Shaft, Declutching, and Bushing Assembly	0803
Shaft, Declutching Shifter	0803
Shaft, Declutch Shifting	0803
Shaft, Disc Brake Cross	1201
Shaft, Distributor Drive	0603
Shaft, Drive	0602
Shaft, Drive, Assembly	0603
Shaft, Driven	0803
Shaft, Drum, Assembly	1900
Shaft, Gear Shift	0805
Shaft, Governor Drive	0305
Shaft, Governor Rocker	0305
Shaft, Governor Valve	0305
Shaft, Hand Brake	1201
Shaft, Idler	0106
Shaft, Idler	0804
Shaft, Lever, with Nut and Lockwasher	1403
Shaft, Main	0802
Shaft, Main Drive	0802
Shaft, Motor Drive	0602
Shaft, Oil Pump	0107
Shaft, Oil Pump Gear	0107
Shaft, Oil Pump Idler	0107
Shaft, Oil Pump Pinion	0107
Shaft, Pedal Lever	0204
Shaft, Power Take-Off Adapter Gear	1911
Shaft, P. T. O. Control Cross	1910
Shaft, Rear Spring Cross	1602
Shaft, Reverse Gear	0704
Shaft, Shifter	0805
Shaft, Shift Lever Rocker	0706
Shaft, Slip Spline Stub (3" Tube)	0902
Shaft, Spline Stub	0902
Shaft, Spur Pinion (6.62 Ratio)	1103
Shaft, Spur Pinion (8.21 Ratio)	1003
Shaft, Spur Pinion (8.21 Ratio)	1103
Shaft, Stub (3" Tube and 1$\frac{1}{2}$" Taper)	0902
Shaft, Take-Off Assembly	1911
Shaft, Thru	1105
Shaft, Water Pump, and Bearing Assembly	0503
Shaft, Water Pump Paddle	0503
Shaft, Water Pump—$\frac{3}{4}$" Diameter	0503
Shell, Connecting Rod Bearing	0104
Shield, Air—between Cowl and Radiator	1800
Shield, Battery, Assembly	0610
Shield, Countershaft Rear Bearing	0703
Shield, Front Brake Dust	1208
Shield, Front Brake Dust—L. H.	1208
Shield, Front Brake Dust—R. H.	1208
Shield, Generator Oil (Steel Washer)	0101
Shield, Intake Manifold	0108
Shield, Rear Brake Dust, Assembly	1208

ALPHABETICAL INDEX—MASTER PARTS LIST

	Group
Shim	1201
Shim, Bevel Pinion Bearing	1003
Shim, Bevel Pinion Bearing Cage—Medium	1003
Shim, Bevel Pinion Bearing Cage—Thick	1003
Shim, Bevel Pinion Bearing Cage—Thin	1003
Shim, Bevel Pinion Forward Bearing	1003
Shim, Bevel Pinion Forward Bearing—Medium	1003
Shim, Bevel Pinion Forward Bearing—Thick	1003
Shim, Bevel Pinion Forward Bearing—Thin	1003
Shim, Brass (.002″)	1403
Shim, Clutch Flywheel Ring Adjusting (.015)	0202
Shim, Connecting Rod (.002)	0104
Shim, Connecting Rod (.003)	0104
Shim, Connecting Rod Bearing	0104
Shim, Connecting Rod Bearing (.002)	0104
Shim, Connecting Rod Bearing (.003)	0104
Shim, Connecting Rod Bearing (.006)	0104
Shim, Countershaft Rear Bearing Cap—.005″	0703
Shim, Countershaft Rear Bearing Cap—$\frac{3}{32}$″	0703
Shim, Crankshaft Bearing—Center (.010)	0102
Shim, Crankshaft Bearing—Front (.010)	0102
Shim, Crankshaft Bearing—Intermediate (.010)	0102
Shim, Crankshaft Center & Rear Main Bearing (.002)	0102
Shim, Crankshaft Center & Rear Main Bearing (003)	0102
Shim, Crankshaft Front Main Bearing (.002)	0102
Shim, Crankshaft Front Main Bearing (.003)	0102
Shim, Crankshaft Intermediate Main Bearing (.002)	0102
Shim, Crankshaft Intermediate Main Bearing (.003)	0102
Shim, Crankshaft Main Bearing—Center & Rear (.002)	0102
Shim, Crankshaft Main Bearing—Center & Rear (.003)	0102
Shim, Crankshaft Main Bearing—Front & Intermediate (.002)	0102
Shim, Crankshaft Main Bearing—Front & Intermediate (.003)	0102
Shim, Declutching Carrier—Medium	0803
Shim, Declutching Carrier—Thick	0803
Shim, Declutching Carrier—Thin	0803
Shim, Distributor Coupling—.005	0603
Shim, Distributor Coupling—.010	0603
Shim, Distributor Drive Shaft—.030	0603
Shim, Driven Shaft Rear Bearing Cap—Medium	0803
Shim, Driven Shaft Rear Bearing Cap—Thick	0803
Shim, Driven Shaft Rear Bearing Cap—Thin	0803
Shim, Driven Shaft Rear Bearing Cover—Thick	0803
Shim, Driven Shaft Rear Bearing Cover—Thin	0803
Shim, Driven Shaft Rear Bearing Cover—Thin	0803
Shim, Drive Pinion Bearing Spacer—Thick	1105
Shim, Drive Pinion Bearing Spacer—Thin	1105
Shim, Drive Pinion Cage (.005)	1105
Shim, Drive Pinion Cage (.007)	1105
Shim, Drum Shaft	1900
Shim, Front Spring	1601
Shim, Front Tubular Cross Member (.037″)	1500
Shim, Front Tubular Cross Member (.062″)	1500
Shim, Idler Shaft Front and Rear Bearing Cap—Med.	0804
Shim, Idler Shaft Front and Rear Bearing Cap—Thick	0804
Shim, Idler Shaft Front and Rear Bearing Cap—Thin	0804
Shim, Jackshaft Bearing Cap (.005)	1105
Shim, Jackshaft Bearing Cap (.007)	1105
Shim, Mainshaft Rear Bearing Retainer	0703
Shim, Mainshaft Rear Bearing Retainer—.005″	0703
Shim, Mainshaft Rear Bearing Retainer—.010″	0703
Shim, Mainshaft Third Speed Gear—.003	0702
Shim, Mainshaft Third Speed Gear—.005	0702
Shim—Medium	0802
Shim, Mounting Side Cap—Medium	1003
Shim, Mounting Side Cap—Thick	1003
Shim, Mounting Side Cap—Thin	1003
Shim, Mounting Side Cap—Medium	1103
Shim, Mounting Side Cap—Thick	1103
Shim, Mounting Side Cap—Thin	1103
Shim, Pinion Shaft Bearing	1003
Shim, Pinion Shaft Bearing	1105
Shim, Pinion Bearing Cage—Medium	1003
Shim, Pinion Bearing Cage—Thick	1003
Shim, Pinion Bearing Cage—Thin	1003
Shim, Pinion Bearing Cage—Medium	1105
Shim, Pinion Bearing Cage—Thick	1105
Shim, Pinion Bearing Cage—Thin	1105
Shim, Rear Spring	1603
Shim, Rear Spring Front Bracket Cross Member ($\frac{1}{16}$″)	1500
Shim, Rear Spring Front Bracket Cross Member (¼″)	1500
Shim, Rear Spring—L. H.	1601
Shim, Rear Spring—R. H.	1601
Shim, Spur Pinion Bearing—Thick	1105
Shim, Spur Pinion Bearing—Thin	1105
Shim, Spur Pinion Bearing Cage—Medium	1003
Shim, Spur Pinion Bearing Cage—Thick	1003
Shim, Spur Pinion Bearing Cage—Thin	1003
Shim, Spur Pinion Bearing Cover—Medium	1003
Shim, Spur Pinion Bearing Cover—L. H.—Thick	1003
Shim, Spur Pinion Bearing Cover—L. H.—Thin	1003
Shim, Spur Pinion Bearing Cover—R. H.—Thick	1003
Shim, Spur Pinion Bearing Cover—R. H.—Thin	1003
Shim, Steel (.003″)	1403
Shim, Steel (.010″)	1403
Shim, Steering Knuckle	1007
Shim, Steering Knuckle Bearing Cap—Thick	1006
Shim, Steering Knuckle Bearing Cap—Thin	1006
Shim—Thick	0802
Shim—Thin	0802
Shim, Thru Shaft Forward Bearing Cover—Thick	1105
Shim, Thru Shaft Forward Bearing Cover—Thin	1105
Shim, Thru Shaft Rear Bearing Cover—Thick	1105
Shim, Thru Shaft Rear Bearing Cover—Thin	1105
Shim, Transfer Case Support Tube	0801
Shim, Upper Cover Brass (.002)	1403
Shim, Upper Cover Brass (.003)	1403
Shim, Upper Cover Brass (.010)	1403
Shims, Anti-Rattle	1800
Shock Absorber, Front Assembly — Monroe — (Complete)	1603
Shock Absorber, L. H. Front	1603
Shock Absorber, L. H. Front and Link Assembly—(Complete)	1603
Shock Absorber, L. H. Rear	1603
Shock Absorber, R. H. Fronk and Link Assembly—(Complete) (Houdaille "BBH")	1603
Shock Absorber, R. H. Front (Houdaille "BBH")	1603
Shock Absorber, Rear Assy.—Monroe—(Complete)	1603
Shock Absorber, R. H. Rear (Houdaille "BBH")	1603
Shoe, Brake	1201
Shoe, Disc Brake—Front	1201
Shoe, Disc Brake—Rear	1201
Shoe, Front Brake, Assembly	1202
Shoe, Front Brake, Assembly—L. H.—Lower	1202
Shoe, Front Brake, Assembly—L. H.—Upper	1202
Shoe, Front Brake, Assembly—R. H. Lower	1202
Shoe, Front Brake, Assembly—R. H.—Upper	1202
Shoe, Rear Brake, Assembly—Lower	1202
Shoe, Rear Brake, Assembly—Upper	1202
Side, Battery Shield, Assembly	0610
Sill, Body—L. H.	1800
Sill, Wood Body—R. H.	1800
Sleeve, Bearing	1911
Sleeve, Bendix—and Shaft Assembly	0602
Sleeve, Bendix Service	0602
Sleeve, Bevel Gear Bearing	1105
Sleeve, Clutch Release	0203
Sleeve, Driving Axle Oil Seal	1102
Sleeve, End Frame	1900
Sleeve, First and Second Speed Slide Gear	0702
Sleeve, Governor Thrust	0305
Sleeve, Housing	1101
Sleeve, Housing—Long	1001
Sleeve, Housing—Long	1101

20

ALPHABETICAL INDEX—MASTER PARTS LIST

	Group
Sleeve, Housing—Short	1001
Sleeve, Housing—Short	1101
Sleeve, Mainshaft Third Speed Gear	0702
Sleeve, Release, Assembly (Including Wick & Spring Post)	0205
Sleeve, Release—Bare	0203
Sleeve, Reverse Gear	0702
Sleeve, Speedometer Cable Extension	2203
Sleeve, Speedometer Driven Gear	0705
Sleeve, Speedometer Driven Gear (S-W)	0804
Sleeve, Steady Bearing Housing	0902
Slinger, Declutching Shaft Bearing Oil	0803
Slinger, Front Brake Dust Shield Oil	1208
Slinger, Front Dust Shield Oil	1402
Slinger, Rear Wheel Brake Drum Oil	1302
Socket, Drag Link	1401
Socket, Gearshift Lever	0706
Socket, Gearshift Lever—and Bushing Assembly	0706
Socket, Trunnion, Assembly	1006
Socket, Trunnion, Assembly (Spindle Pin Pocket)	1006
Spacer, Bearing	1911
Spacer, Bevel Gear Bearing	1105
Spacer, Bevel Pinion Bearing	1003
Spacer, Control Bracket Support	0805
Spacer, Countershaft Front Bearing	0703
Spacer, Deck Plate Bracket	1800
Spacer, Drive Pinion Bearing	1105
Spacer, Fan	0503
Spacer, Fan Blade (⅝")	0503
Spacer, Fan Bracket	0503
Spacer, Front Engine Support Bracket	0110
Spacer, Front Wheel Bearing—Inner	1301
Spacer, Governor Yoke	0305
Spacer, Hand Brake Lever Sector—Front	1201
Spacer, Hand Brake Lever Sector—Top	1201
Spacer, Idler Gear	0804
Spacer, Main Drive Gear Idler Bearing	0802
Spacer, Pinion Shaft Bearing	1003
Spacer, Pinion Shaft Bearing	1105
Spacer, Rear Brake Spider and Brake Flange	1101
Spacer, Rear Spring Pin	1601
Spacer, Rear Spring Shackle	1602
Spacer, Rear Spring Shackle Pin	1602
Spacer, Rear Wheel	1301
Spacer, Reverse Gear Shaft Bearing	0703
Spacer, Second and Third Speed Countershaft Gear	0702
Spacer, Shock Absorber Stud	1603
Spacer, Speedometer Gear	0804
Spacer, Towing Hook—Front	1501
Spacer, Towing Hook—Rear	1501
Spacer, Tubing	1209
Spacer, Universal Drive Inner Race	1007
Spacer, Universal Joint	1007
Spacers, Steel	1800
Spacers, ½"	1205
Speedometer Assembly (Stewart-Warner)	0605
Spider, Differential	1002
Spider, Differential	1103
Spider, Differential	1104
Spider, Governor, Assembly	0305
Spider, Rear Brake	1203
Spider, Rear Brake, and Bushing Assembly	1203
Spider, Rear Brake, and Sleeve Assembly—Long	1203
Spider, Rear Brake, and Sleeve Assembly—Short	1203
Spider, Worm Gear, Assembly	1900
Spindle, Fan	0503
Spindle, Manifold Hot Spot Valve	0108
Spoon, Hand Brake Lever	1201
Spring	0603
Spring	1403
Spring, Accelerator	0303
Spring, Bendix Drive	0602
Spring, Body Hold-Down—L. H.	1800
Spring, Brake	1900
Spring, Brake Control Rod Return	1204
Spring, Brake Shoe	1201
Spring, Breaker Arm	0603
Spring, Breaker Lever—Fastening	0603
Spring, Brush	0601
Spring, Brush	0602
Spring, Brush Arm	0602
Spring, Carburetor Idler Adjusting Screw	0301
Spring, Carburetor Throttle Stop Screw	0301
Spring, Chamber Hose	1209
Spring, Clutch—Inner	0202
Spring, Clutch Pedal Return	0203
Spring, Clutch Pressure	0202
Spring, Clutch Pressure Plate Retracting	0202
Spring, Clutch Release Trunnion Return	0203
Spring, Clutch Throwout Shaft Lever	0203
Spring, Clutch Throwout Shaft Lever, Link	0203
Spring, Contact	0603
Spring, Cowl Side Vent	1800
Spring, Declutching Shifter Shaft Lock	0803
Spring, Declutch Shift Lock	0803
Spring, Disc Brake Lever Arm Release	1201
Spring, Disc Brake Shoe	1201
Spring, Distributor Cap	0603
Spring, Drag Brake	1900
Spring, Drag Link Ball Seat	1401
Spring, First and Reverse Gearshift Latch Plunger	0706
Spring, Front, Assembly	1601
Spring, Front Brake Shoe	1203
Spring, Front, No. 1 and Bushing	1601
Spring, Front, No. 1 and Bushing Assembly	1601
Spring, Front, No. 1 and No. 2 Assembled	1601
Spring, Fuel Pump Diaphragm	0302
Spring, Fuel Pump Rocker Arm	0302
Spring, Fuel Pump Valve	0302
Spring, Gearshift Lever Pivot	0706
Spring, Gearshift Lever Socket	0706
Spring, Gearshift Rod	0706
Spring, Gear Shift Lock	0805
Spring, Gearshift Lockout Plunger	0706
Spring, Gearshift Lockout Plunger Ball	0706
Spring, Gearshift Rod	0706
Spring, Governor	0305
Spring, Governor Bumper	0305
Spring, Hose	1206
Spring, Lever Arm Release	1201
Spring, Oil Filter	0107
Spring, Oil Filter Case Cover (Fram #11583)	0107
Spring, Oil Filter Pressure Relief Valve	0107
Spring, Oil Pressure Regulator	0107
Spring, Oil Pressure Relief Valve	0107
Spring, Oil Relief Valve	0107
Spring, Oil Relief Valve	0110
Spring, Pawl	1911
Spring, Pawl Rod	1201
Spring, Poppet	1900
Spring, Pressure	0205
Spring, Pull Back	0202
Spring, Rear Brake Shoe	1203
Spring, Rear Brake Shoe Return	1203
Spring, Rear, No. 1 and No. 2 Assembled	1601
Spring, Rear, Assembly	1601
Spring, Rear Auxiliary, Assembly	1601
Spring, Release Lever Tension	0205
Spring, Reverse Gearshift Latch Plunger	0706
Spring, Reverse Gearshift Rod Plunger	0706
Spring, Shifter Lock Plunger	0805
Spring, Steering Cross Tube End Ball	1402
Spring, Steering Cross Tube End Cover	1402
Spring, Steering Knuckle Felt Pressure	1006
Spring, Steering Knuckle Pin Felt Retainer	1007
Spring, Trunnion Socket	1006

21

ALPHABETICAL INDEX—MASTER PARTS LIST

	Group
Spring, Trunnion Socket Felt	1006
Spring, Universal Drive Buffer	1007
Spring, Universal Drive Pilot Pin Plunger	1007
Spring, Universal Joint	1007
Spring, Valve	0105
Spring, Water Pump Paddle Seal	0503
Spring, Weight	0603
Spring, Weight, Set	0603
Springs	1800
Staple, Tool Box Door Hasp	1800
Step, Stirrup Assembly	1800
Stove, Intake Manifold	0108
Strainer, Oil Pan Oil, and Water Trap Assembly	0107
Strainer, Oil Pump, Assembly	0110
Strap, Air Reservoir Tank	1205
Strap, Clutch Flywheel Ring Adjusting	0202
Strap, Deck Plate Support—L. H.	1800
Strap, Gas Tank	0300
Strap, Gas Tank, and Anti-Rattle Assembly	0300
Strip, Field Coil Insulation	0602
Strip, Floorboard Insert Retainer	1800
Strip, Terminal Screw Insulation	0603
Strip, Toeboard Retaining—L. H.	1800
Strip, Toeboard Retaining—R. H.	1800
Strip, Toeboard Retaining (Top)	1800
Stubshaft Assembly—17 1/8" Long	0902
Stubshaft Assembly—24 1/8" Long	0902
Stubshaft Assembly—40 7/8" Long	0902
Stubshaft Assembly—40 11/16" Long	0902
Stubshaft Assembly—52 1/8" Long	0902
Stud	0101
Stud, Adapter (Trans. Case)	1911
Stud, Ball	1403
Stud, Bevel Pinion Bearing Cage	1003
Stud, Bevel Pinion Forward Bearing Cage	1003
Studs, Carburetor Adapter	0301
Stud, Carrier to Housing	1103
Stud, Clamping	1209
Stud, Clutch Flywheel Ring	0202
Stud, Clutch Oiler and Coil	0101
Stud, Clutch Oil Line Bracket	0101
Stud, Clutch Pressure Plate	0202
Stud, Cylinder	0101
Stud, Cylinder Block	0101
Stud, Cylinder Head	0101
Stud, Differential Carrier and Cap	1103
Stud, Differential Carrier to Housing	1103
Stud, Differential Carrier to Housing (1/2"—13)	1103
Stud, Differential Support Cap	1003
Stud, Differential Support Cap	1103
Stud, Driving Axle Flange	1102
Stud, Exhaust Manifold Front Flange	0108
Stud, Exhaust Manifold Rear Flange	0108
Stud, Front Brake Diaphragm	1207
Stud, Front Driving Axle Drive Flange	1007
Stud, Front Driving Axle Flange and Hub	1007
Stud, Front Wheel Hub—L. H.	1302
Stud, Front Wheel Hub—R. H.	1302
Stud, Fuel Filter Assembly	0302
Stud, Gear Shaft Bearing Washer	1003
Stud, Gear Shaft Bearing Washer	1105
Stud, Gearshift Connecting Tube	0706
Stub, Gearshift Lever	0706
Stud, Hand Throttle	0101
Stud, Housing and Socket	1001
Stud, Housing (For Upper Cover)	1403
Stud, Housing to Carrier	1001
Stud, Intake and Exhaust Manifold	0108
Stud, Intake and Exhaust Manifold Attaching	0108
Stud, Intake and Exhaust Manifold—Long	0108
Stud, Intake and Exhaust Manifold—Short	0108
Stud, Intake Manifold Heat Box Attaching (To Exhaust Manifold)	0108
Stud—Long	1006
Stud, Magazine Oiler	0204
Stud, Main Bearing	0102
Stud, Main Bearing—Center	0102
Stud, Manifold	0108
Stud, Motor Terminal	0602
Stud, Mounting Side Cap, and Cage	1003
Stud, Oil Pan	0110
Stud, Power Take-Off Adapter	1911
Stud, P. T. O. Control Lever	1910
Stud, Rear Axle Housing and Camshaft Bracket	1203
Stud, Rear Brake Camshaft and Diaphragm Bracket	1203
Stud, Rear Engine Support	0110
Stud, Rear Motor Support	0110
Stud, Rear Spring	1602
Stud, Rear Wheel Hub—L. H.	1302
Stud, Rear Wheel Hub—R. H.	1302
Stud, Shock Absorber (Bottom)	1603
Stud, Shock Absorber (Top)	1603
Stud—Short	1006
Stud, Spare Tire Mounting Padlock	1800
Stud, Spare Tire Mounting—Plain	1800
Stud, Steering Arm Ball	1006
Stud, Steering Arm Ball (1½" Diameter)	1006
Stud, Steering Cross Tube End, and Bearing Assy.	1402
Stud, Steering Knuckle	1007
Stud, Steering Knuckle Bearing Cap	1006
Stud, Steering Knuckle Bearing Cap	1007
Stud, Steering Knuckle Companion Flange	1006
Stud, Spur Pinion Bearing (½" Diameter)	1105
Stud, Spur Pinion Bearing (7/16" Diameter)	1105
Stud, Spur Pinion Bearing Cage	1003
Stud, Spur Pinion Bearing Cage	1105
Stud, Spur Pinion Bearing Cover	1003
Stud, Thru Shaft Forward Bearing Cover	1105
Stud, Valve Plunger Guide Yoke	0110
Stud (½"—13)	1102
Stud (7/16"—18)	1006
Support, Anchor Bracket End	1201
Support, Battery, Assembly	0610
Support, Contact Point and	0603
Support, Control Bracket	0805
Support, Deck Plate—L. H.	1800
Support, Distributor Cap Spring	0603
Supports, Engine Hood	1800
Support Extension Rear Engine	0110
Support, Floorboard Front	1800
Support, Front Engine	0110
Support, Gearshift Lever Bracket	0706
Support, Ignition Wires	0603
Support, Radiator Top	0501
Support, Rear Cab Assembly	1800
Support, Rear Engine	0110
Support, Rear Engine (Bellhousing #3 S.A.E.)	0110
Support, Rear Engine—L. H.	0110
Support, Rear Engine—R. H.	0110
Support, Rear Quarter Fender Rear—Lower	1800
Support, Rear Quarter Fender Rear—Upper	1800
Support, Transfer Case—Front	0801
Support, Transfer Case—Rear	0801
Support, Washer Gasket	1603
Support, Washer Piston	1603
Suppressor, Distributor	0611
Suppressors, Spark Plug	0611
Switch, Blackout (Cole-Hersee)	0606
Switch, Blackout (Delco-Remy)	0606
Switch, Foot Dimmer (Delco-Remy)	0606
Switch, Foot Dimmer (H. A. Douglas)	0606
Switch, Ignition to Filter Jumper Assembly	0611
Switch, Instrument Panel Light (Delco-Remy)	0606
Switch, Magnetic Starting (Delco-Remy)	0602
Switch, Starting (Auto-Lite)	0602
Switch, Starting (Delco-Remy)	0602

ALPHABETICAL INDEX—MASTER PARTS LIST

Group

Switch, Stop Light 1205
Switch, Stop Light (Westinghouse)..................0606
Swivel, Carburetor Air Shutter Lever...............0301
Swivel, Carburetor Throttle Lever...................0301

Tachometer (Stewart-Warner) 0605
Tag, Emergency Line 1209
Tag, Service Line...1209
Take-Off Case 1911
Take-Off, Power Assembly 1911
Tank, Air 1205
Tank, Air Reservoir 1205
Tank, Gas, Assembly 0300
Tank, Gas, 30 Gal..0300
Tank, Gas, 60 Gal..0300
Tappet, Valve 0105
Tarpaulin 1800
Tee, ⅜″ 1209
Tee, Tubing (Manifold) 1209
Terminal 0601
Terminal—Douglas 0607
Terminal—Eyelet 0607
Terminal, Wire 0607
Thermostat 0502
Thermostat (Bishop & Babcock)....................0502
Thermostat (Dole #XT-187) 0502
Thimble 1603
Thrower, Crankshaft Oil 0102
Thrower, Oil 0601
Thrower, Oil, Assembly 0601
Tire, 9.00-20 Spare and Tube.........................2300
Tire, 12.00-20 Spare and Tube.......................2300
Toeboard—L. H. 1800
Toeboard—R. H. 1800
Treadle, Accelerator 0303
Trough, Oil 1103
Tube, Accelerator Cross Shaft.......................0303
Tube, Cam and Wheel, Assembly with Bearings and Wheel Nut 1403
Tube, Cam and Wheel, Assembly with Wheel Nut.....1403
Tube, Carburetor Discharge 0301
Tube, Carburetor Idling By-pass, Assembly...0301
Tube, Drag Link 1401
Tube, End Cover and Oil Seal, Assembly.......1403
Tube, Frame 1206
Tube, Gas Tank Inlet Titeflex........................0300
Tube, Gearshift Connecting 0706
Tube, Gearshift Connecting—Assembly 0706
Tube, Gearshift Connecting—Outer 0706
Tube, Governor Throttle Rod.........................0305
Tube, Housing 1101
Tube, Housing (Long—R. H.).........................1001
Tube, Housing (Short—L. H.).........................1001
Tube, Jacket, and Bearing Assembly..............1403
Tube, Jacket, Jacket Tube Ball Bearing Unit and Upper Cover Assembly 1403
Tube, Jacket, and Jacket Tube Ball Bear. Unit Assy...1403
Tube, Oil Drain 0110
Tube, Oil Filter Center—Assembly (Fram #11733)..0107
Tube, Rear Axle 1101
Tube, Rear Spring 1602
Tube, Shift Rod 1911
Tube, Steering Cross 1402
Tube, Steering Cross, Assembly (Including the following items) 1402
Tube, Steering Cross—Only 1402
Tube, Viscometer Gauge, Assembly...............0107
Tubing, Gas Tank Vent Pipe ⅜″ Bundy...........0300
Tubing, ¾″ O.D. Copper, #20 Ga. (16″ long)........1209
Tubing, ¾″ O.D. Copper, #20 Ga. (40″ long)........1209
Tubing, ¾″ O.D. Copper, # 20 Ga. (106½″ long).....1209
Tubing, Windshield Wiper Only.....................1800

Group

Union, Oil Pump Pipe...................................0107
Unit, Housing Oil Seal..................................1403
Unit, Jacket Tube and Jacket Tube Ball Bearing.....1403
Unit, Jacket Tube Ball Bearing......................1403
Unit, Lever Shaft and Roller Bearing—with Nut and Lockwasher 1403
Unit, Sealed—6-V 0607
Unit, Stud Roller Bearing..............................1403
Units, Lever Shaft, Stud—Roller Bearing, Nut and Lockwasher Assembly 1403

Valve, Air Application, and Stop Assembly.........1205
Valve, Air Supply, Assembly..........................1205
Valve, Brake Application 1205
Valve, Carburetor Check 0301
Valve, Carburetor Fuel (#55)........................0301
Valave, Carburetor Power Jet (#13)...............0301
Valve, Compression Assembly 1603
Valve, Double Check 1205
Valve, Exhaust 0105
Valve, Exhaust (Round Groove Type Latest Design)..0105
Valve, Exhaust (Tapered Groove Type Orig. Design)..0105
Valve, Fuel Pump 0302
Valve, ¼″ Gas Shut-Off..................................0300
Valve, Gas Tank ¾″ Brass (Walworth #95)........0300
Valve, Governor Butterfly 0305
Valve, Hand Control 1205
Valve, Intake 0105
Valve, Intake (Round Groove Type Latest Design)....0105
Valve, Intake (Tapered Groove Type Original Design).0105
Valve, Journal Relief 0902
Valve, Manifold Hot Spot...............................0108
Valve, Oil Filter Pressure Relief.....................0107
Valve, Oil Pressure Regulator.........................0107
Valve, Oil Relief 0107
Valve, Oil Relief 0110
Valve, Pressure Regulating 1801
Valve, Quick Release 1205
Valve, Relay 1205
Valve, Relief Assembly 1603
Valve, Safety 1205
Valve, Single Check 1209
Valve, Tire Inflating, Assembly 1205
Ventilator, Roof 1800
Venturi, Carburetor Main (#33).....................0301
Viscometer (Oil Pan Unit)..............................0107
Viscometer (Visco-Meter Corp.) 0605

Washer 0603
Washer, Advance Control Screw—Bottom..........0603
Washer, Advance Control Screw Spring............0603
Washer, Advance Control Screw—Top..............0603
Washer, Armature Shaft Spacer......................0602
Washer, Bendix Shaft Spacer.........................0602
Washer, Bevel Pinion Nut..............................1003
Washer, Brake Spring 1900
Washer, Brush Holder Rivet...........................0602
Washer, Brush Holder Rivet Insulating.............0602
Washer, Camshaft Gear 0106
Washer, Camshaft Gear Nut Lock....................0106
Washer, Camshaft Gear Thrust 0106
Washer, Camshaft Idler Gear Thrust................0106
Washer, Carburetor Air Shutter Shaft Thrust.......0301
Washer,* Carburetor Cap Jet Fibre.................0301
Washer,* Carburetor Channel Screw Fibre.........0301
Washer,* Carburetor Compensator Jet Fibre.......0301
Washer, Carburetor Discharge Tube Fibre..........0301
Washer,* Carburetor Filter Plug Fibre..............0301
Washer,* Carburetor Fuel Valve Fibre...............0301
Washer,* Carburetor Main Jet Adjusting Fibre......0301
Washer,* Carburetor Main Jet Fibre..................0301
Washer,* Carburetor Power and Accelerator Jet Fibre.0301

ALPHABETICAL INDEX—MASTER PARTS LIST

Part	Group
Washer, Carburetor Swivel	0301
Washer,* Carburetor Union Body Fibre	0301
Washer, Clutch Flywheel Ring Stud Nut Lock	0202
Washer, Clutch Pres. Plate Retracting Spring—Ret.	0202
Washer, Countershaft Front Bearing Locating	0703
Washer, Countershaft Front Bearing Lock Nut	0703
Washer, Countershaft Front Bearing Thrust	0703
Washer, Countershaft Rear Bearing Lock Nut	0703
Washer, Countershaft Rear Bearing Retainer	0703
Washer, Cylinder Water Jacket Pet Cock Cap Screw	0101
Washer, Cylinder Water Jacket Pet Cock Cap Screw—Lead	0101
Washer, Declutching Shaft End	0803
Washer, Differential Bevel Side Gear Thrust	1002
Washer, Differential Bevel Side Pinion Thrust	1002
Washer, Differential Bevel Side Pinion Thrust	1103
Washer, Differential Bevel Side Pinion Thrust	1104
Washer, Differential Cap Screw Lock	1103
Washer, Differential Pinion Gear Thrust	1002
Washer, Differential Pinion Gear Thrust	1103
Washer, Differential Pinion Thrust	1103
Washer, Differential Pinion Thrust	1104
Washer, Differential Side Gear Thrust	1002
Washer, Differential Side Gear Thrust	1103
Washer, Differential Side Gear Thrust	1104
Washer, Disc Retainer	1900
Washer, Distributor Gear Spacer	0603
Washer, Distributor Coupling	0603
Washer, Driven Shaft	0803
Washer, Drive Pinion Bearing Lock Nut	1105
Washer, Drum Shaft	1900
Washer, Fan Bearing Cork	0503
Washer, Fan Bearing Cork Retaining	0503
Washer, Fan Bracket Clamp—Front	0503
Washer, Fan Bracket Clamp Nut Lock	0503
Washer, Fan Bracket Clamp—Rear	0503
Washer, Fan Drive Pulley Lock Nut	0503
Washer, Fan Spindle Nut—Front	0503
Washer, Felt	0501
Washer, Felt	0601
Washer, Felt	0603
Washer, Felt (D.E.)	0602
Washer, Field Terminal Stud	0602
Washer, Flat Steel (⅜")	1800
Washer, Front Brake Shaft	1203
Washer, Front Brake Shoe "C"	1203
Washer, Front Spring Stud Nut Lock	1602
Washer, Front Wheel Bearing Adjusting Nut	1301
Washer, Front Wheel Bearing Felt	1301
Washer, Front Wheel Bearing Lock	1301
Washer, Fuel Pump Diaphragm Alignment	0302
Washer, Fuel Pump Lower Diaphragm Protector	0302
Washer, Fuel Pump Rocker Arm Pin	0302
Washer, Fuel Pump Top Cover Screw Lock	0302
Washer, Fuel Pump Upper Diaphragm Protector	0302
Washer, Gearshift Lever Pivot	0706
Washer, Gear Shaft Bearing Lock	1003
Washer, Gear Thrust	1900
Washer, Generator—Felt	0101
Washer, Idler Shaft	0804
Washer, Idler Shaft Gear Thrust	0106
Washer, Idler Shaft Rear Bearing	0804
Washer, Insulating	0601
Washer, Intake and Exhaust Manifold Attaching Stud	0108
Washer, #6 Lock	0601
Washer, #8 Lock	0601
Washer, #10 Lock	0601
Washer, #14 Lock	0601
Washer, ⅞" Lock	0601
Washer, .669 Lock	0601
Washer, Main Bearing Center Thrust	0102
Washer, Main Bearing Thrust—Center	0102
Washer, Main Drive Gear Idler Bearing	0802
Washer, Mainshaft Drive Thrust	0704
Washer, Mainshaft Overdrive Gear Retaining	0702
Washer, Main Shaft Power Takeoff Clutch	0802
Washer, Main Shaft Rear Bearing	0802
Washer, Mainshaft Rear Bearing Lock Nut	0703
Washer, Mainshaft Second Speed Gear Locating	0702
Washer, Mainshaft Third Speed Gear Locating	0702
Washer, Motor Terminal Stud Insulation (3/32")	0602
Washer, Oil Pump Drive Gear	0107
Washer, Oil Thrower	0602
Washer, Oil Trough	1103
Washer, Pinion Bearing Lock	1003
Washer, ¼ Plain	0601
Washer, ½" Plain Steel	0601
Washer, #10 Plain	0601
Washer, Plain	1201
Washer, Power Take-Off Adapter Gear Thrust	1911
Washer, Pronged	1403
Washer, Rear Brake Shaft	1203
Washer, Rear Brake Shoe Anchor Pin "C"	1203
Washer, Rear Brake Slack Adjuster	1203
Washer, Rear Brake Slack Adj. Retaining, Assy.	1203
Washer, Rear Hub Bearing Felt Retainer	1301
Washer, Rear Spring Shackle—Inner	1602
Washer, Rear Spring Stud	1602
Washer, Rear Wheel Bearing	1301
Washer, Rear Wheel Bearing Felt	1301
Washer, Rear Wheel Bearing Felt Retainer	1301
Washer, Rear Wheel Bearing Lock	1301
Washer, Rear Wheel Bearing Nut	1301
Washer, Retainer—Mainshaft 3rd Speed Gear Roller Bearing	0702
Washer, Reverse Gear Needle Bearing	0703
Washer, Sector Spacer	1201
Washer, Shackle Bracket	1602
Washer, Shim—.005 Thick	0603
Washer, Shim—.010 Thick	0603
Washer, Sleeve Dust Cap Felt	0902
Washer, Sleeve Dust Cap Steel	0902
Washer, Spacer (C.E.)	0602
Washer, Spacer (Between Gear and Motor Drive)	0602
Washer, Spacer (Between Housing and Gear)	0602
Washer, Spacer Support	1603
Washer, Spacer (Under Weight Plate)	0603
Washer, Spacer (.626 I.D. x 1 1/16" O.D. x 3/32")	0602
Washer, Spur Pinion Bearing	1105
Washer, Steady Bearing Housing Lock	0902
Washer, Steering Knuckle Felt	1006
Washer, Terminal Screw	0603
Washer, Terminal Screw Bushing	0603
Washer, Terminal Screw Insulation	0603
Washer, Thrust	0603
Washer, Thrust	1911
Washer, Transmission Driveshaft Lock Nut	0703
Washer (Use with 3TF321A Gear)	0703
Washer, Valve Cover Screw	0105
Washer, Valve Spring	0110
Washer, Water Pump Head Screw Copper	0503
Washer, Water Pump Packing	0503
Washer, Weight	0603
Wedge, Field Terminal	0602
Weight	0603
Weight Assembly	0603
Weights, Governor	0305
Welded, Deck Plate Assembly	1800
Welded, Endgate, Assembly	1800
Well, Carburetor Progressive	0301
Wheel, Budd (B-45520)	2300
Wheel, Budd (#44470—L-Rim)	2300
Wheel, Budd (Type "L" #33287)	2300
Wheel, Front and Rear (Budd)	1301
Wheel, Front and Rear (Budd-L-Rim)	1301
Wheel, Steering	1404

ALPHABETICAL INDEX—MASTER PARTS LIST

	Group
Wick, Cam Sleeve Felt	0603
Wick, Clutch Release Sleeve Oil	0205
Wick, Clutch Release Trunnion Oil	0203
Wick, Crankshaft Oil, Assembly	0102
Wick, Felt	0603
Wick, Hand Brake Lever Oiler	1201
Wick, Magazine Oiler	0204
Wick, Oil—and Spring (C.E.)	0602
Wick, Oil (C.E.)	0602
Wick, Oil (D.E.)	0602
Wick, Oil (in Gear Housing)	0602
Window, Rear, and Screen	1800
Windshield Assembly—L. H.	1800
Windshield Assembly—R. H.	1800
Windshield and Brackets	1800
Windsplit, Radiator Filler Door	1800
Wiper, Windshield Assembly Complete (Trico)	1801
Wire, Fan Bearing Cork Retainer Lock	0503
Wire, Choke Control, Assembly	0303
Wire, High Tension, Assembly	0604
Wire, Ignition	0603
Wire, Ignition—#1	0604
Wire, Ignition—#2	0604
Wire, Ignition—#3	0604
Wire, Ignition—#4	0604
Wire, Ignition—#5	0604
Wire, Ignition—#6	0604
Wire, Jumper—Coil to Distributor	0604
Wires, Ignition, Assembly	0604
Wire, Throttle Control, Assembly	0303
Wire, Water Pump Paddle Seal Snap	0503
Wiring, Spotlight—Dash to Light Location	1800
Worm Assembly	1900
Wrench	1603

	Group
Wrench—$\frac{3}{8}''$—$\frac{7}{16}''$	2300
Wrench—$\frac{1}{2}''$—$\frac{19}{32}''$	2300
Wrench—$\frac{3}{4}''$—$\frac{7}{8}''$	2300
Wrench—$\frac{9}{16}''$—$\frac{11}{16}''$	2300
Wrench—$\frac{5}{8}''$—$\frac{25}{32}''$	2300
Wrench, Auto	2300
Wrench—Crescent Type	2300
Wrench, Front and Rear Wheel Bearing Nut	1301
Wrench, Front and Rear Wheel Nut	1302
Wrench, Front Wheel Adjusting	1301
Wrench, Front Wheel Bearing	2300
Wrench, Front Wheel Bearing Adjusting Nut	2300
Wrench, Front Wheel Bearing Nut	1301
Wrench, Mainshaft Drive Thrust Nut	0704
Wrench, Oil Pressure Crow Foot	0107
Wrench, Oil Pressure Crow Foot (Hercules #2268-A)	2300
Wrench, Oil Pressure "T"	0107
Wrench, Oil Pressure "T" (Hercules #X5800)	2300
Wrench, Rear Wheel Bearing Adjusting Nut	2300
Wrench, Rear Wheel Bearing Nut	1301
Wrench, Socket (Rear Wheel)	2300
Wrench, Spark Plug	2300
Wrench, Water Pump Nut	2300
Wrench, Water Pump Spanner	2300
Wrench, Wheel Bearing Adjusting Nut	2300
Yoke, Clutch, Assembly	1900
Yoke, Fixed Stub	0902
Yoke, Fixed Stub (3″ Tube)	0902
Yoke, Gearshift Connecting Rod (Lever End)	0706
Yoke, Governor	0305
Yoke, Release Lever Adjusting	0205
Yoke, Shift	1911
Yoke, Slip Joint Sleeve	0902
Yoke, Valve Plunger	0110

NUMERICAL INDEX—MASTER PARTS LIST

Part No.	Group	Part No.	Group	Part No.	Group	Part No.	Group
1TE2-10	2300	2H2-39	0103	2N2-127	0107	2BF2-231	0104
1UV2-10	2300	2UU0-39	0103	2UU0-127	0107	2BG2-231	0104
1ZN2-10	2300	2C2-40	0110	2N2-128	0107	2D2-231	0104
1BL0-37	0102	2BL0-43	0106	2UU0-128	0107	2DE2-231	0104
1UU2-37	0102	2UU0-43	0106	2N4-129	0107	2DF2-231	0104
		2BL0-44	0106	2UG0-129	0107	2DG2-231	0104
2B7-01	0102	2UU0-44	0106	2UU0-129	0107	2UUA0-231	0104
2BL0-01	0102	2UUA0-44	0106	2UUA0-129	0107	2UUB0-231	0104
2UU0-01	0102	2T2-44B	0106	2D0-130	0101	2UU0-231A	0104
2SA7-01B	0102	2T2-45A	0106	2J0-130	0101	2T2-239	0110
2SA7-01C	0102	2SA2-48	0101	2BL0-131	0107	2BL2-239A	0110
2UU0-02	0101	2T2-48	0101	2N4-131	0107	2D3-244	0110
2BN0-06	0109	2BL0-49	0110	2UU0-131	0107	2G2-245	0110
2UU0-06	0109	2M1-49	0110	2BL3-132	0110	2G2-246	0110
2RM5-07	0101	2UU0-49	0105	2NJ4-132	0110	2B2-248	0104
2SA5-07	0101	2NB2-49A	0110	2UG0-138	0107	2D2-248	0104
2UU0-07	0101	2SA3-50	0106	2UU0-138	0107	2UU0-256	0101
2SA2-08	0108	2BL0-52	0105	2SA5-141B	0108	2BL2-277	0110
2B2-10	0105	2T2-52	0105	2UU0-142	0108	2BL2-278	0110
2BL0-10	0105	2UU0-52	0105	2SA7-142B	0108	2N2-295	0107
2UU0-10	0105	2UG0-54	0106	2SA7-142C	0108	2SA0-310A	0106
2D2-10A	0105	2UU0-54	0106	2UU0-143	0108	2A0-318	0107
2SA2-10C	0105	2SA6-54B	0106	2T2-145	0108	2J0-318	0107
2C2-12A	0108	2SA3-61	0101	2BL0-148	0107	2UG0-318	0107
2C2-13	0108	2URL4-61A	0101	2UU0-148	0107	2UU0-318	0107
2UU0-13	0108	2SA3-61B	0101	2T2-149	0101	2A0-319	0107
2B2-15	0104	2SA2-62	0106	2N0-150	0107	2J0-319	0107
2T2-15	0104	2SCH2-62	0106	2UG0-150	0107	2UU0-319	0107
2T2-15A	0104	2BL0-65	0101	2UU0-150	0107	2UU0-319	1205
2T2-15A	0110	2UU0-65	0105	2T2-151	0101	2UUA0-319	0107
2N2-19	0107	2SA3-65C	0101	2UU4-152	0110	2UUA0-319	1205
2B4-20	0104	2BL0-67	0101	2BL4-152A	0110	2B0-320	0101
2BL0-20	0104	2T2-67	0101	2B6-153	0110	2SA0-320	0101
2D4-20	0104	2UU0-69	0101	2DK6-153	0110	2UG0-320	0101
2UU0-20	0104	2UU0-72	0105	2UG0-153	0110	2BL0-329	0110
2BL0-21	0105	2SA3-72A	0101	2UU0-153	0109	2M1-329	0110
2M2-21	0105	2T2-74	0101	2C2-158A	0108	2UU0-329	0105
2UU0-21	0105	2UU0-74	0101	2UG0-159	0108	2NB2-329A	0110
2UUA0-21	0105	2BL0-76	0106	2URM2-159	0108	2C2-334	0108
2BL0-23	0102	2SA3-76	0106	2UU0-159	0108	2SA2-342	0108
2UU0-23	0102	2UU0-76	0106	2CF2-159A	0101	2URM3-342	0108
2SA2-23B	0102	2UU0-80	0100	2SA2-159A	0108	2Y1-345	0701
2BL0-24	0110	2SA3-81	0101	2UG0-160	0110	2BN0-353	0103
2T2-24A	0110	2URL3-81	0101	2UG0-161	0107	2D2-353	0103
2BN0-25	0103	2CB2-92	0106	2UU0-161	0107	2UU0-353	0103
2D2-25	0103	2UU0-92	0107	2SA4-161A	0110	2UU0-353S	0103
2UU0-25	0103	2UUA0-92	0107	2BL0-164	0104	2Y1-368	0110
2UU0-25S	0103	2T2-92A	0106	2SA2-164	0101	2B2-371	0102
2BN0-26	0103	2T2-93	0101	2SA2-165A	0110	2D2-371	0102
2D2-26	0103	2SA2-96A	0101	2TH0-168	0110	2T2-371	0102
2UU0-26	0103	2T2-98	0108	2T2-168A	0110	2B2-372	0102
2D2-26A	0103	2T3-98	0108	2CB2-168B	0110	2D2-372	0102
2BN0-27	0103	2G2-99	0110	2SA2-169	0101	2SA2-372	0102
2D4-27	0103	2A2-104	0107	2UU0-178	0101	2B2-373	0102
2UU0-27	0103	2T2-104	0107	2SA4-181A	0101	2D2-373	0102
2D4-27A	0103	2T2-104	0110	2SA4-181B	0101	2SA2-373	0102
2T2-28	0110	2UU0-104	0107	2BL0-183	0106	2B2-374	0102
2N2-29	0107	2BLW0-105	0110	2T3-183	0106	2D2-374	0102
2T2-31	0110	2SA2-105	0107	2UU0-183	0106	2SA2-374	0102
2BL0-33	0106	2URL3-105B	0107	2BL0-195	0107	2UU0-377	0107
2SA3-33	0106	2BL0-109	0104	2UG0-210	0107	2B3-382	0102
2UU0-33	0106	2B4-116	0104	2UP3-210	0107	2D3-382	0102
2N2-34	0107	2BL0-116	0104	2UU0-210	0107	2BL0-397	0102
2UU0-34	0107	2SA2-116A	0104	2URL3-210A	0107	2UU0-397	0102
2B6-35	0101	2B2-118	0104	2BL0-213	0106	2T2-397A	0102
2D6-35	0101	2BL0-118	0104	2SA2-213	0106	2B2-401	0102
2UG0-35	0101	2T2-118	0104	2UU0-213	0106	2D2-401	0102
2URM6-35	0101	2UU0-118	0104	2SA2-213A	0106	2SA2-401	0102
2UU0-35	0101	2BL0-122	0107	2BL0-214	0109	2UU0-401	0102
2UG0-36	0110	2UU0-122	0107	2T3-214	0110	2B2-402	0102
2UU0-36	0105	2BL0-126	0107	2UU0-214	0109	2D2-402	0102
2T2-38	0110	2N2-126	0107	2T2-221	0108	2T2-402	0102
2B2-39	0103	2UU0-126	0107	2B2-231	0104	2UU0-402	0102
2BN0-39	0103	2BL0-127	0107	2BE2-231	0104	2T0-410	0102

1

NUMERICAL INDEX—MASTER PARTS LIST

Part No.	Group	Part No.	Group	Part No.	Group	Part No.	Group
2T0-410	0110	2D4-530	0104	2UG0-651	0102	2T2-855	0603
2UU0-447	0103	2UU0-535	0108	2UU0-651	0102	2URM0-860	0108
2UUA0-447	0103	2UUA0-535	0108	2BL0-652	0106	2D4-862	0103
2UUB0-447	0103	2SA2-530A	0108	2UU0-652	0106	2D4-862A	0103
2SA2-461	0108	2SA2-535B	0108	2B3-653	0102	2D2-863	0103
2UU0-463	0107	2B4-540	0104	2D2-653	0102	2UU0-863	0103
2UU0-464	0107	2D0-540	0104	2SA2-653	0102	2UU0-863S	0103
2UUA0-464	0107	2BL2-542	0110	2UG0-653	0102	2BN0-864	0103
2BL0-477	0107	2BL3-543	0110	2UU0-653	0102	2D4-864	0103
2N2-477	0107	2BL2-544	0110	2B3-655	0102	2D4-864A	0103
2UG0-477	0107	2UU0-544	0109	2D2-655	0102	2UU0-869	0103
2UG0-477	0110	2A2-546	0107	2SA2-655	0102	2UU0-869S	0103
2UU0-477	0107	2BL0-552	0102	2UG0-655	0102	2BN0-873	0103
2T2-477B	0107	2BL0-553	0104	2UU0-655	0102	2D4-873	0103
2T2-477B	0110	2UU0-553	0104	2B3-657	0102	2UU0-873	0103
2B2-489	0102	2UUA0-553	0104	2D2-657	0102	2D4-873A	0103
2D2-489	0102	2UU0-554	0103	2SA2-657	0102	2BN0-875	0103
2SA2-489	0102	2UU0-554S	0103	2UG0-657	0102	2D2-875	0103
2UU0-489	0102	2T2-559A	0104	2UU0-657	0102	2UU0-875	0103
2SA2-492	0106	2B2-571	0101	2T2-659	1911	2UU0-875S	0103
2B2-499	0102	2RE2-571B	0101	2UU3-680A	0107	2BN0-877	0103
2D2-499	0102	2A0-575	0107	2SA2-687	0102	2D2-877	0103
2SA2-499	0102	2H2-575	0107	2TH0-698	0106	2UU0-877	0103
2T2-508	0110	2URL2-575	0107	2UG0-698	0106	2UU0-877S	0103
2TE2-508	0110	2UU0-575	0107	2TH0-699	0106	2BN0-878	0103
2B2-509	0102	2UU0-578	0101	2BL0-710	0105	2D2-878	0103
2D2-509	0102	2UU0-578	0107	2D2-710	0105	2UU0-878	0103
2SA2-509	0102	2UUA0-578	0107	2UU0-710	0105	2UU0-878S	0103
2UU0-509	0102	2UU0-583	0107	2SA2-710A	0105	2D2-879	0103
2B4-510	0104	2UUA0-583	0107	2URL5-727	0107	2UU0-879	0103
2D4-510	0104	2T4-592	0107	2UU5-727	0107	2UU0-879S	0103
2B2-511	0102	2UG4-592	0107	2UG0-728	0107	2UU0-881	0103
2D2-511	0102	2BL0-598	0107	2UU0-728	0107	2UU0-881S	0103
2SA2-511	0102	2UG0-598	0107	2UG0-729	0107	2BN0-888	0103
2UU0-511	0102	2UG0-598	0110	2UU0-729	0107	2UU0-888	0103
2B2-512	0102	2UU0-598	0107	2UG0-736	0107	2UU0-888S	0103
2D2-512	0102	2T2-598C	0107	2UG0-736	0110	2BN0-889	0103
2T2-512	0102	2T2-598C	0110	2UG0-737	0107	2UU0-889	0103
2UU0-512	0102	2UG0-599	0107	2UU0-737	0107	2UU0-889S	0103
2B2-513	0102	2UG0-599	0110	2UG0-743	0107	2BN0-891	0103
2D2-513	0102	2T2-599A	0107	2UU0-743	0107	2UU0-891	0103
2T2-513	0102	2T2-599A	0110	2UG0-745	0107	2G0-910	0110
2UU0-513	0102	2T2-602	0107	2UU0-745	0107	2RL0-910	0110
2B2-514	0102	2T2-602	0110	2A0-746	0107	2ZB0-910	0110
2D2-514	0102	2UGO-602	0107	2E0-746	0107	2BL0-911	0110
2SA2-514	0102	2UG0-602	0110	2J0-746	0107	2J0-911	0110
2UU0-514	0102	2UU0-602	0107	2UG0-746	0107	2SA0-911B	0110
2B2-515	0102	2A0-606	0107	2UU0-746	0107	2D0-912	0101
2D2-515	0102	2A0-607	0107	2TH0-748	0110	2J0-912	0101
2SA2-515	0102	2UU0-607	0107	2BL0-774	0106	2URM0-912	0101
2UU0-515	0102	2A0-608	0107	2UU0-774	0106	2N0-922	0107
2B2-516	0102	2J0-608	0107	2TH0-775	0106	2SA2-937	0110
2D2-516	0102	2A0-611	0107	2BL0-776	0106	2B2-942	0104
2SA2-516	0102	2J0-611	0107	2UU0-776	0106	2CB2-942	0104
2UU0-516	0102	2UG0-611	0107	2UU0-782	0109	2D2-942	0104
2B2-517	0102	2UU0-611	0107	2TH0-783	0109	2BL0-946	0110
2D2-517	0102	2T2-616	0106	2E0-807	0107	2T2-946	0107
2SA2-517	0102	2T2-618	0106	2J0-807	0107	2T2-946A	0107
2UU0-517	0102	2SA2-633A	0108	2UBB0-808	0107	2T2-951	0110
2B4-520	0104	2SK3-634	0107	2UGA0-809	0101	2BL0-956	0107
2D4-520	0104	2T2-634	0107	2A0-828	0107	2UU0-956	0107
2BL0-521	0102	2T3-642	0102	2E0-828	0107	2UUA0-956	0107
2T2-521	0102	2T3-642	0110	2UU0-828	0107	2N2-956C	0107
2UU0-521	0102	2CB2-643	0102	2A0-829	0107	2N2-956C	0110
2UUA0-521	0102	2T2-643	0102	2J0-829	0107	2SD0-957	0110
2BL0-522	0102	2SA2-644	0102	2UG0-829	0107	2UG0-957	0110
2T2-522	0102	2T2-648A	0106	2UU0-829	0107	2UU0-957	0107
2U0-522	0102	2T2-649A	0106	2UG0-831	0107	2UU0-962	0102
2UUA0-522	0102	2DB0-650	0305	2UG0-831	0110	2UUA0-962	0101
2BL0-523	0102	2UK0-650	0305	2UU0-831	0107	2SA2-963A	0108
2SA2-523	0102	2UU0-650	0305	2SA2-832	0706	2SA2-964	0108
2UU0-523	0102	2B3-651	0102	2TH0-837	0107	2BL0-975	0107
2UUA0-523	0102	2D2-651	0102	2N0-845	0107	2N2-975	0107
2B4-530	0104	2T2-651	0102	2BLA0-855	0603		

NUMERICAL INDEX—MASTER PARTS LIST

Part No.	Group	Part No.	Group	Part No.	Group	Part No.	Group
2UU0-975	0107	2UU0-1355	0109	3BLA0-11	0706	3T0-79	0703
2N2-976	0107	2UU0-1356	0101	3T1-11	0706	3S2-82A	0706
2T0-986	0110	2UU0-1356	0102	3RL3-15	0706	3S2-82A	1201
2T2-988	0102	2UU0-1357	0101	3UBL0-15	0706	3S3-82A	1201
2SK2-998	0110	2UU0-1357	0109	3UK0-15	0706	3RL3-83	0706
2T2-998	0110	2UU0-1358	0101	3TF4-15C	0706	3UBL0-83	0706
2B2-1001	0101	2UU0-1358	0109	3RL4-16	0701	3UK0-83	0706
2BL0-1001	0101	2UU0-1359	0101	3UBL0-16	0701	3UU3-83	0706
2D2-1001	0101	2UU0-1361	0101	3UK0-16	0701	3RL3-84	0706
2UU0-1001	0101	2UU0-1362	0102	3UU5-16	0701	3UBL0-84	0706
2SA2-1002	0108	2UU0-1363	0102	3BL0-17	0701	3UK0-84	0706
2C3-1003	0108	2UU0-1364	0102	3BLA0-17	0701	3UU3-84	0706
2RM3-1003A	0108	2UU0-1365	0104	3D3-17	0701	3RL3-85	0706
2TE2-1008A	0101	2UU0-1367	0105	3D3-17	0706	3UBL0-85	0706
2TE2-1008B	0101	2UU0-1368	0105	3TF3-17A	0701	3UK0-85	0706
2TE2-1008B	0204	2UU0-1369	0105	3BLA0-18	0702	3UNF3-85	0706
2TE2-1008C	0204	2UU0-1371	0105	3UK0-18	0702	3UU3-85	0706
2SD2-1011	0108	2UU0-1372	0105	3RL3-18A	0702	3BLA0-87	0706
2SD2-1011B	0108	2UU0-1373	0105	3BL0-21	0702	3RL3-87	0706
2DK2-1051	0110	2UU0-1373	0503	3BLA0-21	0702	3UK0-87	0706
2N2-1051	0110	2UU0-1375	0105	3RL3-21	0702	3DF3-87A	0706
2TE3-1068	0102	2UUA0-1375	0105	3TF3-21A	0702	3UB0-88	0706
2UU0-1109	0104	2UUA0-1376	0101	3BLA0-22	0702	3URL4-88A	0706
2UG2-1111	0603	2UU0-1377	0101	3RL4-22	0702	3TF3-93A	0702
2UK3-1111	0603	2UU0-1378	0109	3UK0-22	0702	3B2-101	1105
2UP2-1111	0603	2UU0-1385	0106	3TF4-22A	0702	3UK0-102	0702
2D3-1131	0101	2UU0-1387	0102	3D3-24	0706	3TF4-102A	0702
2UD3-1150	0107	2UU0-1387	0106	3DF3-24	0706	3B2-103	1105
2SA0-1160	0110	2UU0-1388	0107	3UDF3-24	0706	3BLA0-103	0704
2HD0-1160A	0110	2UU0-1388	0109	3BLA0-26	0702	3DF2-103	0703
2SA4-1161	0108	2UU0-1391	0107	3UK0-26	0702	3Y1-103	0704
2URM4-1161	0108	2UU0-1393	0103	3RL3-26A	0702	3RL0-105A	0205
2D2-1162	0108	2UU0-1394	0107	3TF3-26A	0702	3S2-106	0704
2UU2-1185B	0110	2UU0-1394	0503	3BLA0-27	0702	3Y2-106	0704
2UU0-1188	0103	2UU0-1395	0107	3TFB3-27	0702	3BLA0-108	0702
2UU0-1188S	0103	2UU0-1399	0107	3BLA0-29	0704	3Y1-111A	0706
2BL0-1189	0104	2UU0-1402	0107	3RL4-29	0704	3D3-112	0703
2UU0-1200	0101	2UU0-1403	0107	3UK0-29	0704	3TF3-112A	0703
2UU0-1209	0503	2UU0-1404	0107	3TF4-29B	0704	3RL0-114	0201
2BL0-1212	0107	2UU0-1405	0107	3DA5-30	0200	3ZT0-114	0201
2BL0-1213	0107	2UU0-1406	0107	3UU5-30	0200	3UN2-119B	0203
2UU0-1213	0107	2UU0-1407	0107	3RL4-30A	0200	3T0-124B	0205
2UUA0-1213	0107	2UU0-1408	0102	3BLA0-31	0704	3A0-125	0203
2BL0-1214	0107	2UU0-1409	0102	3RL4-31	0704	3B3-125	0203
2BL0-1214	2300	2UU0-1411	0102	3UK0-31	0704	3RM0-125	0203
2BL0-1215	0107	2UU0-1416	0107	3TF4-31A	0704	3RL0-128	0205
2BL0-1215	2300	2UU0-1417	0107	3D3-35	0706	3RL0-132A	0205
2BL0-1216	0108	2UU0-1418	0107	3TFB4-35	0706	3A0-136	0205
2BL0-1221	0102	2UU0-1421	0107	3UK0-35	0706	3A0-137B	0705
2BL0-1225	0102	2UU0-1424	0107	3D2-36	0703	3A0-137B	0804
2UU0-1225	0102	2UU0-1425	0107	3T2-37	1003	3DW0-140	0902
2BL0-1268	0110	2UU0-1426	0108	3T2-37	1105	3ZKE0-140	0902
2UU0-1268	0109	2UU0-1427	0108	3BLA0-41	0703	3ZUNC0-140	0902
2UU2-1272	0107	2UU0-1428	0108	3RL3-41	0703	3ZC0-140A	0902
2UU3-1274	0107	2UU0-1429	0108	3UBT0-41	0703	3ZD0-140D	0902
2UU0-1327	0107	2UU0-1431	0108	3TF3-41A	0703	3BL0-142	0703
2UG0-1335	0107	2UU0-1432	0108	3RL0-43	0205	3BLA0-142	0703
2UU0-1335	0107	2UU0-1433	0108	3C0-45	0201	3A2-145	0205
2UU0-1338	0101	2UU0-1434	0108	3DA0-45	0201	3A2-145A	0203
2UUA0-1338	0101	2UU0-1435	0108	3RL0-45	0201	3TF2-146	0702
2UU0-1339	0101	2UU0-1436	0108	3NY0-46	0203	3BLA0-147	0703
2UU0-1341	0101	2UU0-1442	0109	3RL0-46	0205	3RL0-147	0703
2UUA0-1341	0101	2UUB0-1447	0107	3B0-46B	0203	3TF0-147	0703
2UU0-1342	0102	2UU0-1448	0102	3RL0-48A	0205	3DC0-150A	0902
2UU0-1343	0102			3T2-61	0703	3ED0-150A	0902
2UU0-1344	0102	3D3-04B	0703	3TF2-61	0703	3D2-152	0706
2UU0-1344	0104	3S3-04B	0703	3T2-62	0703	3DF2-152	0706
2UU0-1344	0107	3BLA0-06	0704	3TF2-62	0703	3B3-154	0701
2UU0-1345	0101	3TFB2-06	0704	3URL3-64	0706	3RL3-154	0701
2UU0-1346	0101	3UK0-06	0704	3RE0-67	1911	3UU3-154	0701
2UU0-1348	0101	3D2-06A	0704	3BLA0-70	0702	3BLA0-156	0703
2UU0-1349	0101	3D2-07	0702	3RL0-70	0702	3D2-156	0703
2UU0-1355	0101	3UK0-07	0702	3UU0-70	0702	3T2-156	0703
2UU0-1355	0107	3BL0-11	0706	3BLA0-79	0703	3UK0-156	0703

3

NUMERICAL INDEX—MASTER PARTS LIST

Part No.	Group	Part No.	Group	Part No.	Group	Part No.	Group
3B3-158	0701	3CD2-291	0902	3B0-682	0202	3TF3-976A	0703
3B3-158	1911	3SA2-293	0902	3B0-683	0202	3BLA0-977	0703
3RL3-158	0701	3SA2-294	0902	3ZT0-683	0202	3UK0-977	0703
3UU3-158	0701	3B3-298	0204	3UU0-685	0202	3TF3-977A	0703
3BLA0-162	0703	3M2-301A	0204	3A0-689	0902	3B0-981	0202
3D3-162A	0703	3CD2-304	0902	3DC0-689	0902	3B0-982	0203
3TF3-162A	0703	3UB0-309	0706	3ED0-689	0902	3NY0-982	0203
3BM2-169A	0203	3UK0-309	0706	3LE0-689	0902	3BLA0-994	0703
3Y1-173	0706	3URL6-309	0706	3ZD0-689	0902	3BL0-1003	0703
3RL3-179	0702	3T1-345	0706	3LE0-689A	0902	3BLA0-1003	0703
3TF3-179A	0702	3UDF2-345	0706	3RG0-694A	0902	3UU0-1004	0203
3D2-187A	0706	3UU2-345	0706	3ZC0-694A	0902	3BL0-1007	0706
3DF4-188	0706	3RL3-346	0706	3DC0-695	0902	3BL0-1008	0706
3RL3-188	0706	3DF4-346A	0706	3RG0-695A	0902	3BL0-1009	0706
3RL3-189	0706	3DF3-347	0703	3CD3-696	0902	3BL0-1011	0706
3UNF3-189	0706	3TF3-347A	0703	3S2-707A	1201	3BLA0-1011	0706
3UU3-189	0706	3D2-351	0706	3DC0-728	0902	3BL0-1012	0702
3BL0-191	0706	3D3-352A	0703	3RG0-728	0902	3BLA0-1012	0703
3BLA0-191	0706	3D3-353	0706	3UN0-728	0902	3BL0-1013	0702
3D2-191	0706	3UU4-353	0706	3DC0-729	0902	3BL0-1014	0702
3DF2-191	0706	3D2-355	0706	3RG0-729	0902	3BL0-1015	0702
3D2-193	0706	3TA4-381	1201	3UN0-729	0902	3BL0-1016	0702
3BL0-196	0705	3SA2-408	0902	3ZT0-749	0202	3BLA0-1016	0702
3D3-196	0703	3SA0-409	0902	3LD0-753	0902	3BLA0-1017	0702
3UK0-196	0705	3A2-413	0705	3LE0-753	0902	3BLA0-1018	0704
3DF3-196B	0703	3SA0-418	0902	3DD0-755	0902	3CA9-1020	0204
3Y2-197	0203	3BLA0-452	0703	3RL2-762	0703	3UU4-1020A	0204
3TF0-200A	0702	3B0-522	0202	3DA0-770	0201	3BL0-1031	0703
3AH0-201	0704	3NY0-522	0202	3RL0-770	0201	3BLA0-1031	0703
3AH3-201	0704	3B0-544	0202	3UU0-770	0201	3BLA0-1035	0706
3CA3-201	0704	3UU0-544	0202	3A0-780	0205	3UB0-1035	0706
3DM3-201	0803	3B0-559	0202	3RL0-781	0205	3UBL0-1036	0706
3EP3-201	1105	3B0-592	0202	3RL0-782	0205	3UK0-1036	0706
3ES3-201	0803	3UU0-622	0203	3RL0-783	0205	3BLA0-1037	0703
3ET3-201	0802	3UN2-625	0204	3RL0-784A	0205	3BLA0-1038	0703
3D2-215	0706	3A2-629	0706	3RL0-785	0201	3BLA0-1039	0703
3D2-216	0706	3BL0-629	0706	3RL0-790A	0205	3BLA0-1041	0703
3DC0-218	0902	3D2-629	0706	3RL0-800A	0202	3BLA0-1042	0702
3LD0-218	0902	3DF2-629	0706	3RL0-816	0701	3BLA0-1043	0702
3LE0-218	0902	3TF2-629	0706	3UU0-816	0701	3BLA0-1044	0702
3RG0-218	0902	3LD0-632	0902	3RL3-863	0702	3BLA0-1045	0702
3UB0-218	0902	3RG0-632	0902	3UU3-863	0702	3BLA0-1046	0702
3A0-219	0902	3UN0-632	0902	3RL0-889A	0205	3BLA0-1047	0702
3DC0-219	0902	3DC0-632A	0902	3B2-914	0702	3UB0-1048	0706
3DF0-219	0902	3CF0-633	0902	3C2-914	0702	3UB0-1049	0706
3RG0-219	0902	3RG0-633	0902	3D2-914	0702	3UB0-1051	0706
3DC0-221	0902	3UN0-635	0902	3E2-914	0702	3UB0-1052	0706
3DF0-221	0902	3RG0-635	0902	3BL0-916	0703	3UB0-1053	0706
3LD0-221	0902	3DC0-635A	0902	3BLA0-916	0703	3UB0-1054	0706
3LE0-221	0902	3RG0-635A	0902	3D2-916	0703	3UB0-1055	0706
3RG0-221	0902	3DF0-637	0902	3D2-917	0702	3UB0-1056	0706
3Y2-222A	1105	3LD0-637	0902	3DF2-923	0703	3UB0-1057	0706
3D2-224	0706	3RG0-637	0902	3DF2-924	0703	3UB0-1058	0706
3TF2-224A	0706	3ZC0-637	0902	3DF2-925	0706	3UK0-1058	0706
3BLA0-236	0705	3LE0-637A	0902	3DF2-927	0706	3UB0-1059	0706
3A0-243	0205	3ZD0-637A	0902	3DF2-928	0706	3TFB2-1071	0702
3B0-243A	0205	3RL0-642	0205	3TF3-948A	0702	3TFB2-1072	0702
3BLA0-248	0703	3RL0-644	0205	3BLA0-949	0702	3TFB2-1073	0703
3HD0-248	0703	3D2-645	0706	3RL3-949	0702	3TFB2-1074	0703
3D2-248A	0703	3DC0-647	0902	3BLA0-951	0702	3TFB2-1075	0703
3BLA0-251	0703	3RG0-647	0902	3TF2-951A	0702	3TFB2-1076	0703
3B2-261	1105	3GE0-647A	0902	3TF2-954	0702	3UU2-1129	0706
3B2-262	1105	3GE0-648	0902	3TF2-955	0702	3UU2-1131	0706
3RL0-270	0702	3UN0-648	0902	3TF2-956A	0702	3UU2-1133	0706
3UK0-270	0702	3B0-673	0203	3BL0-957	0702	3URL0-1350	0706
3BLA0-274	0702	3NY0-673	0203	3RL2-957	0704	3D0-1370	0705
3CKA2-285	0802	3NY0-674	0203	3S2-959	0704	3UBT0-1370	0705
3CKA2-285	0804	3UU0-674	0203	3UB0-961	0706	3UK0-1370	0705
3FK0-285	0803	3DA0-677	0202	3URL2-961	0706	3TF0-1410A	0702
3SA2-285	0803	3ZT0-677C	0202	3UNA3-962	0706	3TF0-1420A	0702
3SA2-285	0902	3DA0-679	0202	3URL2-968	0706	3RL0-1470	0201
3SA2-288	0902	3ZT0-679	0202	3CD4-970	0902	3BLA0-1540	0702
3CD2-289	0902	3DA0-681	0202	3CD4-970A	0902	3TFB3-1580	0702
3D0-290	0706	3ZT0-681C	0202	3DF4-971	0706	3URL0-1630	0706

4

NUMERICAL INDEX—MASTER PARTS LIST

Part No.	Group	Part No.	Group	Part No.	Group	Part No.	Group
4TG7-01C	1101	4UG0-55	1104	4UG0-96	1105	4SA2-139	1006
4DF2-06	1103	4C3-55A	1103	4UGA0-96	1003	4ZDK0-139	1006
4NFG0-06	1103	4C3-55A	1104	4UGA0-96	1105	4FA0-141	1105
4G0-07	1203	4GKW0-57	1002	4UH0-96	1003	4FB0-141	1105
4GEA0-10	1002	4GKW0-57	1103	4UH0-96	1105	4RL0-141	1003
4GEA0-10	1103	4GKW0-57	1104	4UHA0-96	1003	4RL0-141	1105
4NKA0-10	1002	4M4-57	1103	4UHA0-96	1105	4UG0-141	1301
4NKA0-10	1103	4M4-57	1104	4BLA0-96A	1003	4Y2-141	1105
4UG0-10	1002	4NKA0-57	1002	4BLA0-96A	1105	4Y2-141	1301
4UG0-10	1103	4NKA0-57	1103	4BLA0-98	1003	4FA0-142	1105
4UGA0-10	1002	4NKA0-57	1104	4BLA0-98	1103	4NFG0-142	1105
4UGA0-10	1103	4UG0-57	1002	4BLB0-98	1003	4NFG0-142	1301
4TF4-10C	1103	4UG0-57	1103	4GKW0-98	1003	4RL0-142	1003
4TF4-11	1102	4UG0-57	1104	4GKWA0-98	1003	4RL0-142	1105
4Y1-15	1301	4BLA0-58	1002	4NK0-98	1003	4BLA0-143	1003
4BLA0-18	1002	4BLA0-58	1103	4BLA0-98A	1003	4BLA0-143	1103
4BLA0-18	1103	4BLA0-58	1104	4BLA0-98A	1105	4F9-143	1105
4GKW0-18	1001	4M4-58	1103	4FKW0-100	1202	4FA9-143	1105
4NFG0-18	1002	4NK0-58	1002	4GKW0-100	1202	4NFG0-143	1003
4NFG0-18	1103	4NK0-58	1103	4HDA0-100	1202	4NFG0-143	1105
4Y2-18	1103	4NK0-58	1104	4DG0-107	1301	4NK0-143	1003
4ZGEW0-18	1002	4ZGEW0-58	1002	4DK0-107	1301	4DG0-145	1301
4ZGEW0-18	1103	4ZGEW0-58	1103	4FKW0-110	1202	4F2-145	1101
4GA0-20	1102	4ZGEW0-58	1104	4GKW0-110	1202	4G0-145B	1301
4GAA0-20	1102	4BMC0-62	1103	4F2-111	1103	4GKW0-146	1203
4TF0-20	1102	4NFG0-62	1003	4SA2-111A	1101	4GLW0-146	1203
4BP0-21	1302	4NFG0-62	1103	4F4-113A	1105	4HDA0-146	1203
4FKW0-21	1302	4NFG0-62	1105	4F4-114A	1105	4HDAA0-146	1203
4GK0-21	1302	4BMB0-62B	1003	4F4-116	1101	4DG0-147	1101
4FKW0-22	1101	4BMB0-62B	1103	4BMC0-119	1003	4F0-147	1101
4UG0-22	1101	4ZDK0-65	1103	4BMC0-119	1103	4TF2-147	1101
4UGA0-22	1001	4G0-65A	1103	4FKW0-120	1101	4Y1-147	1101
4UH0-22	1001	4GKW0-67	1001	4FKW0-120	1203	4AKB0-149	1203
4ZGEW0-22	1101	4GKW0-67	1101	4GKW0-120	1203	4SK0-149	1202
4TF4-22A	1101	4ZGEW0-67	1103	4UG0-120	1208	4UG0-149	1203
4ZGEW0-22A	1101	4UG0-67	1001	4NFG0-122	1003	4UGA0-149	1203
4D0-25	1301	4AK0-69	1001	4NFG0-122	1103	4SKA0-151	1006
4Y2-25	1301	4AK0-69	1003	4NK0-122	1002	4FKW0-153	1203
4BLA0-28	1003	4NK0-69	1002	4SA2-122	1103	4FKW0-154	1203
4BLA0-28	1105	4SK0-69	1002	4ZGEW0-122	1103	4G0-155	1203
4UG0-28	1003	4AA0-72	1001	4GKW0-123	1002	4G0-156	1301
4UG0-28	1105	4AA0-72	1101	4TF2-123	1105	4NFG0-156	1301
4Y2-28	0703	4G0-79	1203	4ZGEW0-123	1105	4Y2-156	1301
4Y2-28A	1105	4HDA0-79	1203	4BMC0-125	1002	4ZGEW0-156	1003
4NKA0-29	1003	4NKA0-80	1002	4BMC0-125	1105	4D2-157	1002
4NFG0-32	1103	4NKA0-80	1103	4GKW0-125	1002	4D2-157	1105
4DK0-35	1003	4A0-81	0802	4TF2-125	1105	4NFG0-157	1301
4DK0-35	1105	4A0-81	1003	4ZGEW0-125	1105	4Y2-157	1301
4FA4-35	1105	4A0-81	1105	4GKW0-130	1101	4NFG0-158	1003
4FKW0-35	1003	4NKA0-81	1003	4GKW0-130	1203	4NFG0-158	1105
4GKW0-35	1003	4NKB0-81	1003	4GLW0-130	1101	4G0-159	1101
4NFG0-35	1105	4FKW0-84	1003	4GLW0-130	1203	4G0-159	1203
4DK0-38	1102	4FKWA0-84	1105	4NK0-136	1003	4ZGEW0-159	1101
4FKW0-38	1102	4GKW0-84	1003	4SK0-136	1105	4ZGEW0-159	1203
4TF3-38	1102	4UG0-84	1003	4DFL0-137	1003	4TF4-161A	1103
4FA0-40	1105	4UG0-84	1105	4DFL0-137	1105	4TF4-162A	1103
4UG0-40	1003	4A2-87	1103	4H2-137	0703	4A1-163	0902
4UG0-40	1105	4G2-87	1102	4H2-137	0803	4GTC0-163	1105
4BLA0-44	1003	4UG0-87	1102	4H2-137	1105	4NK0-163	0803
4BLA0-44	1103	4G0-89	1202	4H2-137	1301	4NK0-163	1003
4FKW0-44	1003	4UG0-89	1202	4RL0-137	1003	4NK0-163	1105
4GKW0-44	1003	4BLA0-97	1003	4RL0-137	1105	4Y1-163	0704
4ZGEW0-44	1003	4BLA0-97	1103	4Y2-137	1105	4NFG0-164	1203
4ZGEW0-44	1103	4BLB0-97	1003	4Y2-137	1301	4D0-166	1301
4ZGEW0-44	1105	4GKW0-97	1003	4A0-138	1006	4G0-166	1301
4GKW0-51	1002	4BLA0-97A	1003	4DFL0-138	1003	4Y2-166	1301
4G0-51A	1103	4BLA0-97A	1105	4DFL0-138	1105	4BA0-170	1003
4FKW0-52	1202	4BLA0-96	1003	4FD0-138	1301	4BA0-170	1103
4GKW0-52	1202	4BLA0-96	1103	4GKW0-138	1003	4BQ0-170	1103
4UG0-52	1202	4BLB0-96	1003	4NFG0-138	1003	4FC0-170	1103
4NK0-55	1002	4FKW0-96	1003	4NFG0-138	1105	4FKW0-170	1103
4NK0-55	1104	4GKW0-96	1003	4ZDK0-138	1006	4GE0-173	1105
4UG0-55	1002	4NK0-96	1003	4FA0-139	1105	4TF2-174	1102
4UG0-55	1103	4UG0-96	1003	4FD0-139	1301	4FC2-175	1103

NUMERICAL INDEX—MASTER PARTS LIST

Part No.	Group	Part No.	Group	Part No.	Group	Part No.	Group
4FD0-176	1001	4SD3-491A	1302	4XGEW0-695	1102	4N2-1028	1105
4FD0-176	1003	4A0-499	1302	4ZGEW0-695	1007	4TG2-1035	1101
4FD0-176	1101	4SK0-499	1006	4FKW0-696	1301	4F0-1040	1103
4A0-178	1105	4A0-503	1302	4G0-696	1301	4FA0-1050	1105
4U0-178	1003	4A0-503	1800	4G0-696	2300	4FKW0-1050	1003
4CHS2-183	1301	4GKW0-503	1302	4G0-699	1102	4NKA0-1050	1003
4DG0-183	1301	4GLW0-503	1302	4A0-723	1105	4ZGEW0-1055A	1103
4DK0-183	1301	4A0-504	1302	4F0-730	1101	4ZGEW0-1056	1103
4G0-183	1301	4GKW0-504	1302	4SK0-730	1007	4ZGEW0-1057	1105
4EB0-187	1105	4GLW0-504	1302	4UG0-730	1102	4ZGEW0-1058	1105
4Y2-194	1301	4A0-505	1302	4GKW0-743	1208	4ZGEW0-1059	1105
4CGA2-209	1002	4GKW0-505	1302	4FKW0-744	1208	4BMC0-1061	1103
4CGA2-209	1105	4A0-506	1302	4GKW0-744	1208	4GTC0-1065	1105
4FKW0-209	1002	4GKW0-506	1302	4BMA0-751	1002	4GTC0-1066	1105
4CGA2-211	1002	4A0-521	1105	4BMA0-751	1103	4A0-1067	1105
4CGA2-211	1105	4BLA0-521	1003	4BMA0-751	1104	4GTC0-1067	1105
4Y2-211	1105	4BLA0-521	1105	4NK0-751	1002	4ZGE0-1073	1105
4Y2-211	1301	4FA4-521	1105	4NK0-751	1103	4ZGEW0-1073	1105
4SK0-215	1006	4FKW0-521	1003	4NK0-751	1104	4GKW0-1074	1003
4SK0-215	1203	4GKW0-521	1003	4TF2-751	1103	4GLW0-1074	1003
4FD0-219	1003	4CHS2-526	1301	4TF2-751	1104	4GMW0-1074	1003
4BMC0-220	1002	4NFG0-531	1102	4FKW0-762	1003	4ZGE0-1074	1105
4BMC0-220	1103	4D0-550	1301	4UG0-762	1003	4ZGEW0-1074	1105
4NK0-220	1002	4G0-550	1301	4UG0-762	1105	4GKW0-1075	1003
4NK0-220	1103	4NK0-553	1003	4GKW0-763	1003	4ZGE0-1075	1105
4ZGEW0-220	1002	4NKA0-553	1003	4FK0-764	0802	4BLB0-1076	1003
4ZGEW0-220	1103	4NKB0-553	1003	4FK0-766	0802	4GKW0-1076	1105
4H1-227	0102	4UG0-559	1301	4FK0-768	0802	4GKWA0-1076	1003
4FD0-237	1007	4NFG0-567	1105	4FE0-773	1105	4ZGEW0-1077	1105
4FKW0-237	1003	4A0-568	1105	4FKW0-801	1302	4HDA0-1078	1203
4ZDK0-237	1007	4BLB0-576	1003	4GKW0-801	1302	4NFG0-1078	1302
4UG0-251	1203	4GKW0-576	1003	4GKW0-801	1302	4NFG0-1084	1001
4UG0-252	1203	4NK0-576	1003	4FKW0-809	1302	4NFG0-1084	1003
4GKW0-254	1208	4FA3-582	1105	4GKW0-809	1302	4NFG0-1084	1007
4HDA0-254	1208	4NK0-582	1003	4BP0-811	1302	4NFG0-1084	1103
4FKW0-270	1101	4F3-583A	1105	4NFG0-811	1302	4NFG0-1084	1105
4FKW0-351	1105	4UG0-584	1003	4G0-811A	1302	4NFG0-1086	1002
4G0-351	1102	4UG0-584	1105	4UF1-826	0803	4NFG0-1086	1105
4FD0-358	1003	4TF2-597	1103	4FKW0-890	1002	4A0-1087	1105
4GKW0-361	1207	4TF2-597	1104	4FKWA0-890	1103	4A0-1088	1105
4FKW0-395	1101	4G0-624	1007	4GKW0-890	1002	4NFG0-1089	1003
4ZDK0-395	1007	4BM0-638	1002	4ZGEW0-890	1103	4A0-1091	1105
4GE0-395A	1105	4BM0-638	1103	4HDA0-900	1203	4NFG0-1092	1105
4NFG0-404	1302	4BM0-638	1104	4UG0-900	1203	4NFG0-1093	1105
4G0-404A	1302	4ZGEW0-638	1002	4FA0-920	1103	4NFG0-1094	1105
4DG0-412	1301	4ZGEW0-638	1103	4UG0-920	1003	4NFG0-1095	1105
4G0-412	1301	4ZGEW0-638	1104	4UG0-920	1103	4NFG0-1096	1105
4D0-417	1101	4G0-639	1301	4UG0-928	1203	4NFG0-1097	1105
4SK0-417	1001	19FK0-646	0803	4G0-936	1101	4NFG0-1098	1103
4SK0-417	1101	4A0-648	1105	4UG0-936	1003	4NFG0-1099	1103
4NK0-425	1003	4A0-648	1207	4UG0-936	1105	4FA0-1100	1105
4NK0-425	1105	4BLA0-648	1003	4HDA0-957	1203	4GKW0-1141	1203
19FK0-431	0803	4BLA0-648	1105	4FA0-960	1105	4NK0-1141	1003
19SKA0-433	0803	4FD0-648	1001	4ZGEW0-961	1103	4UG0-1143	1102
19SKA0-438	0803	4FD0-648	1003	4ZGEW0-962	1105	4SK0-1154	1001
4UG0-447	1208	4ZGEW0-648	1105	4FKW9-970	1100	4GKW0-1156	1002
4BLA0-460	1003	4FD0-652	0802	4GKW9-970	1100	4GEW0-1160	1002
4BLA0-460	1103	4FD0-652	0803	4AK0-974	1402	4GEW0-1160	1103
4BLB0-460	1003	4G0-652	1003	4ZGEW0-974	1402	4UG0-1165	1102
4FA0-460	1105	4GTC0-652	1105	4ZGE0-992	1105	4BLA0-1178	0804
4GKW0-460	1003	4ZGE0-653	0803	4ZGE0-993	1105	4BLA0-1178	1003
4NK0-460	1003	4ZGEW0-653	1105	4F2-1022	1105	4BLA0-1178	1003
4NK0-470	1003	4GKW0-660	1101	4FKW0-1022	1101	4GTC0-1178	1003
4N3-483	1101	4TGA0-660	1101	4N2-1023	1105	4F2-1199	1103
4N3-483	1103	4UG0-660	1101	4NK0-1023	1003	4NK0-1204	1003
4UG0-483	1002	4FKW0-661	1001	4NKA0-1023	1003	4NK0-1204	1105
4UG0-483	1101	4ZDK0-661	1103	4NKB0-1023	1003	4NKA0-1204	1003
4A3-489	1302	4G0-666	1202	4NKC0-1023	1003	4NKA0-1205	1203
4FKW0-489	1302	4G0-667	1202	4NKD0-1023	1003	4UG0-1232	1003
4FKWA0-489	1302	4F0-671	1203	4NKE0-1023	1003	4UG0-1232	1105
4UK3-489	1302	4G0-674	1102	4NKF0-1023	1003	4A0-1251	1105
4SD3-489A	1302	4G0-674	1103	4NKG0-1023	1003	4UU0-1251	1105
4A3-491	1302	4G0-678	1203	4NKH0-1023	1003	4GKW0-1253	1002
4UK3-491	1302	4G0-683	1301	4NKJ0-1023	1003	4E2-1265	1105

NUMERICAL INDEX—MASTER PARTS LIST

Part No.	Group
4B4-1271	1105
4C4-1271	1105
4D4-1271	1105
4E4-1271	1105
4F4-1271	1105
4B4-1272	1105
4C4-1272	1105
4D4-1272	1105
4E4-1272	1105
4F4-1272	1105
4F4-1273	1103
4FKW0-1273	1302
4GKW0-1275	1105
4HDA0-1317	1203
4GKW0-1321	1202
4UG0-1326	1203
4F2-1334	1103
4NFG0-1344	0802
4NFG0-1344	0803
4NFG0-1344	1003
4UG0-1353	1203
4HDA0-1363	1007
4HDA0-1363	1101
4HDA0-1363	1203
4GKW0-1379	1101
4GKW0-1379	1203
4GKW0-1381	1105
4A0-1420	1105
4NFG0-1420	1105
4UG0-1430	0803
4AA0-1600	1001
4AA0-1600	1101
4BL0-1600	1001
4BL0-1600	1101
5BL0-01	0503
5D4-01	0503
5UU0-01	0503
5BL0-02	0503
5D3-02	0503
5UU0-02	0503
5BL0-03	0503
5D0-03	0503
5UU0-03	0503
5BL0-04	0503
5SA2-04	0503
5UU0-04	0503
5RM4-05	0501
5UP4-05	0501
5UU3-05	0501
5Y3-05	0501
5UU6-06	0505
5A0-07	0503
5UBB0-07	0503
5UU0-07	0503
5ZBLEQ0-7	0503
5BL0-12	0503
5UU0-12	0503
5BL0-13	0503
5UU0-13	0503
5UU2-17	0505
5A0-22	0503
5UU2-22	0503
5SA0-25	0503
5A2-30	0502
5HF0-30	0502
5UU0-30	0502
5BL0-32	0503
5UU0-32	0503
5UUA0-32	0503
5UUB0-32	0503
5UP3-39	0503
5T3-39A	0503
5BL0-40	0503
5D0-40	0503
5UU0-40	0503
5A0-47	0503
5SCM0-47	0503
5SCM0-48	0503
5B0-49	0503
5UU0-49	0503
5T0-49A	0503
5ZBLE0-49A	0503
5B0-51	0503
5UU0-51	0503
5SA0-51A	0503
5ZBLE0-51A	0503
5SA0-53	0503
5T0-53	0503
5RL4-54	0503
5UU0-54	0503
5ZBLE0-54A	0503
5UK0-55	0503
5UU3-55	0503
5ZTL3-55	0503
5T3-55B	0503
5T3-55H	0503
5UK4-61	0501
5UU4-61	0501
5URM4-61B	0501
5RH3-62	0101
5UP3-62	0101
5URM3-62	0101
5UU3-62	0505
5SCM0-71	0503
5SCM0-72	0503
5TL0-72	0503
5SA0-73	0503
5T0-73	0503
5SA0-74	0503
5T0-74	0503
5UU4-80	0503
5BL0-81	0503
5UU0-81	0503
5BL0-86	0503
5D0-86	0503
5BL0-87	0503
5UU0-87	0503
5BA0-88	0503
5UK0-88	0503
5UL0-88	0503
5UU0-88	0503
5SA0-88A	0503
5UK4-90	0501
5UP4-90	0501
5UU5-90	0501
5UL4-90A	0501
5T2-92	0503
5T2-94	0505
5T2-95	0503
5UG4-104	0501
5UK0-104	0501
5UU3-104	0501
5SA0-113	0503
5B4-120	0503
5UK4-120	0503
5UL4-120	0503
5UU4-120	0503
5SA4-120A	0503
5DW0-132	0503
5DW0-133	0503
5UU3-158A	0505
5C2-169	0501
5T2-170B	0503
5UU0-171	0503
5A0-192	0503
5UU0-192	0503
5UUA0-192	0503
5BL0-209	0503
5UU0-209	0503
5TH0-233	0503
5UU0-234	0503
5CB0-242	0503
5UU0-242	0503
5SCM0-256	0503
5UU2-262	0503
5D2-264	0503
5UU0-271	0505
5UU0-272	0505
5SA0-274	0503
5SA0-275	0503
5TL0-275	0503
5C0-300	0501
5C3-300	0501
5UBL2-300	0501
5TL0-341	0503
5B0-369	0501
5B01-369	0501
5UBK2-374	0501
5UKS2-374	0501
5URK2-374	0501
5URM2-374	0501
5D0-378	0503
5UU0-392	0503
5D0-424	0503
5D0-425	0503
5D0-426	0503
5D0-427	0503
5D0-428	0503
5D0-429	0503
5UU4-443	0505
5UU4-553A	1500
5UU0-579	0505
5UUA0-579	0505
5UKS2-582	0501
5UU0-584	0503
5UU0-586	0503
5UU0-594	0107
5UU0-595	0107
5UU0-596	0107
5UKS2-597	0501
5UU0-607	0505
5UU0-608	0502
5UU0-612	0505
6URKA2-11	0300
6URK2-15	0300
6D0-22	0300
6D0-22	0304
6UV3-22	0300
6DA4-50	0301
6UDG0-50	0301
6UG4-50	0301
6UR4-50	0301
6UU3-50	0301
6UR4-50A	0301
6UU0-116	0301
6UU0-117	0301
6UU0-118	0301
6UU0-128	0301
6UU0-131	0301
6UU0-142	0301
6UU0-151	0301
6UU0-156	0301
6UU0-159	0301
6UKSB0-160	0300
6UKSC0-160	0300
6BL0-164	0301
6UU0-167	0301
6UU0-168	0301
6UU0-180	0301
6UU0-190	0301
6UU0-207	0301
6UU0-218	0301
6UU0-250	0301
6UU0-280	0301
6UU3-309A	0301
6CK1-317	1204
6UUA2-317	0304
6UU2-317	0304
6UU2-327	0301
6F0-330	0301
6UKS4-330	0301
6ZBM0-330	0301
6UU0-336	0301
6UU0-343	0301
6UU0-346	0301
6UU0-358	0301
6UU0-374	0301
6N2-375A	0301
6QT3-401	0301
6UU0-434	0301
6UU0-442	0301
6UU0-445	0301
6UU0-449	0301
6BLA0-472	0603
6UU0-475	0301
6UU0-485	0301
6HF4-510	0302
6UG4-510	0302
6URL4-510	0302
6UU4-510A	0302
6N4-510C	0302
6UU0-550	0301
6UU0-553	0301
6UU0-554	0301
6U4-593B	0300
6ZUN0-594C	0300
6UBK3-610	0300
6UBKA3-610	0300
6UN0-610	0300
6URK3-610	0300
6URKA3-610	0300
6UU4-640A	0304
6UKSA4-686	0301
6URK4-720	0300
6URK0-730	0300
6D3-740	0300
6D3-750	0304
6W2-754	2203
6UU0-774	0301
6BL0-775	0301
6UU0-776	0301
6BL0-777	0301
6UU0-781	0301
6UU0-782	0301
6BL0-783	0301
6UU0-785	0301
6BL0-784	0301
6UU0-834	0301
6UU0-891	0301
6UU0-892	0301
6UU0-893	0301
6UU0-894	0301
6UU0-895	0301
6UU0-896	0301
6UU0-897	0301
6UU0-898	0301
6UU0-899	0301
6UU0-901	0301
6UU4-913A	0304
6UKS3-916	0300
6UKSA3-916	0300
6UV0-943	0304
6UVA3-943	0300

NUMERICAL INDEX—MASTER PARTS LIST

Part No.	Group	Part No.	Group	Part No.	Group	Part No.	Group
6URK3-1020	0302	8BLA0-302	0603	9TA0-05A	1301	9N0-288	1007
6UV3-1320	0300	8SA0-302	0603	9NFD0-07	1007	9GKW0-291	1007
		8SCM0-302A	0603	9N0-12	1007	9TA0-294	1103
7UG4-03	0402	8B0-303	0603	9N0-13	1007	9TA0-294	1402
7UK4-03	0402	8SA0-304	0603	9N0-14A	1006	9GKW0-300	1202
7UV4-03	0402	8UU0-306	0603	9N0-16A	1006	9GLW0-300	1202
7BL0-04	0108	8UBA2-308	0603	9B0-17	1301	9SKA0-300	1202
7BL0-04	0402	8UBA2-308	0604	9J0-17	1402	9N0-302	1007
7T2-04	0402	8UU3-308	0604	9K0-17	1301	9SKA0-310	1202
7UU2-04	0402	8UUA3-308	0604	9AA0-20	1402	9SA0-341	1402
7BL3-09	0402	8T2-309	0603	9SK0-21	1402	9SA0-342	1402
7UG3-09	0402	8T2-309	0604	9NK0-22	1402	9A0-344	1402
7N3-09A	0402	8BL2-311	0603	9AK0-23	1006	9BJ0-403	1301
7CBA4-10	0401	8RLA0-311	0603	9TE2-23	1007	9BJ0-403	2300
7TE3-10	0401	8UU0-311	0603	9UGA0-23	1006	9CJ4-403	1301
7BL0-22	0402	8SCM0-315A	0603	9UN2-23	1403	9CJ4-403	2300
7M2-22	0402	8UU0-316	0603	9SK0-24	1402	9KA4-403	1301
7SA3-29	0401	8BLA0-317	0603	9AA0-25	1402	9KA4-403	2300
7UG3-29	0402	8SA0-317	0603	9SK0-25	1402	9GKW0-411	1402
7BL2-42	0402	8UU0-317	0603	9UG0-25	1402	9FKW9-430	1000
7SA2-42	0402	8UUA0-317	0603	9SK0-28	1301	9GKW9-430	1000
7T1-44	0402	8SA0-318	0603	9TA0-28	1301	9FKW0-433	1007
7UG0-320	0402	8SA0-321	0603	9ZDK0-28	1301	9GKW0-433	1007
		8UU0-321	0603	9TA0-28A	1301	9FKW0-434	1007
8BLA0-09	0603	8SA0-322	0603	9N0-38	1301	9GKW0-434	1007
8SA0-09	0603	8BLA0-324	0603	9SK0-39	1301	9AK0-435	1007
8UD0-14A	0604	8SA0-324	0603	9N0-51	1007	9ZDK0-435	1007
8TE0-18	0603	8BLA0-325	0603	9N0-68	1007	9SK0-436	1007
8D0-23	0604	8SA0-325	0603	9FKW0-75	1006	9ZDK0-436	1007
8E0-23	0604	8UU0-325	0603	9GKW0-75	1006	9CK0-437	1007
8CB0-23A	0604	8SA0-326	0603	9A0-76	1402	9SK0-438	1006
8B0-28	0603	8BLA0-327	0603	9TA0-77	1007	9ZDK0-438	1006
8BLA0-28	0603	8SA0-327	0603	9AA0-80	1402	9ZDK0-438	1007
8UU0-28	0603	8UU0-327	0603	9A0-84	1402	9FKW0-441	1001
8BLA0-35	0603	8BLA0-328	0603	9B0-85	1402	9SK0-441	1001
8SA0-35	0603	8SA0-328	0603	9AA0-90	1402	9FKW0-442	1001
8UU0-35	0603	8UU0-328	0603	9SK0-92	1301	9SK0-442	1001
8SA0-57	0603	8B0-329	0603	9UG02-94	1006	9ZDK0-444	1006
8UU0-57	0603	8BLA0-329	0603	9B0-95	1301	9SK0-447	1006
8A0-62	0603	8TE0-449	0603	9DKA0-95	1301	9ZDK0-447	1006
8A0-66	0603	8BL0-455	0603	9SK0-95	1301	9GKW0-448	1006
8A0-68	0603	8BLA0-464	0603	9ZDK0-95	1301	9SK0-448	1003
8B0-69	0603	8BLA0-465	0603	9SA0-109	1007	9SKA0-448	1105
8BLA0-69	0603	8NFS0-465	0603	9FKW0-121	1203	9ZDK0-448	1006
8UU0-69	0603	8BLA0-466	0603	9SK0-121	1203	9ZDK0-448	1007
8SA0-69A	0603	8UU0-469	0603	9SK0-130	1301	9SK0-449	1006
8A0-83	0603	8RLA3-470	0603	9TA0-130	1301	9SK0-449	1402
8BLA0-120	0604	8UG0-470	0603	9TA0-130A	1301	9SK0-451	1006
8SA2-120A	0603	8UK0-470	0603	9SKA0-134	1203	9SK0-452	1006
8A0-179	0604	8URL4-470	0603	9NA0-140	1006	9ZDK0-452	1006
8B0-210	0603	8UU3-470	0603	9SK0-140	1006	9SK0-453	1006
8BLA0-210	0603	8UKS2-506	0611	9ZDK0-140	1006	9UG0-453	1006
8UU0-210	0603	8UKS2-508	0611	9SK0-146	1202	9ZDK0-453	1006
8SA0-210A	0603	8BLA0-524	0603	9AK0-147	1202	9SK0-454	1006
8B0-220	0603	8BLA0-526	0603	9NA0-150	1006	9SKA0-454	1006
8UU0-220	0603	8BL0-527	0603	9N0-168	1007	9ZDK0-454	1006
8BLA0-230	0603	8UUA0-527	0603	9FKW0-170	1002	9ZDKA0-454	1006
8UU0-230	0603	8BL0-528	0603	9GK0-170	1002	9FKW0-455	1006
8SA0-230B	0603	8UU0-531	0603	9N0-173	1007	9SK0-455	1006
8UU0-249	0603	8UG2-537	0603	9SK0-175	1301	9SK0-456	1202
8BLA0-260	0603	8URL2-537	0603	9TA0-175	1301	9SK0-457	1202
8UU0-260	0603	8URL2-538	0603	9ZDK0-175	1301	9AK0-458	1203
8SA0-260A	0603	8URL2-539	0603	9TA0-175A	1301	9BKT0-458	1203
8N0-283	0603	8URL3-541	0603	9SK0-178	1301	9FKW0-458	1203
8UU0-283	0603	8UU0-548	0603	9ZDK0-178	1301	9UG0-458	1203
8UUA0-283	0603	8UU0-549	0603	9A0-216	1302	9AK0-459	1203
8BLA0-284	0603	8UU0-551	0603	9A0-217	1302	9BKT0-459	1203
8BL0-287	0603	8UU0-552	0603	9ZA-223	1403	9FKW0-459	1203
8SA0-287	0603	8UU0-590	0603	9AK0-232	1302	9UGA0-459	1203
8BLA0-288	0603	8BN0-610	0604	9BP0-232	1302	9AK0-461	1203
8SA0-288	0603	8URL0-610	0604	9FKW0-232	1302	9FKW0-464	1203
8UU0-288	0603			9TA0-237	1402	9SK0-464	1203
8SA0-294	0603	9AK0-05	1301	9SK0-251	1203	9NK0-473	1007
8SA0-295	0603	9TA0-05	1301	9A0-261	1402	9AK0-475	1402

NUMERICAL INDEX—MASTER PARTS LIST

Part No.	Group	Part No.	Group	Part No.	Group	Part No.	Group
9ZDK0-475	1402	10UGA3-09	0203	10Y2-79	1204	10M0-254	1404
9SK0-476	1007	10TE3-09A	0203	10ZUBLA2-79	1206	10M2-254	1404
9ZDK0-476	1007	10UG3-10	1401	10CH2-79A	0203	10Y1-264	1401
9SK0-477	1007	10UU3-10	1401	10T1-79A	0203	10URL4-265A	0706
9ZDK0-477	1007	10ZN4-10F	1401	10Y1-84	0706	10A2-266	1007
9AK0-478	1007	10UU3-14	1201	10U2-86	0706	10D2-286A	1201
9ZDK0-478	1007	10Y3-14	1403	10Y1-86	0706	10FE2-292	1204
9SK0-479	1007	10NL2-15	0204	10UG0-93	1006	10UN2-292	0204
9ZDK0-479	1007	10S2-15	0204	10H1-94	1401	10URL4-300A	0706
9SK0-481	1007	10S2-15	0805	10SP0-94	1401	10C2-303	0304
9ZDK0-481	1007	10S2-15	1201	10Y0-94B	1401	10C2-303	1209
9SK0-482	1007	10UA1-15	0204	10H1-95	1401	10D2-303	1209
9ZDK0-482	1007	10UA1-15	0303	10SP0-95	1401	10E2-303	0611
9SK0-483	1007	10UN2-16A	0706	10Y1-95A	1401	10F2-303	1209
9ZDK0-483	1007	10D2-18	0706	10H1-96	1401	10CD0-307	1403
9GKW0-484	1007	10UB0-18	0706	10SP0-96	1401	10D2-311	1201
9ZDK0-484	1007	10UU3-21	1201	10Y0-96A	1401	10SJ0-314	1403
9SKA0-485	1007	10UU4-22A	1201	10ZN0-99	1401	10TE0-314	1403
9ZDK0-485	1007	10DL4-23	0706	10Y0-99B	1401	10CD0-323	1403
9AK0-486	1007	10UK0-23	0706	10DF0-100	1201	10HDA0-323	1403
9AK0-487	1007	10UL0-23	0706	10UU0-100	1201	10UK0-323	1403
9SK0-488	1007	10UU4-23	0706	10A2-126	1201	10UU3-336	0303
9ZDK0-489	1007	10URL4-23A	0706	10D2-126	1201	10UV2-336	0303
9SK0-498	1006	10DF2-24A	1201	10UU2-126	1201	10UVA2-336	0303
9ZDK0-498	1006	10T1-25	1204	10DL0-130	0706	10UVB2-336	0303
9ZDK0-499	1007	10T1-26	0706	10D2-132	1201	10UB3-336B	0303
9FKW0-501	1006	10UB0-26	0706	10UU5-132	0805	10D2-339	0706
9N0-501	1007	10UU2-27	0303	10AD2-135	1601	10UU2-361	0303
9N0-502	1007	10NA9-28	1201	10UG2-135	0805	10A2-383	0204
9ZDK0-504	1007	10UU3-28	0303	10Y1-135	0204	10UB0-391	0706
9N0-506	1007	10Y1-35A	1201	10Y1-136	0108	10UU2-391	0706
9N0-507	1007	10T2-39	1201	10Y1-136	0303	10T2-391A	0706
9N0-508	1007	10D2-41A	1201	10T2-137	1201	10B0-408	0706
9ZDK0-511	1007	10M3-46	1201	10UU3-137	1201	10T2-408	0706
9UG0-513	1302	10M3-46A	1201	10NL4-138	0204	10T2-408	0805
9N0-514	1302	10DFL2-48	0203	10UG5-138	0204	10T2-408	1910
9SK0-514	1302	10UB3-48	0203	10UU5-138	0204	10T1-409	0706
9A0-515	1402	10UU3-48	0203	10S2-147	0203	10URL2-409	0706
9A0-516	1402	10UB5-53	0204	10R1-148	0203	10UU2-409	0706
9N0-525	1301	10NL4-53A	0204	10UG0-162	1401	10DFL0-470	0203
9A0-544	1402	10UB5-54	1204	10UU0-162	1401	10UB0-470	0203
9A0-545	1402	10UB2-55	0204	10ZN0-162F	1401	10UGA3-507	1910
9UG0-750	1208	10NL2-55A	0204	10CF3-175A	1405	10UL3-507	0805
9UG0-750	1402	10A2-56	0203	10UU5-177A	1405	10UU3-507	0805
9AK0-807	1302	10B2-56	1910	10RL3-181	0108	10UUA3-507	0805
9BP0-807	1302	10D2-56	1201	10UF1-193A	1401	10BCA0-520	1403
9BP0-809	1302	10F2-56	0203	10UU3-194	0303	10RJ0-520	1403
9BK0-820	1007	10F2-56	0805	10UU2-198	0303	10SJ0-520	1403
9ZDK0-820	1007	10F2-56	1201	10BCA0-206	1403	10UK0-520	1403
9SK0-825	1301	10F2-56	1204	10J0-206	1403	10D3-545	1201
9SKA0-825	1301	10TE2-56	1910	10J4-210	1404	10RJ0-550	1403
9UF0-825	1105	10UU2-56	0203	10UU5-210	1404	10SJ0-550	1403
9FKW0-830	1001	10UV0-56	1910	10GE0-215	1403	10UK0-550	1403
9SK0-830	1001	10T2-58	0203	10TE0-215	1403	10F2-572	1405
9UG0-830	1001	10UU3-61	0203	10GE0-216	1403	10SA0-592	1403
9FKW0-860	1006	10DF3-61A	0203	10TE0-216	1403	10SA0-593	1403
9SK0-860	1006	10DF2-62	1201	10C0-224	1403	10TA0-593	1403
9ZDK0-860	1006	10UU3-62	1201	10UK0-224	1403	10UT0-593	1403
9FKW0-890	1006	10T2-63	0303	10BM0-227	1403	10ZDT2-596	1007
9FKWA0-890	1006	10T2-63	1209	10SA0-227	1403	10C0-598	1403
9GKW0-890	1006	10URL2-64A	0706	10SCK0-227	1403	10UU2-657	1403
9GLW0-890	1006	10W1-65	0203	10RJ0-228	1403	10NA0-701	1201
9UG0-890	1006	10RJ0-67	1403	10RJA0-228	1403	10NB0-701	1201
9UGA0-890	1006	10UK0-67	1403	10SJ0-228	1403	10S0-701	1201
9LL0-990A	1000	10HDA0-67A	1403	10SJA0-228	1403	10Y0-721	1401
9UG0-1030	1007	10DFJ2-72	0303	10UV3-231	1405	10UB2-729	0303
9UK0-1030	1007	10SA2-72	0303	10UV3-232	1405	10UU4-747	1201
		10AB2-79	1209	10D2-243	1201	10UV3-747	1201
10UE2-04	0204	10D2-79	0203	10D2-244	1201	10UKS3-750	0303
10BB2-05	0706	10D2-79	1204	10D2-249	0706	10UB0-767	0706
10T2-05	0706	10D2-79	1209	10SA0-251	1403	10BCB0-770	1403
10UF1-07	0108	10G2-79	0303	10SA0-251	1404	10RJ0-770	1403
10NL2-08	0204	10SA2-79	0303	10Y0-251A	1403	10SJ0-770	1403
10NL3-09	0203	10Y2-79	0203	10Y0-251A	1404	10UK0-770	1403

NUMERICAL INDEX—MASTER PARTS LIST

Part No.	Group	Part No.	Group	Part No.	Group	Part No.	Group
10NL0-828	0204	10UK0-1143	1403	12UU3-57A	1500	13H3-06	1602
10NL0-833	0204	10CD0-1145	1403	12UU3-57A	1602	13RH3-06	1602
10DF0-872	1201	10CDA0-1145	1403	12UG0-62	1500	13TE4-06	1602
10N0-911	1201	10CDB0-1145	1403	12UGTA0-62	1500	13BF4-09	1601
10TA0-911	1201	10UK0-1145	1403	12UP0-62	1500	13UG4-09	1601
10N3-912	1201	10UKA0-1145	1403	12UU0-62	1500	13UU4-09	1601
10S0-912	1201	10UKB0-1145	1403	12UV0-62	1500	13UV4-09	1601
10NA0-913	1201	10URL2-1166	0706	12UK0-62B	1500	13BF3-11	1601
10S0-913	1201	10BCA0-1176	1403	12UG0-62C	1500	13UG4-11	1601
10S0-914	1201	10UK0-1176	1403	12UG0-63	1500	13UU4-11	1601
10RJ0-920	1403	10HDR0-1178	1403	12UGTA0-63	1500	13UV4-11	1601
10SJ0-920	1403	10UK0-1178	1403	12UP0-63	1500	13A2-12	1602
10UK0-920	1403	10U0-1205	0706	12UU0-63	1500	13Y1-13	1602
10NA0-926	1201	10B2-1245	1201	12UV0-63	1500	13UU2-14	1601
10S0-926	1201	10UKS3-1270	0303	12UK0-63B	1500	13UV3-14	1601
10NA0-927	1201	10UU3-1284	0805	12UG0-63C	1500	13BF0-21	1601
10S0-927	1201	10UU2-1285	1201	12UU5-64	0801	13UG0-21	1601
10S0-928	1201	10DA0-1320	1403	12W1-79	0303	13UU0-21	1601
10NA0-929	1201	10SJ0-1320	1403	12UG3-82	1500	13UV0-21	1601
10S0-929	1201	10UK0-1320	1403	12UK3-82	1500	13BF0-22	1601
10HDA0-930	1403	10L0-1330	0203	12C4-86	0110	13UG0-22	1601
10UK0-930	1403	10UK0-1370	1403	12UU4-86	0110	13UU0-22	1601
10NA0-935	1201	10UK0-1450	0706	12UU5-88	0110	13UV0-22	1601
10S0-935	1201	10UL0-1450	0706	12R4-101	1201	13BF0-23	1601
10NB0-936	1201	10RJ0-1790	1403	12M9-108	1209	13UG0-23	1601
10S0-936	1201	10SJ0-1790	1403	12UT3-108	1501	13UU0-23	1601
10S0-937	1201			12W2-108	1501	13UV0-23	1601
10NB0-938	1201	12UG4-02	1500	12ZUN3-108	0805	13BF0-24	1601
10NC0-938	1201	12UK4-02	1500	12ZDA3-108A	1205	13UG0-24	1601
10S0-938	1201	12UU4-02	1500	12UG3-121	1500	13UU0-24	1601
10S0-939	1201	12UV4-02A	1500	12UUA3-121	1500	13UV0-24	1601
10BCA0-940	1403	12RM5-03	1602	12UVA3-121	1500	13BF0-25	1601
10UK0-940	1403	12UK3-03	1500	12UU3-121A	1500	13UG0-25	1601
10NA0-941	1201	12UV3-03	1500	12UG3-122	1500	13UU0-25	1601
10S0-941	1201	12UU3-03A	1500	12UGA3-122	1500	13UV0-25	1601
10NA0-942	1201	12RM5-04	1602	12UUA3-122	1500	13A1-26	1601
10S0-942	1201	12UK3-04	1500	12UU3-122A	1500	13A1-26	1602
10RJ0-970	1403	12SK4-06A	1602	12CH3-128	1501	13TE2-26	1601
10SJ0-970	1403	12UGTA7-10	1500	12H3-128	1501	13BF0-27	1601
10D2-988	0204	12UP7-10	1500	12W4-128	1501	13UG0-27	1601
10D2-988	1204	12UG7-10B	1500	12CH3-129	1501	13UU0-27	1601
10A2-989	0303	12UK7-10B	1500	12H3-129	1501	13UV0-27	1601
10NR7-990	1201	12UG7-10C	1500	12W4-129	1501	13BF0-28	1601
10SB7-990	1201	12UU4-12	1500	12NJ0-130	1602	13UG0-28	1601
10NC4-995	1201	12A1-25	1602	12RL0-130A	1602	13UU0-28	1601
10NA5-995A	1201	12TE2-25	1602	12UKSC-140	2103	13UV0-28	1601
10S4-995A	1201	12Y1-33A	1201	12UU0-160	1602	13BF0-29	1601
10UU2-996	1201	12NJ4-34	1602	12A0-160B	1602	13UG0-29	1601
10UUA2-996	1201	12RL4-34	1602	12HF3-164	0801	13UU0-29	1601
10NA0-997	1201	12UG4-34	1602	12A0-170B	1602	13UV0-29	1601
10S2-997	1201	12UK4-34	1602	12UU4-201	0801	13BF0-31	1601
10UN2-1005	0706	12UG3-42	1500	12UU9-206	1500	13UG0-31	1601
10UN2-1006	0706	12UL3-42	1500	12UUA9-206	1500	13UU0-31	1601
10BB2-1007	0706	12UU4-42	1500	12RM2-262	1602	13UV0-31	1601
10UNF2-1008	0706	12UVA4-42	1500	12UUA4-360	1500	13BF0-32	1601
10URL2-1014A	0706	12UG3-43	1500	12UU4-360A	0801	13UG0-32	1601
10UN2-1015	0706	12UU4-43	1500	12UU3-399	0110	13UV0-32	1601
10UB0-1017	0706	12UG4-44	1500	12NJ3-401	0110	13BF0-33	1601
10UA0-1018	0706	12UK4-44	1500	12NJ2-402	0110	13UG0-33	1601
10UA0-1019	0706	12UL4-44	1500	12NJ2-402	0501	13UV0-33	1601
10UD2-1045	0706	12UU4-44	1500	12NJ2-403	0110	13BF0-34	1601
10UU3-1052	0303	12UU5-45	1602	12NJ2-403	0501	13UG0-34	1601
10BCA0-1070	1403	12DK4-45A	1602	12RM4-478	1602	13BF0-35	1601
10CDKL0-1094	1403	12A4-45B	1602	12TL2-479	1602	13UG0-35	1601
10UK0-1094	1403	12A4-46B	1602	12RM3-491A	1602	13BF0-36	1601
10RJ0-1095	1403	12N4-54	1602	12P2-492A	0300	13UG0-36	1601
10SJ0-1095	1403	12TE4-54	1602	12UG3-496	1500	13UU0-36	1601
10SJ0-1097	1403	12UG4-54	1602	12UU2-702	1500	13UV0-36	1601
10UK0-1097	1403	12DU2-56	0401			13BF0-37	1601
10B2-1104A	1204	12UB2-56	0401	13DF3-01A	1602	13UG0-37	1601
10UU3-1137	0706	12UG3-56	0402	13RL2-02	1602	13UU0-37	1601
10URL3-1137A	0706	12UK2-56	0401	13RL2-03	1602	13UV0-37	1601
10URL2-1138	0706	12UU3-57	1500	13UU0-03	1601	13BF0-38	1601
10CD0-1143	1403	12UU3-57	1602	13UV0-03	1601	13UG0-38	1601

NUMERICAL INDEX—MASTER PARTS LIST

Part No.	Group
13UU0-38	1601
13UV0-38	1601
13BF0-39	1601
13UG0-39	1601
13UU0-39	1601
13UV0-39	1601
13BF0-40	1601
13UG0-40	1601
13UU0-40	1601
13C3-41C	1601
13ZTE3-41C	1601
13UU3-42	1601
13UUA3-42	1601
13UV3-42	1601
13UVA3-42	1601
13UU3-43	1601
13UV3-43	1601
13BF0-44	1601
13UG0-44	1601
13UU0-44	1601
13UV0-44	1601
13BF0-45	1601
13UG0-45	1601
13UU0-45	1601
13UV0-45	1601
13BF0-46	1601
13UG0-46	1601
13UU0-46	1601
13UV0-46	1601
13BF0-47	1601
13UG0-47	1601
13UU0-47	1601
13UV0-47	1601
13BF0-48	1601
13UG0-48	1601
13UU0-48	1601
13UV0-48	1601
13BF0-49	1601
13UG0-49	1601
13UV0-49	1601
13BF0-50	1601
13UG0-50	1601
13UU0-50	1601
13UV0-50	1601
13BF0-51	1601
13UG0-51	1601
13UV0-51	1601
13BF0-52	1601
13UV0-52	1601
13BF0-53	1601
13UU2-56	1602
13BF0-65	1601
13UG0-65	1601
13UG0-67	1601
13H0-70	1602
13RH0-70	1602
13TE0-70	1602
13GTD3-92	1601
13UU3-92	1601
13Y2-93B	1602
13DK3-95	1602
13UU3-95	1602
13H2-97	1602
13TE2-97	1602
13RH2-98	1602
13TE2-98	1602
13H2-98A	1602
13TE2-99	1602
13H2-99A	1602
13A1-101	1601
13A1-101	1602
13TE2-101	1601
13Y1-101A	1601
13UG2-103	1601
13UU2-103	1601
13B0-107	1601
13TE0-107	1601
13UU0-107	1601
13UU4-111	1601
13UV4-111	1601
13UG3-113	1603
13UU4-113	1601
13UV4-113	1601
13UG3-114	1603
13UU4-114	1601
13UV4-114	1601
13A2-115	1602
13TE2-116	1602
13A2-126	1602
13DF2-126	1602
13H2-126	1602
13TE2-126	1602
13N0-127	1601
13UG0-127	1601
13UU0-127	1601
13UV0-127	1601
13TAS0-128	1601
13UG0-128	1601
13C0-128A	1601
13RH2-129	1602
13RHP2-129	1602
13TE2-129	1602
13RE0-132	1601
13RL0-132	1601
13SK0-132	1601
13UG0-132	1601
13UV0-132	1601
13SK0-135	1601
13UG0-135	1601
13UV0-135	1601
13RH2-149	1602
13UG3-149	1602
13DK3-154	1601
13UU3-154	1601
13UV3-154	1601
13BF4-156	1601
13UG4-156	1601
13UU4-156	1601
13UV4-156	1601
13BF0-157	1601
13UG0-157	1601
13UU0-157	1601
13UV0-157	1601
13BF0-158	1601
13UG0-158	1601
13UU0-158	1601
13UV0-158	1601
13BF0-159	1601
13UG0-159	1601
13UU0-159	1601
13UV0-159	1601
13BF0-161	1601
13UG0-161	1601
13UU0-161	1601
13UV0-161	1601
13BF0-162	1601
13UG0-162	1601
13UU0-162	1601
13UV0-162	1601
13BF0-167	1601
13UG0-167	1601
13UU0-167	1601
13UG2-197	1602
13RM3-209	1602
13RM2-211	1602
13UV0-212	1601
13BF0-214	1601
13BFA0-214	1601
13UU0-214	1601
13UV0-214	1601
13UG3-248	1602
13UGA3-248	1602
14UV3-41	1800
14UVA3-41	1800
14UV3-42	1800
14UVA3-42	1800
14UN2-189	1800
14UV3-211	1800
14UV3-212	1800
14UV3-250	1800
14UKS4-330	1800
14UV4-461	1800
14UV3-463	1800
14UV4-662	1800
15U4-19A	1800
15DLS2-24	1800
15DLS3-26	1800
15DLS3-27	1800
15UKS3-28	1800
15NF2-29	1800
15UKS4-97	1800
15DLS2-146	1800
15DLS2-147	1800
15NF2-182	1800
15UKS2-239	1800
15UKS2-241	1800
15UKS2-244	1800
15T2-274	1800
15T2-274A	1800
15UKS4-284	1800
15T2-293	1800
15T2-294	1800
15DL3-295	1800
15NF2-346	1800
15URK2-395	1800
15UKS3-450	1800
15UKS3-550	1800
15UKSE0-700	1800
15UKS0-890	1800
15U4-901A	1800
15UKS0-910	1800
15UKS3-1939	1800
15UKS3-2069	1800
15UKS2-2310	1800
15UKSA3-3014	1800
15UKS3-3033	1800
16UKS2-06	2300
16URK2-06	2300
16URKA2-06	2300
16URKB2-06	2300
16UKSA3-13	1800
16UKSB0-19	1800
16A0-28	2300
16A0-30	2300
16RE0-33	2300
16SCH0-33	2300
16A0-34	2300
16D0-41	1301
16D0-41	2300
16HD3-41	2300
16UG3-41	2300
16H2-44	2203
16N2-44	1209
16S2-44	0304
16S2-44	2203
16S9-44	1403
16UC2-44	0611
16V9-44	1209
16AK9-50	0610
16AN9-50	0610
16UAP9-50	0610
16UKS3-50	0610
16A0-88	2300
16UKS2-90	1800
16B0-117	2300
16N0-117	1301
16TE2-134	2300
16URK3-165	1800
16B0-273	1209
16UKS3-181	1800
16A0-198	2300
16A0-218	2300
16A0-226	2300
16C0-226	2300
16D0-226	2300
16UBL2-243	1800
16C0-273	1209
16E0-273	1209
16Y0-273	0300
16Y0-273	0304
16URKA0-286	0607
16URKA0-365	0607
16URKB4-400	0608
16URKC4-400	0608
16URKD4-400	0608
16URKE4-400	0608
16URKF4-400	0608
16URKG4-400	0608
16URK3-400A	0608
16URKA3-400A	0608
16BLK0-410	0601
16BLL0-410	0601
16BLM0-410	0601
16BLU0-410	0601
16UU4-414	1800
16UU5-415	1800
16URK2-420	0604
16BL2-440	0610
16TE9-440	0610
16UU0-440	0610
16URK2-443	0602
16UBL2-443A	0602
16UKS4-460	0607
16URKC4-460	0607
16UBK4-460A	0607
16URKA4-460A	0607
16UKS2-465	1800
16GE0-472	0607
16UKS3-480	0607
16UKS2-497	0610
16UKS4-510	0608
16UBK3-510A	0608
16URK3-510A	0608
16UKS3-520	0601
16UKS3-520A	0611
16T0-538	0602
16ZN0-538	0602
16A0-539	0602
16FE0-539	0602
16BLA0-560	0602
16HM0-560	0602
16RE4-560	0602
16REA4-560	0602
16ZN4-560	0602
16UKSD0-562	1800
16UKSE2-562	1800
16UKSF2-562	1800
16UKSG2-562	1800
16RL0-564	0605
16UKS3-564	0605
16D0-569	0705
16S0-569	0804
16B0-569A	0705

11

NUMERICAL INDEX—MASTER PARTS LIST

Part No.	Group	Part No.	Group	Part No.	Group	Part No.	Group
16F0-571	0705	16ZN0-880	0602	16UK2-2285	2200	16HM0-2884	0602
16G0-571	0705	16NC0-936	1201	16UKA2-2285	2200	16HM0-2885	0602
16H0-571	0705	16UKS5-950	1800	16UP2-2285	2200	16HM0-2886	0602
16J0-571	0705	16UKSA5-950	1800	16UPA2-2285	2200	16HM0-2887	0602
16K0-571	0705	16UKSB0-962	1800	16UU2-2285	2200	16HM0-2891	0602
16S0-571	0804	16URKB0-988	0607	16UUA2-2285	2200	16HM0-2934	0602
16RE0-572	0804	16UKS3-990	1800	16UUB2-2285	2200	16HM0-2935	0602
16T2-572	0705	16UKS4-1026	1801	16UUC2-2285	2200	16HM0-2936	0602
16UKS3-580	0602	16UBK2-1029	1800	16UVA2-2285A	2200	16HM0-2937	0602
16AA9-590	0602	16UKS3-1110	0610	16A2-2534	0101	16UU2-3026	2203
16AJ9-590	0602	16URK3-1110	0610	16A2-2535	0101	16UUA2-3026	2203
16UKS3-590	0602	16URK3-1110A	0610	16A0-2555	2300	16UKS3-3050	0607
16UKSA3-590	0602	16UKS0-1113	0604	16A0-2556	2300		
16UKSB3-590	0602	16UKS2-1127	1800	16B9-2556	2300	18UV3-34	1910
16A0-592	0602	16UD2-1127A	1800	16A0-2557	2300	18UV2-46	1910
16F0-592	0602	16UKS4-1128	1801	16A0-2558	2300	18UV0-51	1911
16ZN0-592	0602	16UKS4-1130	1801	16A0-2559	2300	18UV2-99	1910
16A0-593	0602	16UKS4-1135	1801	16UKS2-2587	1800	18UV0-102	1911
16F0-593	0602	16UKS4-1137	1801	16A0-2624	0602	18UV4-105	1910
16ZN0-593	0602	16UKS3-1190	0605	16URKB0-2628	0607	18UV0-108	1911
16A0-594	0602	16UU0-1199	0107	16URKB0-2629	0607	18RL2-127	1911
16ZN0-594	0602	16UKS3-1210	0605	16URKB0-2632	0607	18UV3-127	1911
16F0-595	0602	16UKS3-1210	2203	16UKSA2-2647	1800	18UVA0-127	1911
16T0-595	0602	16UKS3-1219	1800	16URK0-2648	0607	18UVB0-127	1911
16ZN0-595	0602	16UU0-1220	2203	16UKS2-2649	0609	18HF2-128	1911
16T0-596	0602	16A0-1230	0603	16URK2-2649	0602	18RL0-128	1911
16SCM0-603	2203	16UKS4-1241	0606	16URK2-2649	0609	18RL0-136	1911
16BB9-623	0604	16URKB3-1241	0606	16UKS3-2672	0606	18UV0-136	1911
16UKSS2-623	0611	16UKS9-1422	0608	16UKS3-2676	0606	18UVA0-136	1911
16UKST2-623	0611	16UKSA9-1422	0608	16W2-2731	0606	18RL4-146	1911
16UKSU2-623	0611	16UKS0-1428	1801	16WA2-2731	0606	18UV5-146	1911
16UKSV2-623	0611	16UKS0-1449	0602	16WA2-2732	0606	18UV0-169	1911
16RL0-630	0605	16URK2-1449	0602	16WB3-2732	0606	18UV3-171	1910
16UKS3-630	0605	16UBL0-1449C	0602	16UU0-2736	2300	18UVA2-171	1910
16UKSA3-630	0605	16SDL3-1460	0610	16UKS3-2740	0609	18UV2-174	1910
16A0-645	2300	16UKS3-1460	0610	16URK3-2740	0609	18UV0-179	1911
16B0-645	2300	16URKA3-1460	0610	16UKS3-2768	0611	18UVA0-179	1911
16G0-645	2300	16UKS3-1520	0610	16UKSA2-2768	0611	18UV0-181	1911
16UKS2-646	2300	16URK3-1520	0610	16URK4-2775	2300	18UV0-183	1911
16URK2-646	2300	16URK3-1570	0606	16URK2-2793	2300	18UV0-184	1911
16UP0-680	0602	16URK3-1570	2300	16URKA2-2793	2300	18UV0-249	1911
16BL2-680B	0602	16SK0-1705	2203	16URKA3-2793B	2300	18UV0-295	1911
16W2-699	0606	16DK0-1706	2203	16HF2-2794	2300	18HF2-308	1911
16W3-699	0606	16UKS3-1791	0610	16UU2-2842	2200	18RL2-308	1911
16RL2-699A	0606	16UKS3-1844	2203	16UU0-2844	2300	18UVA3-309	0805
16UKS2-721	1800	16UKS0-1875	1800	16UU2-2845	0611	18UVB3-309	0805
16SA2-724	0611	16UG2-1896	2200	16UUB2-2845	0611	18UV0-315	1911
16UP0-730	0602	16UK2-1896	2200	16UUC2-2845	0611	18UV0-388	1911
16UKS3-740	0606	16UP2-1896	2200	16UUD2-2845	0611	18RL0-530	1911
16UU0-743	1302	16UU3-1896	2200	16UUE2-2845	0611	18UV0-530	1911
16SA3-744	1302	16UV2-1896	1800	16UUF2-2845	0611	18UV0-550	1911
16UB3-760	0605	16UV2-1896	2200	16UUG2-2845	0611	18UV3-599	1910
16UKS3-760	0605	16UL2-1896A	2200	16UUH2-2845	0611	18UVA3-599	1910
16UKS3-760	2203	16UKS2-1927	1800	16UU3-2850A	0107	18UV0-613	1911
16UKS4-768	2201	16SA0-2004	0602	16HM0-2857	0602	18UKS2-637	1910
16UKSA3-768	2201	16SCM0-2004	0602	16HM0-2858	0602	18UV3-640	1900
16URK3-768	2201	16ZN0-2004	0602	16HM0-2859	0602	18UV0-641	1911
16URKA2-768	2201	16UKSA2-2016	1800	16HM0-2861	0602	18UV3-646	1900
16URKB2-768	2201	16UKSB2-2016	1800	16HM0-2862	0602	18UV3-650	1900
16URK3-768A	2201	16RL0-2040	0605	16HM0-2863	0602	18UV0-653	1911
16URK3-768C	2201	16UKS3-2040	0605	16HM0-2864	0602	18UVA0-653	1911
16RL0-775	0605	16UU0-2068	0107	16HM0-2865	0602	18UV0-654	1911
16UKS3-775	0605	16AW2-2069	0609	16HM0-2866	0602	18UV0-655	1911
16UKSA4-777	1800	16UKS2-2124	0611	16HM0-2867	0602	18UV0-656	1911
16UKSA0-778	1800	16UKS2-2125	0611	16HM0-2868	0602		
16DFLB9-786	0300	16UKSA0-2147	1800	16HM0-2869	0602	19D0-13	1105
16DFLB3-786-15	0300	16UU2-2192	2203	16HM0-2871	0602	19T2-13	1301
16W2-790A	1502	16UUA2-2192	2203	16HM0-2873	0602	19UU4-19	0805
16UV0-800	2203	16UUB3-2192	2203	16HM0-2874	0602	19UU4-79	0801
16UKS4-823	1801	16UKS3-2210	1800	16HM0-2875	0602	19UV0-89	1911
16UKS5-828	1800	16UKS3-2220	1800	16HM0-2876	0602	19UVA0-89	1911
16UV0-845	1800	16UKS3-2230	1800	16HM0-2877	0602	19SKA2-91	0801
16F0-880	0602	16UG2-2285	2200	16HM0-2878	0602	19UU4-105A	0805
16SA0-880	0602	16UGA2-2285	2200	16HM0-2879	0602	19HF2-111	0801

NUMERICAL INDEX—MASTER PARTS LIST

Part No.	Group	Part No.	Group	Part No.	Group	Part No.	Group
19UU2-111	0801	19SKA0-435	0803	25LE3-03	1205	25FE2-146	1205
19DKB3-131	0804	19FK0-436	0803	25BB4-10	1205	25SD0-173	1203
19FK0-131	0804	19FK0-437	0803	25HB5-10	1205	25C0-187	1209
19UU3-154	0805	19UG0-437	0803	25HC4-10	1205	25N0-194	1206
19DKB0-185	0801	19SKA0-438	0803	25HC5-10	1205	25N0-195	1206
19FK0-185	0801	19SKB0-438	0805	25FK0-21	1207	25N2-215	1206
19FKA0-185	0802	19BK0-440	0800	25KL0-21	1207	25UU0-219	1205
19FK0-188	0805	19FK2-440	0800	25KC0-21	1207	25C2-226	1205
19UG0-188	0805	19FK0-442	0803	25KD0-21	1207	25B2-231	1205
19SK0-189	0805	19SKB0-443	1002	25KE0-21	1207	25E2-231	1205
19FK0-191	0805	19SKB0-443	1105	25KM0-21	1207	25F3-231	1205
19UG0-191	0805	19SKB0-444	1301	25KN0-21	1207	25G0-233A	1203
19UG0-202	0805	19SKB0-445	1301	25KP0-21	1207	25SD0-235	1202
19FK0-205	0805	19SKB0-463	0805	25KQ0-21	1207	25G0-244	1203
19SK0-205	0805	19FK0-468	0804	25KR0-21	1207	25ZT2-287	1206
19DKB0-206	0802	19FK0-469	0804	25KS0-21	1207	25ZT2-288	1206
19FK0-206	0802	19FK0-471	0804	25B0-27	1209	25ZT2-289	1206
19FK0-207	0802	19DKA0-477	0803	25C0-27	1209	25BC0-310	1203
19FK0-208	0802	19UG0-479	0803	25C2-27	1206	25BD0-310	1203
19FKA0-208	0804	19FK0-487	0803	25G0-31	1205	25UU0-317	1205
19DKA0-211	0802	19DKA0-489	0803	25DFL3-32	1209	25LA3-328	1205
19FK0-211	0802	19DKB0-494	0804	25UKS3-32	0605	25A0-350	1209
19DKA0-212	0802	19DKA0-502	0803	25E2-35	1209	25A0-350	2300
19DKA0-212	1007	19DKA0-503	0802	25C0-37	1209	25B0-350	1209
19FK0-212	0803	19DKA0-504	0802	25C0-37	2300	25B0-350	2300
19DKB0-214	0802	19DKA0-505	0803	25TH0-40	1205	25C0-350	1209
19SKB0-214	0802	19DKA0-506	0803	25TH4-40	1205	25T4-380A	1205
19SKB0-214	0803	19DKA0-507	0803	25G2-41	1205	25UU0-417	1205
19UG0-214	0803	19DKA0-508	0803	25A0-43	1209	25A3-630A	1205
19DKB0-215	0804	19DKA0-509	0803	25G0-43	1209	25A4-640B	1205
19FK0-215	0804	19DKA0-511	0803	25A0-44	1209	25A4-650B	1205
19DKA0-216	0804	19DKA0-512	0803	25C0-44	1209	25D0-660	1209
19FK0-216	0804	19FK0-512	0803	25D0-44	1209	25E0-660	1209
19DKA0-218	0803	19FK0-513	0802	25G0-44	1209	25C2-670	1209
19FK0-218	0803	19UG2-513	0803	25B0-45	1209	25E2-670	1209
19FK0-219	0804	19UGA0-513	0803	25C0-45	1209	25C9-730	1209
19FK0-221	0802	19UU3-545	0805	25D0-45	1209	25C9-730	2300
19FKA0-221	0802	19DKB0-570	0802	25E0-45	1209	25DB9-730	1209
19FK0-222	0804	19FK0-581	0803	25G0-45	1209	25DB9-730	2300
19DKA0-223	0804	19FKA0-581	0803	25H0-45	1209	25UU0-780	0505
19FK0-223	0802	19FKB0-581	0803	25J0-45	1209	25UUA0-780	0505
19FKA0-223	0803	19FK0-590	0801	25U0-47	1209	25UKS4-830	1209
19FKA0-223	0804	19UG0-590	0801	25UA0-47	1209	25ZGEW0-874	1203
19DKA0-224	0802	19UU3-613	0805	25UB0-47	1209	25AK0-881	1203
19FK0-225	0802	19UG0-619	0801	25G3-50B	1205	25AM0-881	1203
19FKA0-225	0804	19UG0-622	0803	25G0-53	1209	25AN0-881	1203
19UG0-225	0802	19UG0-623	0805	25G0-54	1209	25AP0-881	1203
19UGA0-225	0802	19UG0-624	0803	25A0-56	1206	25AQ0-881	1203
19C2-305	1105	19UG0-625	0805	25A0-56	1209	25AR0-881	1203
19UV2-352	1911	19UG0-626	0803	25C2-56	1209	25AS0-881	1203
19FK0-356	0802	19UG0-627	0805	25D2-56	1209	25AT0-881	1203
19DKB0-357	0804	19FK0-636	0803	25E2-56	1209	25KH0-881	1203
19FK0-358	0804	19FK0-636	0804	25GF2-56	1209	25A2-1060	1205
19FKA0-358	0803	19FK0-637	0803	25A0-57	1209		
19UG0-358	0804	19FK0-637	0804	25GA0-57	1209	27UV3-08	1500
19FK0-361	0802	19FK0-638	0801	25C2-59	1209	27UV3-09	1500
19DKB0-362	0802	19FK0-643	0801	25C2-59A	1209	27UVA3-09	1500
19FK0-362	0802	19FK0-644	0803	25G2-61	1205	28UV4-10	1603
19A3-391	1911	19FK0-644	0804	25KC0-65	1203	28UV4-20	1603
19UV3-391	1911	19FK0-645	0803	25KD0-65	1203	27UV0-064	1900
19UV0-408	1911	19FK0-646	0803	25AD4-71	1205	27UVB2-217	1800
19FK0-421	0802	19UG0-760	0803	25AE3-71	1205	27UVC3-217	1800
19SKA0-422	0803			25L3-71	1205	27UV5-224	1800
19UG0-425	0803	24CP2-29	1206	25ZGEW0-99	1203	27A2-238	1800
19FK0-426	0803	24CQ2-29	1206	25ZGEW0-99A	1203	27ZA2-239	1800
19SKA0-427	0803	24CR3-29	1206	25G0-103	1203	27UKSB3-241	1800
19UG0-427	0803	24CS3-29	1206	25G0-106A	1203	27UV2-432	1800
19DKA0-429	0803	24UV2-439	1800	25A0-110	1205	27UU0-556	2300
19UG0-431	0803			25A2-110	1205	27UV3-557	1800
19FK0-432	0803	25GC2-03	1205	25N0-112	1203	27UV3-561	1800
19FKA0-432	0803	25GG3-03	1205	25NT2-130	0606	27UV3-567	1800
19FK0-434	0802	25GH3-03	1205	25NT2-130	1205	27UV3-568	1800
19SKA0-434	0803	25GQ4-03	1205	25G3-139	1205	27UV0-573	1900
19DKA0-435	0803	25LA3-03	1205	25G0-145	1209	27UV0-574	1900

13

NUMERICAL INDEX—MASTER PARTS LIST

Part No.	Group	Part No.	Group	Part No.	Group	Part No.	Group
27UV0-583	1900	28UU4-30	1603	DK-23A	0601	GEB-45	0601
27UV0-594	1900	28UU4-40	1603			GEB-1005	0601
27UV0-595	1900	28UU4-100	1603	EA-93	0609	GEB-1005A	0601
27UV0-596	1900	28UU4-200	1603			GEB-1007	0601
27UV0-597	1900			F1-10	0302	GEB-1007A	0601
27UV0-598	1900	30LD4-110	0901	F1-15	0302	GEB-1007B	0601
27UV0-599	1900	30LE4-110	0901	F5-7	0302	GEB-1008	0601
27UV0-601	1900	30ZKE5-232A	0901	F6-12	0302	GEB-1008A	0601
27UV0-602	1900	30ZKEA5-232A	0902	F7-122	0302	GEB-1008B	0601
27UV0-603	1900	30DD5-263	0901	F8-50	0302	GEB-2005	0601
27UV0-605	1900	30DD5-300	0901	F16-27	0302	GEB-2005A	0601
27UV0-606	1900	30ZDE5-302	0901	F26-3	0302	GEH-2054F	0601
27UV0-607	1900	30ZDEA5-302	0902	F36-9	0302	GEH-2090F	0601
27UV0-608	1900	30DD5-330	0901			GK-174	1090
27UV0-609	1900	30ZC5-332B	0901	G-2141	0305	GX-9	0601
27UV0-611	1900	30DD5-340	0901	G-2237	0305	GX-10	0601
27UV0-612	1900	30ZC5-372B	0901	G-2280	0305		
27UV0-613	1900	30CD5-400	0901	G-2281	0305	IG-687	0603
27UV0-614	1900	30ZC5-433B	0901	G-2282	0305	IG-688	0603
27UV0-615	1900	30ZDE5-460	0901	G-2287	0305	IG-688A	0603
27UV0-616	1900	30ZDEA5-460	0902	G-2288	0305	IG-750	0603
27UV0-617	1900	30ZD5-470J	0901	G-2289	0305		
27UV0-618	1900	30ZDA5-470J	0902	G-2290	0305	MN-21	0601
27UV0-619	1900	30DD5-530	0901	G-2291	0305		
27UV0-621	1900	30ZD5-590J	0901	G-2293	0305	RA-12-6	1201
27UV0-622	1900	30ZDA5-590J	0902	G-2335	0305		
27UV0-623	1900			G-2373	0305	S2A-345	0305
27UV0-624	1900	32UV0-500	1800	G-2374	0305	S2B-831	0305
27UV0-625	1900	32UV2-836	1800	G-2376	0305	S2D-170	0305
27UV0-626	1900	32UV2-837	1800	G-2682	0305	SC-150	0305
27UV0-627	1900	32UV3-838	1800	G-2781	0305	SD0-160	0110
27UV0-628	1900	32UV2-839	1800	G-2784	0305	SN-213	0305
27UV0-629	1900	32UV3-844	1800	G-2807	0305	SN-1150	0305
27UV0-631	1900	32UV3-845	1800	G-2811	0305	SP-230	0601
27UV0-632	1900			G-3770	0305		
27UV0-633	1900	A-1870	0305	G-3771	0305	T8S10-9	0301
27UV0-634	1900			G-3978	0305	T41-10	0301
27UV0-635	1900	C120-9	0301	G-3983	0305	T43-6	0301
27UV0-636	1900	C-134	1201	G-4189	0305	T56-5	0301
27UV0-637	1900	C142-28	0301	G-5950	0305	T56-13	0301
27UV0-638	1900	C-241	1201	G-6701	0305	T56-15	0301
27UV0-639	1900	C-270	1201	GAA-32	0601	T56-23	0301
27UV0-641	1900	C-479	1201	GAE-47	0601	T56-24	0301
27UV0-641	1911	C-483-A	1201	GAL-44	0601	T56-27	0301
27UV0-642	1900	C-493	1201	GAR-83	0601	T56-36	0301
27UV0-643	1900	C-506	1201	GAR-95	0601	T56-48	0301
27UV0-644	1900	C-507	1201	GAR-1177	0601	T-317	1603
27UV0-645	1900	C-509	1201	GBD-20A	0601	T-347	1603
27UV0-646	1900	C-513	1201	GBJ-25	0601	TM-10-1116	2300
27UV0-647	1900	C-514	1201	GBM-21	0601	TM-10-1117	2300
27UV0-648	1900	C-535	1201	GBW-34	0601	TM-10-1118	2300
27UV0-649	1900	C-536	1201	GBW-58	0601	TM-10-1119	2300
27UV0-651	1900	C-541	1201	GBW-66	0601		
27UV0-652	1900	C-542	1201	GBW-67	0601	X-11	0305
27UV0-653	1900	C-557	1201	GBY-38	0601	X-17	0305
27UV0-674	2300	C-598-2	1201	GCE-24C	0601	X-23-072	2300
27UV0-680	2300	C-598-4	1201	GC-26	0601	X-74	0305
		C-604	1201	GCE-53	0601	X-153	0601
28M2-01	1603	C-653	1201	GCE-54	0601	X-157	0601
28MA2-01	1603	C-655	1201	GCE-55	0601	X-193	0601
28UV2-01	1603	C-10825	0503	GCE-57	0601	X-194	0601
28UUA2-01A	1603	C-102538	0503	GCE-109	0601	X-195	0601
28UVA2-01A	1603	CGE-91	0601	*GCE-1012	0601	X-196	0601
28UG2-02	1603	CR9-5	0301	GCE-1057	0601	X-198	0305
28UU3-02	1603	CR26-65	0301	GCE-1091	0601	X-203	0601
28UG2-03	1603	CR121-10	0301	GCE-1109	0601	X-234	0305
28UU3-03	1603	CT63-2	0301	GCE-2050	0601	X-239	0305
28M2-05	1603	CT62-3	0301	GCE-3080	0601	X-258	0601
28UU0-09	1603	CT63-4	0301	GDD-25	0601	X-263	0601
28MA0-10	1603			GDD-26	0601	X-295	0601
28UG2-11	1603	DA-39	0601	GEA-30	0601	X-298	0601
28UK2-11	1603	DA-132	0601	GEB-27	0601	X-317	0305
28UG2-12	1603	DG-1144	0601	GEB-29	0601	X-339	0305
28UK2-12	1603	DH-7	0601	GEB-44	0601	X-358	0601

NUMERICAL INDEX—MASTER PARTS LIST

Part No.	Group
X-378	0601
X-404	0601
X-454	0305
X-461	0305
X-489	0601
X-513	0305
X-556	0305
X-572	0305
X-600	0305
X-612	0305
X-715	0601
X-825	0305
X-847	0601
X-864	0601
X-959	0601
X-997	0305
X-32556	0704
5X-177	0601
5X-349	0601
5X-361	0601
5X-1377	0601
8X-62	0601
8X-66	0601
8X-140	0601
8X-146	0603
8X-163	0603
8X-177	0601
8X-305	0601
8X-311	0601
8X-321	0601
8X-349	0601
8X-361	0601
8X-707	0603
8X-794	0601
8X-870	0603
8X-1377	0601
8X-1420	0601
8-0300-24A	0604
8-0300-29A	0604
8-0300-34A	0604
8-0300-42A	0604
8-0300-49A	0604
8-0300-50A	0604
24-241-6	1201
24-241-15	1204
24-241-17	1201
24-241-20	1204
24-241-55½	1204
24-241-57	1204
24-241-59	1204
28M2-01	1603
40-S-5-D	1201
71-65	1603
76-A1	1603
1392-X	1800
1393-X	1800
1395-X	1800
1557-X	1800
4265-A	0503
8533-A	0503
10571	1603
10639-B	1603
10640-B	1603
10641-2	1603
10853	1603
10854	1603
10873	1603
10874	1603
10875	1603
11406	1603
12221	1603
12222	1603
12406	1603
17073B—.010 U/S	0102
40500-B	0503
42297-A	0103
90440	1800
90452	1800
90664	1800
96058	1800
96072	1800
96076	1800
102203	1800
103319	0602
103865	0602
103884	0602
106497	0602
106750	0602
107710	0603
110730	0602
112084	1800
114503	0602
114998	0602
115903	0602
117517	1800
117565	1800
117566	1800
117569	1800
117570	1800
117571	1800
117572	1800
117575	1800
117577	1800
117578	1800
119126	1800
122159	0602
132108	0302
135616	0602
802691	0602
802694	0602
805727	0602
805790	0602
808933	0602
809051	0602
809053	0602
809062	0602
809591	0602
809593	0602
809595	0602
809663	0602
809815	0602
809817	0602
810074	0603
810217	0602
810226	0602
810586	0602
810601	0602
810626	0602
810627	0602
810824	0602
811230	0602
811299	0602
811388	0602
811450	0602
811451	0602
811553	0602
811601	0602
812015	0602
812016	0602
812496	0602
812664	0602
813045	0603
813134	0602
813511	0603
813521	0602
813523	0602
813554	0602
815097	0603
815839	0602
816453	0602
817055	0602
817056	0602
817070	0602
817077	0602
817114	0602
817313	0602
817314	0602
817790	0602
818002	0602
818134	0602
818194	0603
819362	0602
822858	0602
824109	0603
824728	0603
826462	0602
826938	0602
827518	0602
828448	0602
828846	0602
828894	0602
833602	0602
854003	0302
854004	0302
854005	0302
854009	0302
854016	0302
855012	0302
855016	0302
855017	0302
855029	0302
855064	0302
855213	0302
855250	0302
855253	0302
855274	0302
855279	0302
855281	0302
855282	0302
855389	0302
855390	0302
855493	0302
855532	0302
855573	0302
855574	0302
855585	0302
855739	0302
855763	0302
855918	0302
856270	0302
922690	0607
924552	0607
925000	0607
1175161	1800
1175162	1800
1175164	1800
1175168	1800
1175202	1800
1521187	0302
1521288	0302
1521289	0302
1521340	0302
1521720	0302
1521972	0302
1522280	0302
1537204	0302
1537259	0302
1537397	0302
1537710	0302
1835455	0602
1836591	0603
1838409	0602
1838410	0602
1842523	0603
1842827	0602
1844542	0603
1847260	0603
1849774	0602
1857492	0603
1861076	0602
1865182	0602
1867662	0607
1874734	0602
1878192	0607
1880642	0602
1881869	0602
1882978	0603
5903375	0607
5930487	0607
5930502	0607
5931167	0607
5931243	0607
5932419	0607
5932572	0607
5933055	0608
5933056	0608
5933069	0608
5933078	0608
5933104	0608
5933121	0608
5933122	0607
5933156	0607
5933231	0608
5933522	0607
5933524	0607